Introduction to microbiology

Introduction to microbiology

Dean A. Anderson

Department of Microbiology and Public Health, California State University, Los Angeles, Los Angeles, California

With a chapter by **Y. Hokama**, Department of Pathology, School of Medicine, University of Hawaii, Honolulu, Hawaii

With 186 illustrations

The C. V. Mosby Company

Saint Louis 1973

Printed in the United States of America

International Standard Book Number 0-8016-0205-X

Library of Congress Catalog Card Number 72-81116

Distributed in Great Britain by Henry Kimpton, London

Cover photograph courtesy of Professor Wayne Nichols,
Department of Biology, Washington University, St. Louis, Missouri.

TS/CB/B 9 8 7 6 5 4 3 2

To my wife
Elgin Wood Anderson

Preface

Many years of teaching microbiology have convinced me that this subject has the potential of being the most fascinating of the biological sciences. Since it deals with a world of minute forms largely invisible to the unaided eye, the student generally has had only the most superficial insight into this invisible world. He has often seen evidences of the activities of microorganisms and has suffered from their actions as they grow in his body. However, he is generally unable to interpret the significance of many of the microbial phenomena he constantly sees.

The microorganisms are fundamental agents involved in many of our basic environmental problems. Therefore, I have attempted to develop ideas and applications relevant to everyday living and to illustrate the roles played by microorganisms as the cause or solution of many of our problems in the pollution of our environment.

Using microorganisms as tools, scientists have been able to discover and correlate great masses of information that have revolutionized many of our concepts of life processes. In this book I have given the background of some of these remarkable new scientific breakthroughs to afford some insight into the way in which microbiologists in general use microbes in seeking solutions for their specific problems.

Unlike many of the more advanced textbooks of general microbiology, *Introduction to Microbiology* is not based on the assumption that the student will have an extensive background in chemistry. However, the chemistry used is designed to give a simple illustration of the basic changes that are taking place.

A conscious effort has been made to maintain a balance between what a microbe is and what it does without involving the complexities of life cycles and similar highly technical details.

Scientific names and scientific terms are employed without apology, but wherever possible the roots and meanings of each new scientific name or term will be given the first time it appears. This, hopefully, will provide a built-in glossary that will serve to clarify the reading matter.

Although scientific terms are used, where applicable the text has been made as readable as possible by avoiding jargon when more common words will convey the idea as well. Where possible, simple examples are included in place of the broad generalities so often encountered.

Many of the illustrations are original and an attempt is made to make them tell a specific story. I am deeply indebted to those who have generously permitted use of some of their published or unpublished illustrations.

The chapters, as presented, fall into three divisions: Fundamentals, Applied Aspects, and Disease Production.

I am most grateful to Dr. Y. Hokama for the excellent manner in which he wrote Chapter 24, Host Responses to Pathogenic Microorganisms and Other Foreign Agents.

I am most appreciative of the helpful suggestions and encouragement of my colleagues in the Department of Microbiology and Public Health, California State University, Los Angeles. In the photographic work, great assistance has been provided by Dr. Russell E. Duncan, Mrs. Gail Sanoki, Mr. Sirtag Ali, and my son, Mr. Gordon K. Anderson.

To Mrs. Mildred Witbracht, I am deeply indebted for her assistance in typing the manuscript.

Dean A. Anderson

Contents

Part 1 Fundamentals of microbiology

Part 2 Practical aspects of microbiology

Part 3 Microorganisms and the production of disease

Introduction to microbiology

PART 1

Fundamentals of microbiology

1 The science of microbiology

FOUNDATIONS OF THE SCIENCE

TERMINOLOGY USED AND NATURE OF THE INVISIBLE WORLD

ORIGIN OF MICROORGANISMS

INTRODUCTION TO THE MAJOR GROUPS OF MICROORGANISMS

ACADEMIC ASPECTS OF MICROBIOLOGY

FOUNDATIONS OF THE SCIENCE

Because prehistoric men were constantly surrounded by, and often in conflict with, visible plants and animals, they must have quickly evolved a body of observations regarding these forms. They passed these perceptions on to their children in story and legend. With the development of written languages, the more astute men of each civilization began to record these observations along with the often bizarre and erroneous explanations they had devised. Out of these sources, the fields of animal and plant science were formalized. Zoology as a science came into being when Aristotle (384-322 B.C.) published his *Historia Animalium*. This work, based on his own shrewd but not always accurate observations, was generously interlarded with folklore. Theophrastus (370-285 B.C.), with his encyclopedic *History of Plants*, performed a similar service for botany. However, microbiology, dealing as it does with forms largely undiscernible to the eye, could have its inception only after the microscope had extended human vision into this invisible world.

Man was denied all but a most rudimentary knowledge of the invisible world of microbes because of the optical limitations of his own eyes. The physical reason is that in the retina of the human eye, the spatial arrangement of rods and cones is such that the smallest objects that can be clearly discerned must be at least 0.1 to 0.2 mm (1/250 to 1/125 inch) in size. Most microorganisms are far smaller than this limiting dimension.

Although the crude microscopes of Leeuwenhoek (1632-1723) and others opened the door to the new world, microbiology as a science did not develop until more effective microscopes were available. This required nearly two centuries because not only were optically adequate microscopes lacking but many of the so-called scientists of the times, secure in their Aristotelian world of living things, evinced little interest in the newborn science. Even the idea of an invisible world of living things was only reluctantly accepted because it ran counter to many of the religious and philosophical beliefs of the times. The real awakening did not come until Louis Pasteur in the 1860-1880's zealously attacked the idea of spontaneous generation and advanced his, then highly controversial, germ theories of fermentation and disease.

TERMINOLOGY USED AND NATURE OF THE INVISIBLE WORLD

The term microorganisms is derived from the Greek (*mikros* small + organisms). A synonym, microbe (*micro* + Gr. *bios* life) was suggested by Sedillot (1878) and has come into common usage, especially in popular writings.

Microorganisms fall into two groups, the relatively simple one-celled forms and the somewhat more complex multicellular forms. The individual cells of the unicellular forms are mostly far smaller than any objects visible to the unaided eye. A few of the largest protozoa can barely be made out in water as faint, moving objects. Most bacteria are at least 100 times smaller than these comparatively gargantuan protozoa.

The term invisible world is something of a misnomer. Masses of growth of even the smallest bacteria are easily visible. What is invisible are the individual cells themselves. For example, with true bacteria, each cell is a complete and

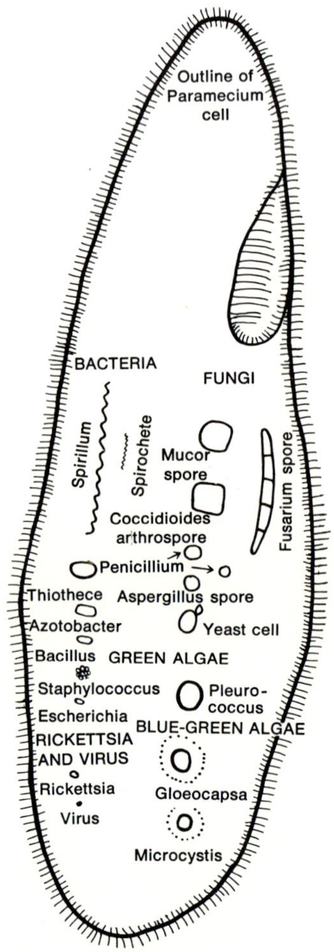

Figure 1-1
Comparative sizes of variety of cells, spores, etc., of fungi, unicellular algae, bacteria, rickettsiae, and viruses, arranged within outline of medium size paramecium (225 μm in length).

independent entity. Hence, bacteria are termed one-celled or unicellular forms. However, when bacteria proliferate in great numbers on a solid growth medium, they form a visible colony representing millions or billions of individual cells. Only when we smear this mass of growth on a microscope slide, stain it with a colored dye or stain, and examine it under a microscope with a magnification of a thousand times or so, can we readily discern the individual cells of the colony.

Great populations of bacteria are often encountered in our daily environment. For example, the slimy coating seen on spoiled fresh meat represents astronomical numbers of bacteria per square inch of slime. The magnitude of these numbers is indicated by the fact that a mass of bacterial cells the size of a drop of water would represent some 50,000,000,000 cells.

The best known example of a mass of microbial cells is the commercial yeast cake. This represents billions of egg-shaped yeast cells, which are each several times larger than bacteria and approximately the same size as a red blood cell (Fig. 1-2).

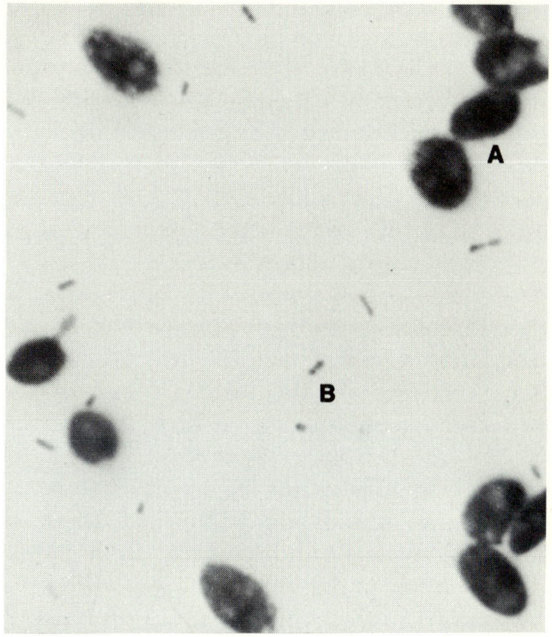

Figure 1-2
Comparative sizes of **A,** *Saccharomyces cerevisiae* (common yeast) cells, and **B,** *Escherichia coli* (common intestinal bacteria) (×2,000).

Molds are simple multicellular fungi. The most familiar are the frequently seen masses of blue or green growth on apples, oranges, cheese, or bread. These are primarily members of the genus *Penicillium.* The mold plant itself is easily seen, but we do not see the intertwining threads of mycelia or the myriads of individual spores borne on the fruiting bodies of the mold plant. However, if we blow gently across the powdery blue mass on a moldy orange or apple, a faint cloud of spores arises.

If a single *Penicillium* spore falls on a moist surface like a break in the skin of an apple, it will germinate and send out threadlike filaments made up of tiny nucleated cells. These spread out in every direction. At first they form a white, faintly fuzzy mass. In a few days upright, fertile threads develop. On the tips of these are formed fingerlike structures and on each finger a beadlike string of spores is formed. A single invisible spore head might have dozens of individual spores, and the whole mold mass might have hundreds of such fruiting heads. These fruiting heads can be made out only with a good magnifying glass. To be seen clearly, the individual spores must be magnified 100 times or more.

ORIGIN OF MICROORGANISMS

Microorganisms, especially simple one-celled bacteria and the closely related blue-green algae, are definitely primitive forms occupying the lowest rungs of the evolutionary ladder. As yet we lack any clear indication of the initial source and manner by which these forms came into being. These minute forms, fragile as they are, have left practically no fossil story that would give even a remote hint as to their origin. The most striking evidence of the antiquity of bacteria upon the earth has recently come to light when the distinctive filaments of *Flexibacter* have been found in sections of rock deposited around ancient hot springs. These ancient filamentous bacteria are almost identical to those still growing in the hot geysers of Yellowstone Park.

Several hypotheses relating to the origin of bacteria have been proposed. The first suggests that the initial life forms (probably bacteria) were borne to the young planet Earth on dust particles or meteorites from some older more advanced planet somewhere in our cosmic environs. Such an explanation only begs the question since it still fails to explain the initial source of these postulated interplanetary visitors. Also the fact that such an interplanetary traveller would have to be able to survive the rigors of extreme heat and cold and the deadly radiation hazard seriously weakens the validity of this theory. A second hypothesis is the autotroph (Gr. *autos* self + *trophē* nutrition) theory. This postulates that bacteria, which were able to secure their energy from sunlight and their carbon source from the carbon dioxide of the atmosphere, grew in the warm, mineral-laden waters of the earth and synthesized the first organic matter out of which to manufacture their own cells and to provide food for subsequent generations of other microbes. The chief difficulty in this concept is that the photosynthetic bacteria, as we now know them, are enzymatically relatively complex forms, which would have been unlikely to suddenly emerge. However, regardless of their origin, such forms must have played an important role in the primordial development of life on the earth.

More recently, the heterotroph (Gr. *heteros* other + *trophē*) hypothesis has been advanced. This postulates that forms that were unable to synthesize their own protoplasm from minerals and carbon dioxide, developed from preformed organic compounds present in the waters of the earth. A Russian scientist, Oparin (1936), extended and applied the heterotroph hypothesis in his book *The Origin of Life on the Earth.* In it he proposed the idea that amino acids and similar compounds were formed from methane, ammonia, hydrogen, and water as a result of lightning discharges in the atmosphere. These molecules would unite to form coacervates (L. *acervatus* heaped or clustered), which became the eventual source of life. Stanley Miller, a young student of Harold Urey, in 1953 reported experiments in which he was able to produce compounds that appeared to be amino acids when he subjected a mixture of methane gas, hydrogen, ammonia, and water to an electric current in a closed system. Following this lead, Sidney W. Fox in 1957 reported that he had produced proteinlike amino acid complexes by heating a mixture of dry amino acids.

These experimental evidences for the heterotroph hypothesis, while attractive, still present certain unsolved and perhaps unsolvable aspects such as the questions (1) were the conditions used by Miller to produce amino acids present in the primordial environment extensively enough to yield appreciable amounts of amino acids, (2) were conditions identical with those used by Fox to form proteinlike complexes present in the early earth environment, (3) by what fortuitous and happenstance operation of chance alone would it be possible to form the initial bacteria-like forms, and (4) could these forms have existed successfully in what was a highly reduced environment possibly completely devoid of oxygen?

As another hypothesis, viruses have also been proposed as the earliest forms of life, but the known complete dependence of viruses on their host cell enzymes for metabolic activity and reproduction casts grave doubt on the validity of this assumption.

At the present time the answer as to the origin of life on the earth is far from settled. The student may well take his choice of any one of the above theories or he may develop his own.

The fact still remains that microorganisms undoubtedly were the earliest forms to appear on the scene. Also, it is evident that many of the present forms of bacteria and perhaps blue-green algae may represent living remnants of that early period when temperature conditions on the newly formed earth first became suitable for the life process to go on.

Regardless of what hypothesis we may espouse, it is attractive to try to visualize the beginning of primitive life on the earth. This probably came as the land masses first began to rise from the primordial seas that no doubt covered the earth with a sheet of warm, mineral-laden water, possibly with very little organic matter being present. A heavy shroud of dense clouds apparently covered the earth, allowing the penetration of little more than infrared rays.

Of the bacteria that we know today, the most likely candidates for growth under these conditions are the purple and green sulfur bacteria. These are no longer widely distributed, but we can still easily isolate them from the oxygen-deficient muds in the bottom of shallow lakes or ponds where reduced sulfur compounds such as hydrogen sulfide or metallic sulfides are present. These forms carry on an anaerobic (Gr. *an* not + *aer* air + *bios*) photosynthesis quite unlike that of higher photosynthetic forms that require oxygen to exist.

It is likely that the newly emergent rocks, splashed by the waves of an almost universal sea, may well have been covered with slimy, red or green masses of these photosynthetic sulfur bacteria. Their dead bodies might have supplied organic compounds to the ancestors of our present day bacteria. The striking feature is that descendants of these primordial sulfur bacteria are still living, largely unchanged over eons of time. Other bacteria that are common today probably evolved somewhat later, but they still appeared on the earth long before the higher plants and animals. The blue-green algae may have preceded the green algae by a substantial period, and the fungi probably came somewhat later. The simplest of the animals, the protozoa, probably were swimming in the warm primordial ponds for many millennia before the fish or land animals appeared. Without doubt, the protozoa of today closely resemble many of those ancient protozoa.

INTRODUCTION TO THE MAJOR GROUPS OF MICROORGANISMS

For convenience microbiologists divide the world of microorganisms into the following major groups. These will be briefly introduced at this point, but discussed in detail in subsequent chapters.

Bacteria

Bacteria are perhaps the most numerous and widely distributed of the microorganisms. They are grouped together in the *Protophyta* or primitive plants under the class Schizomycetes, (Gr. *schizein* to split + *mykēs* fungus). They are so called because the most common method of reproduction is by transverse splitting or fission. They are typically one-celled, very small (usually under 1/5,000 inch in length), generally without photosynthetic pigments, and lacking the well-defined nucleus with its chromosomes found in higher forms. In place of chromosomes enclosed in a nuclear membrane, bacteria have a single, apparently circular chromosome without a confining membrane. They are, therefore, classed as procaryotic (Gr. *pro* or *protos* primitive + *karyon* nucleus) forms.

The three most common shapes are rods (with or without the ability to produce internal or *endo*-spores), spheres, and spirals. The rods may be designated by the term bacterium (Gr. little staff or rod) or bacillus (L. small rod or stick). The term bacterium is also applied collectively to all shapes of bacteria. *Bacillus* is the genus name for the spore-forming rods and is most generally used for the spore formers. However, it is sometimes applied also to nonsporing rods, especially by medical bacteriologists or physicians. The spherical forms are classed as coccus (singular) (Gr. *kokkos* seed or berry) and cocci (plural) (pronounced "cock's eye"). These are further classed as micrococci, diplococci, staphylococci, streptococci, depending on the kind of cell group or cluster formed during growth. The spiral forms are placed mostly in the genus *Spirillum* (Gr. *speira* coil) (plural *spirilla*).

Only a few of the bacteria can carry on photosynthesis. Most of them are like the fungi, which must depend on preformed organic matter such as plant or animal tissues or products to furnish energy and material for cell metabolism.

Algae

Algae represent one of the great groups of plants. They are characterized by a lack of roots, stems, and leaves. However, they range in size and complexity from minute one-cell forms, through the multicellular green algae, to the great kelp plants (brown algae) of the oceans, which occasionally attain a length of 150 feet.

The microbiologist is concerned with only the one-celled or simpler multicellular blue-green or green algae and the diatoms.

The blue-green algae represent a very primitive group of forms apparently closely related to the bacteria in shape, mode of reproduction, and possession of a procaryotic chromosomal structure. In the scheme of classification proposed by Bergey and co-workers, they are placed with the bacteria in the great plant division, Protophyta (*proto* + Gr. *phyton* plant), in the class Schizophyceae (*schizo* + Gr. *phykos* alga). Other classifications designate the blue-green algae as the Cyanophyceae (Gr. *kyanos* blue) or Myxophyceae (Gr. *myxa* slime).

Blue-green algae possess chlorophyll, but it is not enclosed in special structures, the chloroplasts, a unique characteristic of green algae and higher plants. In the blue-greens, chlorophyll is distributed throughout the cell. The blue-green color stems from the presence of a second distinctive pigment, phycocyanin (Gr. *phykos* alga + *kyanos* blue). The habitat of the blue-green algae includes the waters and soils of the earth. They are widely found and are especially notable because of their ability to grow in such harsh environments as deserts, glaciers, and hot springs, or as pioneer plants after a stream bed is badly eroded by floods.

Green algae and diatoms, together with closely related unicellular or multicellular forms, represent the other great algae group of special interest to the microbiologist. Diatoms are sometimes classed with the golden brown algae. These forms are all higher on the evolutionary scale than the blue-green algae and possess a true nucleus, with nuclear membranes and chromosomes. They divide by mitosis as do higher plants. Because of this they are classed with the higher plants as eucaryotic (Gr. *eu* true + *karyon* nucleus) forms. Their chlorophyll is present in membrane-enclosed organelles called chloroplasts. They are widely distributed in fresh and marine waters and in soils. In spring or fall they may produce a striking picture, growing in such profusion in small ponds or lakes that the water becomes a deep green. This green pond scum may contain millions of algae cells per liter of water. The brown coating on rocks in the bed of a freshwater stream is formed primarily by mats of diatoms and green or blue-green algae. Under the microscope the algae and diatoms present a fascinating array of bizarre shapes and forms.

Fungi

The fungi (L. *fungus* a mushroom) represent a great group of plants without chlorophyll, the photosynthetic pigment, and devoid of roots, stems, and leaves. They range in size and complexity from the one-celled yeasts to the complex mushrooms and related fungi. The yeasts and molds are of primary interest to the general microbiologist, while the mildews, rusts, and smuts are of special interest to and are studied by the microbiologist concerned with plant diseases. The more complex forms such as mushrooms and puff balls represent the domain of the botanist rather than the microbiologist.

Fungi, regardless of their complexity, share several characteristics: (1) none of them is photosynthetic, because none of them possesses any photosynthetic pigment (chlorophyll), (2) nutritionally, all of them are heterotrophic, requiring preformed organic matter in order to proliferate, and (3) all of them are eucaryotic, possessing nuclei with a varying number of chromosomes, surrounded by nuclear membrane.

The yeasts represent a convenient rather than

a botanical group of plants. They are commonly one-celled and generally reproduce by budding. Included in this group are the familiar commercial yeasts that ferment sugars to ethyl alcohol and carbon dioxide. Other important forms are the spoilage yeasts that grow in fruit juices and fruits, and certain pathogens that cause desert fever or valley fever (coccidioidal granuloma), blastomycosis, or similar diseases.

The molds are also a convenience grouping and include forms that are botanically dissimilar. In addition to the familiar *Penicillium*, other powdery yellow, olive, or occasionally black molds are often seen. A distinctly different group of molds are cottony rather than powdery. These large cottony masses may be white, gray, or even a dirty black. The most familiar of these is *Rhizopus*, the black bread mold.

Protozoa

Protozoa are the simplest and smallest of the animals. The term *protozoa* means primitive animals. They are designated one-celled, but some of them may possess many nuclei within a single protozoan body. They may range in size from that of the larger bacteria to comparatively huge forms, the largest of which might be barely discernible to the unaided eye. Some may develop into colonies in which some of the cells display a degree of special function. The protozoa are widely distributed in nature, especially in water and less frequently in soil. A few cause important diseases such as amebic dysentery and malaria. They are animal-like in that they possess motility and show irritability to external responses. They are characteristically nonphotosynthetic and, as such, require preformed organic matter for their metabolism. However, one group of flagellated photosynthetic forms are claimed by both protozoologists and the algologists or botanists.

Slime mold or myxomycetes

The myxomycetes represent a peculiar group of microorganisms most commonly encountered on rotting wood or forest soils. They have both animal and plant characteristics. At some stages of their life cycle they form bodies resembling amebas, which ooze across surfaces. In their more mature stage they form a solid mat of growth from which arise fruiting bodies at least superficially similar to the fruiting bodies of some molds.

Rickettsiae and viruses

The final groups of organisms that concern the microbiologists are the rickettsiae and viruses. The rickettsiae are pathogenic forms, apparently bacteria, that have lost many of their enzyme systems. As a result they are able to grow only in the tissues of a host animal. Most of them are found only in the intestines or other tissues of such insects as ticks, fleas, and lice, or in the tissues of animals bitten and thus infected by these arthropods. They are mostly very small, plump rodlets, and are the cause of such diseases as typhus (not typhoid), Rocky Mountain spotted fever, and Q fever.

Viruses represent a very distinctive group of agents. Whether they can be termed "living forms" depends on one's definition of life. They are very simple in structure, consisting of a tiny circular chromosome of DNA or RNA surrounded by a protein coat. In shape they may be long filaments, spheres, or octagons; or may possess a distinctive octagonal "head" and "tubelike tail." They are known only because they are able to destroy their host cell or can be detected inside a host cell when viewed with the electron microscope. They range in size from 12 to 200 nm (Fig. 1-1), which makes them too small to be seen with the light microscope. They cause diseases in animals, plants, and bacteria and are known primarily by the diseases they produce.

ACADEMIC ASPECTS OF MICROBIOLOGY

Once a minor appendage of botany or zoology, microbiology has become a full-fledged science, its ramifications now extending into botany and zoology. The botanist whose primary interest is in the simple fungi and yeasts emerges as a mycologist. He may, however, regard himself also as a biologist or a microbiologist, depending upon his academic background. The botanist with a primary interest in the simple algae would rank as an algologist. The field of zoology provides several microbiologically oriented specialists. The protozoologist, concerned with the simplest animals, might be classed as a microbiological zoologist. The microbiologist enamored with the slime molds or mycetozoa may have come through the ranks of zoology, botany, or the broader field of biology. Today, many schools do not separate botany and zoology but their programs may encompass broad training in both fields, often including microbiology as one of the disciplines.

On the other hand, microbiology includes a number of specialties peculiar to this aspect of science. The oldest of these is medical bacteriology. Originally, primary emphasis was placed on the bacteria as causative agents of many of the major diseases of man. However, with the advent of public health control measures the importance of many of these bacteria diminished; indeed some have completely disappeared from the scene. Chemotherapeutic agents and antibiotics once were hailed as the miracle drugs, the panacea of all ills. Although miracles have been achieved, the problem is far from solved. The emphasis has shifted. We are now confronted with the threat of penicillin-resistant staphylococci and with intestinal bacteria that have become resistant to several antibiotics. Such multiple resistance was not considered a major threat until it was found that this resistance could be transferred genetically to strains formerly sensitive to all of the antibiotics. Due in part to the false security conferred by antibiotics, hospital infections have increased rather dramatically in the last few years. With the microflora of the intestines changed by oral antibiotic treatment, bacteria that we formerly considered nonpathogenic are now being found in a variety of infectious conditions. Although the emphasis has thus changed, medical bacteriology still remains an important field. It has expanded to include the study of diseases caused by fungi, rickettsiae, and viruses and is now termed medical microbiology.

Serology, the study of the production of antibodies in the blood, has moved far beyond the mere production of vaccines for immunization against diseases. Experiments with antibody production have revealed that noninfectious diseases can develop from the action of immune agents we develop against our own tissues. Also, immunology has become a useful tool for the botanist, zoologist, geneticist, biochemist, and even the biophysicist for the elucidation of similarities in tissues and cell products.

From applied microbiology, procedures have been developed for the control of water pollution and safety of water supplies, the transmission of diseases in foods, or the prevention of microbial impairment of quality in foods, dairy products, or cosmetics. The plant pathologist is concerned with the control of bacterial, fungal, and viral diseases of the plants that provide our food, clothing, and shelter. The crop-producing power of our soils depends largely on the microflora and fauna of arable soils—the special domain of the soil microbiologist.

In the last two decades a whole new branch of science has emerged from breakthroughs provided by the dogma of DNA as a carrier of genetic messages, the electron microscope, labelling of compounds with radioactive molecules, and a host of remarkable new instruments for measuring microquantities of cell constituents. These concepts and tools have advanced us from the study of cells to the study of the role of the individual molecules in life processes. This field of science is known as molecular biology. It has given great impetus to interdisciplinary sciences such as biochemistry, biophysics, and biomathematics.

Microorganisms have played the major role in the growth of molecular biology. Because of their morphological, genetic, and metabolic simplicity, they provide ideal subjects for the amazing and elegant experiments that have helped to explain many of the secrets of the processes of life. The future possibilities of the impact of these discoveries on our daily lives are breathtaking to consider.

Concomitant with, or serving as the basis for, advances in molecular biology, a number of specialized microbiological disciplines have developed, such as microbial physiology and metabolism, microbial genetics, and the study of microbial cell morphology and ultrastructure. With this newer knowledge available, the general microbiologist is taking a new look at the whole field of microorganisms. Using the remarkable new tools of the molecular biologist, the taxonomy of microorganisms is being reassessed. Microbial ecology, concerned with the interrelations of microorganisms to their environment, has recently evolved as a full-fledged discipline in microbiology.

The following outline summarizes the various academic fields of microbiology and their relationships:

I. Botany related
   A. Algology—simpler and one-celled true algae
   B. Mycology—molds, rusts, mildews, yeasts
   C. Zymology—commercial applications of yeasts

II. Microbiology
   A. Medically related (oldest branch)
      1. Medical bacteriology
      2. Medical virology-rickettsiology
      3. Medical mycology
      4. Serology and immunology
   B. Applied fields
      1. Sanitary microbiology (water, sewage)
      2. Dairy microbiology
      3. Food microbiology
      4. Industrial microbiology
      5. Soil or agricultural microbiology
      6. Microbiology of air and environment
      7. Plant pathology
   C. Basic or theoretical microbiology
      1. General microbiology
      2. Taxonomy
      3. Microbial ecology
      4. Physiology and metabolism
      5. Microbial genetics
      6. Morphology and ultrastructure
      7. Microbial applications of molecular biology

III. Zoology related
   A. Protozoology (general)
   B. Medical parasitology
   C. Study of myxomycetes (slime molds)

2 Historical evolution of microbial knowledge

The growth of microbiology as a science is a story of an invisible living world, revealed through the medium of the microscope. Therefore, it is logical to begin this history with an account of the discovery and development of the lens and the microscope.

DEVELOPMENT OF THE MICROSCOPE

The concept of lenses is not a new one. Excavation of Ninevah (1845-1855) unearthed a planoconvex lens apparently dating back to about 720 B.C. Pliny, writing about A.D. 75 mentioned the burning power of lenses. The optics of the simple lens were discussed by Roger Bacon in 1267. While various accounts have been given of the development of spectacles, a tomb in Florence bore the date 1317 and the inscription, "Here lies Salvino d'Armato—inventor of spectacles; may the Lord forgive him for his sins." Still, nearly three centuries passed before the compound microscope was invented.

Historians do not agree, precisely, as to who invented the first double-lens, that is, compound microscope. Between 1591 and 1608 a Dutch spectacle maker, Zaccharias Janssen* (also spelled Jansen or Jensen) is credited with placing two lenses in a tube to form the first compound microscope. This was apparently a crude affair with no provisions for focusing.

However, in 1610 Galileo Galilei, who had already made a workable telescope, was told of the existence of a compound microscope in Holland. Accordingly, he put together a double-lens system in a tube and mounted it so it could be focused. This he called his "occiale." Under it, according to his description, "a water flea looked as large as a hen."

In 1625 the term microscope, which means "to see small" (Gr. *mikros* small + *skopein* to spy or watch), was coined by a young Italian, Faber or Fabri. He was a member of the Society of Lynxes (Lincei), a group of inquisitive, bright young men who met and discussed their own observations and new scientific knowledge that came to their attention. Galileo was a member of this group.

With the principle of the compound microscope established, amateur lens grinders began here and there to make such instruments. One of these was Christian Huygens, who put two lenses in the eyepiece of the microscope. This ocular system still bears his name and, until recently, Huygenian eyepieces were found on nearly all microscopes.

Athanasius Kircher, a Jesuit priest, during the plague epidemic in Rome (1656) examined blood from victims with a simple microscope and reported seeing "small worms." However, with the low magnification available, the worms he saw could not have been bacteria.

Robert Hooke (1635-1703) was essentially an experimental physicist. He was also one of the first serious microscopists and in his *Micrographia* published in 1665, he first applied the term cell to the boxlike structures he observed in thin sections of cork.

FIRST REPORTED OBSERVATION OF MICROBES

The time was now ripe for revealing microorganisms to the world. It all began when the Dutchman, Antony van Leeuwenhoek (1632-1723), owner of a Delft drapery shop, began to pursue his hobby. About 1670 he designed and began to build simple single-lens microscopes, hardly as large as a cigarette lighter. How he was

*One biographer suggests a father and son team—Hans and Zaccharias Janssen.

able to detect bacteria with these instruments is still something of a mystery. The greatest magnification he obtained was apparently not much more than 160 (some commentaries say 270) times. It is thought by some that he might have gained enhanced magnification as a result of a drop of water inserted between the object and the lens. In the course of his lifetime he built perhaps 200 or more of these simple microscopes. The simplicity of his microscopes is illustrated by the photographs of an exact replica made a few years ago by a Dutch instrument maker (Fig. 2-1).

Without doubt Leeuwenhoek's instruments (although for some reason they were not widely copied or used after his death) gave better resolution and perhaps greater magnification than the compound microscopes of the period. Furthermore, the early compound microscopes were plagued with chromatic aberration, a rainbow of colors around the object being observed. This would have been sufficient to obscure such tiny forms as bacteria.

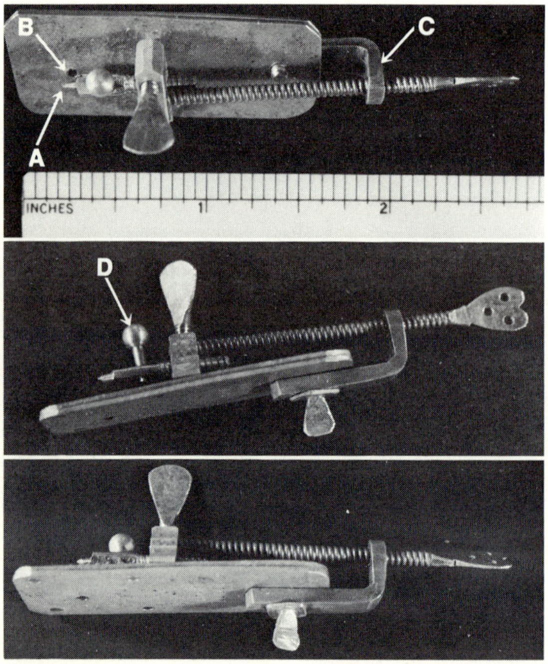

Figure 2-1
Replica of Leeuwenhoek microscope from Leyden, Holland, showing various views of instrument. **A,** Pin on which object is placed for viewing; **B,** lens; **C,** screw for coarse adjustment; **D,** fine adjustment. (Enlarged 2×.)

Leeuwenhoek was a man of infinite curiosity, blessed with most acute vision. For him, nothing was too sacred for observation. He placed on the observing pin of his microscope a great variety of substances. These included sediments from water, bits of rotting peppercorns, scrapings from teeth, and even materials from the cavity in a decayed tooth that he pulled from his own mouth. He was the first to describe scales of the outer skin, the crystalline lens of the eye, spermatozoa, and striations of the voluntary muscles. All of these observations he described in homely Dutch in rambling letters that he sent periodically to the Royal Society of London.

The first description of what are now recognized as microorganisms (including bacteria) came in his eighteenth letter (dated October 9, 1676), a long somewhat desultory description of a series of experiments covering nearly a year. In it, he told of examining water in which peppercorns and other materials had been soaked. The animalcules that he described included some protozoa and algae, plus minute rods, spheres, and spirals that must have been bacteria.

In his thirty-ninth letter (September 17, 1683), he described and drew animalcules (bacteria) in materials scraped from the teeth.

While Leeuwenhoek may or may not have been the first to see bacteria under the microscope, certainly he was the first to see and describe them. He could well have been one of the first to recognize the value of the microscope as a laboratory instrument.

Even though Leeuwenhoek demonstrated the existence of microbes, few among those interested in science caught the significance of his observations. Also, they were reluctant to accept ideas at variance with abiogenesis or spontaneous generation, the popular biological theory of the time. Indeed, for nearly 200 years microbiology as a science made slow progress against this fallacious current of thought.

SPONTANEOUS GENERATION OR ABIOGENESIS

Spontaneous generation or abiogenesis, which apparently might have been part of human thought even in prehistoric times, held that living forms such as fish and frogs might arise spontaneously from mud or similar sources without parents. In 384 B.C. Aristotle, by reducing the theory to writing, misled scientific minds for more than two millenia. Even though today this theory is advanced to explain the origin of life eons ago, it does not satisfactorily explain the origin of the individual.

By the time of Leeuwenhoek, men of vision were beginning to note certain changes in nature that did not square with this stifling theory. Francesco Redi was one of these. He was irked by the belief that blowflies originated spontaneously in rotted meat. In 1668 he performed a simple experiment to test this theory. He placed pieces of meat in a series of jars. Some he covered with parchment, others with gauze or wire netting, while others were left open. Attracted by the stench of the rotting meat, swarms of flies began to fly over the ripe meat, depositing their eggs bomber-fashion on the parchment, gauze, or unprotected meat. No maggots appeared in the protected containers although the unprotected meat was soon crawling with them.

This experiment, and similar ones that followed, forced philosophers to rescind the application of abiogenesis to higher forms. Many, however, continued to apply it to microorganisms. Some of them cited Leeuwenhoek's observations as proof that in the invisible world life was being generated from lifeless matter.

In 1711 Louis Joblot boiled a hay infusion, afterward placing one-half in a covered, sterilized bottle and the other half in an open pan. The covered portion remained unchanged while that in the open pan soon teemed with life. When he removed the parchment cover, that portion, too, showed evidence of much microbial activity, with the theory being that life arose from "eggs" floating in the atmosphere.

Buffon and Needham were active proponents of microbial abiogenesis. Buffon felt that vitality was composed of two parts, interaction of which spontaneously produced organic molecules.

In 1749 Needham conducted an experiment in which he heated mutton gravy, later observing cloudy growth in the heated portion. Although the answer was not obvious then, we know today that the heat probably killed the vegetative bacteria, but heat-shocked the more resistant spores into rapid germination and growth. An Italian priest, Abbe Lazzaro Spallanzani, questioned these results and in 1765 repeated his own version of the experiment. Spallanzani, however, boiled the sealed tubes for a much longer time. As a result, the heated material remained sterile, and the unboiled evidenced growth.

The controversy resulting from the simple variation of Needham's original experiment raged for almost a century. Needham argued that Spallanzani had killed some "vital" force by the prolonged heating, and when Spallanzani countered with another set of experiments in 1776, Needham used Lavoisier's discovery of oxygen the previous year by saying the oxygen had been destroyed.

In 1810 Nicholas Appert heated foods to the boiling point and then sealed them hermetically, discovering the principle of canning, a major by-product of Spallanzani's original work. Following this, and in answer to Needham's oxygen argument, in 1836 Franz Schulze and Theodore Schwann independently performed experiments that proved air alone was not the cause of growth. Schulze bubbled air through an acid solution, then into boiled infusions of meat and vegetables. The flasks receiving the washed air remained sterile. Instead of acid, Schwann passed the air through heated coils of glass tubing to achieve the same results. Proponents of Needham's theory, however, maintained that the vitality was destroyed by the acid and the heat. Schroeder and von Dusch carried the experiments one step further when, in 1854, they plugged flasks with cotton wool and boiled them. Even though air could continually pass into the plugged flasks, they remained sterile indefinitely.

In spite of the mounting evidence from such experiments, Pouchet, a French naturalist, continued to espouse the theory of spontaneous generation. Spurred by this continuous controversy, Pasteur devised his famous experiment with swan-necked flasks to counter the arguments, although not everyone was convinced. In some flasks, Pasteur boiled a milk or yeast-mineral-sugar solution to kill the "seeds" and sealed the tips of the flask necks. When the tips were broken and air rushed in, microbial growth ensued. However, unsealed, curved flask necks allowed no growth, even when the flasks were allowed to cool and air could pass freely into the solution. Microbes apparently remained on the walls of the neck and never passed into solution.

Tyndallization, a process of intermittent sterilization developed by Tyndall in 1877, and Lister's antiseptic surgery with phenol or carbolic acid sprays supported Pasteur's work and caused the ultimate defeat of the theory of spontaneous generation, paving the way for modern developments in microbiology.

GERM THEORY OF FERMENTATION

Even Saint Paul recognized the process we term fermentation as he spoke of the common practice of mixing a bit of unbaked sour dough to flour: "A little leaven leaventh the lump" (I Cor. 5:6). Leeuwenhoek saw yeast cells in beer dregs, although he probably did not realize their role in fermentation. In 1820 Thaer, describing the blue color in milk, attributed it to a fungus; he was one of the first to credit these changes to microbial action.

In 1836 several investigators made significant discoveries on the cause of fermentation. Working independently, Cagniard-Latour and Schwann concluded that yeast cells, acting on sugars, cause alcoholic fermentation. At the same time Kützing described yeasts as living organisms. However, these discoveries were not unopposed, as the famous Liebig maintained that fermentation is caused by "molecular instability induced in the albuminoid fraction of the beer." The controversy continued until 1857 when Pasteur proved yeasts produced alcohol and carbon dioxide from sugar in fermentation.

GERM THEORY OF DISEASE

As the germ theory of fermentation grew in acceptance, it paved the way for recognition of the germ theory of disease. However, this concept had its beginnings earlier, since the Old Testament (Leviticus), Galen, Aristotle, and other writers of antiquity relate the spread of disease to contact with those already infected. Varro in the first century B.C. and Columella in the first century A.D. suggested that *animalia minutae,* which entered the body when eaten or inhaled, could cause disease.

Others, much later, continued to suggest that invisible forms caused disease. Robert Boyle in 1663 felt that a better understanding of fermentation would lead to better understanding of disease mechanisms. Leeuwenhoek's observations led Benjamin Marten (1720) to write that diseases could be caused by "animalicula or wonderfully minute creatures," varying according to the disease, and Plenciz (1762), an Austrian physician, to state his belief in *living* disease agents, *contagium vivum,* specific for different diseases.

In the late eighteenth and nineteenth centuries, many investigators, working with plant and animal diseases, supplied the experimental proof to support this disease theory. Major figures include Dr. Oliver Wendell Holmes and Dr. Ignaz Philipp Semmelweiss of Vienna, both of whom denounced physicians for transmitting puerperal fever to uninfected patients as they examined them without first cleansing their hands. In 1840 Henle, affirming the theory that microorganisms cause disease, established rules for proving that a given form causes a given disease. Pasteur became a convert to the germ theory and militantly supported it. His research in this field led to the anthrax (1881) and rabies (1885) vaccines. Robert Koch was another major proponent of the theory. Koch's famous *Postulates* were based on Henle's concepts. As more people began to accept the theory, research interests were channelled into the search for disease producers. This, at last, established microbiology as a distinct science.

THE GOLDEN AGE OF BACTERIOLOGY

The period 1870-1890 has been called the Golden Age of Bacteriology. The remarkable scientific strides made during these two short decades stemmed from a series of fortuitous developments and impelling national pride.

Two intellectual giants, the fiery Pasteur and the methodical and determined Koch, largely dominated the field. The intense rivalry between France and Germany spilled over into their laboratories and infused them with a great sense of the urgency and importance of their research.

Out of their two laboratories came most of the basic bacteriological techniques that we use today and that provided the foundation for many of the discoveries made during this remarkably fruitful period.

In Pasteur's laboratory, the ingenious Chamberland developed the bacteria-tight unglazed porcelain filter, which opened the door to the study of viruses. He also adapted the hermetically sealed retort, long used for softening bones, as an autoclave for the sterilization of media and glassware.

Out of Koch's laboratory (1883) came liquefiable solid media made for the first time with the remarkable algal component, agar-agar. This permitted the ready isolation of bacteria in pure cultures, an amazing scientific breakthrough hailed even by Pasteur. When this was coupled with the Petri dish, designed by J. R. Petri (1887), bacteriologists were given a powerful tool for isolating the causative agents of disease or finding strange new bacteria in a variety of environments.

Two British workers contributed new methods of sterilization. These were Tyndall's intermittent or fractional sterilization and Lister's dry-heat oven.

Early workers found bacteria notoriously difficult to visualize under the microscope until Weigert (1878) first successfully stained them with the newly discovered aniline dyes and Koch developed the microscope slide heat-fixed–smear method that we still use.

Out of these basic developments came the Gram stain (1884), basic to all bacterial identifications, the Ziehl-Neelsen stain for the diagnosis of tuberculosis by examination of sputa, and Loeffler's methylene blue stain for the recognition of the diphtheria organism in throat smears.

Concurrent with the development of these new techniques came marked improvements in the compound microscope, largely as a result of the activities of Zeiss and Abbe in Germany. The oil-immersion objective provided the additional resolving power and magnification necessary for the ready visualization of bacteria. The apochromatic objective largely eliminated the chromatic aberration that had plagued the earlier microbiologists, while the new substage condenser greatly improved the illumination when examining preparations of bacteria.

When Lister (1867) conceived his new antiseptic surgery, it provided great impetus to the use of chemical disinfectants for the control of infections. Based on this beginning, aseptic surgery came into fruition during the Golden Age.

Vaccines that immunized against rabies and some of the bacterial diseases were also developed. With the recognition of the powerful toxins produced by the diphtheria and tetanus bacteria, therapeutic antitoxins came into being.

Metchnikoff (1884) was the first to recognize the antibacterial action of the white cells of the blood, and a similar property was soon found in the antibodies in the serum of the blood.

Table 2-1. Important disease-producing bacteria discovered between 1870 and 1890

YEAR	DISEASE	CAUSATIVE AGENT
1872	Blue-pus organism	*Pseudomonas aeruginosa*
1872	Anthrax	*Bacillus anthracis*
1873	Relapsing fever	*Borrelia recurrentis*
1874	Leprosy	*Mycobacterium leprae**
1877	Lumpy jaw (actinomycosis) in cattle	*Actinomyces bovis*
1879	Gonorrhea	*Neisseria gonorrhoeae*
1880	Typhoid	*Salmonella typhosa*
1881	Lobar pneumonia	*Diplococcus pneumoniae*
1882	Tuberculosis	*Mycobacterium tuberculosis*
1882	Pneumonia-associated	*Klebsiella pneumoniae*
1884	Serious pus infections	*Staphylococcus aureus*
1884	Strep throat, scarlet fever	*Streptococcus pyogenes*
1884	Diphtheria	*Corynebacterium diphtheriae*
1884	Tetanus	*Clostridium tetani*
1887	Cerebrospinal meningitis	*Neisseria meningitidis*
1887	Malta fever (brucellosis)	*Brucella melitensis*
1887	"Blackleg" of cattle	*Clostridium chauvoei*
1888	Farcy in horses	*Nocardia farcinica*
1888	Rat-bite fever	*Spirillum minus*
1889	Soft chancre (chancroid)	*Haemophilus ducreyi*

*Demonstrated but not isolated.

As a result of the increased interest in bacteria, many of them were identified to be the causative agents of disease (Table 2-1).

While these medically-related discoveries were being made, many important observations concerning bacteria were reported in the nonmedical field. In 1877 Burrill of the University of Illinois demonstrated that a bacterium caused a serious plant disease, the fire-blight of pears. The legume root nodule bacterium *(Rhizobium)* was discovered and its role as a symbiotic nitrogen fixer established. The nonsymbiotic nitrogen-fixing bacterium, *Azotobacter*, was also isolated together with the vital soil bacteria, *Nitrosomonas* and *Nitrobacter*, that oxidize ammonia to nitrates.

What may have been the first bacteriology course in the United States was taught at the University of Wisconsin. Within the next few decades, bacteriology courses were beginning to appear in many of the major universities.

MICROBIOLOGY SINCE THE GOLDEN AGE

With the firm foundation provided by the discoveries of the Golden Age, the science of microbiology, in spite of hiatuses created by two world wars, has made astounding advances. Since World War II, the rate of acquisition of new knowledge has become almost geometric and its impact is amazing to contemplate.

Earlier highly significant applications of microbiology included filtration, chlorination and testing of water, sewage treatment, and pasteurization of milk. These, coupled with newer immunization procedures, have made typhoid, diphtheria, and tetanus rare indeed. Microbiological surveillance procedures have kept the dread cholera out of the United States for some eight decades.

Knowledge of the viruses has made enormous gains within the last two or three decades. The result is that vaccines for poliomyelitis, measles, mumps, and several related diseases have made these diseases almost completely preventable.

The influenza pandemics of 1918 and the 1950's caused by new strains of the virus are not likely to recur. When new outbreaks are found almost anywhere in the world, surveillance teams of virologists immediately isolate and identify the new strains. Vaccines can be prepared immediately and widespread immunization can extinguish a potential worldwide threat.

Antibiotics and chemotherapeutic agents have changed the entire picture of bacterial disease control. Although often remarkably effective, they are not the panacea we once thought. A two-edged sword, they have also created the specter of new bacterial strains resistant to many of these agents.

The electron microscope has provided a remarkable new insight into bacteria, viruses, and their infinitesimal organelles.

Using bacteria as tools has made it possible to establish the remarkable new biological dogma that DNA is the repository of genetic information. This has opened the amazing new field of molecular biology—life processes defined and explained on the basis of the very molecules involved. The future impact of this newer knowledge on human life and well-being staggers the imagination.

SUGGESTED READINGS

Bulloch, W.: The history of bacteriology, Oxford, 1960, Oxford University Press.

Clark, P. F.: Pioneer microbiologists of America, Madison, 1961, University of Wisconsin Press.

Conant, J. B., editor: Pasteur's study of fermentation. (Harvard case histories in experimental science), Cambridge, Mass., 1952, Harvard University Press.

Conant, J. B., editor: Pasteur's and Tyndall's study of spontaneous generation. (Harvard case histories in experimental science), Cambridge, Mass., 1953, Harvard University Press.

Dobell, C.: Antony van Leeuwenhoek and his "little animals," New York, 1960, Dover Publications, Inc.

Lechevalier, H. A. and Solotorovsky, M.: Three centuries of microbiology, New York, 1965, McGraw-Hill Book Co.

Vallery-Radot, R.: The life of Pasteur, translated by Mrs. R. L. Devonshire, New York, 1960, Dover Publications, Inc.

Zinsser, H.: Rats, lice and history, Boston, 1935, Little, Brown and Co.

3 Natural distribution of microorganisms

COMMON HABITATS RICH IN MICROBIAL LIFE

HUMAN BODY AS A MICROBIAL HABITAT

UNUSUAL HABITATS

ROLE OF WINDS IN DISTRIBUTION OF MICROORGANISMS

Microbiologists accept as a truism the statement that microbes are found essentially everywhere on the earth even in the most unlikely environments. The scientist is hard pressed to find in nature, surface areas or substances that are devoid of microorganisms. Even the harshest, most inhospitable environments provide a habitat for some forms. Many of these forms are so highly specialized enzymatically that they rarely compete with the commonplace microflora of the more hospitable environments and seldom are found outside their own peculiar habitat.

The environment in which a microorganism is most commonly found is termed its habitat, while a study of the relation of living forms, including microbes, to their environment is termed ecology (Gr. *oikos* house + *logos* word).

Except under rare conditions, the natural microbial population of any area is mixed rather than pure, that is, consists of one species only. However, the mixed population itself often represents a group of forms peculiarly adapted or environmentally selected to live in the same habitat.

COMMON HABITATS RICH IN MICROBIAL LIFE

Soil

In terms of the number and variety of microorganisms present, the soil of the earth is probably the most densely populated of all natural habitats. A pinch of rich garden soil may have in it more individual bacterial cells (including actinomycetes) than there are people in the United States. This same pinch of soil may also contain many thousands of fungi (mycelial fragments or spores) and many hundreds of yeast cells. Variable numbers of active or encysted protozoa, algae (blue-green, green, diatoms), and bacteriophages (viruses capable of attacking bacteria, algae, or yeasts) may also be found. Cysts of slime molds may be present together with microbial forms that are as yet undiscovered.

The number of microbes in a given soil is never static but may suddenly burgeon a thousandfold when organic matter is added, or the soil is moistened by a rain, or both, and the day becomes warm. Thus, the first balmy days of spring may see a relatively quiescent soil population suddenly produce a great burst of reproductive life and physiological activity.

However, even the soil has areas of dense or meager population. Around roots (the rhizosphere) the numbers of bacteria and other forms may be enormous, resembling a great metropolitan city, while an inch or so away the population may be only a hundredth as great as that of the rhizosphere.

Rotting plant materials

Decaying plant remains are alive with great numbers of microorganisms. The predominating forms will be determined by moisture content, cellulose and nitrogen content, waxes, oils, and other conditions. Straw and high-cellulose grasses may be permeated with threads of the cottony molds, especially *Mucor,* while very moist plant materials, especially if high in protein, may become slimy with growing bacteria. By the time the plant remains have been broken down into brown leaf mold, much of the earlier teeming microbial growth will have subsided into a steady state of relatively slow growth.

At times the growing microbes will produce so much heat as a waste product of their violent physiological activity that the mass of organic matter may become noticeably hot. On rare occasions spontaneous combustion of a mass of plant material (such as a stack of hay) may occur. Among the microorganisms so involved are the thermophiles (Gr. *thermē* heat + *philein* to love).

Much is known about the bacteria and algae that grow at high temperatures. However, information is limited about the fungi (or molds) that grow at temperatures higher than those usually encountered in nature. Cooney and Emerson in *Thermophilic Fungi* define them as forms that are able to grow at 50° C (122° F) or higher temperatures, while their lowest or minimum growth temperature is 20° C (68° F). They usually make their best growth at 25° to 35° C (77° to 95° F). A few common molds, such as *Aspergillus fumigatus*, are thermotolerant and able to grow at temperatures approaching 50° C and also at temperatures below 20° C. Most common molds make their best growth at lower temperatures. As an example, the common black bread mold, *Rhizopus nigricans*, has a maximum growth temperature of 30° C and a minimum of about 8° C. The common green mold, *Penicillium*, grows moderately well in the household refrigerator where temperatures are most commonly 5° to 10° C.

The first thermophilic mold was described in 1886 by Lindt and named *Mucor pusillus* (L. very small). Only five more species were described by the late 1950's. Since that time additional species have been reported, bringing the known thermophilic molds to approximately twenty.

Thermophilic fungi have been isolated from "heating" leaves, straw, grass, and rotting guayule. Environmental conditions must be right if heating is to occur. The organic matter must be relatively moist (at least 30%) and fairly high in cellulose. Microorganisms that yield unusual amounts of heat during the decomposition of plant materials are termed thermogenic (heat-producing) forms.

However, nearly all biological oxidation results in the production of appreciable heat energy as a by-product. If growth of microorganisms is not inhibited by the accumulated heat they are then termed thermophilic. Thermogenic forms are generally quite thermophilic or at least thermotolerant.

Water

Freshwater may have few or many bacteria, algae, fungi, and protozoa, depending on the amount of organic matter present, temperature, degree of acidity or alkalinity, and the amount of dissolved oxygen. In naturally unpolluted waters these microbes are largely native inhabitants that are always present but generally in relatively small numbers. Many of these forms will not grow on common bacterial media; and their presence can be demonstrated only by submerging glass slides in the water. The bacteria then attach to and grow on the slides and can be studied under the microscope.

Pollution of water with sewage, other highly organic wastes, or soil may cause relatively pure water to suddenly abound with microorganisms, especially bacteria. If the pollution is not continuous, then as soon as the added organic matter has been mineralized (that is, changed to carbon dioxide and water plus nitrates, sulfates, and phosphates) the invading microbial horde dies out, the native population now predominates, and the water becomes clear and pure once more.

Water of the ocean differs materially from freshwater in numbers and kinds of microbes. The open sea has its own particular flora made up largely of enzymatically versatile bacteria able to grow at the salinity of the ocean and at relatively low temperatures. This population is often sparse. Along the shore, washings from the soil or industrial or human wastes introduce large numbers of alien microorganisms and quantities of organic matter. These exotic forms are generally short-lived in the ocean due to the saline nature (about 3% salt) of the ocean water and because the growth products of normal flora, particularly some of the algae, appear to have a toxic effect on the introduced bacteria.

Sewage

Sewage, because it is rich in protein, carbohydrates (including cellulose), fats, and other nutrients, is literally teeming with microorganisms, especially bacteria. A gram of sewage may have in it billions of bacteria plus some fungi, protozoa, and related forms. These all serve as purifiers of the sewage. Thus, if raw sewage is dumped into a pond and allowed to stand for a month or so, it will change from a gray-black, foul-smelling fluid into an almost clear, odorless water. During this time the bacteria will have decomposed the putrescible matter and converted it into carbon dioxide, water, nitrates, and sulfates, and protozoa may have eaten many of the bacteria. When the food supply is exhausted, the microorganisms settle to the bottom (often as dead bodies) together with nondecomposable matter, leaving clear sparkling water. This process of complete oxidation of putrescible material is called stabilization, or mineralization.

Rumen of cattle and other ruminants

The ruminants, or cud-chewing, cloven-hoofed animals, (for example, cattle, sheep, goats, camels, llamas, and caribou) have in operation in the rumen, the first of their four stomachs, a remarkable process of food digestion in which microorganisms play the major role. Because of this, microbiologists have taken an interest in this vigorous microbial activity and the interactions of the microbial population involved.

The rumen with its microflora and microfauna represents an evolutionary adaptation peculiarly fitted to the feeding habits and types of food consumed by ruminants. Most of the plant roughage that they eat would be indigestible for us or many other animals.

Cattle, for example, when feeding, rapidly gulp down great quantities of coarse, woody plant material. This, representing several pounds of hay, grass or leaves, is ingested into the rumen until the animal is satiated. The animal lies down and begins the contented process of regurgitating a bolus of the gulped plant material, chewing it some 30 to 85 times for a half minute or so, then swallowing it. This rumination process may be repeated at regular intervals for several hours and is commonly called cud-chewing. The plant material thus broken into bits is returned to the rumen, where vigorous microbial action takes place. Great numbers of bacteria and protozoa carry on the fermentative breakdown of the seemingly indigestible and unnutritious cellulosic plant remains. The rapidity and effectiveness of this enzymatic attack is amazing.

The existence of such microflora and microfauna has been known for more than a century. Ciliated protozoa were first found in the rumen by Grunby and Delafond in 1843. They suggested that the protozoa enabled the animals to survive on plant remains. It was not until 1879 that bacteria were found that fermented the plant materials anaerobically, converting them into acids and gases. We now know that the products of bacterial fermentation include acetic, propionic, and butyric acids, plus quantities of carbon dioxide, hydrogen gas, and methane. Since a cow may take up great quantities of plant materials, great amounts of fermentation products, including gases, may be produced. This gas may be quietly eructated, or a part of it may be absorbed into the blood and expelled through the lungs. The acids and salts may be absorbed directly into the bloodstream or serve as nutrients for other bacteria. The protozoa apparently disintegrate in the second and third stomachs, serving as a source of rich animal protein for the host. Dead bacteria also serve as a source of proteins and other vital nutrients.

The rumen fluids contain billions of bacteria per milliliter and many protozoa. Many of the bacteria are similar to forms that are encountered elsewhere but some seem to be indigenous to the rumen.

In addition to the rumen the other three stomachs, the reticulum, the omasum, and the abomasum, possess characteristic microbial populations; however, neither in number nor activity do they approach that of the rumen.

Intestines of the horse

In the intestinal tract of the horse a similar situation prevails. The woody stems of hay or grass, essentially resistant to the digestive juices, furnish food for certain cellulose-digesting protozoa and bacteria. As these die, their bodies become prime sources of protein and amino acids.

HUMAN BODY AS A MICROBIAL HABITAT

The various cavities and structures of the human body serve as habitats for microorganisms, especially bacteria. Microbes that inhabit the human body without causing any appreciable damage to the host and are characteristic of a person in good health are spoken of as the normal flora of a given region.

The skin

The skin, for example, carries a normal flora. The predominating organism appears to be *Staphylococcus epidermidis.* (This was once erroneously called *S. albus.*) This can be isolated from the surface of the skin even after the skin has been washed. Apparently it is lodged around hair follicles and sweat glands, especially if the skin is oily. For this reason the bacteria are not readily completely removed.

Normally *S. epidermidis* causes little damage to the unbroken skin, but if the organism is carried through the epidermis by a splinter the area around the splinter may fester. When pus obtained from such a festered or infected area is streaked on a plate of nutrient or blood agar it often yields a pure culture of *S. epidermidis.* The bacteria may infect a hair follicle or a skin abrasion caused by shaving, resulting in a small pimple. The relatively mild nature of such infections is due to the fact that the organisms lack the enzymes and toxins necessary for marked invasiveness and pathogenicity.

On the other hand, severe skin infections such as boils and abscesses are caused by the pathogenic *S. aureus,* which we seem to acquire from persons who have active infections or who carry the organism in their nose or throat. A few individuals appear to carry *S. aureus* as a constant inhabitant of the skin. Here, a delicate balance seems to exist between the pathogenicity of the bacterium and the resistance mechanism of the host.

Another frequent inhabitant of the skin is *Corynebacterium acnes.* Because it grows well only under anaerobic conditions, it has been missed in surveys of the skin flora that have used only aerobic cultures. The species name is something of a misnomer. While often associated with or isolated from acne lesions, it does not appear to be a primary cause of acne.

In addition to the previously mentioned organisms, the skin carries surface contaminants from soil, air, water, or other sources. They generally do not remain for any period and rarely grow on the skin. More commonly they die out rapidly, due, perhaps, to certain antibacterial compounds or agents secreted from the skin.

Microflora of the mouth

The human mouth is inhabited by a natural flora made up primarily of apparently harmless parasitic forms such as *Streptococcus salivarius.* Some of these produce antibiotic-like substances that may protect us against pathogens. Others are scavengers that decompose the food particles between our teeth and at gum margins and keep our mouths relatively clean. Still others, as a result of their growth products, may contribute substances responsible for undesirable mouth odors.

In studies of the saliva and tooth surfaces, *Staphylococcus epidermidis,* the alpha hemolytic *Streptococcus mitis, Neisseria catarrhalis,* as well as the similar gram-negative anaerobic coccus, *Veillonella alcalescens,* all have been found to be consistently present. *Lactobacillus* species are also generally isolated and may, in fact, be a contributory factor in dental caries because of the ability of these bacteria to produce quantities of lactic acid that dissolve the tooth enamel.

Anaerobic bacteria, actinomycetes, and spirochetes are frequently found in the tooth-gum margins where anaerobic conditions prevail.

The intestinal tract

Our intestinal tract begins at the mouth and ends at the anus. Although located within the body it may be considered external rather than internal because the tissues are directly in contact with the outside through the foods and liquids we ingest.

Just as plants or animals often grow predominantly in one place, so our intestinal tract provides a variety of habitats, each with its predominating normal flora. In persons in good health, the forms appear to be universally present. However, they tend to fluctuate in numbers and predominating types as a result of the kinds of foods taken into the body. Thus a high concentration of sugar, like that produced by quantities of candy at Christmas, would stimulate the growth of one group of bacteria, while a meal, such as a large steak, high in protein and amino acids, would cause other organisms to burgeon.

In general, even though the population is dynamic and variable, certain kinds of bacteria seem to be constantly present. It is likely, although good experimental proof is lacking, that various nationalities with different food habits might well show definite differences in the normal intestinal flora and fauna.

Other external factors, such as heavy dosages of oral antibiotics, often produce dramatic and sometimes disastrous changes in the normal flora. Wholesale and selective killing or inactivating of certain bacteria that inhibit other forms may give rise to an overgrowth of undesirable microbes. A classic example is the uncontrolled overgrowth of the yeast, *Candida albicans*, following a partial sterilization of the intestinal tract by antibiotics or sulfa compounds.

This organism, which may cause thrush or white canker sores in the mouth, especially in infants, is normally present in small numbers in the intestinal tract. However, it is kept in control by the other microbes normally growing in the intestines. When this inhibitory population is destroyed by antibiotics the balance in nature is destroyed and *Candida albicans* may quickly gain the ascendancy. The result may be a serious intestinal diarrheic disturbance, which in infants or debilitated older persons may lead to dangerous or even fatal consequences. This man-made upset of a natural microbial balance is analagous to that which develops when stockmen who have wiped out coyotes by shooting or poison are suddenly faced with a scourge of jackrabbits. Without the biological control exercised by the coyotes, the rabbits multiply astronomically and become competitors for the forage feed of the stockmen's herds.

The stomach has a variable population. When food is initially ingested into the stomach, it may carry with it large numbers of bacteria. However, the body pours quantities of hydrochloric acid and active proteolytic enzymes into the stomach. This creates an environment inimical to many of the microorganisms. Most of those coming in contact with this enzyme-acid mixture may be quickly killed. Some, however, inside the large food particles, are effectively protected and may move unscathed through the duodenum into the small intestines. Secretions of bile at the duodenal juncture change the digestive juice–food fluid from a highly acid to alkaline mixture. Some bacteria, especially the intestinal streptococci (enterococci) immediately begin to grow.

When the partially digested food enters the large intestines, bacterial growth becomes vigorous. As a result the fecal material that is finally voided from the body is made up of about one-third microorganisms, mostly bacteria. A good proportion, but not all of them, are dead.

In the seething growth that characterizes the large intestine, organisms may compete with each other for food materials, produce metabolic products (antibiotics) that kill those around them or form products that stimulate growth of their neighbors. Many bacteria die from attacks by phages (Gr. *phagein* to eat) in the intestines while the slow growers may be literally crowded to death by the enzymatically active rapid growers. Also, the mixture is completely devoid of dissolved oxygen. As a result, only obligate anaerobes (requiring the absence of oxygen) or facultative anaerobes (able to adjust to the presence or absence of oxygen) can survive.

A recent study of the bacterial content of feces presents an interesting picture. Bacteria were found to make up about one-third of the mass of the fecal material. Microscopic counts of all cells showed about 300,000,000,000 bacteria per gram of wet feces. However, only 8,500,000,000 of these bacteria were found to be viable. These consisted mostly of obligate anaerobes of the genus *Bacteroides*. The ratio of obligate to facultative anaerobes was about 40:1. The facultative anaerobes included such forms as the coliforms (*Escherichia coli* and related bacteria) and the fecal streptococci. A variety of other bacteria and yeasts were present, but in smaller numbers. Of colonies isolated at random from blood agar, 80% were *Bacteroides* species, 8% *E. coli*, and about 1% or less *Streptococcus faecalis*. These values are representative rather than absolute because numbers and kinds are in a state of constant flux as the food intake varies or as protozoa, phages, or antibiotics take their toll.

UNUSUAL HABITATS

Under unusual natural conditions the environment may be so harsh that one would suppose that microorganisms simply could not grow. Contrary to popular belief, microorganisms can and do grow, often rather vigorously, under the most improbable conditions.

Hot springs

As an example, one could consider the hot springs and pools of Yellowstone National Park unlikely habitats. Here the water as it comes steaming from Vulcan's furnaces may be at boiling point or even hotter. Yet in these waters, microorganisms continue to grow, producing striking pink, yellow, and white masses. Certain algae have been observed growing at temperatures of 56° to 60° C (133° to 140° F). Some of the blue-green algae are even more heat tolerant, growing at temperatures up to 73° to 75° C. However, their optimum growth temperature is about 55° C.

Certain filamentous bacteria, probably *Flexibacter*, produce visible whitish, pinkish, or yellowish masses of growth at 88° C. (The boiling point of water at the altitude of the park is 92° C.) In one pool, with a steady temperature of 91.5° C, glass slides immersed in the water for 48 hours showed a microscopically visible growth of filamentous bacteria. Slides immersed in superheated pools (93.5° to 95.5° C) when removed 7 to 10 days later showed adherent growth of filamentous bacteria, often visible to the eye. Thus, these bacteria were growing at temperatures near or above boiling point of water at that altitude.

In the hot springs of Iceland a blue-green alga, *Mastigocladus laminosus*, was found to live and grow at temperatures even higher than those reported for the blue-greens of the Yellowstone pools. Growth at such temperatures represents something of an enigma when compared with our daily experience. However, the logical explanation appears to be that their enzymes are stable at the temperatures at which these forms grow or they may be produced more rapidly than they are heat inactivated.

The origin of such heat-loving forms invites conjecture. Certain rock formations that formed from hot springs of pre-Cambrian times show the presence of great numbers of what appears to be flexibacteria similar to, if not identical with, those found in Yellowstone Park.

Brock suggests that the present bacteria might be merely surviving remnants of a ubiquitous primordial population that inhabited the once widely distributed hot springs of a much warmer earth. In fact, he suggests that our present mesophiles might be mutants from these original heat-loving organisms.

Such thermophilic bacteria also might well be the ancestors of the common spore-forming aerobic rods that grow in heating organic matter (hay and grass), milk during long-time pasteurization, and freshly canned meats or vegetables during the prolonged period required for cooling.

Inhabitants of cold environments

An opposite but equally harsh environment is that represented by glaciers, refrigerators, and other low-temperature situations. Bacteria that grow under these conditions may be termed psychrophiles (cold-lovers). Again, they may be obligate psychrophiles, able to grow only at low temperatures (15° to 10° C down to −4° C), or facultative, growing at a maximum of about 30° C but able to grow quite rapidly at 0.5° C. A classic example of such psychrophiles is the algae that grow on glaciers and impart pink patches on the icy surfaces.

Psychrophilic bacteria have been found in the bitter cold of the Antarctic waters. A particularly striking observation was made of a highly saline lake, so salty it never froze although its temperature ranged from −5° to −24° C. It was found that these bacteria were apparently closely related to, if not identical with, some of the common micrococci and bacilli. Such forms would be forced to grow as both halophiles (salt-lovers) and psychrophiles. Under such conditions the nature and functioning of their enzymes invite some interesting conjectures.

Bacteria that grow at refrigerator temperatures are responsible for spoilage of meats, milk, and many other products stored in the household refrigerator. The main reason that these products keep as long as they do in a refrigerator is not because the cold always prevents growth of the bacteria but because the cold causes the generation time (time for one cell to become two) to become relatively long. Thus, bacteria that may double in number every 30 minutes at room temperature might require 10 to 15 hours for a doubling of their numbers in a refrigerator. To illustrate this point, suppose that two bottles of milk, each initially containing 10,000 bacteria per milliliter, were used in an experiment. One was left at room temperature on a warm day, while the other was placed in the refrigerator. At the end of 10 to 11 hours, the population of the one at room temperature would have risen to about 20 billion bacteria per milliliter and that of the refrigerated bottle would have been about 20,000 bacteria per milliliter. Milk does not become detectably sour until it contains 10 to 40 billion bacteria per milliliter. Therefore, milk from the room temperature bottle would probably taste sour, while the refrigerated bottle might require several days to reach this state.

Psychrophilic bacteria were first described by Forster in 1887. The common growth character of all psychrophiles is their ability to grow at 0° C or lower temperatures. Mesophiles (middle-temperature lovers) rarely grow at temperatures much below 10° C. The psychrophiles run the whole gamut of bacteria—gram negative, gram positive, spore formers, nonspore formers. Stokes and Ingraham define psychrophiles as those bacteria that produce colonies on agar dish cultures at 0° in 14 days. In addition, their optimum growth is about 20° C (68° F).

Barophilic bacteria

Barophilic bacteria have been found growing in sediments at the bottom of the Mindanao Trench or Deep of the Pacific Ocean. At this point, the water depth is over 7 miles. The weight of the water (1 cubic foot of water weighs 62.4 lb) pressing down on a square foot of the ocean floor would be approximately 1,200 tons. Bacteria that have become adapted to growth under such enormous hydrostatic pressures are known as barophilic (pressure-loving) bacteria. They can be grown successfully only in a test tube of suitable medium and incubated in a heavy steel cylinder into which water is forced until a pressure of several thousand pounds per square inch is reached. Also, such cultures must be incubated at low temperatures because they have become habituated to the low temperature of the ocean floor.

Mars-like environments

Recently, as part of the Space Program, an attempt was made to find a spot on the surface of the earth comparable to the supposed climatic conditions on Mars. The nearest approach to such an environment was the Atacama Desert of northern Chile. This is a band, 18 to 100 miles wide, of unbelievably dry land, which runs for some hundreds of miles through the arid part of Chile. Rain falls only once every 10 years or so and then only a quarter of an inch or less. Many years go by without a drop of rain. The landscape shows a total absence of visible vegetation or animal life. The relative humidity rarely rises above 10%.

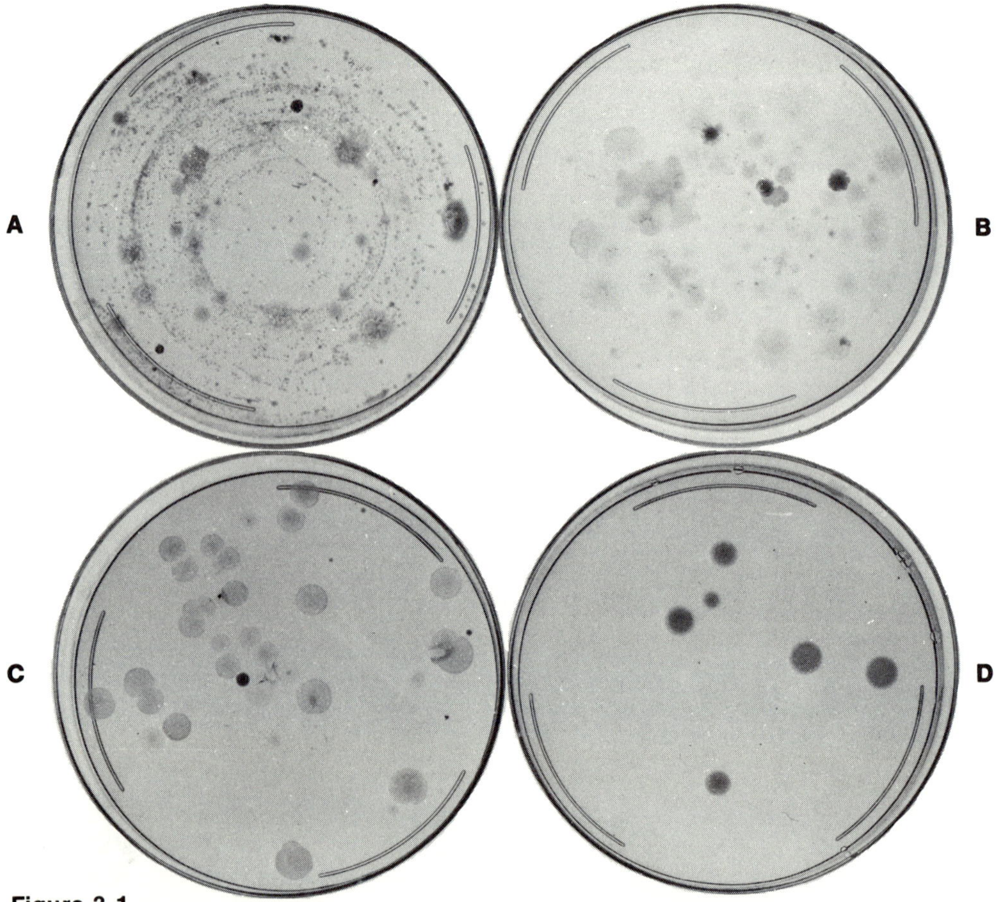

Figure 3-1
Halophilic bacteria in ocean water as shown by colonies on plates of nutrient agar or nutrient agar plus salt (NaCl), when seeded with 0.1 ml of sample. **A,** No salt; **B,** 5%; **C,** 10%; **D,** 15%. (Corresponding plates seeded with fresh pond water, showed no growth on any of the salt-containing plates.)

The surface is a rocky desert pavement held together at lower levels by salt crystals. Soil samples from this forbidding environment were taken under sterile conditions and handled with meticulous care to prevent contamination. These were then plated on a nutritionally meager soil extract medium with a trace of yeast extract. Surprisingly enough, the plates showed a hundred to several thousand colonies per gram. Many were spore formers and could have lain quiescent in the soil for centuries. However, some were nonspore-forming bacteria and others were actinomycetes. A few blue-green algae and occasionally fungi were found. In the most hostile areas, the highest concentration of microbes was found at a depth of 4 to 10 inches. It was thought that solar radiation destroyed most of the surface forms.

The Great Salt Lake and the Dead Sea

Scattered in arid regions are the remnants of ancient lakes. Centuries of evaporation have concentrated the salt levels almost to the saturation point. The Great Salt Lake in Utah and the Dead Sea are classic examples. In addition, many playas or undrained desert basins become temporary salt lakes after desert thunderstorms.

The Great Salt Lake contains up to 25% sodium chloride and the Dead Sea is even higher in salt content, but both are far from being devoid of life. Although shores are forbiddingly barren, an examination of the shallow waters along the shores of the Great Salt Lake often shows windblown masses of the tiny, fragile brine shrimp, *Artemia gracilis*, the main inhabitant of the lake. These accumulate in quiet spots, settle to the bottom, and are decomposed by the bacteria that are present. Because the highly saline water will hold little dissolved oxygen, the shrimps are decomposed anaerobically, setting up a "Stygian stink," as Fremont, the explorer, described it. The water itself contains hundreds of obligate halophilic bacteria per milliliter. These forms can be grown only in media containing at least 12% sodium chloride (Fig. 3-1). In addition, glass slides immersed in the water are covered with many unusual bacteria that have never grown artificially. The black, foul-smelling mud has in it many thousands of bacteria per gram. Thin, slimy layers containing many blue-green algae and a few green algae may be visible at the bottom of shallow pools.

The Dead Sea is equally alive with bacteria, mostly gram-negative rods plus a few pigmented micrococci. These halophiles, *Halobacterium* species, and micrococci have been isolated not only from the Dead Sea but from the Great Salt Lake, and shallow, temporary salt lakes from many parts of the world, indicating a universal distribution.

Waters rich in hydrogen sulfide (H_2S) furnish a habitat for bacteria that require sulfur for their metabolism, while waters laden with iron salts favor the growth of such iron bacteria as *Ferribacterium*. Deposits of such iron bacteria may well have been responsible for great iron ore deposits such as the bog iron deposits of the Masabi Range.

ROLE OF WINDS IN DISTRIBUTION OF MICROORGANISMS

At relatively high altitudes, 20,000 to 40,000 feet, riverlike bands of air flow along through the atmosphere. These are localized, almost vagrant masses, perhaps 10 to 30 miles wide, which may extend up and down several hundred or thousand feet. They do not flow continually in the same region but move north or south. They blow always in approximately a west to east direction. Their speed, usually 100 to 200 miles per hour, may occasionally reach 300 miles per hour.

In addition to these jet streams, and local winds such as the trade winds, the boras, and the mistrals, hurricanes, and tornadoes all move great masses of air capable of carrying microbe-laden dust. With such errant air masses continually winging their way around the world, it is easy to understand why the halophilic bacteria found in the waters of the Great Salt Lake, the Mono Lake of California, and the Salton Sea (which is less than 70 years old) all resemble closely the halophiles of the Dead Sea of the Holy Land. It also explains why *Azotobacter* strains found anywhere in the world are so similar to the organisms originally isolated by Beijerinck from the mud of a Delft canal.

SUGGESTED READINGS

Alexander, M.: Introduction to soil microbiology, New York, 1961, John Wiley & Sons, Inc.

Alexander, M.: Microbial ecology, New York, 1971, John Wiley & Sons, Inc.

Brock, T. D.: Principles of microbial ecology, Englewood Cliffs, N. J., 1966, Prentice-Hall, Inc.

Hungate, R. E.: The rumen and its microbes, New York, 1966, Academic Press, Inc.

Kudo, R. R.: Protozoology, ed. 5, Springfield, Ill., 1966, Charles C Thomas, Publisher.

Marples, M. J.: Life on the human skin, Sci. Am. **220**: 108-115, January 1969.

Postgate, J.: Microbes and man, Baltimore, 1969, Penguin Books Inc.

Rosebury, T.: Microorganisms indigenous to man, New York, 1962, McGraw-Hill Book Co.

4 The higher microorganisms—plants

Of special interest to the microbiologist is a varied and not easily delineated group of living forms, much smaller and structurally less complex than higher plants and animals but larger and more complex than the simple, one-celled bacteria and their relatives. These are arbitrarily termed the higher microorganisms. The plant forms include the smaller, simpler algae and fungi while the animal forms are represented by the protozoa. Size and ease of cultivation make many of these higher microorganisms amenable to study with the methods and tools of the microbiologist.

THE PROTISTS

The life forms included in the higher microorganisms do not fit neatly into the plant or animal kingdoms. Because of this, Haeckel, as early as 1866, proposed that all living forms be classed in three kingdoms: Planta, Animalia, and Protista. The term Protista is from the Greek and implies first (in terms of time), or primitive. Accordingly, this third kingdom includes primitive forms, low on the evolutionary scale, and sometimes not clearly definable as plants or animals. The designation, Protista, while adopted by a good many authors, has not been given official acceptance by any of the biological disciplines. The protists generally include the simpler algae and fungi, the bacteria and sometimes the viruses. The term microorganisms is almost synonymous. In the next few chapters the characteristics of the various protists will be described.

EUCARYOTIC AND PROCARYOTIC MICROORGANISMS

With one exception these higher protists have one distinctive character in common with higher plants and animals. They possess a true nucleus, enclosed in a nuclear membrane and complete with chromosomes, nucleolus, and karyosome. For this reason they are eucaryotic forms. Cell reproduction involves mitosis and meiosis, depending on the type and function of the cell. Within the cells of both plant and animal eucaryotic microorganisms are many functional organelles such as the mitochondria. Some of these eucaryons are one-celled while others are multicellular.

The blue-green algae represent the exception among the higher plantlike microorganisms. They do not have a true nucleus, and their chlorophyll is not confined within a membranous sac, the chloroplast. (In the true algae the chlorophyll is confined within a chloroplast.) Rather the photosynthetic pigments are localized in certain membranous areas. Their nuclear structure appears to be similar to that of the bacteria, essentially a single, very long, circular strand of nuclear material resembling a long, closed string of beads. Microorganisms (blue-green algae, bacteria, rickettsiae) that possess such a simple primitive type of chromosome are classed as proto- or procaryotic forms.

Thus far no definite intermediates have been found, presumably making the line of demarcation between the procaryotic and eucaryotic forms a distinct one, that is, an "either or neither" division. This suggests that the evolutionary development of each group may have been abrupt rather than gradual.

ALGAE

The algae represent a large and varied group of living forms. They range from the huge kelp plants of the ocean, which occasionally reach a length of 150 feet, to the one-celled forms, a few of which are almost as small as the larger bacteria. The unicellular forms are of definite interest to the microbiologist because of their practical importance in water supplies, their possible future use as sources of food for protein-poor human populations, and the possibility that they might be used to provide food and oxygen for future astronauts in outerspace travel.

The larger algal forms, such as kelps, while superficially resembling some of the higher plants, differ in that they lack the true roots, stems, and leaves with their complex of phloem and xylem conducting tubes, which characterize the higher plants. The unicellular algae, on the other hand, represent independent, self-sufficient entities.

About 70% of the earth's surface is covered with water, a natural habitat for algae. Add to this the algae that grow in or on moist soils, trees, or rocks, and it becomes apparent that algae are ubiquitous forms, the most widely distributed of the green plants. They play a vital part in the economy of nature. In fresh and sea waters of the earth they become pasturage, the green grass as it were, feeding the hosts of aquatic animals from minnows to whales, as the grassy pastures feed cattle. Indirectly, the carnivores feed on algae when they consume algae-fattened forms.

Just as the cattleman makes his pastures more lush by fertilizing them, fish farmers often increase their fish yields by adding artificial fertilizer to their ponds to increase the growth of algae.

As photosynthetic forms, during the day they may take up quantities of carbon dioxide from the air or from that dissolved in the surrounding water. Substantial amounts of oxygen may be given off, increasing the dissolved oxygen content of the water. However, during the night, dark-respiration results in the addition of appreciable quantities of carbon dioxide to the water or to the atmosphere. Although a net reduction in carbon dioxide may result, the amount will depend on such factors as light intensity, temperature, and soluble nutrients.

Algae represent the oldest of the aerobic, photosynthetic plants. They probably are the forerunners of the present higher green plants.

They have been classified into 9 to 15 phyla divided in turn into some 1800 genera and 21,000 species. Only those phyla rich in unicellular forms will be given more than cursory notice in the discussion that follows.

Two degrees of evolutionary development are found in the algae: the more advanced eucaryotic forms (green, brown, golden, and red algae), and the simpler, much more primitive procaryotic blue-green algae. Members of this latter group will be treated separately because of their apparent close resemblance to the bacteria.

Major groups

All of the algae possess as their primary photosynthetic pigment, chlorophyll *a*. In addition, some contain chlorophylls *b*, *c*, *d*, or *e*. In all but the blue-green algae, the chlorophyll is distributed in chloroplasts throughout the cell. Also, all algae have various carotenoid pigments that give them their distinctive coloration. Some phyla are made up primarily of one-celled forms, while others may be partially or completely devoid of such forms.

PHYLUM I—Chlorophycophyta or Chlorophyta. The Chlorophycophyta (Gr. *chloros* green + *phykos* seaweed + *phyton* plant) are the green (or sometimes grass-green) algae, a large and important multicellular group containing many unicellular or simple forms. Some 325 genera and 6,500 species have been identified.

Chlamydomonas represents a large and widely distributed genus. It has an ellipsoidal cell with two whiplike flagella that assume a parallel position, thrust back stiffly, then return to the original position. This motion has been described as the breast stroke. The cells also have a light-sensitive red spot or stigma, thought to function in orienting the cell in relation to the light for the most efficient photosynthesis. Because of its motility, zoologists often consider this to be a genus of the protozoa.

The most frequent mode of reproduction is by asexual means. When this occurs, the parent cell becomes nonmotile. Two to eight small daughter cells are then formed internally and continue to grow until they reach a genetically determined size. The cell wall bursts, releasing the daughter cells, which again become vigorously motile. Under some conditions sexual reproduction may occur, involving the union of gametes. This property has made *Chlamydomonas* useful in genetic studies.

Chlorococcum is a widely distributed unicellular, spherical, nonmotile form. This genus in the course of vegetative reproduction at times produces zoospores. These may unite to give a type of sexual reproduction.

Chlorella, a small, rapidly growing, spherical form reproduces only asexually. Its vigorous and rapid growth habit has made it ideal for studies of photosynthesis, and it has been grown experimentally as a source of human and animal food.

Scenedesmus is a striking form that occurs as four or eight cells attached laterally. Such a colony resembles a pile of divan cushions. Hornlike structures occur on the end cells.

The Desmidaceae represent a unique family of the green algae. Desmids are also often seen in freshwater. These are single cells, constricted in the center to form mirror-image semicells resembling two buns with the bottoms pressed together. Most of them have bizarre, fingerlike protuberances.

PHYLUM II—Charophyta (stoneworts). The Charophyta are fairly large, multicellular, filamentous forms. Their long growing strands represent a common nuisance, water grass, which impedes the flow of water in streams and canals. The phylum contains no unicellular forms.

PHYLUM III—Euglenophycophyta. Characteristically, the Euglenophycophyta (Gr. *eu* true + *glenē* eyeball) are unicellular with 1 to 3 flagella and a red eyespot or stigma. These algae have a cell membrane or pellicle that in some forms is rigid enough to maintain the shape of the cell. In other forms, without the rigid cell wall, the shape changes continuously during cell movement. The photosynthetic pigments are similar to those of the green algae. Cells reproduce by longitudinal cell division.

The Euglenophycophyta bear a close relationship to certain of the pigment-free, flagellate protozoa, and such forms suggest the tenuous nature of the boundary between simple plants and animals. Colorless members of the *Euglena* group can be grown heterotrophically in the dark if supplied with suitable nutrients. These forms are often classed as protozoa by the zoologists. The known members of the phylum comprise some 25 genera and 400 species.

PHYLUM IV—Xanthophycophyta (yellow-green algae). The phylum Xanthophycophyta (Gr. *xanthos* yellow) comprises some 70 genera and 360 species. These resemble the green algae in many respects but differ in the type of storage products (oils), pigments, and the possession by motile forms of two unequal flagella. The genus *Botrydiopsis,* a spherical unicellular form commonly found in the soil, closely resembles *Chlorococcum.* Carotenoid pigments predominate over chlorophyll. Among the larger and more complex of this group is the rather common *Vaucheria* or water felt.

PHYLUM V—Chrysophycophyta (golden algae). Chrysophycophyta (Gr. *chrysos* gold) is a widely distributed and numerous group found especially in freshwater but also in seawater and soil. It is primarily composed of unicellular, flagellate forms. The described forms comprise some 75 genera and 300 species. Cellulose is not found in the cell wall and the membrane may be silicified or calcified. One of the forms, *Synura,* is frequently responsible for the production of bad taste in water supplies.

PHYLUM VI—Bacillariophycophyta (diatoms). The Bacillariophycophyta, or diatoms, instead of representing a distinctive phylum, may also be placed in the Bacillariophyceae, a class in the Chrysophycophyta. The diatoms constitute an important, unique, and widely distributed group of algae. Some 700 genera and 5,000 species have been described. They are found in great numbers in the oceans and freshwaters of the world where they may be planktonic (free-living forms), benthic (bottom-living forms), or epiphytic (growing on stems of water plants). In freshwater streams they often form a slippery olive-brown coat on submerged rocks. The cells are generally elongated, often cigar-shaped, and they consist of two halves. Since one half is smaller and the other larger, the algae are often described as pillbox algae. The overlapping walls form the boxlike *frustules* and are involved in reproduction. A heavy deposit of siliceous materials forms in the cell wall. This serves as a resistant shell that remains intact

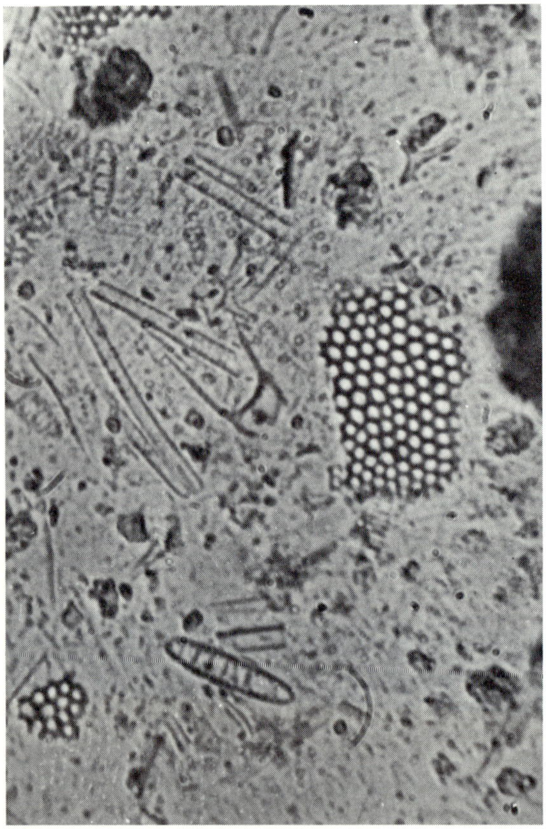

Figure 4-1
Photomicrograph of Lompoc diatomaceous earth showing siliceous skeletal remains of diatoms deposited under prehistoric sea several million years (×1,000).

after the diatom has died and bacteria have decomposed organic portions of the cell.

Because of the almost eternal nature of such skeletons, great quantities of such remains, accumulated as sediments during the Tertiary and Quaternary periods, are still to be found. At Lompoc, California, a huge bed of diatomite, probably the largest in the world, is being dug with power equipment and chemically processed for commercial use. The finished product contains 85% silica. Although diatomite is made up largely of fragments, a good many intact diatom skeletons (frustules) are still visible under a microscope (Fig. 4-1). Diatomite finds wide use as filtering material, polishing powders, and insulation.

The diatoms have no flagella but motility can be observed. The presumption is that water is drawn in through the front end of the cell and forced out through the rear end. This constitutes a kind of jet propulsion, a type of locomotion not often encountered among living things.

Storage materials are oils. In some forms, these may impart undesirable tastes and odors to water supplies.

PHYLUM VII—Phaeophycophyta (brown algae). The Phaeophycophyta (Gr. *phaios* dusky) represent a group of large marine algae, attached or benthic forms. Kelp *(Macrocystis)* is a classic example. Also included are the common sea or rock weeds, *Fucus* and the *Sargassum* of the open sea for which the Sargasso Sea is named. Some 230 genera and 1,500 species are known.

PHYLUM VIII—Rhodophycophyta (red algae). The Rhodophycophyta (Gr. *rhodon* red) are usually large marine forms of warmer waters. A small number are found in freshwaters and a few in soil. Some 400 genera and 1500 species are known. The algae species that furnish agar for bacterial media belongs to this group. Very few of these forms are unicellular. One genus, *Porphyridium,* may grow on moist soils producing deep red patches.

PHYLUM IX—Pyrrophycophyta (includes the dinoflagellates). The Phyrrophycophyta forms, found in fresh, brackish, and sea waters are one-celled flagellates with a red eyespot. They fall into two classes. In the first, the cells have two watchglasslike halves. The second group have a furrow that divides the cell into two parts from which flagella emerge.

The phylum is made up of some 100 genera and 900 species. Similar flagellates are known that have no photosynthetic pigments. These are classed as protozoa. This again emphasizes the thin line dividing simpler plants and animals.

The most notorious member of the dinoflagellates is *Gonyaulax* (Gr. *gony* knee + *aulax* furrow), which at certain seasons of the year causes the "red tide" in ocean waters. Mussels feeding on this form take up a highly toxic substance that makes them poisonous to man. Red tides may also result in the death of large numbers of fish. Zoologists often classify this genus as a protozoon.

PHYLUM X—Cryptophycophyta (Cryptomonads). The Cryptophycophyta represent a small group made up primarily of one-celled flagellates. They divide by longitudinal fission and lack the sexual stage. They have not been widely studied.

Blue-green algae

The blue-green algae occupy a unique position among the algae. They are more primitive, nearer the bottom of the evolutionary scale than the other algae. They are commonly studied with the other algae in courses in cryptogamic botany. However, inclusion of these forms with the algae has been questioned since they have so many characteristics in common with the bacteria. Like bacteria they are procaryotic forms in that they lack the organized nucleus (with its chromosomes and nuclear membrane) of the higher forms. In addition, the blue-green algae have a good many other characteristics peculiar to the bacteria, a fact emphasized by Stainer and van Niel (1941). These distinctive characteristics are as follows:

1. Blue-green algae have a simple bacteria-like chromosome without an accompanying nuclear membrane or other chromosomal structures.
2. Cell walls contain mucopolysaccharides, found also in bacteria but not in higher forms.
3. The cell walls, like bacterial cell walls, can be dissolved by lysozyme.
4. Bacteriophage-like viruses attack the blue-green algae.
5. Unlike the other algae, chlorophyll is not organized into membrane-limited chloroplasts, but similar to the photosynthetic bacteria, is distributed through the cells.
6. In common with the bacteria, cell division is accomplished by a transverse division of the cell.
7. The fairly complex sexual reproduction, common to other algae, is lacking in the blue-greens.
8. Other algae contain sterols in their cells. These are lacking both in the blue-green algae and the bacteria.
9. Nitrogen-fixation is a property peculiar to the blue-green algae and the bacteria.
10. One group of blue-green algae produces endospores similar to those produced by the spore-forming rod-shaped bacteria.

One characteristic peculiar only to the blue-green algae is their apparent lack of flagella. Cell movement is accomplished by a gliding motion, the mechanism for which is still unexplained.

Blue-green algae may be encountered often if one is aware of their existence. They may be seen as a light, greenish-blue froth along the wave-washed shore line. On limestone rocks, constantly moist with seeping waters, the genus *Nostoc* may grow, forming round, slimy balls resembling cooked tapioca. On rare occasions similar tiny *Nostoc* masses are found along lake shores. On wet acidic rocks, blackish slimy growths of *Chroococcus* or *Gloeocapsa* may develop. In tropical mist forests blue-green algae may grow throughout the year on tree bark and leaves. The brightly colored growths surrounding some of the geysers in Yellowstone Park may be rich in blue-green algae. Occasionally slimy masses of blue-green algae form on moist soil. In the deserts temporary pools left by summer thunderstorms develop, after a few days, a bottom film of blue-green algae.

While generally blue, blue-green, or a pale green, the growing algae may exhibit a whole gamut of colors depending upon the amount of light, temperature, pH of soil or water, nutrients, or salinity. The color range may be blue, blue-green, yellow, olive-green, red, violet, or even black. As an example, *Oscillatoria*, a filamentous form, produces blue growth when grown in green

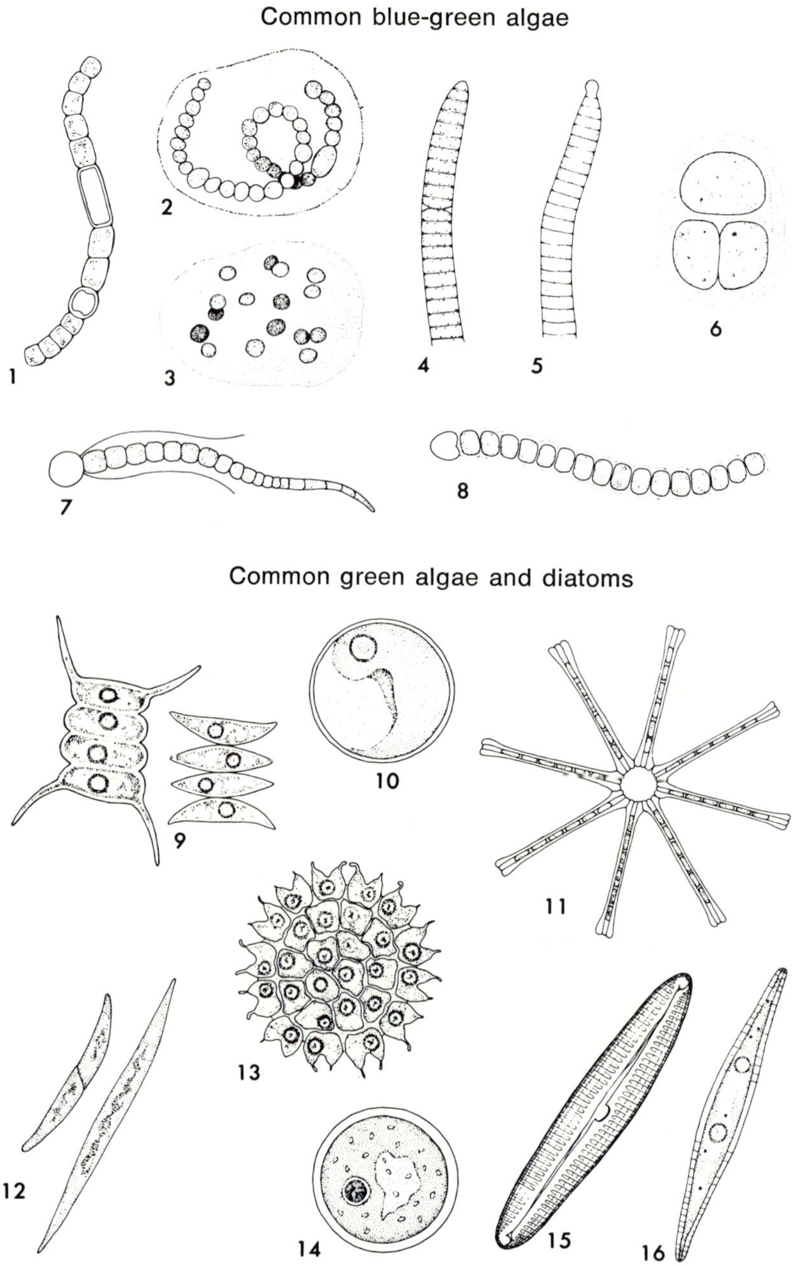

Figure 4-2
Representative types of blue-green algae, green algae, and diatoms. Key to genera: **1,** *Anabaena;* **2, 8,** *Nostoc;* **3,** *Anacystis;* **4,** *Oscillatoria,* **5,** *Phormidium;* **6,** *Chroococcum;* **7,** *Rivularia;* **9,** *Scenedesmus* (two growth forms); **10,** *Chlorella;* **11,** *Asterionella;* **12,** *Ankistrodesmus;* **13,** *Pediastrum;* **14,** *Chlorococcum;* **15,** *Pinnularia;* **16,** *Fragilaria* (cells often lie side-by-side like books on a shelf). (From Anderson, D.: Laboratory instructions in microbiology, St. Louis, 1969, The C. V. Mosby Co.)

or yellow light, a yellow growth in blue light, and green growth in red light.

Botanists place the blue-green algae in the phylum Cyanophyta (or Cyanophycophyta), a term proposed by Sachs (1874). Earlier (1860) because of the slimy growth habit, Stitzenberg had proposed the name Myxophyceae, but because of the possibility of confusing it with myxomycetes (slime molds), this term was not generally accepted. Cohn in 1879, on the basis of the apparent bacteria-like characteristics, suggested the name Schizophyceae. *Bergey's Manual* places the blue-green algae (Schizophyceae) and the bacteria (Schizomycetes) (von Naegeli, 1857) in the division Protophyta (primitive plants) (Sachs, 1874).

Currently the blue-green algae are placed in five orders. The first includes one-celled forms with no endospores or trichomes (growing as slimy threads). The second includes mostly unicellular forms that produce endospores. The remaining three orders are filamentous forms.

Most forms, even the one-celled types, are embedded in a slimy coat. *Nostoc* resembles a string of beads with an occasional large heterocyst. When grown in the dark with glucose as a carbon source, the chain may break up into individual cells.

Oscillatoria is filamentous, with its thread cell embedded in slime. In water, its threadlike trichomes oscillate regularly back and forth, justifying the name *Oscillatoria.*

Merismopedia occurs as a flat plate made up of blocks of four essentially spherical cells. As many as 64 cells may be present in one colonial mass.

Certain strains of *Nostoc* growing in the soil are able to carry on active nitrogen fixation. By this process, gaseous nitrogen of the air is incorporated into the body protein of the cell. In addition, ammonia or related compounds are formed. A portion of these compounds may be released as waste products into the surrounding soil.

Occasionally, large masses of blue-green algae may produce "water blooms" in fresh water ponds. If certain *Anabaena, Microcystis,* or *Aphanizomenon* strains are the predominating forms, animals that drink the water may become seriously ill or even die. One spectacular epidemic of poisoning in Storm Lake, Iowa (1952), due to an *Anabaena* bloom, killed some 5,000 to 7,000 Franklin gulls, 560 ducks, 400 coots and 200 pheasants, 500 squirrels, 11 muskrats, 15 dogs, 4 cats, 2 hogs, 2 hawks, 1 skunk, and 1 mink. Many poisonings of cattle have resulted from drinking water during a blue-green algae bloom. Some cattle have died in as little as a half-hour. Swimmers in water with blue-green algae blooms may develop unpleasant skin rashes. Blue-green algae are also notorious as producers of bad tastes or odors in drinking water.

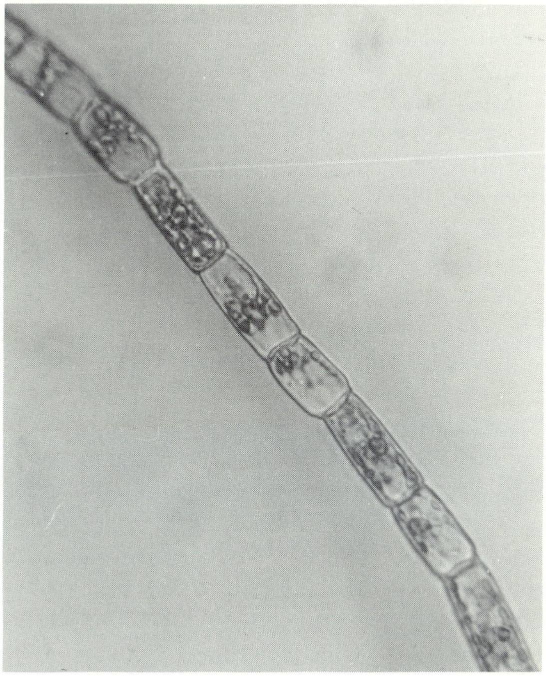

Figure 4-3
Green alga filament showing tubelike morphology and internal organelles.

FUNGI

The fungi (pronounced fun'jī) (Eumycetes) represent a large and varied array of chlorophyll-free plants ranging from mushrooms and the large bracket fungi seen on dead trees, to one-celled yeasts, some as small as large bacteria. In terms of size the mushroom plant is deceptive. The fleshy body is merely the fruiting body. Unseen is a vast array of underground filaments that makes up the vegetative body of the plant. Rarely noticed are the airborne spores.

The term eumycetes (Gr. *eu* true + *mykēs* fungus) is used to distinguish fungi from the so-called slime molds or myxomycetes, which do not fit neatly into either plant or animal categories. However, both the eumycetes and the myxomycetes have two traits in common that separate them from algae and higher plants: (1) they do not possess the photosynthetic pigment, chlorophyll, and (2) because they cannot synthesize the organic compounds necessary for the building of body tissue or for metabolic processes, they must rely on preformed organic compounds to supply these essential life needs.

Nutritionally, like ourselves, fungi are heterotrophic. As such they must obtain nourishment from outside themselves. In contrast, the algae are autotrophic because through photosynthesis, using light as the source of energy, they can build within their bodies the compounds needed for growth and other metabolic processes.

The fungi can feed on a great variety of preformed organic compounds of which carbohydrates seem to be the most readily utilizable. These compounds range from the simple sugar, glucose, to the complex carbohydrate (polysaccharide), cellulose, which is the major component of woody plant tissue. Many fungi can directly utilize proteins, fats or other lipids, organic acids, higher alcohols, and other complex compounds.

Most of the fungi are saprophytes (Gr. *sapros* rotten). As such they attack dead plant and animal remains. The cellulose molecule, resistant to attack by most bacteria and most of the protozoa, is readily broken down by a great group of wood-rotting fungi. The rapidity with which the fallen giant in a forest disintegrates is a silent witness to the enzymatic vigor of the "wood-rotters."

A few fungi are parasitic on living plants or produce plant diseases. A few others attack man, higher animals, fish, insects, and protozoa. These pathogenic (Gr. *pathos* disease + *gennan* to produce) fungi will be discussed in more detail in a subsequent chapter.

One small but amazing fungus possesses a lassolike organ by which it captures nematodes or eelworms in the soil. The fungus then quickly sucks the life substance out of the captive.

The fungi show remarkable enzymatic versatility. In addition to proliferation of their own tissues in the course of growth, they are able to synthesize organic materials into a wide range of important compounds that may accumulate as by-products. These include fats, polysaccharides, organic acids (such as citric acid), enzymes, and antibiotics (especially penicillin). Because of the intrinsic value of many of these growth products, fungi (mostly molds) have found wide commercial application in industrial fermentations.

Since fungi can disintegrate cellulose with great rapidity, they pose a threat, especially in the tropics, as destroyers of the timbers of houses, of paper products, tent cloth, and many other cellulose-containing compounds. They are also economically important as spoilers of fruits, vegetables, meats, and other foods.

The major habitats of the fungi are the soil, the forest floor, or any other areas where quantities of dead organic matter are present. This organic matter becomes the reservoir from which spores of fungi, especially the molds, may be wafted by every vagrant breeze. Lighting on any suitable moist organic surface, the spores may quickly germinate and grow into fungus plant masses that lead to decay or spoilage. Fungus spores in the atmosphere may be the cause of allergic reactions such as hay fever or asthma. Indeed, they may run a close second to pollens as allergens.

Some disease-producing fungi apparently

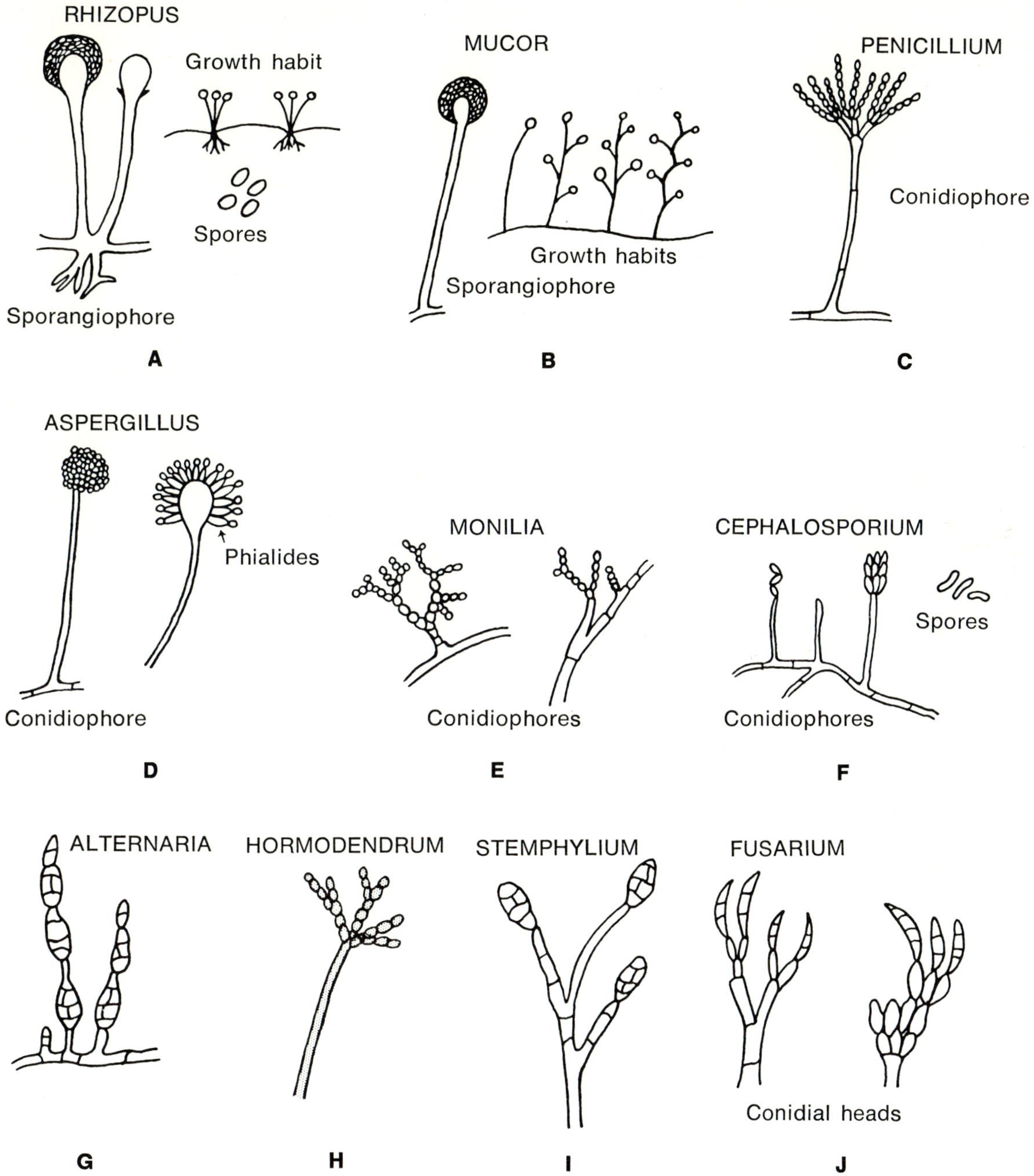

Figure 4-4
Fruiting bodies of commonly encountered molds. (From Anderson, D.: Laboratory instructions in microbiology, St. Louis, 1969, The C. V. Mosby Co.)

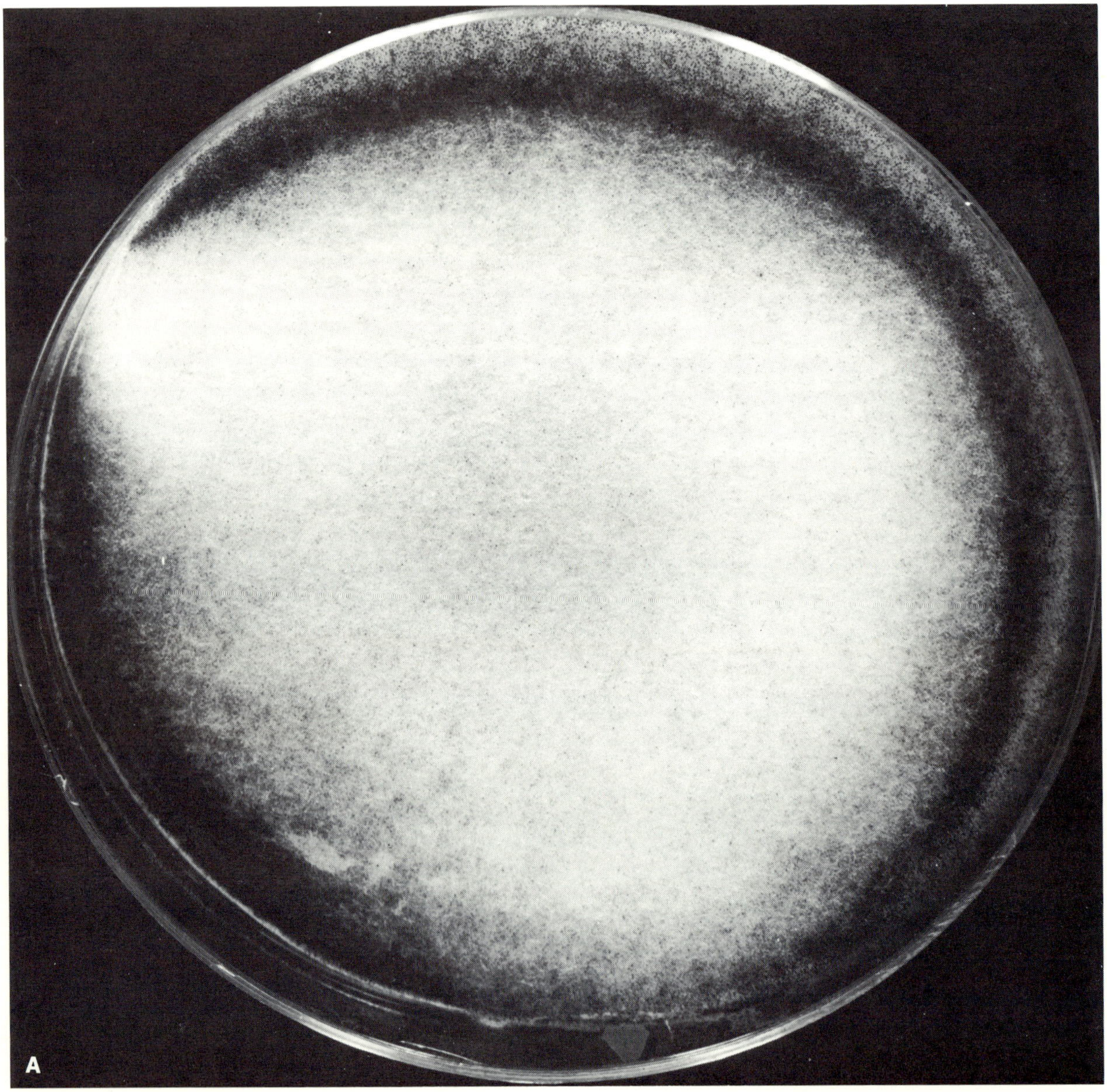
A

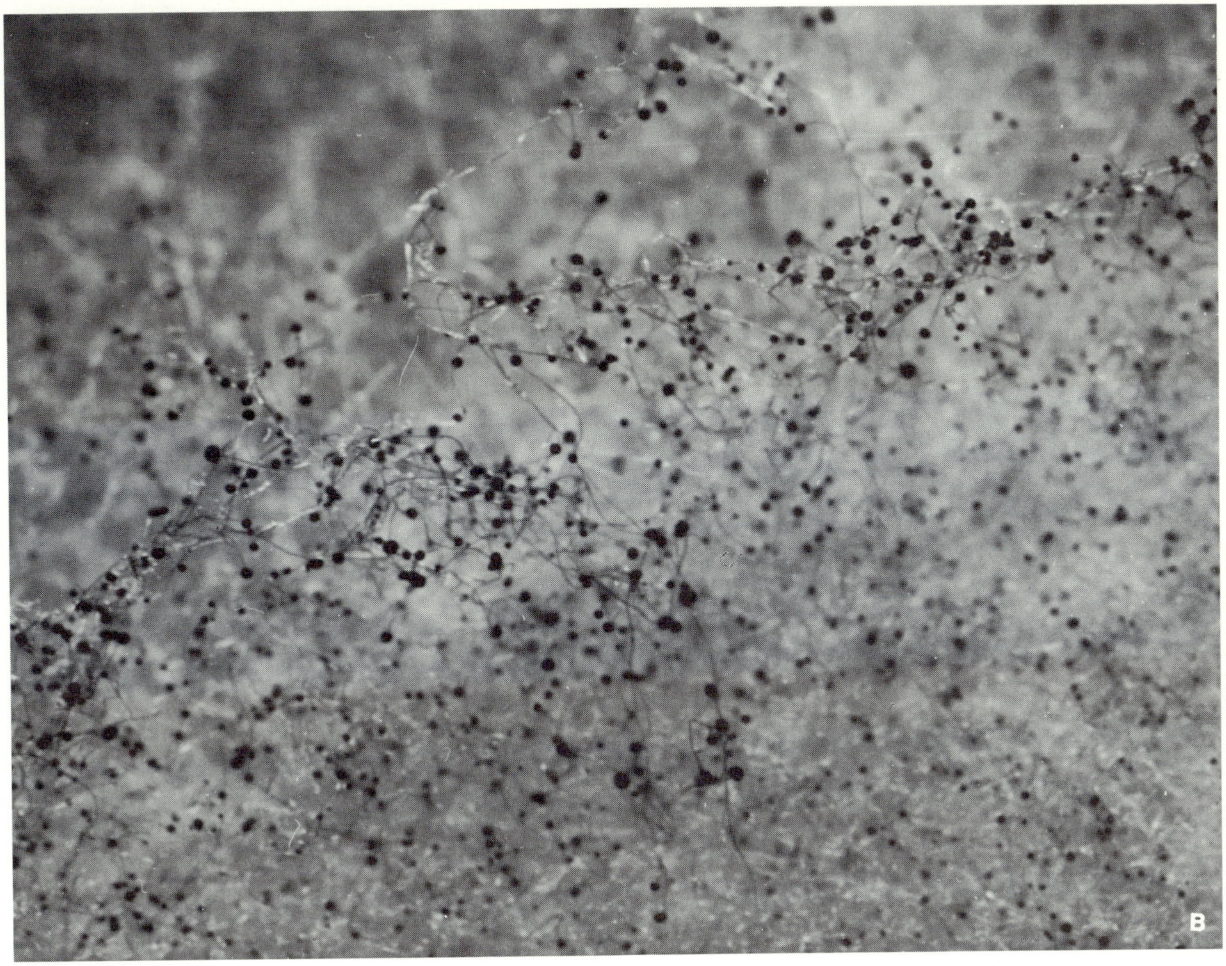

Figure 4-5
A, Petri plate completely filled with *Mucor* mycelia. Note concentration of sporangiophores (sporing bodies) around edge of plate, probably due to more available oxygen.
B, Looking down on a growing culture of *Mucor* sp. showing aerial distribution of mycelia and sporangia (×50).

spend most of their lives in the soil, with the pathogenic phase an incidental one. A small number of forms are found in freshwater or seawater. These are grouped together as water molds but they represent a varied crowd related primarily by their environment.

The specialist who devotes his life to a study of the whole group of the fungi is termed a mycologist.

The microbiologist is primarily concerned with the simpler molds and yeasts that might be classed as microorganisms. Primary attention will, therefore, be given to the molds and yeasts in this discussion.

The investigator whose major interest involves the mildews, smuts, and rusts that cause plant disease is termed a plant pathologist. A subsequent chapter will be devoted to plant diseases.

MOLDS

Mold is a trite or convenience term with no real botanical meaning. It comes from an Old English word meaning dust. The common molds are representatives of three different major classes of the fungi, the Phycomycetes, the Ascomycetes and the Fungi Imperfecti. Characteristics of the fruiting bodies of these forms are compared in Fig. 4-4.

Molds in the major groups of the true fungi

LOWER OR PRIMITIVE FUNGI

CLASS I—**Phycomycetes (The filamentous fungi).** The vegetative filaments of the Phycomycetes (Gr. *phykos* seaweed or alga + fungi) have few cross walls and vegetative spores generally are formed in a spore case or sporangium. Sexual spores when present develop as zygospores (Gr. *zygon* yoke). Mold growth is usually cottony. The most commonly encountered molds in this class are *Rhizopus* and *Mucor*.

HIGHER FUNGI

CLASS II—**Ascomycetes.** The second class of fungi, Ascomycetes (Gr. *askos* sac), encompasses some 15,000 species including large, fleshy fungi (some edible), plant pathogens, and common powdery molds. The common yeast also falls in this group. The characteristic that all have in common is the ability to produce sexual spores in an ascus or sac. Asexual spores are borne naked, not in a spore case. The mycelia generally show numerous cross walls or septae.

Included in this class are various plant pathogens such as powdery and sooty mildews, causative agents of slime flux of elms and black knot of plumes, as well as *Claviceps purpurea*, which causes ergot in rye.

The most common genera of molds of this class are the powdery blue, green, yellow, or pink, or occasionally brownish forms. They include *Penicillium* (most commonly encountered of all the molds), *Aspergillus*, and *Neurospora* (pink bread mold).

CLASS III—**Fungi Imperfecti or Deuteromycetes.** The class Fungi Imperfecti is so named because no perfect or sexual stage has been found. Some 20,000 species have been described. Many of the dark-colored, powdery molds fall in this group. Spores are borne naked, not in a spore case. Colors most frequently seen are olive drab, brown, dark green, and black. Common molds in this class are *Monilia (Candida)*, *Alternaria*, *Cephalosporium*, *Hormodendrum*, and *Stemphylium*. The mycelia are septate—possessing cross walls or septae.

CLASS IV—**Basidiomycetes.** The name, Basidiomycetes (Gr. *basis* foundation or base), derives from the fact that spores, often four in a group, are borne on a base cell or basidium. The major members of the group are mushrooms, toadstools, and related fleshy fungi. The smuts and rusts, with their complicated life cycle, also fall in this group. Members of this class are not commonly encountered or cultivated by the microbiologist, except by the plant pathologist.

Growth habits of molds in four major classes

CLASS I—**Phycomycetes.** Members of the class, Phycomycetes grow either as single cells (often with a motile stage) or as cottony filaments that tend to be nonseptate or coenocytic (Gr. *koinos* common + cyte or cell). These filaments are essentially tubes with numerous nuclei, hence they are termed common or shared cells because nuclei are not separated by a septum or cross wall. Because of the filamentous nature of the Phycomycetes and their preference for a moist habitat, they are classed as the algae-fungi. They are probably lower in the evolutionary scale than other molds.

A number of the forms in the class Phycomycetes are found in water, including the following:

1. Chytridiales or the chytrids produce few if any mycelial threads but have motile zoospores. These are often claimed as fungi by the botanists and as protozoa by the zoologists. They are mostly saprophytes but some attack algae or even higher water plants. A few plant diseases are attributed to this group.
2. The Saprolegniales (Gr. *sapros* rotten + *legnon* border and ales or of the order) are considered true molds. The name is applied because filaments of the mold often grow out as a white beard or fringe from the infested area on a fish, piece of wood stem, leaf fragment, or water plant. They often kill fish in home aquariums. These forms can be grown quite readily by baiting water with hemp seeds. The filaments grow as a fringe on the seeds and can be isolated from this source and cultivated on a suitable agar medium. They can also be isolated from moist soils adjacent to freshwater. *Pythium*, a plant pathogen that causes "dampening off" of seedlings, is related to this group of lower fungi.

From the viewpoint of the microbiologist the most important member of the Phycomycetes is the subclass Zygomycetes. These forms produce a branched mycelium (fungus nail), made up of threadlike filaments largely devoid of cross walls. Asexual spores are borne primarily in a spore case or sporangium, no motile cells are formed, and occasionally sexual spores (zygospores) are produced. The sexual spores are formed by the tip-to-tip union of unlike (essentially male and female) mycelial threads. Where the tips come together, the two nuclei are both walled off by cross walls. The cell walls between them dissolve and the two nuclei then unite. These first undergo diploid mitotic division, then meiosis, that is, reduction-division occurs, with the formation of haploid spores. Reassortment of genes occurs at this time.

The most familiar member of the Zygomycetes is *Rhizopus nigricans*, the black bread mold. This is distinguished by its growth, which resembles dirty cotton, and the many visible, tiny, pepperlike spore cases. Another commonly seen member of the group is *Mucor*. In the laboratory *Rhizopus*, *Mucor*, or other related members of the zygomycetes can be isolated readily by placing a small amount of moist soil or moist organic matter in an agar plate of suitable nutrient medium such as Sabouraud's agar. The mold plants begin to grow rapidly and soon fill the plate with a cottony mass of growth. If dry soil is used, vigorous *Mucor* or *Rhizopus* growth seldom develops because actively growing mycelia are absent and the spores seem to germinate more slowly than those of some other molds.

The asexual fruiting body develops from the vegetative mycelia as a round spore case filled with spores and attached to fairly long threads, the fertile hyphae (Gr. *hyphos* web). Because of this fruiting structure in which the sporangiophores resemble pins, *Mucor* and *Rhizopus* are often called "pin molds." The individual spores, released when the spore case is ruptured, are termed sporangiospores. The case with its enclosed spores, the sporangium (Gr. *sporos* seed + *angeion* vessel) and the spore case together with its supporting fertile hypha is termed a sporangiophore.

The primary difference between *Rhizopus* and *Mucor* is that with *Rhizopus*, two or three sporangiophores arise from the same spot on the vegetative hypha and at the base a rootlike structure, the rhizoid (Gr. *rhiza* root + *eidos* form), with the appearance but not the cellular make-up of true roots, also develops. In *Mucor* the sporangiophores arise singly and no rhizoids are formed. The fruiting habits of these forms are shown in Fig. 4-6.

Figure 4-6
Fruiting bodies (sporangia) and fertile hyphae of *Mucor* sp. (×200).

CLASS II—**Ascomycetes.** The molds of the Ascomycetes include the forms most commonly encountered in the laboratory—*Penicillium* (L. *penicillum* brush), and *Aspergillus* (L. *aspergere* to scatter or sprinkle). *Penicillium* is the most ubiquitous of all molds. It is seen as a blue, green, or occasionally yellow powdery growth on cheese, oranges, apples, jams and jellies, hams or other smoked meat. It is the ripening agent of the blue cheeses and the producer of penicillin. Its distinguishing characteristics are color, powdery appearance of growth, and the production of long chains of conidiospores (Gr. *konis* dust) like a paint brush at the tips of fingerlike sterigmata (Gr. *sterigma* support). When the long filaments of vegetative hyphae of a colony of *Penicillium*, growing on the surface of an agar plate, are examined under the microscope through the bottom of the plate, frequent cross walls or septae may be seen.

The growth habit of *Aspergillus* is similar, but in the sporing head, chains of spores grow out in every direction from spikelike sterigmata that arise from a round central structure (Fig. 4-7). When devoid of spores, the sporing head resembles a spiked war club. Colonies may be blue, green, yellow, or brown. The aspergilli are primarily saprophytes.

Neurospora grows as a powdery but less compact colony than *Penicillium* or *Aspergillus*.

However, conidiospores are formed in profusion, and an occasional thin mycelial filament may be seen. The most common color of the colony is a light pink although occasionally it may be a fairly deep red, orange-yellow, or almost pure white. *Neurospora crassa* has been widely used in genetic studies.

Class III—Fungi Imperfecti. The class Fungi Imperfecti is something of a paradox. It is made up of a very large crowd of powdery molds that produce neither ascospores, zygospores, nor basidiospores. Many of these forms probably should belong to the Ascomycetes. For example, until relatively recently when ascosporogenous forms were first found, *Penicillium* and *Aspergillus* were included in the Fungi Imperfecti.

The most frequently seen molds of this group are *Monilia*, *Oospora*, and members of the family Dematiaceae, a group of dark-colored molds in which the conidiospores or fertile hyphae, or both, are dark or smoky.

The *Monilia* colonies are generally olive drab, relatively small, compact, and velvety. Conidia arise from short, fertile hyphae (see Fig. 4-4). Occasionally white or purple forms may be identified as *Monilia*, but these may be instead imperfect stages of *Neurospora*.

Oospora may be isolated from dairy products, especially spoiled cottage cheese. It grows on culture medium to produce a flat, tight growth made up of fine filaments growing close to the medium. The nature of growth depends on the isolation medium. The conidiophores are short and made up of a short line of single ovoid spores like a strand of elliptical beads that vary somewhat in size.

The most frequently encountered genus of the Dematiaceae is *Alternaria*. Its colony is dark brown to blackish with sparse vegetative filaments and very distinctive pear-shaped conidiospores divided into irregular segments. It often grows on tomatoes or strawberries.

The pink, flat colonies of *Fusarium* with its sickle-shaped conidia are found occasionally, especially on plates seeded directly with soil. This genus belongs to the family Tuberculariaceae and is a fairly common soil inhabitant. One species causes a serious potato disease. See Fig. 4-4 for typical fruiting bodies.

By using selective media such as Sabouraud's dextrose or maltose agar, or other media high in sugars and sufficiently acid to inhibit bacterial growth, a wide variety of interesting molds can be isolated from soil and the environment in general.

Some molds that do not produce fruiting bodies are difficult to identify. These are termed Mycelia Sterilia (Fig. 4-8).

YEASTS

Like the term mold, the term yeast has no specific botanical significance. It is a convenience term, given stature because of its long usage in the English language. While the origin

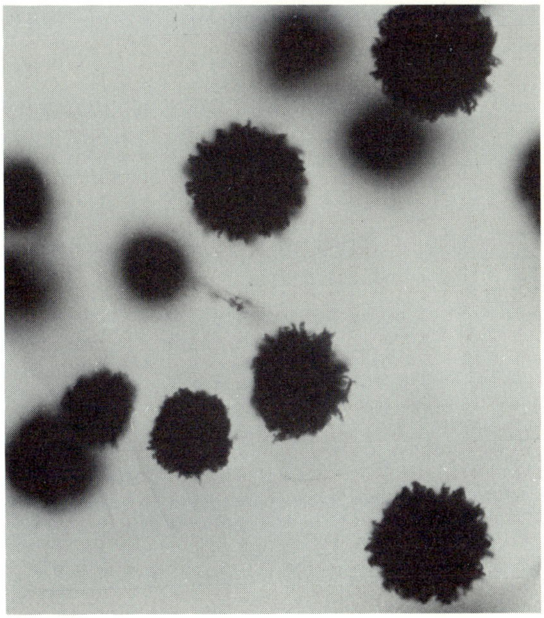

Figure 4-7
Conidiophores (sporing heads) of *Aspergillus niger* as seen when looking down on petri plate culture (×200).

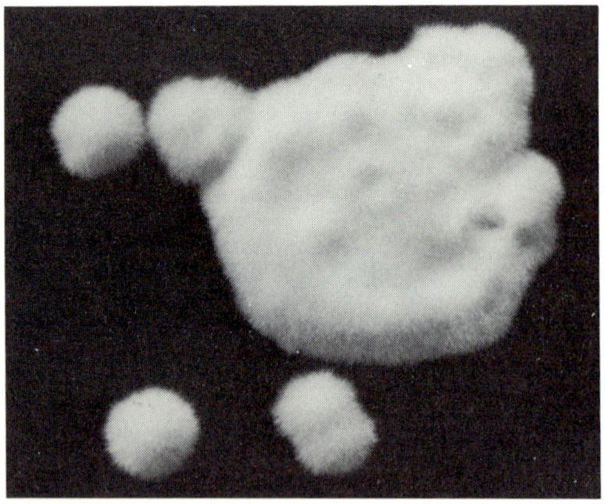

Figure 4-8
Mycelia Sterilia colony, so called because of absence of identifying fruiting bodies (×2).

of the term yeast is not clearly known, it apparently came from the Old English or Dutch *gist,* meaning to foam or boil, as seen in fermenting beer.

The yeasts are true fungi but differ from the molds in that they normally grow as single cells, rarely forming the filaments characteristic of the molds. They apparently contain representatives of three different classes of fungi. A few rare forms are directly related to the Basidiomycetes. These forms shoot out bulletlike ballistospores similar to the basidiomycete, *Sporobolus.* However, most of the yeasts fall into one of two classes—those that produce ascospores (ascosporogenous forms) are placed in the class, Ascomycetes (the familiar commercial yeasts fall in this group), while those that do not produce ascospores (anascosporogenous forms) are grouped with the Fungi Imperfecti. As with the molds, nutritional requirements, growth habits, and other physiological characteristics suggest that these two common groups of yeasts are closely related.

Yeasts are found on the surface of fruits, in the nectar of flowers, the flux that flows from trees, soils (especially vineyard or orchard soils), water, in the mouth and intestines of humans and animals, and in a number of other habitats.

The nutritional requirements of the yeasts include relatively high concentrations of sugar plus some preference for acids. This explains the large numbers in vineyard and orchard soils on which acid fruits have fallen. The yeast cells on fruit surfaces and possibly in the flower nectar may well come from the adjacent soil. Bees carry yeast cells from flower to flower.

The production of ethanol (ethyl alcohol) and carbon dioxide from sugars is characteristic of some but not all yeasts. We are prone to consider yeasts as the only microorganisms that produce alcohol. This misunderstanding arises from the fact that, commercially, certain yeasts have been selected for this ability. Their value lies in the fact that they are homofermentative. The primary product of the fermentative breakdown of sugar is ethanol + carbon dioxide. With certain strains, more than 90% of the sugar that is fermented to produce energy (not utilized to form body tissue) may be converted into these products. In beer fermentations, the percent of alcohol may reach 7% to 8%. The actual concentration will depend on the tolerance of a particular yeast strain to ethanol, its own waste product. Fermentation is an anaerobic process. Its value to the yeast is primarily as a source of energy for life processes.

Alcoholic fermentation is only one of a wide variety of microbial fermentative processes. Also included are the lactic acid fermentation by bacteria in milk and a number of other familiar or unfamiliar growth reactions.

Some yeasts are heterofermentative, producing not only ethanol but also a variety of other products that may be undesirable in terms of taste, odor, or toxicity. Hence, when these yeasts grow they produce spoilage. Some spoilage yeasts may be primarily oxidative; that is, they convert sugars directly into carbon dioxide plus water, producing undesirable clouding and taste in fruit juices, or film on the surface of fruit juices, dill pickles, or other products.

The most common method of yeast cell reproduction is by budding (Fig. 4-9). In budding, the mother cell begins to form a protuberance

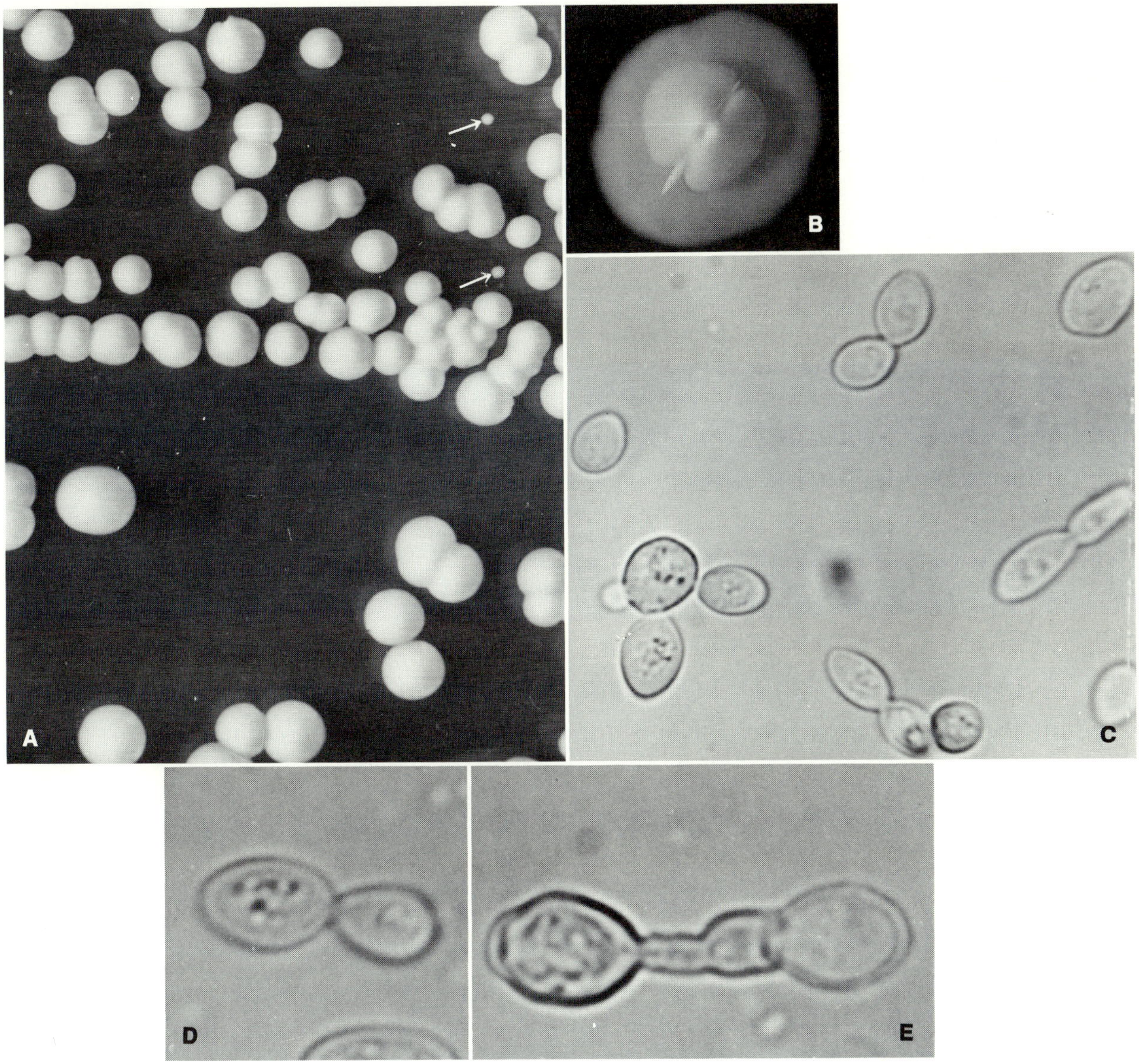

Figure 4-9
A, Typical raised, domed colonies of *Saccharomyces cerevisiae,* directly isolated from commercial yeast sample on nutrient agar plus 4% sucrose. Small colonies shown microscopically to be haploid strains (×3). **B,** Unusual giant colony of *S. cerevisiae* on Sabouraud's agar after 14 days growth (×3). **C-E,** Types of budding seen in *S. cerevisiae* (×1,600).

from a point on the cell wall. This continues to enlarge and a constriction develops. In the process, at a certain size usually considerably smaller than the mother cell, this protuberance breaks off at the point of constriction leaving a scar. Occasionally a bud will form another bud while still attached to the mother cell, and sometimes a third bud may develop from the second. When this is repeated several times, a chain of buds resembling a filament will develop. Counts of bud scars suggest that a given cell can produce only a certain number of buds.

In some yeast strains, instead of budding the cells grow into long rodlike cells. When these reach a certain size they divide into two equal-sized cells by forming a septum or cross wall. This resembles cell division in rod-shaped bacteria.

The ascosporogenous yeasts, such as the common commercial yeast, *Saccharomyces cerevisiae* (Gr. *sakcharon* sugar and L. *cerevisiae* beer), grow as elliptical cells about the size of a red blood cell and several times larger than a bacterial cell. They are diploid, apparently with 14 to 15 pairs of chromosomes. The cells normally can continue to reproduce by asexual budding without producing ascospores. Anaerobic conditions and the production of alcohol tend to inhibit ascospore formation. Cells can be stimulated to grow in a medium high in sugar enriched with plant extract such as beet root and top extract. If a drop of such vigorously growing cells is placed on a small block of plaster of Paris, ascospore formation will occur. This is apparently due to the sudden increase of oxygen and the rapid loss of nutrients, which are absorbed into the plaster. The ascospores, when completely formed, will be slightly larger than the parent vegetative cell and will contain four round haploid cells. These will contain one set of chromosomes (14 to 15 single chromosomes) instead of the 14 to 15 pairs of chromosomes in the parent cell.

Single spores can be isolated by using a microscope equipped with a micromanipulator. When the spores are grown, they will generally form haploid cultures instead of the diploid cells of the parent culture. However, in some cases when the ascospores germinate, adjacent haploid cells extend out protuberances that then unite with a protuberance of the adjacent cell. The nucleus of one cell enters the other cell and a diploid is formed. Whether or not the isolated culture remains haploid or immediately becomes diploid depends on genetic factors.

In haploid cultures, the cells are round and smaller than the large ellipsoidal cells of the diploid strains. The haploid strains grow less vigorously and are less effective fermenters than the diploid strains. Occasionally, in examining commercial yeasts microscopically, one sees both the diploid cells and an occasional haploid cell. Because haploid strains can be isolated, it is possible to produce genetic crosses in tame yeasts to yield strains with special desirable characteristics much as higher plants are crossed to produce superior strains.

The commercially important yeast strains are sometimes arbitrarily classified as tame and wild types. The tame types are essentially domesticated microbes. Some of these strains have been carried by brewers, distillers, or yeast manufacturers for many years and through hundreds of transfers. They may be selected and kept with great secrecy to produce a particular type of beer or baking yeast. Yeast geneticists are often employed to produce new and superior strains.

With wines, while tame yeasts may be added to the grape juice to produce a certain type of wine, some wineries depend on the wild yeasts that are present on the grapes to produce the desired bouquet of the finished product. This explains in part why wines of some seasons are superior to those of others. Most of the commercially important fermentative yeasts are diploid but occasional haploid strains are encountered, especially with spontaneously fermented wines.

The presence of alcohol-fermenting yeasts in the atmosphere is witnessed by the fact that when syrup or honey is diluted to a 5% to 10% solution and exposed in an open container, an alcoholic fermentation may occur before mold growth can cover the surface. However, both honey and syrup may harbor osmophilic yeasts. These grow slowly in the product. Therefore, to prove that the fermenting yeasts came from the air, the diluted honey or syrup must be sterilized by boiling before exposing it to the air. Most of the anascosporogenous strains of yeasts associated with spoilage or disease production are haploid. They are discussed in a subsequent chapter.

LICHENS

As a classic example of mutualism or symbiosis, lichens are of interest to the microbiologist. They represent an associative type of growth in which blue-green or green algae, primarily unicellular forms, grow enmeshed in fungus mycelia of higher fungi, mostly ascomycetes (not too far removed from the puff balls) or more rarely basidiomycetes (relatives of the mushrooms).

The algae, being photosynthetic, furnish the energy and carbon sources for the fungi, while the mycelia of the fungi provide mechanical support and hyphal bodies serve to absorb water, which becomes available to the algae. The relationship appears to represent a mild parasitism on the part of the fungi.

The lichens are familiar as flat-growing bodies with leaflike structures or granular surfaces seen on tree bark, shingles, posts, or rocks. They range in color from grayish or yellowish green to orange, yellow, brown, black, or white. Although erroneously called moss, they lack the distinct bright green color and gross surface structure of true mosses.

They grow well in adverse environments such as high mountains, deserts, the Arctic or Antarctic. They are, in fact, the major form of plant life in the Antarctic. Although their growth is usually inconspicuous, in the Arctic they form the so-called reindeer moss, which may sometimes grow up to four feet or so in height. They have some practical value. In parts of Lapland, Iceland, and the Orient they are harvested, dried, and used as human food. They were a source of dyes in early times, yielding bright blue, red, yellow, and orange dyes.

Litmus, now largely synthesized, was first obtained from two different lichens that grew in Holland. Lichens have been used as a source of tannin for tanning hides. In parts of Russia and Siberia they have also been substituted for hops in the brewing of beer.

Most of the lichens can resist adverse dry periods. They may move from one place to another as wind-borne fragments of algae cells and fungi mycelia, or the component algae and fungi may in some cases be transported independently.

They apparently served as one of the foods of the ancient Egyptians and are thought by some to be the manna of the fleeing Israelites.

The lichens are a widely varied group, of which some 15,000 species have been described. This great variety of species is due primarily to the many possible combinations of fungal and algal types that may make up the lichen bodies.

SUGGESTED READINGS

Alexander, M.: Introduction to soil microbiology, New York, 1961, John Wiley & Sons, Inc.

Christensen, C. M.: Molds and man, Minneapolis, 1951, University of Minnesota Press.

Hale, M. E.: The biology of lichens, London, 1967, Edward Arnold (Publishers) Ltd.

Jackson, D. F., editor: Algae and man, New York, 1964, Plenum Publishing Corp.

Pfaff, S. A., Miller, M. W., and Mrak, E. M.: The life of the yeasts, Cambridge, Mass., 1966, Harvard University Press.

Waksman, S. A.: Actinomycetes, nature, occurrence and activities, Baltimore, 1959, The Williams & Wilkins Co.

Waksman, S. A.: Soil microbiology, New York, 1957, John Wiley & Sons, Inc.

5 The animal-like higher microorganisms

PROTOZOA
MYXOMYCETES
ACRASIALES OR CELLULAR SLIME MOLDS

In the world of microorganisms the line of demarcation between plants and animals is not as distinct as it is in the higher forms. However, the protozoa and mycetozoa (myxomycetes) can conveniently be considered as the animal representatives of the microbial world if the interrelated forms are ignored.

PROTOZOA

The Protozoa (Gr. *protos* first or primitive + *zoon* animal) (singular—protozoon) represent the simplest members of the animal kingdom. This simplicity, together with their relatively small size and primitive nature, justifies their inclusion in the Protista. They are often described as one-celled (unicellular) animals; however this is something of an oversimplification. Although generally functioning as one-celled forms, they vary greatly in size and complexity. Their sizes range from 2 μm (about the size of the larger bacteria) to about 3,000 μm (large enough to be barely discernible with the naked eye).

Most of the protozoa possess a single nucleus, while others may have 2 to 4 nuclei. A few of the very large forms may have hundreds of nuclei scattered throughout their bodies. In addition, a few produce colonies made up of many cells in which some cells exhibit a degree of specialized function.

Protozoan cells are metabolically and structurally fairly complex, often possessing a considerable number of organelles or tiny organs, each with a specialized function. In one large form, some 42 different organelles have been described.

Taxonomically, the protozoa represent a large and varied group. Some 30,000 species have been described and some 25 of these species have been found as parasites or pathogens of man.

Habitats of protozoa

In nature protozoa are found primarily in water, especially freshwater. Soil is another important habitat. The numbers of protozoa may be fairly high in moist soil, but some are able to grow well even in the thin water film surrounding particles of relatively dry soil.

Lakes, sluggish streams, and the seas furnish a permanent abiding place. Even any temporary body of water, such as a puddle, holes in stumps, or water in a tree crotch at times teems with protozoa. Melting snow from glaciers or snow banks may often provide a home for some protozoa. Indeed, the snow itself may possess protozoa (or algae) in such numbers that they impart color to it. Although a few of these forms can grow at low temperatures, the optimum temperature for most free-living protozoa is between 16° and 25° C (61° and 77° F). A few rare forms have adapted to life in hot springs. Japanese workers found some of these unusual forms in a hot spring with a temperature of 56° C (133° F).

In oceans and seas, some protozoa are characteristically found in the shallows and others in the depths. Great numbers occasionally develop in open areas of the Arctic or Antarctic. Here the waters almost become a protozoan "soup." Forming, as they do, the lowest link in the food chain, such protozoan swarms may account for the great catches of fish in such areas. Tropical seas are generally lower in protozoan numbers and many of these protozoa are bizarre in appearance. Unique forms develop in such harsh environments as the Great Salt Lake or Dead Sea that have salt concentrations of 20% or more.

Many common forms are ubiquitous. Thus the same protozoan species are found in freshwater ponds of New Zealand, Germany, or California, or in soils from widely divided areas of the earth.

Waters low in organic matter and rapidly moving waters generally show a meager protozoan fauna. On the other hand, the first 6 inches of soil may harbor a good many protozoa per gram. Typical counts range from 10,000 to 100,000 and even sometimes as high as 300,000. These estimates are made by adding soil dilutions to plates of plain agar heavily seeded with bacteria such as *Enterobacter aerogenes.* No doubt counts arrived at by this method may be considerably lower than the numbers actually present in the soil.

The universal distribution of protozoa in nature is due largely to their ability to form tough, resistant cysts that remain viable for long periods under adverse environmental conditions. Thus, when a puddle of water once abounding with protozoa dries, many of these forms remain alive as cysts, continuing in a state of quiescence until the next rain or they are swept up into the air by a violent wind and redeposited miles away by another downpour of rain. In oceans, protozoans rarely encyst because this offers no advantage for survival.

When cysts germinate or excyst, a series of steps are involved. Moisture must first be taken in to permit a regeneration of organelles that degenerated during encystment. This is followed by regeneration of cilia or flagella and the active vegetative life or trophozoite (Gr. *trophē* nutrition + *zoon* animal) of the protozoon begins. Reserve food stored at the time of encystment furnishes the energy and building materials necessary for excystment. A period of dormancy is usually required between cyst formation and excystment. This is commonly a few days or weeks, but may be as long as 2 to 9 months.

Factors that cause free-living water (or soil) forms to encyst are evaporation of moisture, change in pH, sudden increase or decrease in oxygen, raising or lowering of temperature, accumulation of certain products of metabolism, the sudden growth of certain objectionable bacteria, or overpopulation. In fresh waters and soils, definite seasonal fluctuation of the protozoan population occurs. Excystment of the protozoa follows renewed moisture, adequate oxygen, optimum temperature, addition of certain plant materials, and the resultant burgeoning of the bacteria that provide nutriment.

Nutrition of protozoa

Protozoa vary greatly in their nutritional requirements. A few forms possessing chlorophyll may carry on active photosynthesis or perhaps live a partially saprophytic existence. These represent a disputed group that the botanists claim as plants by virtue of their photosynthetic habit and the zoologists insist are protozoa because of their motility and other animal-like characteristics.

In contrast with the self-sufficient photosynthetic forms, many of the protozoa are carnivore-like. They dart to-and-fro, pursuing and devouring other smaller protozoa, bacteria, algae, and even the tiny metozoans. Some of the bacteria-devouring forms appear to have a definite preference for certain kinds of bacteria. Occasionally, bacteria-eaters have been seen to writhe in apparent agony or even die after ingesting certain bacteria; a probable case of microbial food poisoning.

Some protozoa, holozoic types, engulf and ingest particles of organic matter. Others, the saprozoites, resemble the saprophytic bacteria in absorbing nutrients in solution from the growth environment through their external membranes. Many forms secure their nutrition in more than one way.

Some forms that ingest whole microorganisms do so through permanent mouthlike structures or gullets, that often are surrounded by a fringe of hairlike cilia that sweep the particulate matter or other microbes into the gaping mouth (Fig. 5-2). Other forms, such as the ameba, may send out pseudopodia (false feet) that engulf the food or may form cuplike depressions that function as a temporary mouth.

Reproduction of protozoa

Reproduction in the protozoa is accomplished in a number of ways. In many of the true unicellular forms, it involves asexual binary fission. In this case the mother cell divides longitudinally into two similar-sized daughter cells, leaving no parent cell. This resembles, but is not identical with, the transverse fission of bacteria since the rigid cell wall and septum formation of bacteria are lacking in the protozoa.

Among the multinucleate protozoa, binary fission may occur, but the size of the daughter cells and number of nuclei may vary considerably. This is called plasmotomy (Gr. *plasma* molded + *temnein* to cut).

In other forms, the parent cell may produce a smaller daughter cell or bud with the same sized nucleus but less cytoplasmic material. This is similar to the budding observed in yeasts. In some, multiple budding may occur.

In schizogony a series of fairly rapid nuclear divisions occurs and a number of small uninucleate buds are split off. These grow quickly into new organisms. Such reproduction is seen in the malarial parasite.

Sexual reproduction may occur in which cells first form a uninucleate haploid gamete by meiosis. Among the ciliates, such as *Paramecium*, after haploid gamete formation, conjugation occurs between compatible mating types. This is analogous to male-female sexual contacts in high animals. At the point of contact between the two conjugation cells, the pellicles fuse and pores apparently open to allow the haploid gametes to unite to form a diploid nucleus. With some forms such sexual reproductions rarely occur. With others, sexual conjugation occurs at intervals, interspersed with a fairly large number of asexual reproductions. Such periodic sexual reproductions seem to be required by some strains to prevent the strain from degenerating and dying out.

Behavior of protozoa

A major animal-like characteristic of the protozoa is their marked response to external stimuli. This results in their swimming toward or away from light, oxygen, or low or high temperatures much as fish would do. They often pursue their prey or move away to avoid danger. Cilia may serve as organs of feeling (tactile organs) as well as organs of propulsion. Some larger forms such as *Paramecium* may even have a neuromotor system that resembles the nervous system of higher animals. The photosynthetic *Euglena* have a red spot or eyespot or stigma containing some 50 tiny granules made up mostly of β-carotene.

Major classes of protozoa

CLASS I—Flagellata or Mastigophora. The Flagellata (L. whip) or Mastigophora (Gr. *mastix* whip + *phoros* carrying) are the most widely distributed of the protozoa. They are found in a wide variety of environments, particularly in water. Their habitats may range from glacial waters to hot springs, and they are often encountered in moist soils. Some have even become parasitic (or pathogenic) in animals. They are generally spindle- or fish-shaped with one to four long, whiplike flagella inserted in a sort of depression at one end of the cell. The flagella impart very active motility. Others have stages that are almost ameba-like and others develop a shell-like coating. Asexual reproduction is by longitudinal fission. In some forms, divisions may occur during the encysted state.

A subclass, the Phytomastigina represents a biological enigma in that its members do not fit neatly into our definitions of either plants or animals. They are highly motile flagellates (an animal characteristic) but many strains possess chlorophyll and are able to carry on photosynthesis. (Some mutant strains lack chlorophyll and live a saprophytic or saprozoic existence.) Pigmented members of this group are claimed as motile algae by the algologists or motile protozoan flagellates by the protozoologists.

The major photosynthetic flagellates are the unicellular forms, *Euglena* and *Chlamydomonas*, and the colony-forming *Volvox*, an actively motile ball-like arrangement of cells, with each outer cell being flagellate. These *Volvox* balls are often as large as 1/10 inch in diameter, making them readily visible to the naked eye. At times, either the two unicellular or the one colonial flagellate (or all three) may grow in such profusion in freshwater that they form a green pond scum.

Many of the dinoflagellates that have a shell-like covering are also able to carry on photosynthesis. These again are claimed by both algologists and protozoologists. *Gonyaulax* species are the most notorious members of this group.

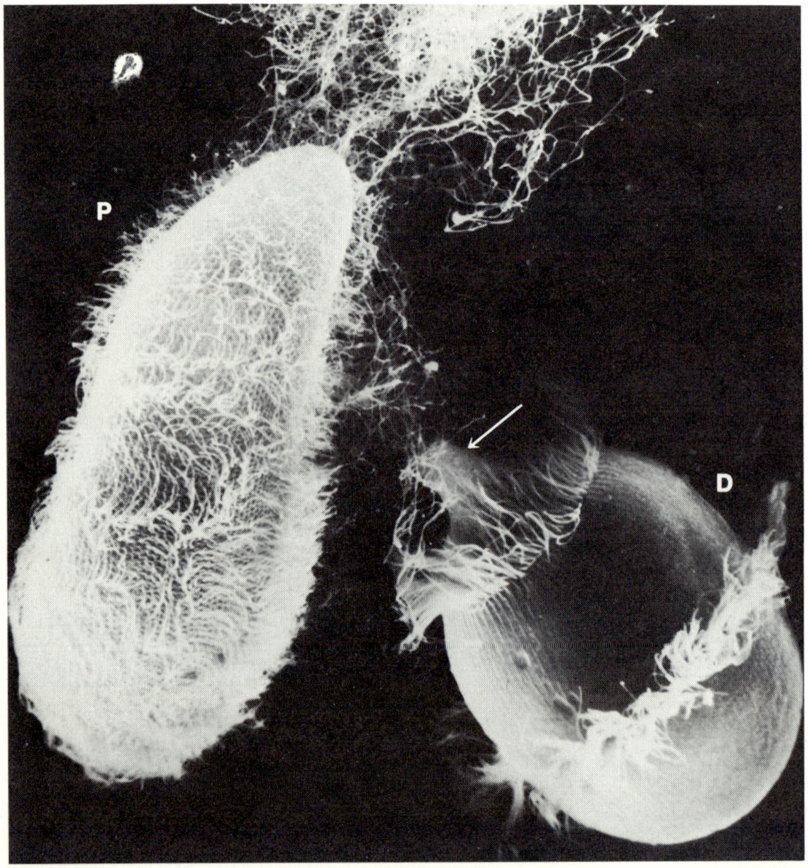

Figure 5-1
Comparative sizes of *Paramecium* **(P),** and the predacious protozoan, *Didinium nasutum* **(D).** Note proboscis indicated by arrow. (Courtesy Dr. Gregory Antipa, The University of Chicago.)

Zoomastigina, a second flagellate subclass, is distinguished by the absence of photosynthetic pigments and the inability to store starch. Most are free-living water and soil forms, but some are parasitic inhabitants of animal and man.

Of the pathogens, the trypanosomes that cause African sleeping sickness and Chagas' disease are of major importance. Likewise, *Leishmania* species, which cause kala-azar and tropical sores, are significant pathogens. *Trichomonas vaginalis* is a common and often irritating inhabitant of the vagina. In one study, 20% to 40% of all women studied carried the organism, while among those who suffered from leukorrhea 50% to 70% were infected.

Flagellates are active inhabitants of the hindgut of termites, where they take an active part in the breakdown of cellulose. A single termite was found to be infested with about 25,000 large and 500,000 small flagellates. The large forms were found to be primary wood digesters.

Class II – Sarcodina or Rhizopoda. Sarcodina (Gr. *sarkodes* fleshlike) or Rhizopoda (Gr. *rhiza* root + *podos* foot) are distinctive in that they are literally an ever-changing "blob" of protoplasm. They move by extruding from their mass a fingerlike structure, the false foot or pseudopodium. When this is anchored they move forward by literally flowing into it with an oozing motion. The pseudopodium may also serve as a type of food-gathering organelle, engulfing particles of organic matter or microorganisms.

The Sarcodina or Rhizopoda are commonly found in fresh or marine waters or in soil. When floating in water, their pseudopodia may extend like long pointed fingers. Some forms live in the intestinal tract of animals. Among those found in man, some appear to be harmless; others are important pathogens. Chief among these is *Entamoeba histolytica*, the cause of amebic dysentery.

Ameba proteus, the most common of the harmless bottom-living water forms is rather large, being up to 600 μm in its largest dimension. On the other hand, *Pelomyxa carolensis* is monstrous, 5 mm (5,000 μm) in length when fully extended. It is a remarkable form with from 100 to 1000 nuclei and many contractile granules often filled with ingested bacteria. It is seldom seen in nature, but in the laboratory it may be observed gulping one paramecium after another—a striking display of carnivorous action.

Ameba with a kind of exoskeleton known as a *test* or shell are found, often in profusion, in fresh or marine waters. Some secrete siliceous plates or calcium carbonate shells. In others, the shell is formed by sand grains or diatom shells that adhere to sticky external secretions. In ocean waters, these forms may grow in abundance. Their dead bodies then settle to form thick deposits resembling gray, claylike sediments. In highly calcareous waters, these sediments, sometimes a thousand feet or more deep, are termed globigerina ooze after the major genus. In the less calcareous waters of the Pacific and Indian Oceans these deposits are termed radiolarian ooze after the primary source organism. The white cliffs of Dover are made up primarily of the fossil remains of protozoan tests.

CLASS III—Ciliata. The Ciliata (L. eyelid or eyelash) are mostly free-swimming water forms characterized by being covered with many short cilia. They feed chiefly on bacteria or other protozoa or to a lesser extent on particulate organic matter. Their size may vary from a few micrometers to 2 mm or more. Because their numerous flagella make them extremely motile, microscopic examination of living forms is often difficult. They are perhaps the most morphologically complex of the protozoa and may possess both micro- and macronuclei.

The best known member of the group is *Paramecium* (Fig. 5-1). While they vary in size, some species are large enough to be barely discernible with the unaided eye.

Many ciliates are voracious feeders. The relatively huge trumpet-shaped *Stentor*, 1 to 2 mm in size has been observed to gulp as many as a hundred *Euglena* cells in a minute. These were swept into the gullet by the fringe of cilia that surrounded the trumpet-shaped opening. *Didinium* is another example of a rapacious protozoan. This barrel-shaped form is much larger in diameter than *Paramecium* but only about half as long. However, it has a cavernous mouth and, as one snake swallows another, can engulf a paramecium by degrees (Fig. 5-2). A few *Didinium* cells can soon wipe out a paramecium culture.

The ciliates are primarily freshwater or marine forms. Some, however, are parasites of fish and amphibia and a few inhabit the intestines of animals such as pigs, cattle, and camels.

CLASS IV—Sporozoa. All of the Sporozoa are parasitic, lack organelles of locomotion, and form spores late in their life cycle. In the host red blood cell, multiple divisions of the sporozoan take place, producing great numbers of individual cells. The host cell ruptures, and these are all released at one time. This process, schizogony, is the cause of the chills and fever characteristic of malaria.

Some sporozoa are parasites of arthropods; others produce diseases in fish (especially salmon or halibut), animals, or man. Being completely parasitic, they are never exposed to an

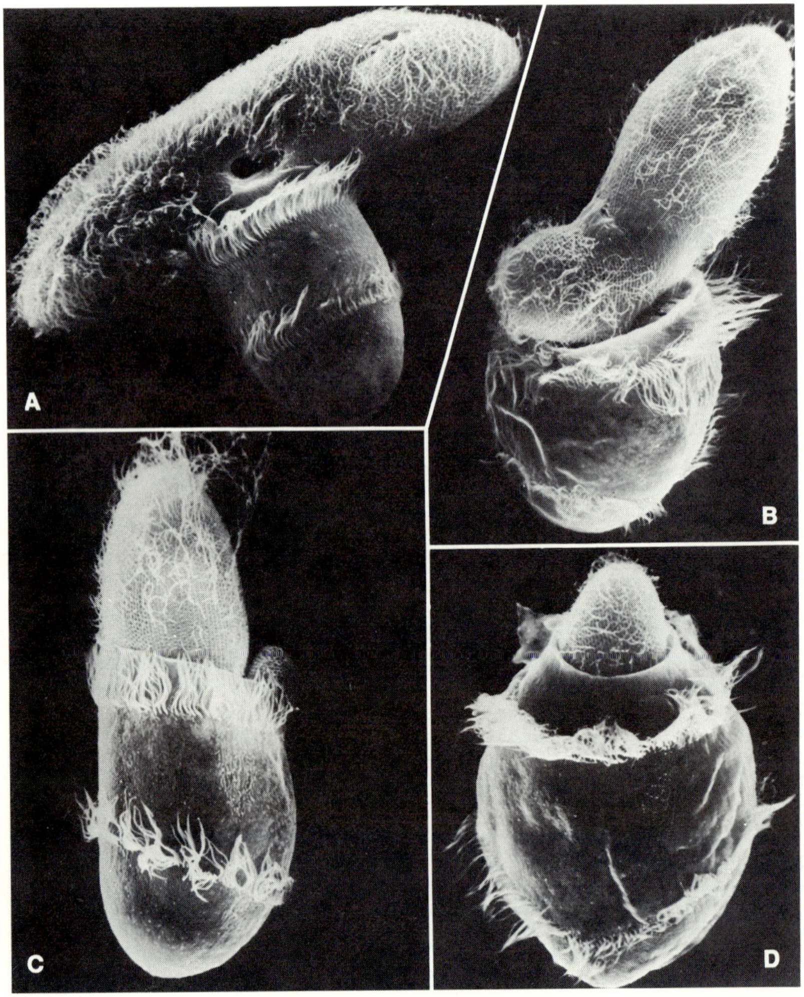

Figure 5-2
Four different paramecia in stages of ingestion by *Didinium nasutum*. **A,** Initial phase; **B,** slightly later, *Paramecium* bent, ingestion beginning; **C,** *Paramecium* half swallowed; **D,** *Paramecium* almost swallowed, *Didinium* greatly extended. (From Wessenberg, H., and Antipa, G.: J. Protozool. **17**(2): 250-270, 1970.)

external environment except as resistant spores. Nutrients are absorbed directly from the animal or insect host through cell membranes.

Although many of the sporozoa live as parasites without serious injury to the host, others may produce serious illness or death. Malaria in man is a classic example. Caused by any of four species of *Plasmodium*, it is one of the most serious infectious diseases in man.

Isospora hominis may cause sporadic cases of intestinal infection. It produces a mild, often prolonged but self-limiting diarrhea. Coccidiosis is a highly fatal disease in chickens that can literally wipe out poultry flocks. Some species of *Emeria* not only infect chickens and other domestic poultry but may cause serious diseases in rabbits, cattle, dogs, cats, and canaries.

Babesia bigemina is the cause of hemoglobinuric fever, Texas or red-water fever in cattle. This disease producer can also infect deer, zebu, and water buffalo. The insect vector is the cattle tick, and the disease is not transmitted if the infection chain is broken by killing the ticks, even if cattle sick with the disease are present in the herd. The protozoan attacks the red blood cells producing anemia and injury to the kidneys, spleen, and liver because of the large amount of red-cell debris that must be cleared from the circulation.

Details of protozoan diseases are given in Chapter 27.

MYXOMYCETES

The myxomycetes (slime fungi) represent a unique, uncommon, and almost bizarre group of living forms. Because of their funguslike habit of spore formation and the superficially funguslike appearance of their fruiting bodies, botanists have long considered them to be fungi. Protozoologists, on the other hand, consider them to be protozoa because of their flagellated zoospore stages and their ameba-like vegetative growth forms. Other biologists consider them to be distinctive life forms, neither plants nor animals but members of the Protista.

The origin of the class name is of some interest. The first known drawings of the fruiting forms were made by Panckow in 1654. Fries placed these forms in a suborder, Myxogastres (*myxo* + Gr. *gaster* stomach), of fungi. In 1833 Wallroth substituted Myxomycetes and De Bary (1860-64) coined the word Mycetozoa emphasizing both the fungus and the animal nature of the forms. Both terms are used today, with botanical-minded mycologists using the term myxomycetes, while textbooks of protozoology designate the organisms mycetozoa.

Classification of the myxomycetes

Although a number of classifications are currently in use, the one that follows has been arbitrarily adopted:

I. Myxomycetacae
   These are sometimes termed the true slime molds because after sexual fusion of the ameboid cells a true plasmodium develops from which the fruiting bodies arise. They are of two types.
   A. Endosporae—bears spores within a sporangium
   B. Exosporae—spores are borne on short stalk

II. Acrasieae (Gr. unmixed)
   In the Acrasieae, an incomplete, pseudoplasmodium is formed. There is not any evidence of a sexual stage. Myxamoeba remain as distinct individuals but unite to form a stalked spore mass.

III. Labyrinthulae
   These are rare water forms. The myxamoeba cause disease of eel grass and are parasitic on algae. The life cycle is not well defined.

The distinctive feature of the slime molds in general are these:

1. They are chlorophyll-free.
2. Part of their vegetative life is passed in an ameba-like state; in this respect they resemble the protozoa.
3. The ameba-like cells may fuse or merely clump to form a multicellular slimy protoplasmic mass, the plasmodium.
4. Fruiting bodies are formed that bear at least a superficial resemblance to the fruiting bodies of molds.
5. Spores are relatively tough and are released readily from the fruiting bodies. The myxomycetes inhabit the moist soils of forests and and the adjacent decaying leaves, stems, and fallen trees of the forest floor. They also grow on damp sawdust and compost or manure. Their primary food source is apparently the bacteria and fungi growing on or in the decomposing organic matter. For germination of the spores and subsequent growth into a plasmodium, the environment must be relatively moist. Myxomycetes play only a minor role, if any, in the decomposition of plant remains.

The plasmodia of true slime molds are seen

occasionally in moist forest environments. Here they may form a slimy, white, red, yellow, or purple hand-sized patch on a rotting fallen tree or, growing between the wood and the bark, they may be exposed when the bark is knocked off. These same slimy masses may appear on the surface of a leafy mat on the soil or on fallen branches.

Fruiting bodies can sometimes be seen as stemlike structures with a pinlike head, or they may resemble a minute puffball or mushroom. When dry, the plasmodia degenerate into a leatherlike thin mat, a sclerotium (Gr. *scleros* hard). The spores are relatively tough, usually retaining viability for 3 to 4 years. However, in one case, spores grew after being stored for 61 years.

Starting with a spore the life cycle is as follows: When the spore is deposited in a moist area or in a drop of water it quickly absorbs moisture. A naked swarm cell is then expelled. This has 1 to 4 flagella and a single nucleus. It is highly motile and swims about in the water film. At this point the swarm spores are able to take up nutrients, such as bacteria or fungi, in solution. Soon the flagella disappear and the zoospore rounds out becoming an ameba-like form, the myxamoeba. This, in turn, may produce more swarm cells that form more myxamoebas, which ooze across the moist surface ingesting quantities of bacteria, yeasts, and fungi. They apparently move by chemotaxy to suitable food sources.

After a period, as the environment becomes less moist or the food supply diminishes, the swarm cells or myxamoebas pair and the nuclei are fused. These diploid zygotes may then continue to divide. Eventually they coalesce to form the plasmodium, which is capable of creeping or oozing about. If the environment becomes dry or extremely cold, the plasmodium may break up into cell masses containing 10 to 12 nuclei that encyst to form a microcyst.

Or, when the food supply diminishes, the plasmodium may creep toward an adjacent drier area. Projecting small masses then become apparent on the surface of the plasmodium. These projections push intermittently upward with a peculiar pulsating motion. At this point, upright stalks may rise and the protoplasm may flow into these stalks forming sporangia, which may assume any one of several shapes but are characteristic for a given genus. The spores that form in the sporangium are haploid. After the sporangium dries the spores are readily released by wind or rain.

Isolation of the myxomycetes

Myxomycetes can be quite easily isolated by use of an oatmeal agar petri dish culture, the surface of which has been thickly seeded with *Enterobacter aerogenes* cells either living or dead. Bits of sclerotia, rotting wood, or soil may then be seeded on the agar as the myxomycetes source. In a few days, spores in the inoculum germinate and filaments of growth may develop eventually covering the surface. If the medium is too rich in nutrients, the bacteria may overgrow the surface and crowd out the myxomycetes.

ACRASIALES OR CELLULAR SLIME MOLDS

Acrasial slime molds differ from the true slime molds in two important respects. (1) When the spore germinates it immediately grows into an ameboid form without being preceded by the development of a flagellated swarm cell. The vegetative stage is, therefore, made up of ameboid cells. (2) When the cells finally clump together, the mass formed is a pseudoplasmodium in which individual cells still retain their identity. Apparently sexual reproduction does not occur.

The Acrasiales are seldom observed growing in nature because their fruiting bodies are minute, inconspicious, and ephemeral. However, students of these forms have found them to be abundant under laboratory conditions. The method of isolation is similar to that used with the true slime molds, but the isolating medium must be low in nutrient (other than the added bacteria) to prevent overgrowth of competing forms.

SUGGESTED READINGS

Alexopoulos, C. J.: The myxomycetes II., Botan. Rev. 29:1-78, 1963.

Curtis, H.: The marvelous animals—an introduction to the protozoa, Garden City, N. Y., 1968, Natural History Press.

Hall, R. P.: Protozoa, New York, 1964, Holt, Rinehart & Winston, Inc.

Kudo, R. R.: Protozoology, ed. 5, Springfield, Ill., 1966, Charles C Thomas, Publisher.

Lister, A.: Monograph of the Mycetozoa, New York, 1925, Johnson Reprint Corp.

Manwell, R. D.: Introduction to protozoology, New York, 1961, St. Martin's Press, Inc.

Milne, L. J., and Milne, M.: Animal life, Englewood Cliffs, N. J., 1959, Prentice-Hall, Inc.

6 The bacteria

THE PROTOPHYTA

DISTRIBUTION

RELATIONSHIP TO OTHER FORMS

SHAPES AND SIZES

FLAGELLA

ENDOSPORE FORMATION

PHOTOSYNTHETIC BACTERIA

PREDATORY BACTERIA

The bacteria as a group represent the most widely distributed and the most primitive of all living forms. Over the eons of their evolutionary development they have become adapted to almost every conceivable natural environment. As a result they are found in every water environment from oceans to puddles, hot springs to glacial lakes, acid waters from coal mines to briny waters of salt lakes. They inhabit plant surfaces, man and animal bodies, and every kind of soil.

THE PROTOPHYTA

Bergey's Manual of Determinative Bacteriology (7th edition), in spite of its limitations, provides the only complete assemblage of descriptions of all of the various genera and species of the bacteria. Because of the primitive nature of the bacteria, in preparing *Bergey's Manual* various contributors argued that the bacteria should be assigned to a third and distinct kingdom of living things, that is, animals, plants, and bacteria. However the more conservative attitudes prevailed and the bacteria as a group were assigned to a major division of the plant kingdom, the Protophyta (a term suggested by Sachs, 1874).

The Protophyta division in turn was divided into three classes, the Schizophyceae (the blue-green algae already discussed), the Schizomycetes, and the Microtatobiotes (smallest living things). For convenience in the discussion that follows, the Schizomycetes (bacteria) are divided into two major subdivisions, the simple bacteria and the higher bacteria.

Schizomycetes

THE SIMPLE BACTERIA. In *Bergey's Manual* the simple bacteria are assigned to two great orders. The first, the Pseudomonadales, includes two suborders, 10 families, 72 genera and 427 species.

These are all gram-negative forms. They have a distinct cell wall that imparts rigidity to the cells. They are primarily nonsporing rods, curved rods, or spirals. The major habitats in which they are found are fresh or marine waters or soils. Many of them are plant pathogens, but only a few of them cause diseases in man or other mammals; a number cause diseases in fish and other inhabitants of natural waters.

The remarkable photosynthetic bacteria, relics of the primordial past, are included in this order along with the chemosynthetic bacteria. These latter bacteria possess the unique property of growing in a medium completely devoid of organic matter. They use, as the carbon source for the synthesis of their cells, carbon dioxide or carbonates. Energy for their life processes is provided by the oxidation of inorganic compounds such as ammonia or sulfur to nitrates or sulfates.

The Eubacteriales, or true bacteria, represent the second order. It includes a great array of forms distributed among 13 families, 4 tribes, 63 genera, and 602 species. They are the most widely distributed of all of the Schizomycetes. Included are gram-negative or gram-positive rods or spherical forms. No spiral forms are included. The gram-positive property is unique to this order.

Two widely distributed genera, *Bacillus* and *Clostridium* produce remarkably tough endospores, the most heat resistant of all living forms.

The Eubacteriales are found in abundance in the soil, where they play a major role in maintaining its fertility—a role essential to our continued existence.

Members of this order are also responsible for most of the diseases of man, higher animals, and insects. Many plant pathogens fall in this group. Also included are the bacteria found in great numbers in and on our bodies.

Most of the problems of the spoilage of foods and milk products are caused by the true bacteria. On the other hand, they are also the bacteria that ripen our cheeses and aid in the production of dill pickles, sauerkraut, and other fermented foods.

The higher bacteria. The higher bacteria comprise the orders Chlamydobacteriales (sheathed bacteria), Hyphomicrobiales (budding bacteria), Caryophanales (shining-nucleus bacteria), Actinomycetales (filamentous, funguslike bacteria), Beggiotoales (algaelike water forms), Myxobacterales (slime bacteria), Spirochaetales (protozoan-like spirals; includes the syphilis organism), and the Mycoplasmatales (unique forms with characteristics of both bacteria and viruses).

Microtatobiotes

To the third distinct class, Microtatobiotes, have been assigned the rickettsia (completely parasitic minute bacteria-like forms), the so-called larger viruses (chlamydia), and the true viruses.

DISTRIBUTION

Similar environments over the earth have a similar or often identical bacterial flora. As previously mentioned, over eons of time, the jet streams of the upper atmosphere, movements of great air masses in turbulent weather fronts, and ocean currents have all contributed to the universal bacterial seeding of like environments of the earth. Migrating birds (especially those inhabiting waters) and migrating animals might have played an important role. Even the intestinal bacteria common to all of us are remarkably similar regardless of our nationality or geographical location. A recent survey of the distribution of *Escherichia coli* strains revealed that the periodic variations in the strains carried by an individual were as great as those observed in strains isolated from widely separated populations. How or why these strains are so distributed in widely divergent populations is still an enigma of microbial ecology.

RELATIONSHIP TO OTHER FORMS

The question are bacteria plants or animals is not readily answered. Indeed, in many respects bacteria fit into neither category but form their own primitive life group. In general, however, bacteria are more plantlike than animal-like. For this reason they are classified as Protophyta.

Plantlike characteristics are as follows:

1. Rigid cell walls similar to those of plants are present in the majority of the bacteria, although they are missing in the spirochetes, slime bacteria, and the mycoplasma group.
2. Energy and carbon sources may be:
   a. Preformed organic matter that must be in solution or solubilized outside the cell by extracellular enzymes and absorbed in solution through the cell membrane.
   b. In the case of some bacteria, inorganic compounds may be oxidized to provide energy with carbon dioxide serving as the carbon source. This physiological pattern (chemosynthesis) is found only among these bacteria.
   c. Photosynthetic action is carried on by a few bacteria. However, bacterial photosynthesis differs from photosynthesis in algae in that it is an anaerobic process, and sulfur rather than oxygen is its by-product.
3. Multiplication is generally by binary or transverse splitting rather than longitudinal division as seen in one-celled animals.

4. In plants, reserve carbohydrates are stored as starch. This is also true of some but not all of the bacteria. In many, reserve foods are stored as glycogen or animal-starch.

SHAPES AND SIZES

The simple bacteria grow characteristically as single cells or clusters of cells in which each cell is a self-sufficient individual. Three basic cell shapes characterize these forms, spherical, rod-shaped (with or without endospore), and the rigid spiral forms or strongly curved rods in which several cells may unite to form a spiral.

Cocci

Cocci are spherical forms further separated into genera on the basis of the cell grouping that occurs. This grouping depends on the plane at which cells divide and is as follows:

1. *Micrococcus* grows as single, slightly elliptical cells in small or large irregular masses resulting from cell division on several different planes and the tendency of cells to remain attached after division.
2. *Staphylococcus* (Gr. *staphylē* a cluster of grapes) has essentially the same grouping and appearance as the micrococci, but it is parasitic or pathogenic and able to ferment glucose anaerobically with acid production. Pimples and boils are caused by staphylococci.
3. *Diplococcus* (Gr. *diplous* double) cells divide only on one plane and remain in twos, flattened at the point of contact.
4. *Streptococcus* (Gr. *streptos* twisted or pliant; not chain; so-called because a long string of these cells often coils in a pliant fashion) cell division is in one plane only, the grouping resembling a string of beads. Cells are often somewhat elliptical.
5. *Sarcina* (L. *sarcina* packet) cells divide in two or three directions at right angles, forming cubical cell clumps of 4 or 8 cells (sometimes 16).

Bacteria or bacilli

Rod-shaped bacteria are of two types, bacteria (nonspore-forming rods), and bacilli (small rods that form endospores within the cells). If any rod-shaped forms occur characteristically in two's, they are termed diplobacilli and, if in a longer chainlike arrangement, streptobacilli, regardless of the presence or absence of endospores. Shapes of various bacteria are shown in Fig. 6-1.

Sizes

The exact size of growing bacteria is difficult to determine because the process of heat-fixing and staining prior to measurement may cause them to shrink as much as 20%. However, in a survey of descriptions of rod-shaped pseudomonads and true bacteria from *Bergey's Manual*, of 80 species taken at random, 70 were 1 μm or less in diameter and less than 5 μm in length. Few of them exceeded 0.7 μm in diameter or 2.5 μm in length.*

Aerobic spore-forming rods were generally somewhat larger than nonsporing forms but most were still less than 1 × 5 μm. *Bacillus anthracis*, (cause of anthrax), largest of the aerobic spore-forming rods, was described as being 1.0 to 1.3 × 3.0 to 10.0 μm. The anaerobic spore formers were slightly larger than the aerobes. Largest of these is the gas-gangrene organism, *Clostridium perfringens*, with dimensions of 1.0 to 1.5 × 4.0 to 9.0 μm.

Individual cells of the cocci generally have diameters of 0.6 to 1.5 μm. A few are as small as 0.2 μm and some as large as 4.0 μm. Sarcinae are generally larger than the micrococci or staphylococci.

*Metric system units employed in the measurement of bacteria and other microorganisms:
1 meter = 1,000 millimeters (mm)
1 mm = 1,000 microns (μ) or micrometers (μm)
1 micron (μ) or micrometer (μm) = 1,000 millimicrons (mμ) nanometers (nm)
1 micron or 1 micrometer = 1 millionth meter
1 millimicron or 1 nanometer = 1 billionth meter
Abbreviations used in this text are μm and nm.

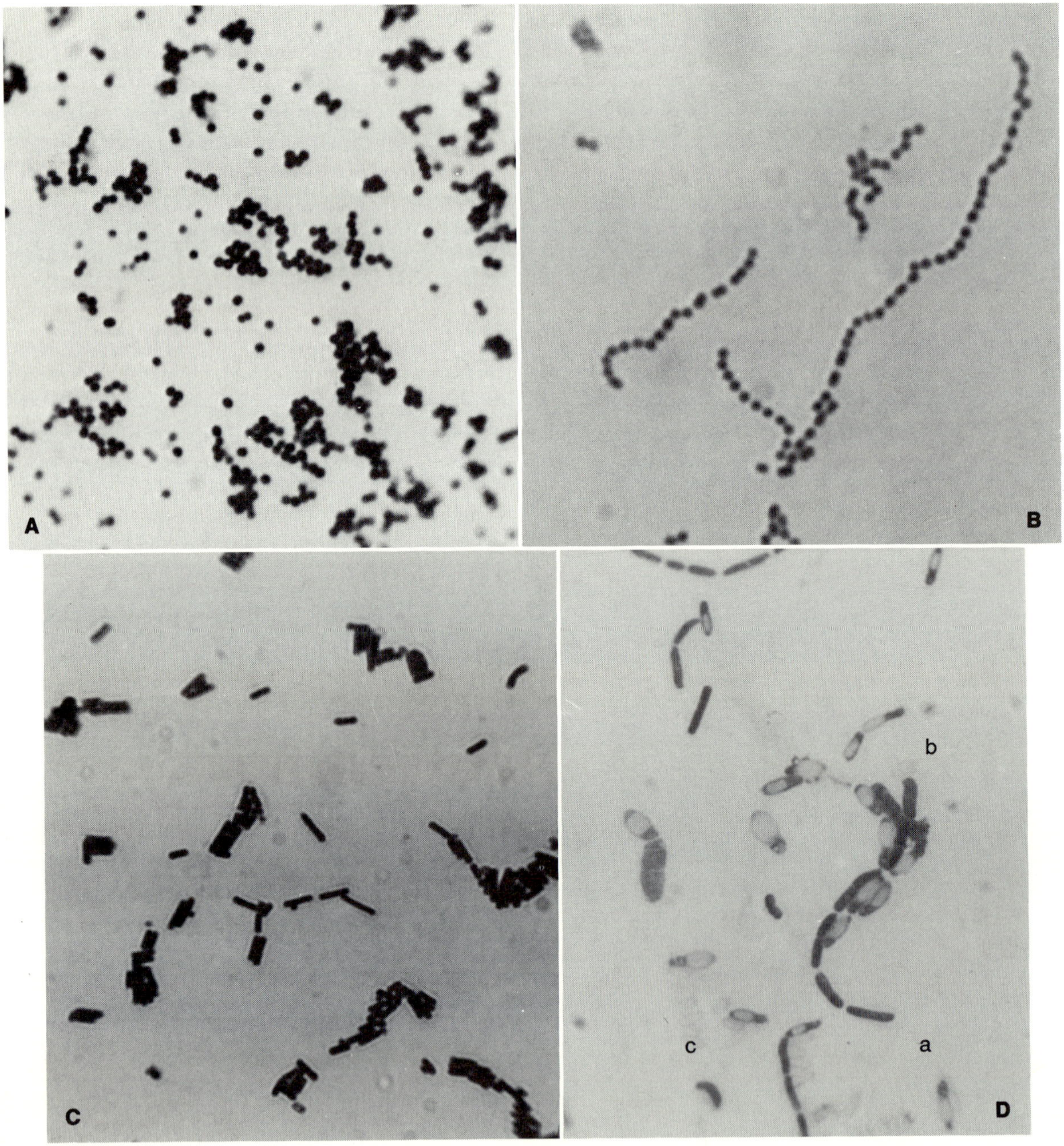

Figure 6-1
A, *Staphylococcus aureus,* cause of boils (×3,000); **B,** *Streptococcus pyogenes,* pus former (×3,000); **C,** *Bacillus polymyxa* in vegetative state (×3,000); **D,** *B. subtilis* showing **a,** vegetative cells, **b,** sporulating cells, **c,** naked spores (×3,000).

Most spirilla are within the size range of the rod-shaped forms but a few water forms are several times longer than the large rods of the true bacteria. Cells of the photosynthetic bacteria are often rather large. This extra size may be necessary in order to provide space for the special enzymatically active organelles involved in photosynthesis.

FLAGELLA

Bacterial cells often possess flagella (Fig. 6-2). These whiplike structures are found on actively motile bacteria and may range in number from a single flagellum to as many as 100 per cell. They are very tiny with a diameter of 20 nm or less. Because of their small size and lack of optical density, they cannot be resolved with the light microscope unless the cells are stained with special stains that plate the flagella, rendering them coarse and dense enough to be seen with the light microscope. The electron microscope, with its greatly enhanced resolving power and its high magnification, is an ideal instrument for the study of flagellation.

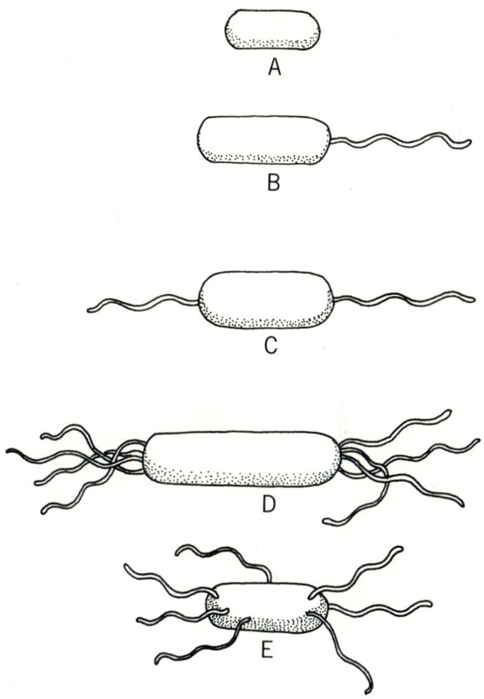

Figure 6-2
Different types of bacterial flagellation. (From Swatek, F. E.: Textbook of microbiology, St. Louis, 1967, The C. V. Mosby Co.)

Flagella are often very long in comparison with the size of the cell. Flagella more than 50 μm in length may originate from bacteria with lengths of less than 2 to 3 μm. Flagella are usually curved and undulating, often with a characteristic number of waves (Fig. 6-3).

Specific terms are used to designate the various types of flagellation. Cells without flagella (Fig. 6-2, *A*) are termed atrichous (*a* neg. + Gr. *thrix* hair). Monotrichous cells *(B)* have a single flagellum that usually extends from one end of the cell. If single flagella extend from both ends of the cell *(C)*, the term amphitrichous is applied (Gr. *amphi* on both sides). When a tuft of several flagella protrudes from one or both ends of the cell *(D)*, flagellation may be termed lophotrichous (Gr. *lophos* small crest or tuft). If several flagella with two or more curves originate from one end of the cell they may be termed cephalotrichous (Gr. *kephalē* head), which is essentially a synonym for lophotrichous. Sometimes for cells having several flagella with two or more waves or undulations protruding from one or both ends, the term multitrichous is used.

The various types of polar flagellation are encountered in and characteristic of the Pseudomonadales. With the true bacteria, Eubacteriales, a few or many flagella may extend out from both the sides and ends of the cell *(E)*. Such cells are said to be peritrichous. (Gr. *peri* around).

The typhoid organism may have a large number of flagella and be almost violently motile. On the other hand, the legume root module organism, *Rhizobium*, may be sparsely flagellate with one to four flagella extending promiscuously from the sides of the cells. However, both are peritrichous.

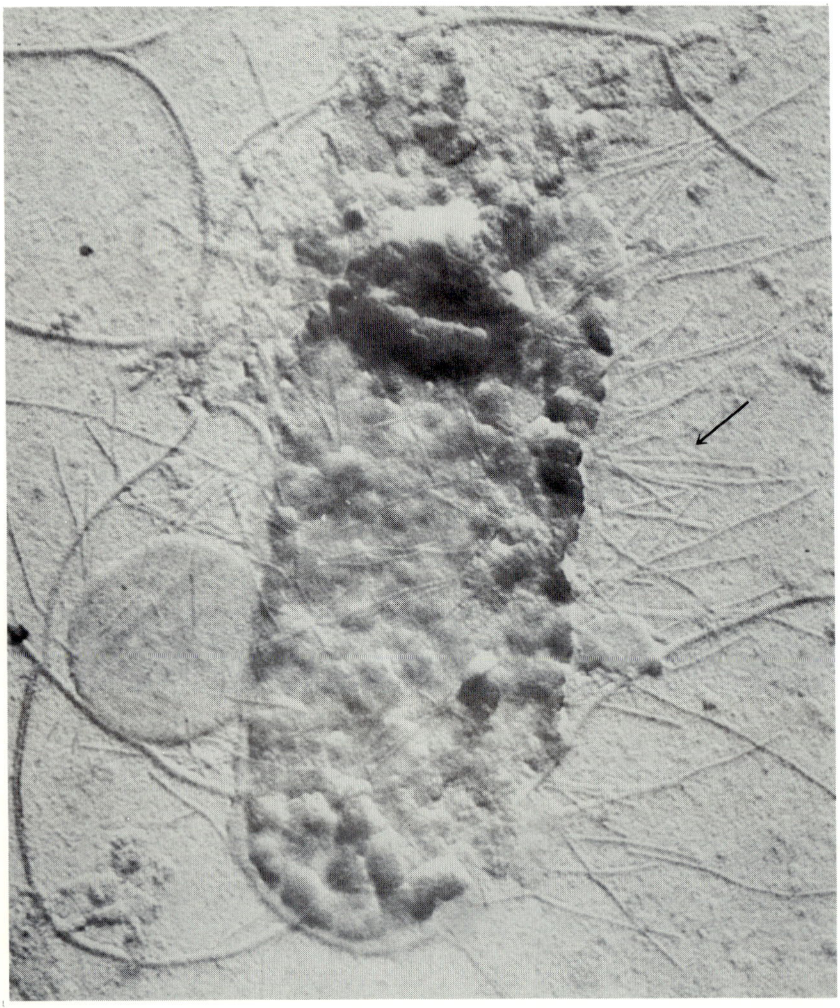

Figure 6-3
Partially disrupted *Escherichia coli* showing flagella (large filaments), and pili (small filaments indicated by arrow) (×90,000).

Nature of flagella

Motility of bacteria has long been known. Even Leeuwenhoek observed motion among his "animalcules." Most of these forms were probably protozoa but a few of his "very tiny animalcules" might well have been bacteria. Although some had motion, it would have been impossible with his microscope to see the flagella.

Ehrenberg in 1838 described volitional movement in bacteria and suggested that the bacteria might possess "muscles." This must have been a shrewd or fortuitous guess. Ehrenberg saw motion, but the flagella, well below the resolving power of his microscope, must have been invisible. Only in 1872 when Cohn succeeded in staining flagella to increase their thickness and light density was it possible to actually visualize them. With the advent of electron microscopy we are now able to secure enough resolution to discern the fine structure of these organelles. Astbury and Weibull (1949) ascertained that they were made up of a distinctive type of protein that they termed flagellin. The suggestion has been made that flagella are primordial muscles.

Fine-structure studies indicate that the flagella have a diameter of about 10 nm and are made up of filaments with a diameter of about 4.5 nm. These filaments apparently occur as helixes. Some workers visualize the entire flagellum as resembling a rope with several main strands each made up of even finer strands. One mathematical model proposes 3 major filaments each made up of 19 minute helixes. Another suggests 7 major filaments each composed of 7 helixes. The entire flagellum may be surrounded by a thin sheath. At the end of each flagellum is a hooklike structure termed the basal granule or blepharoplast. It is found within the cell barely inside the cell membrane. These granules are thought to contain RNA that probably plays a part in energy transfer responsibile for the vigorous motion of the flagella.

X-ray diffraction studies of isolated flagellar preparations yield patterns resembling those of human muscle preparations. The energy required for flagellar movement appears to be derived from the energy-rich ATP (adenosine triphosphate) bonds. Proof that ATP is involved is shown by the simple procedure of placing on a glass slide flagella that have been broken off the bacterial cell by agitation in an electric blender. The broken ends will attach to the glass surface leaving the flagella free to move in the suspending medium. Adding ATP causes the flagella to beat rapidly with a motion similar to their normal movement (see Chapter 15). This same reaction occurs with various muscle fibers when ATP is added. On the basis of these observations it appears likely that flagella may be a primordial muscle or muscles as Ehrenberg had guessed.

In addition to removing flagella by exposing the cells to the shearing action of a high-speed blender, they may be dissolved off with certain surface active agents or detergents. However, they may, to some extent, be regenerated by growing cells. This may require 30 minutes to an hour, and new flagella may not appear on all of the cells.

Flagellar movement

The rate at which flagella move is almost unbelievable. Metzner (1940) used a *Chromatium* species, one of the relatively large, spiral-shaped, purple sulfur bacteria, for motility studies. He calculated that the cell was rotating, corkscrew fashion, at about 40 turns per second. The flagella were revolving about 3 times for every turn of the cell or a speed of about 7200 revolutions per minute. This would be comparable to the rotating of a small high-speed electric motor.

The forward motion of motile bacteria is also amazingly rapid. *Bacillus megaterium* has an estimated speed of 27 μm per second while *Vibrio comma* moves at 200 μm per second or 200 times its own length. By comparison, if a fairly tall man were to move as rapidly in water he would have to swim about 1200 feet per second. The rapidity with which bacteria move is especially evident when one attempts to follow microscopically the movement of some bacteria as they dart about in a wet mount.

ENDOSPORE FORMATION

The rod-shaped bacteria that produce endospores are grouped in the family Bacillaceae and are represented by two genera, *Bacillus* (the aerobic spore-forming rods) and *Clostridium* (the anaerobic spore-forming rods). One other species of bacteria, *Sarcina ureae* (or *Sporosarcina ureae*), a spherical form occurring singly, in twos, or fours, produces endospores. However, these endospores are not as resistant as most of those produced by the Bacillaceae. *Azotobacter* may have a cyst stage. These cysts are tough, resistant bodies but differ in a number of respects from true endospores. Molds produce asexual and sexual spores. Although they resist drying and similar adverse conditions their resistance to heat is far less than bacterial endospores.

The endospores resist drying for long periods of time. Bacterial counts made on a sample from the center of a sun-dried adobe, which had been inside a second-story wall for over 80 years, still contained more than 2,000,000 viable spores per gram of the clay.

Bacterial spores are the most heat resistant of all life forms. Some of the bacillus spores withstand boiling temperatures for several minutes. Occasionally strains are encountered whose spores are still viable after being boiled for an hour or more in a neutral liquid medium. However, in the vegetative state, cells of spore formers are as sensitive to heat as the cells of nonspore formers. In addition to marked resistance to heat, the individual spores show only half as much DNA per spore as in the vegetative cell.

Spores, when placed in a complex medium such as nutrient broth, will germinate into vegetative cells. Rapid vegetative growth will then occur. This growth, however, will continue for only a limited time. If the medium is slightly alkaline, these vegetative cells often begin to form spores within 4 to 5 hours after germination and the beginning of vegetative growth. Within 7 to 8 hours, as many as 75% to 96% of the cells may have sporulated. In fact, in media with special combinations of nutrients and pH, 100% sporulation may occur in as little as 5 hours. In other media only a small proportion of the cells shows early sporulation, and in some simple media they may even fail to form spores.

The newly sporulated cell is generally still surrounded by the cell wall of the mother cell. At this state of sporulation the structure is known as a sporangium. The shape of the sporangium will be determined by the position of the spore in the mother cell, that is, central, subterminal, or terminal.

In the anaerobic spore formers, if the spore is larger than the mother cell it may be swollen in the center to produce a spindle shape known as a clostridium or it may resemble a drumstick (plectridium). This latter type of sporulation is characteristic of *Clostridium tetani.*

Endospore formation in bacteria, first described by Cohn (1873), occurs in the following steps. An hour or two before spores are formed, a portion of the inside structure of the vegetative cell becomes increasingly dense, then rounds up into a highly refractile body. This body becomes the spore and is surrounded by a dense tough coating. The cell wall of the mother cell initially surrounding the endospore then sloughs off, leaving a naked spore. The rapidity with which such sloughing off occurs varies with different species and in different media.

Factors involved in spore formation are rather complex. They include a pH of 7.5 to 8.0 in the medium; presence in proper concentration of Mg^{++}, Ca^{++}, Fe^{++}, and in some cases Mn^{++} ions; temperature generally close to the growth optimum; for *Bacillus* species, adequate oxygen (well aerated cultures will sporulate more quickly than static cultures in tubes); and presence of an available carbohydrate, such as glucose, stimulates spore formation.

Just what triggers spore formation at a certain time in a culture is not well understood. It may involve exhausting of nutrients essential for vegetative growth, removal of some agent inhibiting sporulation, or synthesis of some factor required for sporulation. Various batches of commercial media have different effects on spore formation due apparently to presence or absence of adequate amounts of such growth factors as thiamine.

Some strains, especially of *Clostridium*, may require as long as 2 weeks to produce 100% sporulation. With recently formed spores, germination may depend in many cases on heat-treating or heat activation of the spores. *Bacillus* species vary in the degree to which heat activation is required for optimum germination. Mild sonic (high frequency sound) treatments have also been found to enhance spore germination while vigorous sonification kills the spores.

The exact point at which a spore becomes a vegetative cell is often hard to establish. However, if the spore has germinated it will be killed by heating to the temperature normally lethal to vegetative cells. When mixed with a methylene blue stain, ungerminated spores remain unstained while the vegetative cells readily take up the stain. When examined in a wet mount under the phase microscope, spores appear as refractile bodies while the vegetative cells have a homogeneous bluish or blackish cast. When spores are placed in a Warburg respirometer flask and respiration studied, a sudden marked increase in uptake of oxygen signals the beginning of active spore germination.

In the process of germination two major steps are involved: a new cell wall is produced as an extrusion from the spore, and chromosome (nucleoid) replication occurs.

PHOTOSYNTHETIC BACTERIA

The photosynthetic bacteria assigned to the order Pseudomonadales represent a unique segment of that great array of bacterial forms. They are able to carry on the photosynthetic process by which carbon dioxide (or carbonates) are converted into carbon compounds that become building material for new cells. However, unlike green plants and algae, they carry on the process anaerobically and give off sulfur rather than oxygen in the process. Therefore, they can be found only in habitats that furnish anaerobic conditions plus sulfur (or other reducing compounds).

Photosynthetic bacteria can be isolated quite readily from constantly wet soils, the mud in shallow ditches, ponds, lakes, slow flowing streams, sulfur springs, and oceanic mud flats. In fairly deep lakes they may grow in numbers sufficient to produce purple or green growth layers. These layers form immediately below the thermocline, that portion of the lake where the water is overturned and oxygenated by seasonal changes in temperature. At these levels, essentially anaerobic conditions are present. In one European lake such a layer of growth has been found at a depth of about 100 feet.

Light energy is furnished by wavelengths of 450 to 550 nm that are able to penetrate to considerable depths and that are absorbed by the carotenoid pigments of the photosynthetic bacteria.

The photosynthetic bacteria have been divided into the following three major families.

1. Athiorhodaceae (Gr. *a* neg. + *theion* sulfur + *rhodon* rose, red) are the nonsulfur purple photosynthetic bacteria. If grown in the absence of oxygen and in light they can carry on photosynthesis even in the absence of hydrogen sulfide or other reduced sulfur compounds. In fact, the presence of such compounds may be inhibitory. For reducing carbonates to the precursor molecule for forming glucose, these bacteria use as hydrogen donors such simple compounds as ethyl alcohol, acetates, and propionates, whereas the purple and green sulfur bacteria use hydrogen sulfide (H_2S) and other reduced sulfur compounds as hydrogen donors. The purple nonsulfur bacteria are facultative anaerobes in that they grow well under anaerobic conditions if light is present but also grow well in air on ordinary media in the dark. The bacteria in this group are spherical or rod shaped and motile by means of polar flagella. Growing cultures range in color from brownish yellow to deep red.
2. Thiorhodaceae are the purple sulfur bacteria. In the presence of light and under anaerobic conditions they use hydrogen sulfide (H_2S) as a hydrogen donor to reduce carbon dioxide or carbonates to the glucose precursor (CH_2O) and ultimately to glucose. The H_2S in turn is oxidized to sulfate (H_2SO_4). Some of these forms resemble the purple nonsulfur bacteria in that they can use simple carbon compounds, in addition to reduced sulfur compounds, as hydrogen donors in the photosynthetic process.
3. Chlorobacteriaceae (the green photosynthetic bacteria) are primarily obligate photoautotrophs utilizing only inorganic sulfur compounds to reduce CO_2 in the photosynthesis of glucose. However, one species, *Chloropseudomonas*

ethylicum, can use ethyl alcohol as a hydrogen acceptor in photosynthesis. Generally the green sulfur bacteria are found in waters or muds high in hydrogen sulfide and exposed to the light. Some form cell aggregates and the cells often have intracellular globules of sulfur. At least one genus grows in symbiotic aggregates with other bacteria or protozoa. They are generally more difficult to grow in pure culture than the purple sulfur or nonsulfur bacteria. In fact, much of our present knowledge is based on microscopic examination of crude cultures of the bacteria.

The photosynthetic bacteria have an interesting and important role in nature's mineralization of organic matter. When we visualize the black mud at the bottom of a shallow pond we become aware that it is rich in partly decomposed organic matter. Constant additions are made as dead algae, protozoa, and bacteria fall to the bottom as sediments, or as dead water plants and their roots add additional new organic remains. With each addition of fresh organic matter a vigorous bacterial population begins to burgeon. Many of these forms are obligate or facultative anaerobes that gain their energy by phosphorylations (see Chapter 15). They leave, as end products of their anaerobic fermentations, carbon dioxide, hydrogen gas, acetic, propionic, and other simple organic acids, ethyl alcohol, and amines. These generally accumulate, but never in high concentrations because other bacteria that can obtain oxygen from sulfate and nitrate ions complete the breakdown of the acids, alcohols, and similar compounds into carbon dioxide and water. The activity of these bacteria depends on the amounts of nitrates and sulfates that are available.

At this point the methane bacteria may make their entry. These are the bacteria that can use the accumulated CO_2 as a hydrogen acceptor. The molecule is reduced to methane (CH_4) or marsh gas. Such quantities may be produced that pockets of gas will be formed in the mud. When sufficient pressure is produced the gas is literally burped out of the mud and rises as a bubble to the surface of the water. This bubble rises into the air and is rapidly oxidized back to CO_2. This yields considerable energy, enough so that on a dark night the bubble of gas becomes a weirdly glowing ball. This has been called a "will o' the wisp."

Near the end of the chain of anaerobic and fermentative respiratory processes, the photosynthetic bacteria appear on the scene. The other bacterial performers have had to rely on energy stored in the organic matter. The photosynthetic bacteria, actuated by light energy, can utilize CO_2 as materials out of which to build their constituents. Carbon dioxide may also be reduced to methane by the hydrogen produced in the cell metabolism. As hydrogen sulfide is utilized it may be oxidized to sulfates, which in turn are used by some facultative anaerobes. On the other hand, sulfur may be precipitated to ultimately be oxidized by the sulfur-oxidizing bacteria. A few types of sulfur bacteria are able to carry on nitrogen fixation, converting the gaseous nitrogen (N_2) of the atmosphere into amino acids that form the proteins of their cells.

PREDATORY BACTERIA

Not only are many bacteria susceptible to attack by viruses (phages), or poisoned by growth products of other bacteria (antibiotics), but recently (1962) predatory bacteria were discovered. These attack other bacteria in a way much like a small weasel killing a much larger chicken. These minute ravenous rodlets plunge headlong into the bacteria host, often driving it several lengths by the force of the impact. The host cell begins to swell and the tiny bacterium works its way inside. It then begins to multiply and in a short time the host cell disintegrates. The predatory bacteria are tiny (0.3 μm wide $\times$ 1.0 μm long), curved rods with a single polar flagellum. They have been given the name *Bdellovibrio bacteriovorus* (Gr. *bdella* leech + L. *vibrare* to vibrate; L. *bacteria* rodlet + *vorare* to devour).

Predatory bacteria appear to be quite abundant in nature but are extremely difficult to grow in the absence of the host cell. Counts have indicated 40 to 50 *Bdellovibrio* cells per gram of soil or milliliter of water. Sewage, on the other hand, may contain a thousand or more per milliliter. *Salmonella* species and *Escherichia coli* are frequently attacked although some strains are resistant. When growing on a host-cell plate culture, the predators produce clear areas similar to phage plaques. However, 48 hours are required for development of the *Bdellovibrio* plaque, and the plaques continue to increase in size for a long period. Phage plaques develop within 24 hours and do not increase appreciably after this time.

The suggestion has been made that these bacterial wolves may keep down the population of certain bacteria much as animal predators keep down deer populations.

SUGGESTED READINGS

Gunsalus, I. C., and Stanier, R. Y., editors: The bacteria, vol. I, New York, 1960, Academic Press, Inc.

Hawker, L. E., Linton, A. H., Folkes, B. F., and Carlile, M. J.: An introduction to the biology of the microorganisms, New York, 1960, St. Martin's Press, Inc.

Lamanna, C., and Mallette, M. F.: Basic bacteriology, ed. 3, Baltimore, 1965, The Williams & Wilkins Co.

Salton, M. R. J.: The bacterial cell wall, Amsterdam, 1964, Elsevier Publishing Co.

Sistrom, W. R.: Microbial life, ed. 2, New York, 1969, Holt, Rinehart & Winston, Inc.

Thimann, K. V.: The life of bacteria, ed. 2, New York, 1963, The Macmillan Co.

7 Higher bacteria

CHLAMYDOBACTERIALES

HYPHOMICROBIALES

ACTINOMYCETALES

CARYOPHANALES

BEGGIATOALES

MYXOBACTERALES

SPIROCHAETALES

MYCOPLASMATALES

Microorganisms termed higher bacteria resemble both the true bacteria and pseudomonads in that they are procaryotic and cell diameters are generally similar to those of the true bacteria. Also flagella, when produced, are of the bacterial type, and rigid cell walls, when present, are chemically similar to those of the true bacteria. On the other hand, many of these forms are limited almost entirely to special habitats such as soil and fresh- or seawater. They may remotely resemble either the blue-green algae, molds, or protozoa. Only a few are important as disease producers, spoilage organisms, or fermenters. Thus, while all of them occupy special and important niches in the economy of nature, only a few of them are important in the economy of man, and they are rarely encountered in the laboratory.

Some forms have never been grown in pure culture, and others do not grow readily in the media and under the conditions usually employed for the cultivation of true bacteria. In many cases, the media routinely employed are too rich in nutrients or may lack certain required growth substances. The usual incubation temperatures are often too high for their growth or the microorganisms may not grow well in the presence of atmospheric oxygen.

The higher bacteria are represented by the following orders: Chlamydobacteriales, Hyphomicrobiales, Actinomycetales, Caryophanales, Beggiatoales, Myxobacterales and Mycoplasmatales and will be considered in this order.

CHLAMYDOBACTERIALES

The Chlamydobacteriales are primarily inhabitants of freshwater. In many cases, the presence of fairly high concentrations of iron or manganese may favor growth or at least sheath formation by these bacteria. The sheaths often contain deposits of iron or manganese hydroxides.

As encountered in nature, the common growth forms are trichomes (Gr. *trichoma* a growth of hair). These are thin filaments made up of a long chain of rods frequently enclosed in a sheath. They are generally attached to some surface such as a dead branch or the stem of a water plant. Often these trichomes may be several millimeters in length and when more than one trichome arises from the same area, growth often forms a visible whitish-gray tassel. This tassel may wave gently back and forth as the water moves. Many of the trichomes are difficult to grow in pure culture. They have been isolated by submerging glass slides in water to provide a surface for trichome attachment. After several days the slides are removed from the water and the attached forms studied microscopically.

Two forms, *Sphaerotilus* species (Gr. *sphaira* sphere + *tilos* down; so called because when the bacteria grow in dirty water, heavy tassel formation may occur), and *Leptothrix* species (Gr. *leptos* fine + *thrix* hair) are often classed as nuisance bacteria because in sewage or polluted waters they form tassels in such profusion that the flow of water through pipes may be impeded.

The nuisance forms can be isolated quite readily. Pieces of alfalfa straw or pea stems are boiled and drained several times to remove excessive amounts of soluble nutrients. These pieces of stem are then added to a polluted water sample. After incubation for several days at room temperature, tassels attached to the straw begin to appear. These can be carefully removed and used to seed a plate culture of special agar media, low in nutrients, high in minerals. The bacteria will not grow on ordinary media.

Leptothrix species are often called the iron bacteria because of the deposits of iron or manganese in their sheaths. For microscopic examination of cells of the sheathed bacteria, the metallic deposits must be removed with hydrochloric or similar acids. The cells may then be stained and examined. Under some conditions, cells can also be seen quite readily with the fluorescent microscope. Fine structures are revealed by means of the electron microscope.

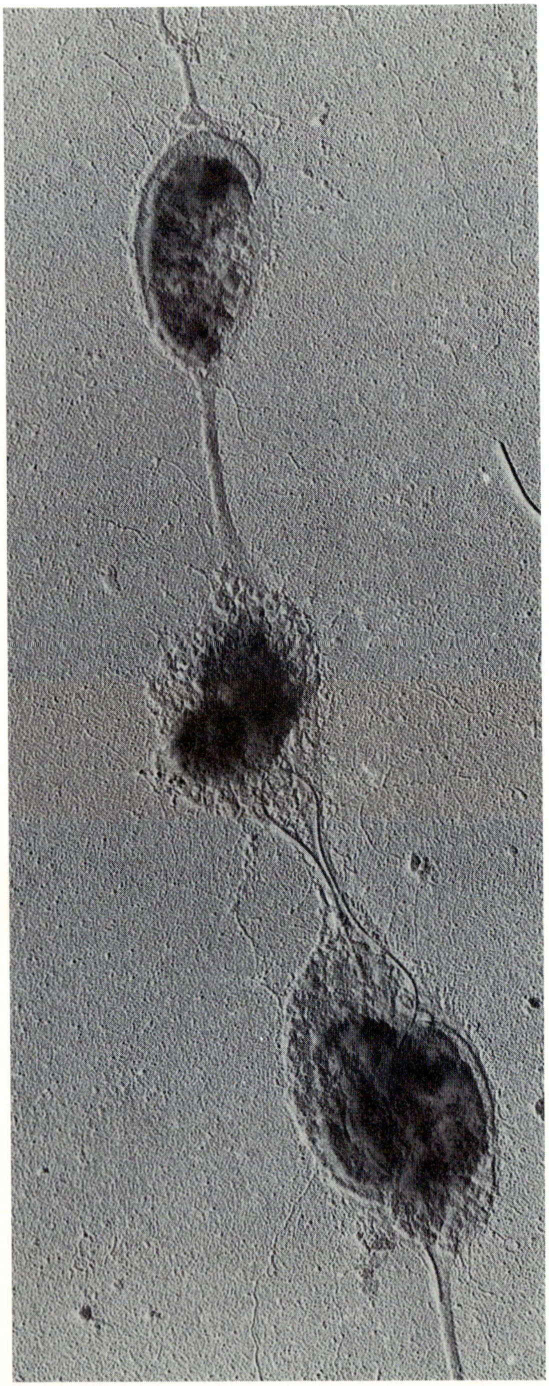

Figure 7-1
Unidentified, marine, agar-liquefying species of Hyphomicrobiales, showing filamentous, budding type of growth. (×30,000.) (Courtesy Margaret Hultgren and Michael Tomasulo.)

Figure 7-2
Electron micrograph of single cell (Fig. 7-1) showing structure and attached filaments. (×70,000.) (Courtesy Margaret Hultgren and Michael Tomasulo.)

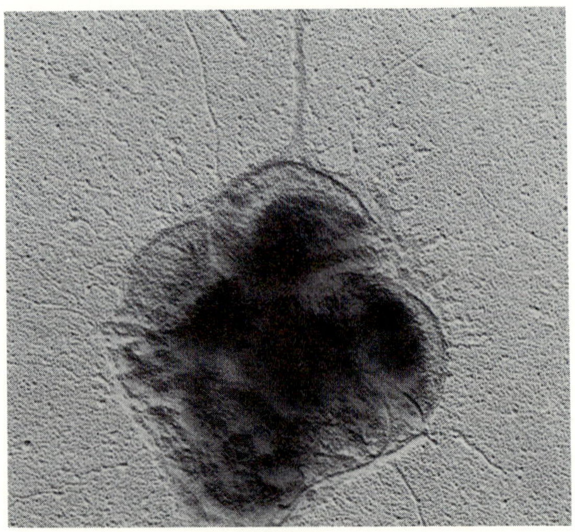

HYPHOMICROBIALES

The most distinctive feature of the Hyphomicrobiales is their distinctive method of reproduction. This is by budding, or budding and longitudinal fission. Buds may be borne at the tip of a mature cell or at the end of a filament arising from the mature cell (Fig. 7-1). Sometimes filaments may branch and several buds may be borne on different filaments. These often form a mass of filaments and buds and may arise from a single holdfast. Free-floating groups of budding filaments may be encountered. Freshwater and mud adjacent to or in streams and ponds are the major habitats of the Hyphomicrobiales. Some are parasites of *Daphnia* (water fleas) or other water forms. Surface growths made up of several hundred filaments and buds are sometimes seen.

Individual cells are often ovoid, measuring 0.5×1.0 μm, or they may be spherical or ellipsoidal (Fig. 7-2). Some species are photosynthetic. Others are chemoheterotrophic growing on mineral-salts media with an added carbon source.

ACTINOMYCETALES

The Actinomycetales represent a borderline group ranging from thin, long rods or spherical forms to long branched filaments resembling, in many respects, miniaturized molds. They are rather widely distributed in soils and decaying organic matter and are occasionally found in waters. Both the mycelia and spores can be isolated from the air. Actinomycetales may cause earthy odors in water supplies. The pleasant aromatic odor of freshly plowed soil is due to the vanillin (a vanilla-like compound) that they produce.

The rod-shaped forms include *Mycobacterium tuberculosis,* the tuberculosis organism, and *M. leprae,* the cause of leprosy. These are acid-fast bacteria (see Chapter 9).

The actinomycetes show varying degrees of filament formation. Representative forms of actinomycetes include the causes of lumpy jaw in cattle (actinomycosis in man); Madura foot (a foot infection in India); tuberculosis-like infections of men, cattle, and birds; human mouth and tooth infections; and a few plant diseases.

The first description of an actinomycete was published by Ferdinand Cohn (1875). This was an organism found in a cheesy mass in a tear duct. Harz (1877) described the organism causing lumpy jaw in cattle *(Actinomyces bovis)* and Beijerinck (1900) first demonstrated the presence of actinomycetes in soil. However, these microorganisms received little real attention until their antibiotic-producing ability was demonstrated by Selman A. Waksman (co-discoverer of streptomycin) and his co-workers in the 1930's. Prior to this time the actinomycetes were grouped along with the molds as fungi. In recent years, microbiologists have concluded that these forms are more bacteria-like than moldlike. As a result they are classed as an order of bacteria.

The bacteria-like characteristics include:

1. The diameter of the mycelial filaments and spores is generally 0.5 to 1.0 μm. This is the diameter of most of the true bacteria and many times smaller than fungal filaments.
2. In many forms, the filaments break up into fragments (oidia) similar in shape to bacteria.
3. The actinomycetes in terms of their genetic mechanism are procaryotic in that they possess a nucleoid similar to that of bacteria rather than the true nucleus of the molds.
4. The chemical composition of cells is similar to that of bacteria in that they lack sterols and the cell walls contain mucopeptides.
5. Like many bacteria, the actinomycetes are susceptible to the action of lysozyme.
6. Like bacteria, they are subject to phage attack and their susceptibility pattern to antibiotics is similar to that of the bacteria rather than fungi.
7. Lysine synthesis is through the diaminopimelic acid pathway, a typical bacterial growth pattern.
8. Flagella, when produced, are of the bacterial type.

Figure 7-3
Micrograph of *Streptomyces spinichromogenes* made with scanning electron microscope, showing three-dimensional nature of spiral. (×10,000.) (Courtesy JEOL [U.S.A.], INC.)

9. Some actinomycetes, like some bacteria, are strict anaerobes. Strict anaerobiosis is unknown in the fungi.
10. Whether or not the filaments have cross walls is still not clearly demonstrated.

The Actinomycetales are divided into four families:

MYCOBACTERIACEAE. The Mycobacteriaceae are thin rods. True filaments are rarely seen. This is the family to which the tuberculosis and leprosy organisms belong. A number of species are widely distributed in the soil or as parasites of man.

ACTINOMYCETACEAE. One genus of the Actinomycetaceae, *Nocardia* (named for Prof. E. Nocard), is aerobic. A few species cause lung infections resembling tuberculosis. Many are soil forms.

The second genus, *Actinomyces*, is anaerobic to microaerophilic. It has true mycelia that reproduce by fragmentation. In addition to lumpy jaw in cattle, species of this genus are responsible for actinomycosis in men, cats, and dogs.

STREPTOMYCETACEAE. The Streptomycetaceae (pliant fungus family) forms produce true mycelia. Reproduction is not by fragmentation but by conidia borne on sporing structures. They are primarily soil forms.

Of the more than 500 different antibiotic compounds discovered, most are produced by species of *Streptomyces*, the major genus of this family (Fig. 7-3). Classification of the streptomyces group is in a state of flux. Isolates that have produced antibiotics are often inadequately described, or have been patented then discarded when the antibiotic was found to be useless. An international committee is currently engaged in an extensive study designed to unravel this tangle.

ACTINOPLANACEAE. The Actinoplanaceae is a family consisting of two genera. One genus produces no aerial mycelia, the other produces them in abundance. In the first, the sporangiospores are motile, in the second, nonmotile.

CARYOPHANALES

The Caryophanales are rarely encountered forms that have been found in the feces and saliva of cattle, in intestines of termites, cockroaches and other insects, the buccal cavity of man, and in decaying organic matter, apparently as harmless saprophytes. They are seen as many-celled filaments or trichomes. The distinctive feature is the presence in the individual cells of a large, bright, disklike body seen in the center of the cell. It stains with the Feulgen stain, indicating its DNA content, hence it appears to be a true nucleoid. This feature explains the order name (Gr. *karyon* nucleus + *phaneros* bright or conspicuous). Some species can be cultivated on a special soil extract medium, but most of the described species have not been grown in pure culture. Three families, six genera, and twelve species are described in *Bergey's Manual.*

The most extensively studied species is *Caryophanon latum* (L. *latus* broad). It is one of the largest of the bacteria, growing in trichomes 3 μm or more in diameter and some 20 μm in length with 4 to 22 discoid units per trichome. Long filaments are also sometimes found.

The first member of this order was described by an American as early as 1849. These forms received little further attention until 1940 when the Russian worker, Peshkoff, published the first extensive study laying the foundation for the present order.

BEGGIATOALES

The Beggiatoales are made up of algalike forms with no flagella or other organs of motion. They grow as long trichomes that can bend and flex. Motility, when it occurs, is in the form of a gliding or a jerky rolling motion along a solid surface. The name, *Beggiatoa,* was first applied by Trevisan in 1842, in honor of an Italian physician. The organisms grow in freshwater especially in the presence of decomposing algae, in seawater, and in soil. They appear to be closely related to the blue-green algae, *Oscillatoria.* On the other hand, some of their characteristics are definitely bacteria-like. They often produce white or creamy growth masses. A few species have been isolated in pure cultures, but most of them have been described and studied microscopically only from natural materials. Some forms produce slender trichomes as long as 100 μm.

MYXOBACTERALES

The Myxobacterales, slime bacteria, were so named by Thaxter (1892) when he observed bacteria growing in slimy colonies from which minute but visible fruiting structures developed. He considered them to be analogous to the slime molds (myxomycetes), and termed them *Myxobacteria.* The common name now used is the myxobacters.

They fall into two groups—those forms that from older colonies produce fruiting bodies filled with a large number of resting cells that are not especially heat resistant but withstand dessication for several years, and those forms that do not produce fruiting bodies. In the vegetative state the latter are rods with tapering or blunt, rounded ends, and flexible cell walls.

Myxobacters move by a gliding motion similar to that seen in some of the blue-green algae and in *Beggiatoa* species. Cells divide by binary fission similar to the true bacteria. Also, they have nucleoids similar to those found in the simple bacteria.

The Myxobacterales may be isolated from the feces of herbivores such as rabbits or sheep. They are also found in soils. Singh estimated 2,000 myxobacters per gram of soil, but this estimate is probably conservative. Two aquatic species are known that are parasitic on algae, and some marine forms have been described.

The methods of isolation are of interest. The original method consists of placing sterilized pellets of sheep or rabbit dung on moistened filter paper in a petri plate. The pellet is then seeded with a small amount of soil suspension and incubated at 26° to 30° C for 5 to 10 days. The myxobacters form a slimy film of growth that spreads out from the pellet. Fingers of growth are often formed due to the rapid movement of the highly motile bacteria that exude large quantities of slime. After 5 to 10 days, minute fruiting bodies are barely visible. They often appear as knobs on a short stem. (Occasionally slime molds, rather than the slime bacteria, develop as contaminants. These not only overgrow the slime bacteria but often may ingest them.)

The second method consists of pouring plainwater agar without any added nutrients into a petri plate. This is then heavily seeded with *Enterobacter aerogenes,* either living or dead, to furnish nutrients for the myxobacters. The plate is then inoculated with a considerable number of soil granules well spaced over the surface. After a week or so at 25° to 30° C, the myxobacters begin to grow out as a slimy, film-like colony from the soil particle. Fingers of slimy growth are often seen. The slime bacteria can be isolated by transferring a loopful of growth from the tip of one of these fingers to another bacteria-water agar plate. (If killed bacteria are used, a pure culture made up only of myxobacter can be secured.) The chief disadvantage of this method is that myxobacters that grow on the sheep pellet medium may not always grow on the bacteria-water agar plates. However, once secured in pure culture, the myxobacters can often be grown on a mineral-salts agar to which has been added cellulose in the form of shredded filter paper or chitin in the form of ground up lobster exoskeletons.

SPIROCHAETALES

The Spirochaetales occur as thin, flexuous spirals or helixes having at least one complete turn but usually many complete spirals. They may be 6 to 500 μm in length. The Spirochaetales should not be confused with the simple bacterium, *Spirillum,* which grows as a rigid spiral and has true flagella. Spirochetes have no rigid cell walls and no flagella. They may be highly motile and move by whirling rapidly along their axis or sometimes by a rapid bending motion. They can move either forward or backward. Unlike the spirilla, which can easily be stained with common bacteria stains, many of the spirochetes are difficult to stain and are often studied unstained in wet mounts under the dark-field or phase microscope. They have a lower refractive index than the simple bacteria, which renders them difficult, if not impossible, to see with the ordinary light microscope.

SPIROCHAETACEAE. The Spirochaetaceae are very long slender forms found in fresh- or salt-water (especially polluted water) or living as parasites in shellfish. One species has been reported from a hot spring. The cell diameter is generally about that of the rod-shaped true bacteria, but the organisms may be 100 to 500 μm in length. Most of these forms have not been successfully grown in pure culture on the usual laboratory media. Three genera have been described: *Spirochaeta, Saprospira,* and *Cristispira* (distinguished by a distinctive crista or crest, a thin membrane on one side of the body following along the helixes of the entire body).

TREPONEMATACEAE. Treponemataceae (Gr. *trepein* to turn + *nema* thread) are usually parasitic or pathogenic in man and other higher animals. Three genera comprise the group:

1. *Borrelia* (named for A. Borrel) is a genus with some 28 species. Several species of *Borrelia* cause relapsing fever in man and may be transmitted by insect vectors. Others are parasites in the mouth or genital mucosa of man. One, *Borrelia vincentii,* is associated with a bacterium, *Fusobacterium fusiforme,* as the cause of trench mouth or Vincent's angina. All of these spirochetes can be stained with ordinary stains.
2. A second genus, *Treponema,* is made up of forms that are difficult to grow and then only under strict anaerobiosis. They are also difficult to stain except with silver stains and are generally examined only by means of the dark-field microscope. *Treponema pallidum* (L. *pallidus* pale) is the cause of syphilis. *Treponema pertenue* (L. *pertenuis* very thin or slender), while undistinguishable from the syphilis organism by dark-field examination, causes yaws and may be transmitted by flies.
3. *Leptospira* (Gr. *leptos* slender) is long, finely coiled, with a hook at one or both ends. *L. icterohaemorrhagiae,* the type species, is the cause of Weil's disease, an infectious jaundice of man for which wild rats are the apparent reservoir.

MYCOPLASMATALES

The Mycoplasmatales order represents an enigmatic group of microorganisms having unusual growth forms different from other bacteria. The name Mycoplasmatales suggests that they may be related to the fungi. This is an unfortunate choice of terminology that has been applied because some strains, when examined microscopically, appear as long, irregular filaments. For example, *Mycoplasma pharyngis,* a form isolated from the human throat, develops as a very thin, irregular filament 0.12 $\mu m \times 1.5$ to 8.0 μm in length. Other forms appear as tiny spheres (0.25 to 0.30 μm) that appear in clusters. Many other intermediate shapes, often bizarre and irregular in appearance, have been described. They are sometimes designated myceloid forms and may be quite long.

The mycoplasma are difficult to grow, requiring media heavily enriched with blood serum or ascites fluid. When grown on such a medium, the colonies are extremely minute, often requiring study under a dissecting microscope. Colonies grow into a typical "fried egg" form, thick in the center and thin around the edges. The cell bodies, if they can be called cells, are very soft and fragile without walls as such, but having a retaining membrane.

They are difficult to stain because they may be destroyed in the staining process. Even with the electron microscope, the structures may be distorted in preparing mounts.

Under some conditions, minute bodies known as elementary bodies may be released. These pass freely through bacteria-tight filters. This is amazing because these tiny bodies, 125 to 150 nm in diameter, are about the size of many of the viruses. While all viruses require living host tissue for growth, the mycoplasma elementary bodies are remarkable because they can be grown on nonliving media although they must be greatly enriched with body fluids. The significant point is that this minute body provides space enough to house the vital enzymes necessary for reproduction. These are the smallest forms that can maintain an independent existence.

The mycoplasma organism was originally discovered in 1898 by Nocard and Roux in tissue taken from the lungs of cattle suffering from a type of pleuropneumonia. They found that the exudates were infectious but could demonstrate no bacteria. To prove the infectious nature of the agent, they filled two collodion sacs with an enriched liquid medium. One sac was then inoculated with infected exudate and both were placed in the peritoneal cavity of a rabbit. The contents of the inoculated sac became cloudy and highly infectious. The uninoculated sac remained sterile. Later Nocard and Roux were able to demonstrate similar growth in a greatly enriched liquid medium. The organism acquired the name pleuropneumonia-like organism, shortened to PPLO. Techniques were developed for the microscopic examination of PPLO's. Subsequently, a similar organism was isolated that caused contagious agalactia in sheep and goats.

In recent years the organism causing primary atypical pneumonia (the Eaton agent), formerly considered to be a virus, has been shown to be a mycoplasmal form. It is now known as *Mycoplasma pneumoniae.* A number of other mycoplasmas have been isolated from the genitourinary tract in man where they appear to be either mildly pathogenic or merely parasitic. Others have been isolated from the human mouth and throat. *M. pharyngis* was found to be present in 12% of the throat cultures examined. One strain, *M. laidlawii,* appears to be a strictly saprophytic form and can be quite easily isolated from sewage, manure, humus, and soil. Recently mycoplasmas have been reported as the cause of some plant diseases.

The exact place of the mycoplasma group in the microbial world is still problematical. One school of thought maintains that they are an independent assemblage of microorganisms. Others feel that they probably have developed from the L-forms of bacteria. These are bacteria that have lost the ability to synthesize cell walls but continue to reproduce with only a plasma membrane as an outer coating. One such organism, a strain of *Streptobacillus moniliformis,* was first isolated by Klieneberger-Nobel of Lister Institute in 1935. The "L" in L-form is derived from Lister. Since that time a number of bacteria have been found to produce L-forms.

In spite of extensive studies, the true microbial position of the mycoplasma group still remains an enigma. Indications are that the microorganisms are something more than bacteria that have permanently lost their cell walls.

Currently, workers in the field recommend that the order name, Mycoplasmatales, be abolished and that these forms be considered as an entirely new class of microorganisms, the Mollicutes (L. *mollis* soft + *cutis* skin).

SUGGESTED READINGS

Breed, R. S., Murray, E. D. G., and Smith, N. R.: Bergey's Manual of determinative bacteriology, ed. 7, Baltimore, 1957, The Williams & Wilkins Co.

Hawker, L. E., Linton, A. H., Folkes, B. F., and Carlile, M. J.: An introduction to the biology of microorganisms, New York, 1960, St. Martin's Press, Inc.

Klieneberger-Nobel, E.: Pleuropneumonia-like organisms (PPLO) Mycoplasmataceae, London, 1962, Academic Press, Inc.

Panos, C., editor: A microbial enigma; mycoplasma and bacterial L-forms, Cleveland, 1967, World Book Publishing Co.

Waksman, S. A.: The actinomycetes, vols. I and II, Baltimore, 1961, The Williams & Wilkins Co.

8 The rickettsiales and virales

The third great class of the Protophyta, the Microtatobiotes, are, as the name implies, the smallest of living things and represent an unusual array of forms. They are distributed among two major orders, the Rickettsiales and the Virales.

THE RICKETTSIAE

The name of the order, Rickettsiales, honors Dr. Howard T. Ricketts, one of the discoverers of the Rocky Mountain spotted fever agent and an investigator of other organisms that bear his name. He died as a result of a laboratory infection acquired while working with the rickettsia that causes typhus.

The rickettsiae, while having a few of the characteristics of viruses, bear a close resemblance to the bacteria in the following ways:

1. In structure they resemble very small gram-negative bacteria. Although often pleomorphic (variable in shape) they may be coccus-like, short rods, or diplobacilli.
2. They divide by binary fission.
3. Growth is inhibited by several of the drugs and broad spectrum antibiotics that are active against bacteria, especially the gram-negative forms.

4. Only a few of the rickettsiae can pass through bacteria-tight filters.
5. They are stainable with some types of stains and visible under the light microscope.

Rickettsiae resemble viruses primarily because both forms, with the exception of one species of rickettsiae, can be grown only in the presence of living host cells. With rickettsiae, host cells are provided by living animals, chick embryos, animal tissue cultures or single-cell cultures of animal cells. They have reached this highly dependent state because they have lost the ability to produce many of the enzymes necessary for growth and reproduction on simple, cell-free media. Thus they have become complete parasites. Like some of the viruses, they are transmitted primarily by insect vectors. In most cases they live and reproduce in the insect gut. Like the viruses, the rickettsiae are known primarily by the diseases they cause. The major rickettsial diseases, briefly described below, will be considered in more detail in a subsequent chapter.

Epidemic typhus (Gr. *typhos* smoke, stupor arising from fever) is perhaps the most important of the rickettsial diseases. Transmitted by human body lice, it has had tremendous impact on history and may have won or lost more wars than the most famous and skillful of the generals.

Endemic or *murine typhus* is a much milder and rarer disease than the classic epidemic typhus. The disease occurs as epizootic in rats but may be transmitted to humans by bites of the rat fleas.

Rocky Mountain spotted fever, often a severe disease, was first described in Montana in 1873. It was thought at one time to be peculiar to the Rocky Mountain area and to be transmitted only by the wood tick *(Dermacentor andersoni).* However, in recent years, sporadic cases have been reported in most of the states of the continental United States. In many cases, the dog tick has apparently been the vector.

Scrub typhus or *tsutsugamushi fever* is an important disease in Japan. During the occupation of Japan after World War II, a number of American soldiers suffered from the disease. It is borne by the harvest mites known familiarly as "chiggers."

Q fever was first identified by Derrick in 1937. A previously unrecognized disease that appeared among packinghouse workers in Queensland, Australia, it was termed Q fever—"Q" standing for "query" (not Queensland), just as physicians sometimes speak of an unknown disease agent as "virus X."

It is an enervating but not highly fatal disease, characterized by high fever, headache, and often by pneumonia-like symptoms. The skin rash, characteristic of the typhus and spotted fever group, is lacking. Also, the disease agent is capable of passing bacteria-tight filters. Since the initial recognition of the disease, sporadic cases have been found in the western United States. During World War II, outbreaks of Q fever caused pneumonitis (lung inflammation) among American troops in the eastern Mediterranean areas. Axis troops also suffered similar outbreaks. In 1946-1947 small epidemics occurred among packinghouse workers in Amarillo, Texas, and Chicago. In 1947 a major outbreak occurred in the Los Angeles area. The source of infection was traced to dairy herds. Most of the cases were among dairy workers. Also, raw milk appeared to have been another source of the infection. The important point is that the disease not only is endemic in Australia but also occurs in a number of other parts of the world.

Chlamydiaceae

In the Chlamydiaceae, a family in the Rickettsiales, are found a number of important disease producers formerly considered to be "large viruses." Like the viruses they require living tissue in order to grow. However, they have both DNA and RNA; they are coccoidal, and in the smallest phase (elementary bodies) may be 200 to 350 nm in size, although inclusion bodies made up of a number of elementary bodies may be up to 2 μm in diameter; they stain with some aniline dyes, are gram-negative; and they are susceptible to many of the antibiotics and the

Table 8-1. Disease, organism, and vector of the most important rickettsial diseases

DISEASE	CAUSATIVE ORGANISM	VECTORS
Epidemic louse-borne typhus	*Rickettsia prowazekii*	Human body lice, human head lice
Murine typhus	*Rickettsia typhi*	Rat fleas, rat lice
Rocky Mountain spotted fever	*Rickettsia rickettsii*	Wood ticks, dog ticks, rabbit ticks and others
Rickettsialpox	*Rickettsia akari*	Mites from wild house mice
Scrub typhus	*Rickettsia tsutsugamushi*	Harvest mites in Japan
Q fever	*Coxiella burnetii*	Ticks, human body lice, contact with infected cattle, sheep, goats, infected meat, raw milk

sulfonamides. They grow intracellularly in various warm-blooded animals.

The most important diseases caused by the Chlamydiaceae are as follows:

1. *Trachoma* or *granulated eyelid* is the disease of antiquity that still constitutes a major public health problem in many parts of the world such as Egypt and the Middle East. If not controlled, it often causes blindness. However, sulfonamides and antibiotics have proved quite effective in its control.
2. *Inclusion conjunctivitis* or *inclusion blennorrhea*, often called swimming-pool conjunctivitis, is a self-limiting but unpleasant irritation of the conjunctival membrane of the eye and may persist for 3 months to a year. It may occur in newborn infants but also may cause a prolonged conjunctivitis in adults. It appears to be transmitted among bathers in some poorly controlled swimming pools.
3. *Lymphogranuloma venereum*, also known as the "fourth" or "fifth" venereal disease or climatic bubo, is another disease caused by this group. It is usually characterized by the development of shallow, grayish, raw lesions with clean-cut edges on the external genitalia. This may be followed by involvement of the adjacent lymph nodes. Scar tissue may form. Occasionally the infection becomes generalized. It is encountered less frequently than gonorrhea and syphilis, but it appears to be a significant disease.
4. *Psittacosis* or *parrot fever* is another of the diseases of the chlamydia group. Although the disease originally was called parrot fever, the disease agent may also infect a good many other birds such as ducks, turkeys, chickens, and pigeons. The more generalized term, ornithosis (disease of birds), is synonymous and more desirable. The disease is pneumonia-like in its symptomology. It may range from a mild, influenza-like disease in which the patient, although ill, may remain ambulatory, to a full-blown acute toxic fulminating pneumonitis. It generally responds to antibiotic therapy. Like the viruses and the rickettsiae, the chlamydiae are known primarily because of and by the disease they produce.

THE VIRUSES

The origin of the term virus is not entirely clear. It came from the Latin meaning a slimy liquid or poison. Early Roman writers applied it to the unseen causes of disease, which they envisaged as poisons, or to the venoms of snakes or spiders. Louis Pasteur early in his career applied the term both to bacterial pathogens and the nonbacterial, unidentified causes of disease such as the "rabies virus."

The series of events that led to the final discovery of viruses as distinct entities began in 1885 when Mayer, a Dutch scientist, described a yellow mottling of tobacco leaves as "mosaic-like." When sap from diseased leaves was placed in a capillary tube and scratched into a healthy leaf, it produced the symptoms of tobacco mosaic, as the disease became known. Microscopic examination failed to reveal a causative agent.

In 1890 Iwanowsky, a Russian, began a study of tobacco diseases in the Crimea. Seeking the causative agent of tobacco mosaic, he crushed infected tobacco leaves and found that the juices produced the disease symptoms when applied to the leaves of uninfected plants. When heated to boiling, the leaf juice lost its infectivity. However, when he filtered the fluid through the bacteria-tight, unglazed porcelain filters developed a few years earlier by Chamberland, the juice still retained the ability to produce mosaic spots on normal tobacco leaves. He concluded that the disease agent was either a toxin produced by bacteria in the diseased leaf or by bacteria that passed through the pores of the Chamberland filter.

In 1898, Beijerinck, who was working in the bacteriological laboratory at the Polytechnical School in Delft, Holland, reported on his repetition of Mayer's work, which he had seen while they were associates at the School of Agriculture at Waneningen. He performed a faultless series of inoculations proving that the sap of infected plants was able to produce the disease even after passage through a bacteria-tight porcelain filter (presumably a Chamberland filter). The unfiltered juice contained no bacteria. The filtrate diluted a millionfold still caused the infection. Filtered juice from the experimentally infected plant could again be diluted a millionfold and would again cause the infection. He concluded that the disease was not caused by microbes but a living liquid virus. He questioned that the agent was particulate in nature but did feel that it was "living."

The same year (1898), Loeffler and Frosch, assigned to a commission for the study of foot-and-mouth disease, secured an exudate from infected cattle. After passing it through a diatomaceous filter, which had held back experimentally added *Pseudomonas aeruginosa*, the filtrate still retained high infectivity and produced foot-and-mouth disease in calves. They found that they could dilute this "active toxin," as they called it, to 1/3,000,000,000 of the weight of the infected calf and still produce the disease. They guessed correctly that the agent was only about one-tenth the size of pathogenic bacteria.

By the turn of the century, the existence of plant and animal viruses had been established. During the next decade a number of other filter-passing plant and animal pathogens (viruses) were discovered. Because of their filter-passing ability they soon came to be called filterable viruses to distinguish them from other viruses, that is, disease-producing agents.

Soon the third large group of viruses, those that infect bacteria, were found. This discovery came about in an indirect, but fortuitous fashion. In 1915 Twort, a young English bacteriologist, was given the task of growing vaccinia (cowpox) virus on a bacterial medium with the hope that this process could be used in the wholesale production of much needed smallpox vaccine for British troops. He was unsuccessful, but on one of his plates, which was overgrown with *Staphylococcus aureus*, he noted watery looking areas. The unknown agent, when washed from these glassy colonies, would produce the same type of clearing on other plate cultures. Also, the washings could be diluted a millionfold, passed through a bacteria-tight filter and would still produce "transparent points." He suspected the agent might be a virus.

It remained for d'Herelle, a French-Canadian working at the Pasteur Institute, to clearly establish the existence of viruses capable of attacking bacteria. While following the course of convalesence in several patients suffering from bacillary dysentery, d'Herelle discovered a filter-passing, anti-Shiga lysing agent that would quickly clear a turbid broth culture of rapidly growing *Shigella*. This could be diluted a billion times and still cause lysis. Later he demonstrated the formation of cleared plaques or virus colonies on an agar plate culture heavily seeded with susceptible bacteria. He applied the name bacteriophage (bacteria-eater) to this agent. This term is now generally shortened to phage.

Thus, in less than two decades, three major groups of filterable viruses had been shown to exist—plant, animal, and bacterial viruses. While such a neat but arbitrary grouping generally holds true, some exceptions have been observed. For example, some plant viruses, such as curly leaf in sugar beets, are apparently able to proliferate in insects; some phages are reproduced in plants; some animal viruses are apparently able to multiply in bacteria.

Recently it has become apparent that viruses specific for insects may be widely distributed. Although insects are animals, the virus disease agents of insects differ in many respects from the viruses and virus diseases of higher animals. Insects play an important role in the transmission of virus diseases of plants. In some cases, the plant viruses may themselves attack the insect vectors. The larvae are most commonly infected; however, adult insects at times succumb to viruses. Although the picture is not yet clear, while the DNA viruses seem to predominate, RNA viruses of insects have been found. Some recent experiments have suggested that virus preparations hold promise as agents for the biological control of insect pests. Many practical problems must be solved before this can become a reality.

The question, are viruses living forms, is a perennial but unanswerable one because it requires a clear and specific definition of the term, "living form." Thus far, an unequivocal definition of living has yet to be evolved.

A listing of the distinctive characteristics of viruses provides the easiest approach to the question. The primary distinguishing characteristics of viruses are as follows:

1. Most of the active virus particles or virions range in size from 10 to 250 nm. This places them below the resolving power of the usual light microscope, which is about 200 nm.
2. In their infective stage, viruses are strictly intracellular pathogens and their presence is generally recognized only by the diseases they produce.
3. Unlike the cells of bacteria and all higher forms that contain both DNA and RNA, the viruses have only one nucleic acid. In the phages, this is most often DNA but RNA phages are also known. In plant viruses, RNA types seem to predominate. With the animal viruses, DNA and RNA viruses are about equally represented.
4. Viruses multiply in the "form of their genetic material." In other words, their reproduction is dictated by the genetic message inherent in their own genome.
5. Viruses can neither grow independently of the host cell nor undergo binary fission.
6. Viruses are devoid of energy-producing enzymes although some meager enzyme systems have been found. Some phages, for example, apparently have proteolytic enzymes that permit the phages to eat their way into the bacterial host cell. The influenza viruses possess the enzyme, sialidase, that attacks sialic acid, a vital constituent of many cell mucoproteins. Possession of this enzyme may impart to the influenza virus much of its characteristically high infectivity.
7. This high specificity of viruses for particular host cells is dictated by the genetic message of the virus itself. Indeed, this property has been described as parasitism on a genetic level.

Lwoff has proposed what may be a logical answer to the question are viruses living forms by his intriguing comment, "Viruses are viruses."

On the question, where did viruses originate, no satisfactory answer has yet been found. One school of thought believes that viruses are degenerate bacteria. In such event, intermediate forms should be found. Some rickettsia, such as the agents of psittacosis, lymphogranuloma venereum, or trachoma have been suggested as intermediate forms. They were formerly called

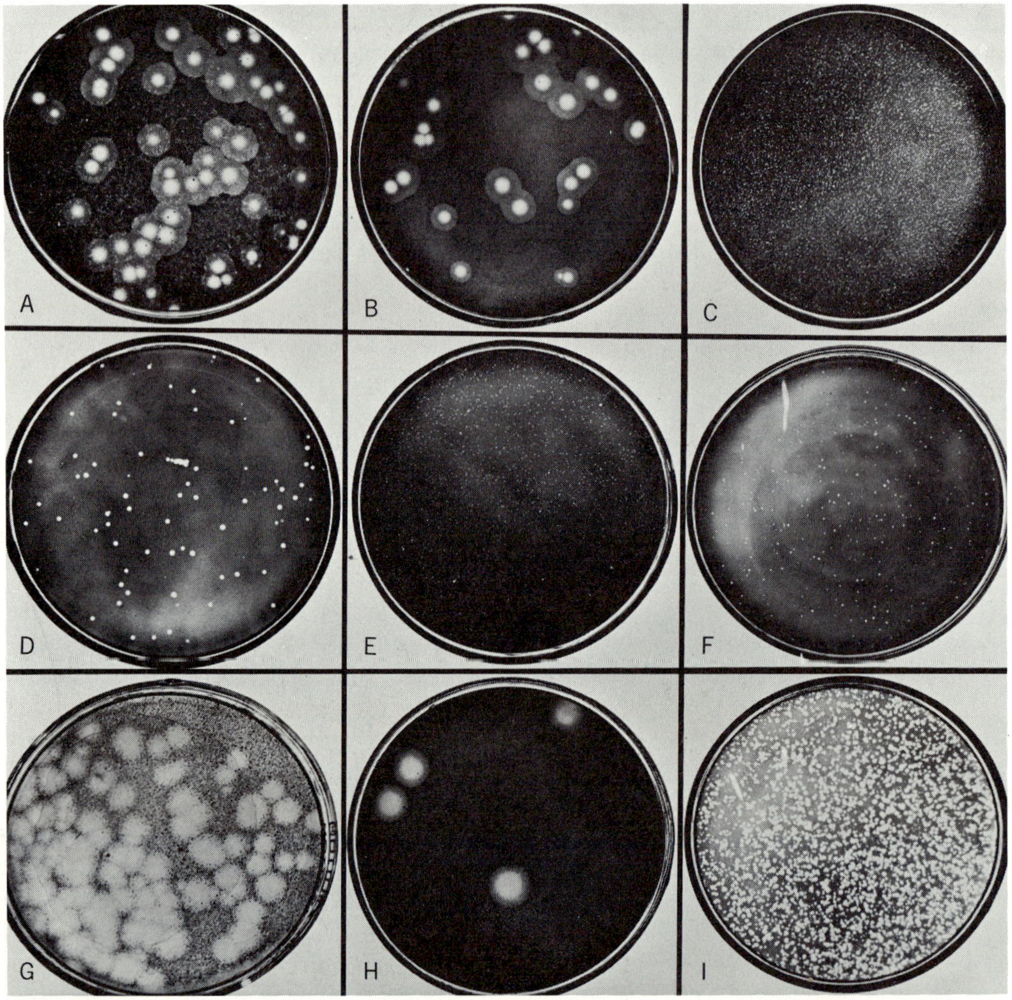

Figure 8-1
Types of plaques formed by different bacterial virus-host systems. **A,** T_1 on *Escherichia coli.* **B,** T_2 on *Escherichia coli.* **C,** ϕ42 on *Staphylococcus aureus.* **D,** ϕ52 on *Staphylococcus aureus.* **E,** ϕ81 on *Staphylococcus aureus.* **F,** ϕ80 on *Staphylococcus aureus.* **G** and **H,** D29 on *Mycobacterium smegmatis.* **I,** ϕ29 on *Staphylococcus aureus.* All plates were flooded with a tincture of iodine before photographing in order to darken the agar background. (From Swatek, F. E.: Textbook of microbiology, St. Louis, 1967, The C. V. Mosby Co.)

large viruses. However, recent evidence suggests that these forms more nearly resemble true bacteria than intermediates. They divide by forming cross walls, having an envelope much like the cell wall of gram-negative bacteria, and are inhibited by penicillin and D-cycloserine that function by inhibiting cell wall formation. Muramic acid, a component of bacterial cell walls, has been isolated from some of these forms. They also possess both DNA and RNA and an energy-producing metabolism. In late stages of growth these forms become extracellular. Again Lwoff has a logical answer, "A large virus is a virus, a small bacterium is a bacterium."

The other concept, that viruses have come from simple primordial forms, is difficult to defend philosophically, since viruses lack completely the ability to reproduce themselves independently.

Virus terminology

The individual virus particle is termed a virion. It is made up of an inner core of genetic material (the template for the reproduction of additional virions by the host enzymes) and surrounded by a coat of protein, the capsid (L. *capsa* box). The capsid is made up of structural units termed capsomeres (Gr. *meros* part). Each virus capsid appears to be composed of a characteristic number of capsomeres with characteristic shapes. In addition to the capsid, some viruses have a second or outer shell termed a peplos or mantle. The capsid structure gives the virus an appearance resembling a naked ear of corn except that the kernels (capsomeres) are arranged in a helical fashion.

How viruses are grown

Viruses are completely dependent upon living host cells for their reproduction. Plant viruses are grown chiefly by grinding up infected leaf or stem tissue and passing it through a bacteria-tight filter or centrifuging it at high speed to throw down plant debris and any bacteria that may be present. The plant-virus material is then inoculated into the host plant by scratching or rubbing the filtrate into the leaves. A more detailed discussion of the plant viruses is given in Chapter 29.

Animal viruses can be grown in any one of the following ways:

1. By injecting virus-bearing material from the blood, spleen, spinal fluid, feces, and so on, of the infected animal or man into suitable host animals. The sample may be injected into the skin (intradermal), through the skin, (subcutaneous), into the intestinal cavity (intraperitoneal), into the muscle (intramuscular) or a number of other ways. For many years the only method for growing the poliovirus was to inject infected material into the brains of monkeys after boring a tiny hole through the skull.
2. Fertile eggs (about halfway through the incubation period) may be inoculated with material containing infective viruses. The viruses may be grown in the yolk sac, or the allantoic membrane, in the brain or other tissues of the developing chick embryo.
3. Bits of animal tissue may be grown in a suitable enriched liquid medium. This is termed a tissue culture, and virus-bearing substances may be inoculated directly into such tissue cultures.
4. Animal tissue may be ground up, the cell debris digested with proteolytic enzymes, and single animal cells separated out. These may then be grown as a layer, one cell in thickness, to form monolayer cell cultures. To count the number of virions in an infected exudate or to compare the growth of various viruses, dilute virus material is inoculated into special petri plates or tissue culture dishes containing a monolayer of cells. Wherever a single virus is present it may destroy the adjacent cells, producing a cleared spot, or plaque, as a visible evidence of cell destruction. By counting the number of plaques, a fairly accurate estimate can be made of the number of virus particles (virions) present. Also, different strains of viruses produce different kinds or sizes of plaques. This enables the virologist to separate out pure plaque strains and to study genetic variation in viruses.

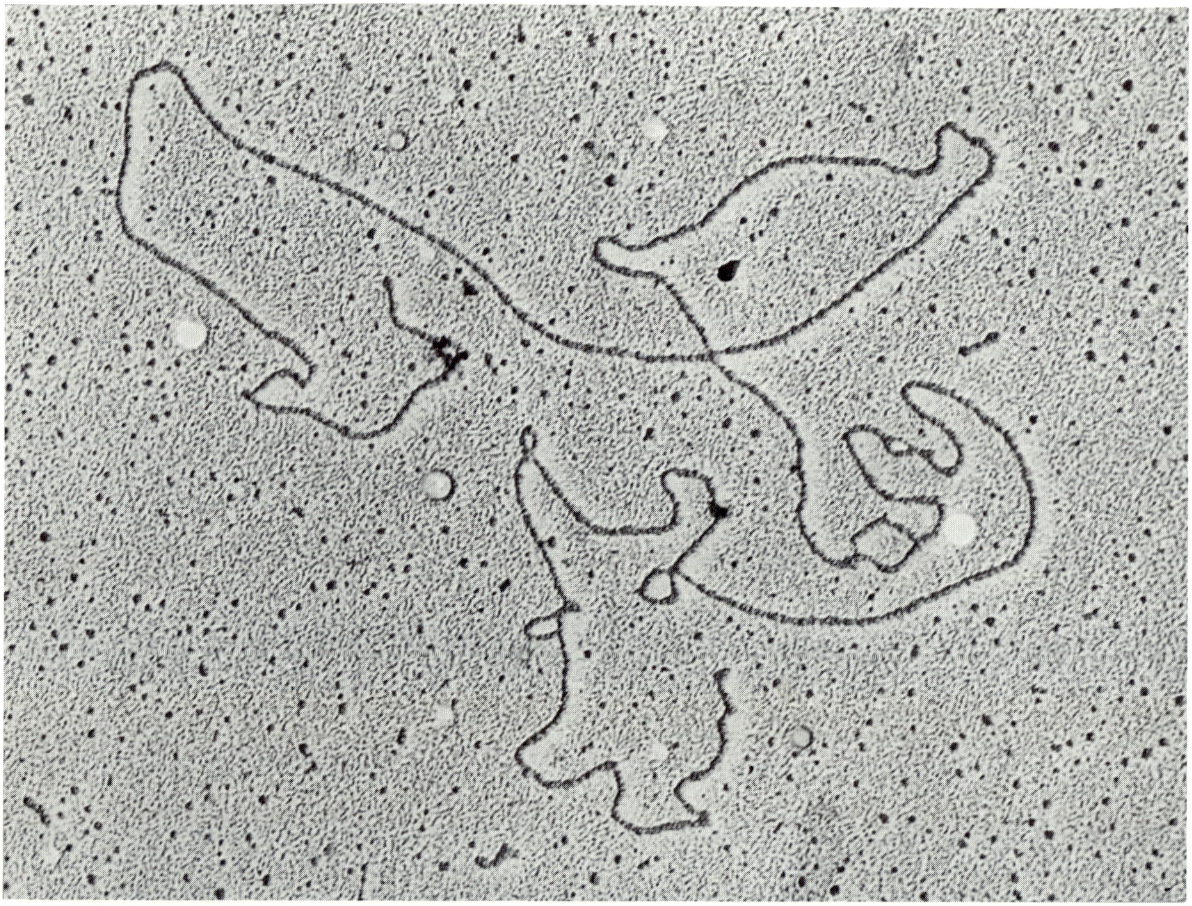

Figure 8-2
Escherichia coli-lambda-phage DNA, showing circular nature of phage chromosome. (×70,000.) (Courtesy K. K. Li and D. Mohr.)

Tissue cultures or monolayer cell cultures may come from a variety of sources such as embryonic chick tissue, monkey tissue, or even human tissue such as the HeLa cells coming originally from a human tumor. Phages growing on a lawn of bacteria in a petri dish produce plaques similar to those formed by animal viruses in cell cultures (Fig. 8-1).

Animal viruses

The animal viruses are often cuboid or hexagonal in shape. Some animal viruses producing visible damage to tissue or cell cultures (the so-called cytopathogenic effect) may not be associated with any specific human or animal disease. These are sometimes termed orphan viruses.

The animal viruses have been arbitrarily classified as RNA viruses and DNA viruses.

The RNA viruses are the following:

1. The small RNA viruses include the enteroviruses (intestinal viruses) represented by poliovirus, Coxsackievirus and the echoviruses (enteric cytopathogenic human orphans), the reoviruses (respiratory enteric orphans), and the foot-and-mouth disease viruses.
2. The myxoviruses include the influenza, parainfluenza, Newcastle, measles, mumps, and rabies viruses.
3. Arbor (arthropod-borne) viruses, of which there are over 200 types, are primarily carried by arthropods (joint-footed insects such as mosquitoes), and may cause encephalitis (western equine encephalomyelitis, eastern equine encephalomyelitis, St. Louis encephalitis) and other diseases, such as yellow fever and dengue or break-bone fever. In addition, indications are that a large number of poorly defined members of this group may exist in South America.

The DNA viruses are the following:

1. Variola (smallpox), cowpox, and rabbit myxoma viruses are considered DNA viruses.
2. The herpes group includes herpes simplex (cold sore virus), herpes zoster (shingles virus), and varicella (chicken pox virus).
3. Adenoviruses represent large arrays of poorly defined viruses that produce diseases of the upper respiratory tract and the conjunctiva. Included in the group is the common cold agent that is apparently a symptomology (rather than an entity) involving a number of different viruses rather than a single virus. Many of these viruses produce coldlike symptoms. Some 28 (or more) serotypes of adenoviruses have been described.
4. Small DNA viruses include a group of viruses that cause abnormal cell growth such as warts, Shope papilloma in rabbits, and other animal tumors.

Bacterial viruses or phages

Bacterial viruses or phages differ considerably from the plant and animal viruses. They are often tadpolelike structures with a large hexagonal head and a long tail that is apparently a hollow tube. At the end of the tail is a "tail plate" with spikelike structures plus long tail fibers. These apparently perform the function of attaching the phage to the host bacterium. Many phages have a collar resembling a coiled spring with a tube in the center not unlike the needle of a hypodermic syringe. This tube is apparently hollow and seems to provide a channel through which the DNA strand of the phage (Fig. 8-2) may be injected into the bacterium.

A few tiny phages, the male specific RNA phages, consist essentially of hexagonal or similar head structure with or without a short spike. Filamentous forms similar to plant viruses have also been found. These are apparently specific for the pili of male bacterial cells.

The classic course of phage infection of bacterial cells appears to be as follows: The phage adheres to the bacterial cell by its tail and the fibers or hooks then attach to the cell. After a brief period the enzymes of the tail appear to eat through the cell wall and cell membrane. The coil-like collar then retracts sharply, forcing the tube into the cell and discharging the DNA strand as a long stringlike fiber.

Two courses of action may follow: the lytic

(or productive) cycle, or the less common lysogenic cycle.

Lytic cycle. In the lytic cycle the genetic message of the virus DNA supercedes the genetic instructions of the DNA in the host cell chromosome. The cell machinery is then instructed to "make phage DNA" or, as Cairns suggests, special phage factories are set up. In a relatively short time the cell manufactures a large number of strands of phage DNA. However, since this DNA is naked it cannot function as a virulent phage. If the cell is ruptured at this time the inactive phage DNA is released. This period is spoken of as the eclipse phase since completed or virulent phages are not present. The cell machinery then begins rapidly to produce capsomeres, forming capsids or phage coats. In addition, tailpieces, fibers, and spikes must also be formed, probably by a separate series of processes. The phage completion process goes on rapidly and in a short time the bacterium is filled with 50 to 300 completed, virulent phages. The cell then bursts due to the action of a phage-produced, lysozyme-like enzyme that dissolves the cell wall. The virulent phages are released into the medium and the next cycle of phage adsorption, virus replication, and cell-burst begins. This may proceed rather rapidly and be completed in as little as 15 minutes or it may require as long as 1 to 2 hours.

Both rapidity of replication and size of burst are determined by the genetic information carried by the DNA (or RNA) of the phage, influenced in turn by genetic information in the host cell, temperature, and components of the medium.

The phage that habitually gives rise to the lytic cycle is said to be virulent. Some phages damage the cell irreparably during the process of cell adsorption. These highly virulent phages have been called intemperate phages.

When lysis occurs, cultures in tubes of broth seeded with host bacteria and phage will first become turbid as the bacteria begin to grow, then, due to rupture of the bacterial cells, clear as the lytic process goes on.

Lysogenic cycle. With sensitive (susceptible) strains of bacteria, attachment of a phage particle is followed by rapid proliferation of phage particles within the host cell. When the number of complete phage particles has reached a certain level, the host cell bursts, releasing the phages.

In some bacterial strains, the DNA of the phage is injected in the same manner as with the lytic strains. However, due to genetic information carried by the host cell, the phage is attached to the bacterial nucleoid, possibly incorporated within the ring itself, and is subsequently replicated along with the DNA of the host cell. This naked DNA is noninfectious and is termed the prophage. This state is called lysogeny and the host cells are said to be lysogenic. Phages that give rise to this unique relationship are said to be temperate. The reason for the term lysogenic—meaning dissolving-producing—is that in a lysogenic population a very small number of cells will apparently mutate to the lytic state, and complete, infective phage particles will be released. These can be detected when the culture itself or filtrates of the culture are mixed with a sensitivity host strain.

With many lysogenic strains of bacteria, exposure of the bacteria to ultraviolet light will cause essentially all of the cells to revert to the lytic state. Immediately, rapid replication of the phage DNA occurs along with formation of protein coats. When a large number of complete particles are thus formed, the cell bursts. In this manner, most, if not all, of the cells are lysed. This process of converting from the lysogenic to the lytic state is termed induction.

Occasional strains of bacteria have been found in which a mutation of the prophage occurs that prevents formation of complete phages following exposure to ultraviolet light. Such phages are spoken of as defective lysogenic strains.

SUGGESTED READINGS

Adams, M. H., et al.: Bacteriophages, New York, 1959, Interscience Publishers, Inc.

Andrews, C. H.: The natural history of viruses, New York, 1967, W. W. Norton & Co., Inc.

Fraser, D.: Viruses and molecular biology, New York, 1967, The Macmillan Co.

Fenner, F., and White, D. O.: Medical virology, New York, 1970, Academic Press, Inc.

Horsfall, F. L., Jr., and Tamm, I., editors: Viral and rickettsial infections of man, ed. 4, Philadelphia, 1965, J. B. Lippincott Co.

Luria, S. E., and Darnell, J. E., Jr.: General virology, New York, 1967, John Wiley & Sons, Inc.

Smith, K. M.: Plant viruses, ed. 4, London, 1968, Methuen & Company, Ltd.

Stent, G. S.: Molecular biology of bacterial viruses, San Francisco, 1963, W. H. Freeman & Co. Publishers.

9 Microscopic methods for the study of microorganisms

THE NEED FOR MICROSCOPY

OPTICS OF THE LIGHT MICROSCOPE

PHASE MICROSCOPE

FLUORESCENT MICROSCOPE

DARK-FIELD MICROSCOPE

TRANSMISSION ELECTRON MICROSCOPE

SCANNING ELECTRON MICROSCOPE

PREPARING MICROORGANISMS FOR LIGHT MICROSCOPE EXAMINATION

BACTERIAL STAINS

Microorganisms, representing as they do a great world of living things largely invisible to the unaided human eye, can be effectively studied only by microscopic methods. Many of these forms, particularly the bacteria, are relatively transparent when mounted on the microscope slide and must be stained before they can readily be seen. Special staining procedures may reveal such structures as spores, capsules, and flagella, or may serve to differentiate and sometimes identify certain kinds of bacteria.

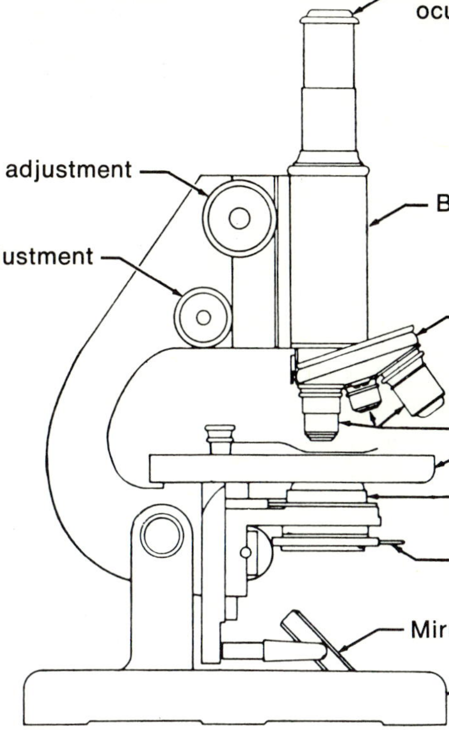

Figure 9-1
Compound, monocular microscope showing basic parts. (From Anderson, D.: Laboratory instructions in microbiology, St. Louis, 1969, The C. V. Mosby Co.)

THE NEED FOR MICROSCOPY

Until Leeuwenhoek first trained his simple microscope on water containing rotting matter, saliva, or the dregs of beer, the whole world of microorganisms remained invisible. The reason lay in the optical limitations of the human eye. These limitations are resolving power and focusing distance.

Resolving power is the ability to separate two objects so that they will appear as two distinct objects and not as one blurred spot. The limit of resolution of most human eyes is about 0.2 mm. In other words, if two separate objects are closer than 0.2 mm they will appear to the eye as a single blurred object, or if an object is smaller than 0.2 mm it will be seen indistinctly, if at all. The theoretical reason for this limitation lies primarily in the spacing of the light receptors (rods and cones) in the retina of the eye. Thus, if two objects are so close that light reflected from them falls on only one receptor, the two will appear as a single object. If they are far enough apart that light reflected from the object falls on two receptors we are able to distinguish them as two objects.

The second limiting factor is focusing distance. With the normal eye, the closest we can bring an object to our eye is about 250 mm (approximately 10 inches). If we hold the object closer, the lens of the eye cannot bring the image into focus on the retina.

OPTICS OF THE LIGHT MICROSCOPE

The light microscopes of the last two centuries have been termed compound microscopes (Fig. 9-1) because they have a dual or compound set of lenses. These consist of the objective (lens nearest the object) and the eyepiece or ocular. The lens systems of each usually consist of two or more separate lenses containing layers of two or more kinds of glass to correct for chromatic aberration.

Objectives perform two functions—they resolve the objects being examined and they magnify this apparent image. Each objective has a limit of resolution, that is, the smallest

Table 9-1. Useful information relative to microscope objectives

NAME OF OBJECTIVE	WORKING DISTANCE*	NUMERICAL APERTURE*	MAGNIFICATION*
Low power	6.8 mm	0.25	10×
High power	0.73 mm	0.66	43×
Oil immersion	0.12 mm	1.25	97×

*Exact values may vary with different makes of microscopes.

object that can be resolved, and an inherent magnification. Oculars or eyepieces in turn magnify the image produced by the objective. The total magnification obtainable with a microscope is the magnification of the objective lens times the magnification of the image by the eyepiece.

Engraved on the objectives and oculars of a compound microscope are values that are most useful to the trained microscopist. By applying them, he can estimate the resolving power of an objective and also the magnification obtainable (Table 9-1).

Working distance is essentially the clearance between the bottom of the objective lens and the microscope slide. This becomes progressively less as the magnification and resolving power of the objectives increase. This distance is very small with the oil-immersion objective. As a result, careless focusing may result in the objective's being jammed into the slide with such force that the lower lens of the objective or the microscope slide or both may be cracked. Also, if a thick cover glass is placed over the object on the slide, it may be impossible to bring the objective close enough to the object itself to bring it into focus.

Numerical aperture (N.A.) is a mathematical value derived as shown below:

$$\text{N.A.} = i \sin \Theta$$

If you visualize the light entering a microscope objective as a cone, Θ equals one-half the angle of this cone, and i equals the index of refraction of the medium that surrounds the lens. With the low-power and the high-power objectives, the medium is air, which has a refractive index of 1.0. With that of glass being about 1.5, the beams of the cone of light are bent abruptly when the light passes from the surface of the microscope slide into the air. As a result, a portion of the rays are dissipated into the surrounding area.

With the low- and high-power objectives, the value of Θ cannot be greater than 1.0; hence their rated N.A. values are less than 1.0. However, the oil-immersion objective is very shortsighted since it comes into focus at only about 0.12 mm from the object. To avoid the marked light loss of the cone of light illuminating the object on the slide, as it bends abruptly on entering the air, the objective must be immersed in a liquid (immersion oil) that has approximately the same refractive index as glass (about 1.5).

Also, the N.A. of the substage condenser should be the same as that of the objective. One of the prime functions of the iris diaphragm of the condenser is to vary the N.A. Most condensers have a rated N.A. of about 1.30 when wide open. When half open the N.A. is about 0.6 to 0.7 (equivalent to the high power) and when one-eighth open it approximates that of the low-power objective (about 0.25).

Although the rated N.A. of the condenser when wide open may be about the same as that of the oil-immersion objective, when the light passes directly into the air from the top lens the rays are bent sufficiently to reduce the functional

N.A. to 1.0 or less. Therefore, for maximum resolution with the oil-immersion system, a drop of oil should be placed both on the top of the condenser lens to produce contact with the slide, and on the slide to permit immersion of the objective.

Because of the presence of immersion oil, the cone of light passes through the condenser, microscope slide, and into the objective lens without the rays being bent and dissipated into the atmosphere. It is as if the entire observation were being conducted through solid glass.

A good approximation of the resolving power of your microscope can be calculated by use of the formula below:

$$\text{Resolving power} = \frac{\text{Wave length of light (in nm)}}{2 \times \text{N.A.}}$$

Different authors list different formulas, but the above applies if the microscope is equipped with a condenser that has the same numerical aperture as the objective, and if the whole aperture is filled with light. Other formulas may be applied if the microscope is not equipped with a condenser or if a thin pencil of light is used.

Suppose we want to know if bacteria can be seen with the low-power objective. A simple calculation will supply the answer. If we use a single wavelength of light at about the middle of the visible spectrum, that is, 5,550 Å or 550 nm, the values would be as follows:

$$\begin{aligned}\text{Resolving power} &= \frac{550 \text{ nm}}{2 \times 0.25}\\ &= \frac{550}{0.5}\\ &= 1{,}100 \text{ nm} = 1.1\ \mu\text{m}\end{aligned}$$

A bacterial cell that is $0.8 \times 1.0\ \mu$m would not be resolvable. Even one that is $1.5 \times 5.0\ \mu$m would be barely discernible. However, this problem would not exist with the oil-immersion objective because the values would be:

$$\text{Resolving power} = \frac{550 \text{ nm}}{2 \times 1.25} = 220 \text{ nm or } 0.22\ \mu\text{m}$$

Thus even the smallest bacterium can be seen by using the oil-immersion objective and a 10× ocular. However, an internal structure in a bacterial cell such as a granule or organelle 0.1 μm or less in diameter is invisible or seen only as a blur. In the same way, unstained flagella are not resolvable. On the other hand, a yeast cell $3 \times 7\ \mu$m can be discerned easily with the low power objective.

The total magnification attainable with a given microscope represents the magnification of the objective times the magnification of the ocular or eyepiece. Thus, with a lower power objective that magnifies 10× and a 10× eyepiece, the total magnification is 100×. With a 5× eyepiece the total magnification is only 50×.

You might surmise that a much higher magnification could be obtained by the simple expedient of using a 20× or 30× ocular. However, this is not optically feasible. While 20× eyepieces are available, they are limited in value because only about the center fourth of the image will be in clear focus. The remaining portion is likely to be fuzzy. The rule of thumb is that the maximum practical magnification is about 1,000× the N.A.

PHASE MICROSCOPE

A major contribution to light microscopy came in 1932 when F. Zernike produced the phase or phase-contrast microscope. This earned him the Nobel Prize in physics in 1953.

While the optics of this system are relatively involved, they consist essentially of phase condensers equipped with a metallic coating, and annular rings in the lenses that reduce the intensity of the undeviated rays and bring the refracted rays into phase by a one-fourth wavelength retardation. The coating reduces the glare that obscures fine details, and the reinforced, deviated rays now in phase result in increased contrast that reveals relatively transparent structures within or around the living cell.

The objectives have similar annular rings and built-in ring plates, which absorb about three-fourths of the direct light from the object. In use, accurate optical alignment of the condenser and objective is required. Two light combinations are possible: dark (positive) contrast, used especially for living cells that appear dark, and (negative) contrast, most useful with bacteria and similar minute objects that appear light surrounded by a darker halo.

The great advantage with phase-contrast microscopy is that living cells in suspension can be examined without staining and many details invisible in the conventional microscope can be revealed.

FLUORESCENT MICROSCOPE

The fluorescent microscope, which utilizes ultraviolet light instead of visible light, was first designed by Lehmann (1911) but was not highly useful until the development of the high pressure mercury vapor arc in the 1940's and the discovery of the fluorescent antibody technique by Coon and co-workers in 1950. Since that time it has become a valuable instrument for scientific research. Its value lies in the fact that ultraviolet light causes many types of living material to fluoresce, that is, the invisible ultraviolet rays are changed into wavelengths emerging as luminous blues, greens, yellows, oranges, or reds.

The applications are threefold.

1. Many naturally occurring microorganisms, especially salt- and freshwater forms, will fluoresce naturally and can be recognized in specimens because of this reaction.
2. Specimens containing bacteria, and so on, can be stained with fluorescent dyes such as acridine orange, and these stained microorganisms may fluoresce in a distinct fashion. Also, the DNA in stained tissues will show a green fluorescence while RNA fluoresces red. When sputum from a tuberculosis patient is stained with auramine, the *Mycobacterium tuberculosis* cells can be identified by their fluorescence and counted.
3. When antibodies are developed in experimental animals by injecting certain bacteria, these antibodies can be labelled or tagged with dye, fluorescein isothiocyanate. When such labelled antibodies are mixed with material containing these specific bacteria, the antibodies will attach to the specific bacteria. When the specimens are examined under the fluorescent microscope, the labelled bacteria will fluoresce strongly. Other bacteria that are untagged will show little or no fluorescence.

DARK-FIELD MICROSCOPE

The dark-field microscope represents a relatively simple modification in the method of illumination of the light microscope. It was originally designed for the examination of colloidal particles in suspension and for forms such as the syphilis organism, *Treponema pallidum*, which is extremely fragile and difficult to stain.

The simplest form of dark-field illumination can be achieved by placing in the condenser a metal disk with narrow slits around the outside. These slits admit light rays from the sides that are bent by the microscope slide to illuminate the minute particles or microorganisms in suspension causing them to scintillate against the dark background.

However, the so-called cardioid dark-field condenser is much more effective and is used routinely for syphilis dark-field examinations. This condenser has a stop that blocks out transmitted light from the field but reflects surfaces inside the condenser forming a hollow cone of refracted light that illuminates the object from the sides.

TRANSMISSION ELECTRON MICROSCOPE

Although the light microscope can resolve nothing smaller than about 0.2 μm, we are still able to study the whole cells of algae, yeasts, protozoa, bacteria, and even rickettsiae. The light microscope filled the void between the eye limitation of 0.2 mm (200 μm and 0.2 μm) (200 nm). However, with the development of the electron microscope, a whole additional field of view has been opened to man because the limit of resolution has been lowered to 2 nm or even 1 nm or less. Just as the light microscope gave us a hundredfold increase in our ability to see small objects, so the electron microscope has increased our apparent visual resolution another 100 to 200 times.

Instead of being able to see only bacteria, we can now see viruses (including their finer structures), and the tiny external structures, such as pili or fimbriae, of bacteria. In extremely thin sections of bacteria and other microorganisms we are able to see tiny organelles and relate these structures to the metabolic activities of the cell. We are now also able to see macromolecules such as DNA.

The first transmission electron microscope (TEM), the prototype of the modern instrument, was described by Knoll and Ruska in 1932. The resolution of this first instrument was poorer than that of the light microscope but it yielded fairly good pictures. Ruska in 1934, by adding a condenser lens, increased the resolving power beyond that of the light microscope. Subsequent improvements by von Borries and Ruska (1938) increased the resolving power to 100 Å (10 nm).

Really successful electron microscope examinations of bacteria, viruses, and other biological material were not made until 1950 when three vital improvements in technique appeared (Fig. 9-2):

1. Discovery of the glass knife, with an edge sufficiently sharp and thin, made it possible to section bacteria. (Recently diamond knives have become available.)
2. Coupled with this was the development of an ultramicrotome with which to use the glass knives. Mechanically, the ultramicrotome is designed to advance the glass knife at each stroke only an infinitely short distance. This limited advance is necessary if a section is to be sliced off a tiny bacterium. The older microtomes cut sections generally much thicker than the bacterial cell.
3. To hold the bacteria for sectioning, something stronger than the paraffin blocks used in conventional histology was required. Nonshrinking resins, the epoxy, Araldite, or the polyester, Vestopal, met this requirement. Bacteria are now embedded in resins in capsules and sections are made of this embedded material.
4. For examination of bacteria or tissues with the electron microscopes, the common bacterial or tissue stains are useless. Instead, electron-dense salts of lead, uranium, or other heavy elements are utilized as stains.

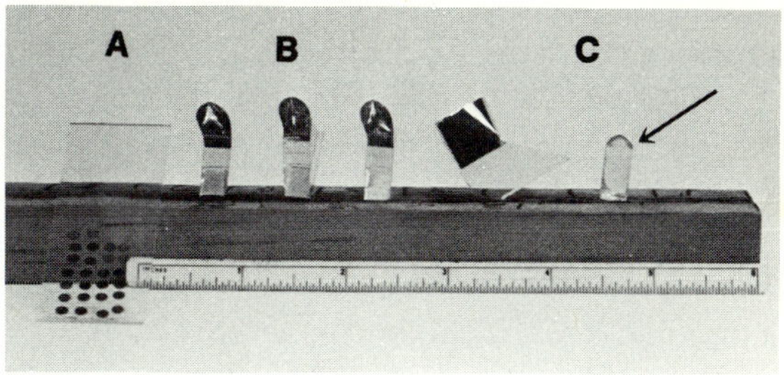

Figure 9-2
Special devices used in electron microscopy. **A,** Grids of brass screen used for mounting specimens; **B,** glass knives ready to cut ultra-thin sections (cutting edge surrounded by heavy foil to form well for collecting sections); **C,** capsule for embedding bits of tissue in plastic prior to sectioning; arrow indicates location of specimen.

Optics of the electron microscope

In many respects the electron microscope is similar to the light microscope. The chief difference is that a beam of electrons (with an extremely short wavelength) replaces the beam of light, and electromagnetic lenses take the place of the glass lenses of the light microscope.

The mechanism and operation of the electron microscope can best be visualized by comparing it to a light microscope turned upside down. At the top of the electron microscope is located the electron source called the electron gun. This is made up of a filament (electron source), shield, and anode. Below the gun is an aperture, a tiny metal disk with a fine hole that allows a thin stream of electrons to pass through. Immediately below that is the condenser lense that functions like the condenser on a light microscope focusing the electron beam on the object being examined. This object is placed on a tiny disk generally made of a fine copper or brass mesh with a very thin coating of Formvar (collodion-like) and carbon. This is held in a small thimblelike holder that is placed in the microscope with special tongs. Below the object are two lenses, the objective and projector lenses, which are much like the objective and ocular of the light microscope. The electron beam falls on a fluorescent screen that is activated to give a visible image, which allows focusing and searching the field. When an object is found suitable to photograph, the screen is dropped out of the way and the beam falls on a camera aperture, allowing a photograph to be made. This is the negative from which an electron micrograph is made.

Unlike the light microscope, the electron microscope must operate under a high vacuum, otherwise the metal filament would be quickly oxidized and the apertures and electromagnets contaminated with metal oxides.

Preparing a specimen of whole bacteria is considerably more involved than smearing a slide for the light microscope. It is usually desirable to fix the bacteria, then metal shadow them to increase the details. This is done by drying bacteria on a number of grids on a microscope slide and evaporating a metal, such as platinum, gold, or uranium, in a vacuum so that a stream of metal ions passes at an angle over the grids depositing metal on the bacteria and causing them to stand out strongly on the photograph.

Bacteria also may be negatively stained. Bacteria dried on the grid are flooded with phosphotungstate solution and allowed to dry. This compound is electron dense, and does not penetrate tissues themselves but fills in the spaces between the fine surface structures causing them to stand out sharply in electron micrographs.

Magnification possible

Theoretically, magnification directly obtainable with the electron microscope would be about 100,000 times. However, when working with whole bacteria the highest useful magnification is about 8,000 to 10,000. With viruses, thin sections of bacteria, or tissue preparations, the magnification used is about 30,000 to 40,000. Even at this magnification, sharply focused pictures are difficult to obtain because the images are often extremely faint. A recently developed accessory, the image intensifier, makes it possible to see images too faint to be seen on the fluorescent screen. This is essentially a television amplifier. Enlarged pictures can be viewed on a television screen or recorded on video tape. Also, the image can be studied in a distant classroom by using remote monitors. The image intensifier increases the intensity of the image some 2,500 times. Thus, specimens that would quickly disintegrate in a high intensity beam can be studied at low electron intensity for much longer periods.

A negative made of an electron microscope image can be enlarged 5 to 10 times by photographic enlargement. Thus a 20,000× image could be enlarged to give the equivalent of a 100,000 to 200,000. If a magnification of 100,000 was achieved in the electron microscope, the image from the negative might be then blown up 10 times to give a magnification of up to a million.

SCANNING ELECTRON MICROSCOPE

In 1953 McMullan conceived the idea of combining the design principles of the electron microscope with the image-scanning ability of the cathode ray tube of the television. This recently perfected instrument, known as the scanning electron microscope (SEM) (Fig. 9-3), makes it possible to examine surfaces and whole objects as large as a tribolium beetle or as small as a bacterial spore. While images are similar to those of the light microscope, the depth of field is 300 to 500 times greater than that of the light microscope. This results in beautiful three-dimensional images or pictures, revealing minute fine details never before detected (Fig. 9-4).

The instrument itself involves a vacuum col-

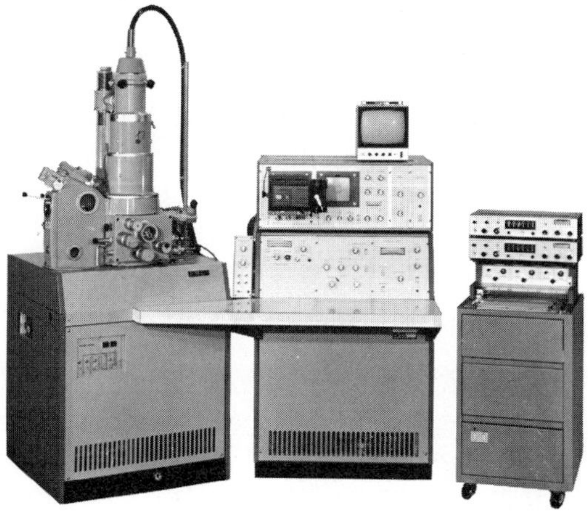

A

Figure 9-3
A, Units comprising JEOL scanning electron microscope. Electron microscope is on left, scanning and amplifying units on right. (Courtesy JEOL [U.S.A.], INC.) **B,** Schematic diagram of electron microscope, illustrating mode of operation, (Courtesty JEOL [U.S.A.], INC.)

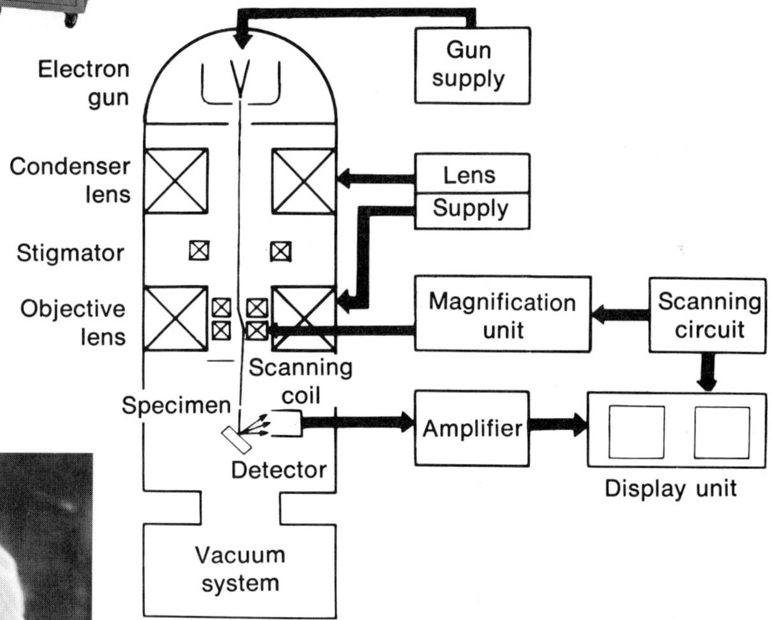

B

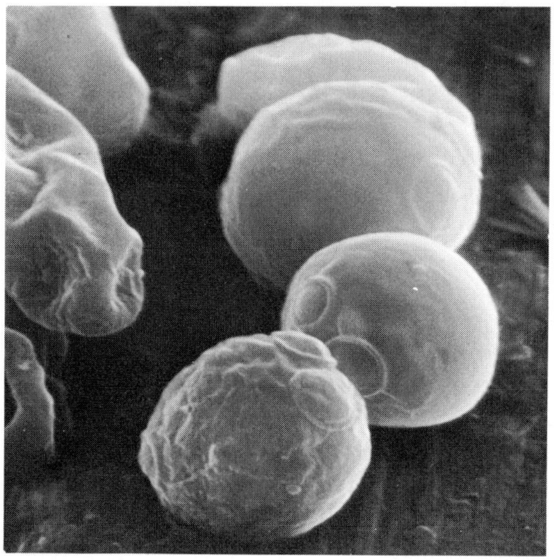

Figure 9-4
Yeast cell, *Saccharomyces cerevisiae,* illustrating three-dimensional effect achieved by scanning electron microscopy. (Courtesy JEOL [U.S.A.], INC.)

umn like that of an electron microscope with an electron gun and magnetic lens. Instead of the electron beam's passing directly through the object being studied, it strikes the preparation at an angle and the surface is scanned by the electron beam or probe a small area at a time. Many of the rays are absorbed, but some are reflected from the surface and fall on a scintillator creating an image that is transmitted to a photomultiplier tube. Variations in signal voltages are reflected in variations in brightness of the image displayed by the cathode (TV) tube. This image may be photographed or visually observed. If scanning moves at a rate that allows more than 30 images per second, the observer sees a complete image on the screen.

The resolving power of the SEM is lower than that of the TEM, being about 100 Å (10 nm) compared with 10 Å (1 nm) for the TEM. The practical range of magnification is about 14 to 50,000 times although magnifications up to 130,000 are possible.

For examination, specimens are mounted on stubs with a special cement, then thinly plated with gold, or gold and palladium, in a vacuum evaporator.

PREPARING MICROORGANISMS FOR LIGHT MICROSCOPE EXAMINATION

The microscopic examination of microorganisms requires special techniques for mounting and staining specimens. Smaller, unicellular microorganisms are often examined by placing a drop of a suspension of the microbes on a slide and covering with a thin coverslip. This is known as a wet mount. A second method is to place a drop of suspension in the center of a cover glass and suspend it over a concave depression in a special type of microscope slide. This is known as a hanging drop.

Wet or hanging drop mounts are often used to determine whether or not bacteria are motile. Motile forms move in a purposeful manner, often changing and even reversing direction. Nonmotile bacteria may display a quivering nondirectional motion, known as brownian motion, that is sometimes confused with true motility. This movement is due to the impact of rapidly moving, but invisible, water (and other) molecules in which the particles are suspended. By counterbalancing the force of gravity, these tiny particles are kept from settling out. Motion ceases when the particles reach a diameter of about 4 μm.

BACTERIAL STAINS

Bacteria in a liquid suspension, such as broth or water, are difficult to see under the microscope, because their refractive index (ability to bend light rays) is not too far removed from that of the glass and the suspending medium. The use of chemical dyes that have a special affinity for the protein of bacteria makes them stand out from the background. A single stain used for this purpose is termed a simple stain. If two or more stains are used in succession, as with the Gram stain or the acid-fast stain, this process is termed counterstaining. A mixture of stains (such as the Giemsa or Wright's stains used for blood cells) is termed a polychrome stain.

Simple stains

Weigert in 1871 tried carmine as a bacterial stain with indifferent success. In 1875 he successfully stained cocci in tissues using one of the newly-developed coal-tar dyes, methyl violet. Koch in 1877 found that he could spread bacteria on a cover slip, and that on drying little or no change occurred in their shapes. He used alcohol to fix them to the slides and stained them with methyl violet and fuchsin. Paul Ehrlich, in Koch's laboratory introduced methylene blue as a bacterial stain (1881). This same year, Koch found that he could fix bacteria to glass slides by flaming. Ehrlich added aniline to methyl violet as an intensifier and was able by prolonged contact to successfully stain the tuberculosis organism, a bacterium notoriously difficult to stain because of the high lipid content of its outer layer. Ehrlich's discovery was, therefore, an important one. He also found that once these bacteria were stained, the stain remained inside the bacteria even when washed with a weak nitric acid solution in alcohol. Ziehl and Neelsen, working independently, developed parts of what we now term the Ziehl-Neelsen acid-fast stain. With some slight modification this has become the standard acid-fast stain.

Christian Gram, a Danish physician, used the aniline-water-methyl-violet stain of Ehrlich, plus Lugol's solution (iodine and potassium iodide) as a mordant and ethyl alcohol as a decolorizing agent. He attempted to stain the pneumonia organism in lung tissue from a pneumonia patient and found that the bacteria were stained a blackish purple while the tissues were decolorized. He counterstained the tissues with Bismarck brown. This has since been replaced by safranin, and this staining procedure constitutes the highly useful Gram stain.

Following an observation made earlier by Ehrlich, Loeffler added potassium hydroxide (1:10,000) to methylene blue stain and found that it would deeply stain granules in the diphtheria organism *(Corynebacterium diphtheriae)* while the body of the organism stained a lighter blue. This proved to be a highly useful method for identifying the diphtheria organism in throat smears or in cultures grown on coagulated blood serum. These intracellular granules are termed metachromatic granules.

More recently, a number of techniques have been developed for the differential staining of spores, capsules, and flagella.

The Gram stain

The Gram stain is the most useful of all staining procedures. It enables microbiologists to readily divide various bacteria into two major groups, gram positive (those retaining the initial stain when washed with the decolorizing agent, ethanol and acetone) and gram negative (losing the stain and showing the color of the counterstain). The significance of the Gram stain lies in the fact that gram properties arise from true basic differences in the two classes of cells. The gram-positive property is inherent only in certain types of true bacteria (mostly cocci and spore-forming rods) and in actinomycetes and yeasts. Many of the bacteria and all tissues are basically gram negative. The major differences between gram-positive and gram-negative bacteria are outlined in Table 9-2.

Table 9-2. Some differences between gram-positive and gram-negative bacteria

PROPERTY	GRAM-POSITIVE	GRAM-NEGATIVE
Cell wall	Relatively thick	Much thinner
Variety of amino acid in in cell wall	Small number	Large number
Lipid in cell wall	Low or absent	Always present
Nutrient required	Often complex, no autotrophs	With exceptions, less complex, some autotrophs
Susceptibility to antibiotics	Relatively high	Relatively low
Dissolving effect (1% KOH)	Not dissolved	Dissolved
Effect of anionic detergents	Growth generally prevented, cell lysed	Generally grow, lysis moderate
Lysis by lysozyme	Easily lysed	More resistant to lysis
Bacteriostatic effect of crystal or gentian violet	Markedly susceptible	Little susceptibility
Magnesium content	High	Low

The acid-fast stain

The acid-fast staining procedure of Ziehl and Neelsen is employed for demonstrating the presence of the tuberculosis organism *(Mycobacterium tuberculosis)* in sputum or other body materials, or the presence of the leprosy organism *(M. leprae)* in scrapings from infected portions of the body. These bacteria are distinctive in that they retain the initial stain (hot carbolfuchsin) even when washed with alcohol containing 3% hydrochloric acid or other acids. A counterstain, usually methylene blue, is used after decolorization to provide a contrasting background.

Acid-fast bacteria will retain the color of the first stain (pink or red) while the nonacid-fast bacteria will be blue.

Other differential stains

Spore and capsule stains are similar to the Gram and acid-fast stains in that counterstains are used. The spores or capsules are stained a color different from that of the rest of the bacterial cell.

Capsules can also be demonstrated by *negative stains.* In the simplest of these, the India ink method, bacteria and ink are mixed on the slide and allowed to dry. The capsule, which is unstained, stands out as a clear zone around the body of the bacterium against the black background of the dried ink.

SUGGESTED READINGS

Aaronson, S.: Experimental microbial ecology, New York, 1970, Academic Press, Inc.

Bradbury, S.: The microscope, past and present, London, 1968, Pergamon Press, Inc.

Gray, P.: The use of the microscope, New York, 1967, McGraw-Hill Book Co.

Harrigan, W. F., and McCance, M. E.: Laboratory methods in microbiology, New York, 1966, Academic Press, Inc.

Meynell, G. G., and Meynell, E.: Theory and practice in experimental bacteriology, Cambridge, 1965, Cambridge University Press.

Sirockin, S., and Cullimore, S.: Practical microbiology, London, 1969, McGraw-Hill Book Co.

10 Pure culture methods for cultivating microorganisms

NEED FOR PURE CULTURES

In such populous habitats as our intestines, the soil, decaying organic matter, sewage and polluted waters, microorganisms exist as mixed cultures. Only when we can isolate a given bacterium, yeast, or mold as a pure culture, free from all other forms, can we gain a true picture of its nature, characteristics, metabolic activities, and importance.

Indeed, the vast amount of fundamental knowledge embodied in the science of microbiology is based largely on the study of pure cultures. Some forms have never been grown successfully in pure culture, so the little we know of them is limited largely to microscopic examinations.

The medical microbiologist can prove that a given microorganism causes a given disease only when he has isolated it. To produce a specific vaccine or determine the sensitivity of a given pathogen to antibiotics or disinfectants, a pure culture is generally required. The applied microbiologist who is attempting to control the spoilage of foods or the microbial deterioration of other materials can work most effectively when the offending microbe (or microbes) has been grown alone. Desirable fermentations or the microbial production of valuable industrial organic compounds are most effective when pure cultures of enzymatically active forms are available. These may even be crossed genetically to yield highly productive strains not encountered in nature.

Various strains of bacteria and other microorganisms have become the tools of molecular biologists in their basic studies of physiology and microbial genetics.

METHODS FOR OBTAINING PURE CULTURES

The early microbiologists were forced to work largely with mixed cultures of bacteria and other forms. This led many of these early workers to accept the convenient theory of pleomorphism; that is, that under different conditions a given organism could develop as a rod or coccus and become the cause of an intestinal infection or a sore throat.

Lister was one of the first to conduct a fundamentally sound experiment that challenged this theory. In 1877 he isolated the common milk souring bacterium, *Streptococcus lactis,* by diluting the milk stepwise and inoculating it into sterile milk. Microscopic examination of the culture revealed only one kind of organism, a short-chained coccus. Today the dilution method is rarely used, but it is still the method of choice for those microorganisms that will grow in a liquid culture but not the surface of an agar medium.

Robert Koch was convinced that different forms of pathogenic bacteria were distinct and constant species. This prompted him in 1881-1883 to add agar-agar to liquid media as a solidifying agent. This marine algal product has the following important properties:

1. When suspended in water or a liquid medium, it can be liquified by heating to near the boiling point.
2. It gels or solidifies at about 45° C to form a firm gel.
3. When solidified, its consistency remains unchanged at higher temperatures until the boiling point is again reached, permitting the growth of heat-loving bacteria at temperatures of 80° C or even slightly higher.
4. Very few bacteria are able enzymatically to attack and decompose agar; thus it is essentially an inert solidifying agent.

Koch's pure culture technique was the breakthrough needed to advance microbiology by making pure cultures readily available. Basically this is accomplished by the use of pour, streak, or spread petri plate cultures.

For pour plates, natural material, such as soil, milk, or feces, containing a mixed population is diluted in sterile water or saline solution to provide a suspension containing a hundred bacteria or so per milliliter. A 1.0 ml portion is measured with a sterile pipette into a sterile petri plate. Over this is poured a sterile, liquid agar medium that has been melted then cooled to 45° C. The plate is now rocked to mix thoroughly the bacteria-containing suspension with the agar. As soon as the agar has cooled a few degrees it forms a firm gel. Wherever a bacterium is caught in the gel, it begins to reproduce itself. In a matter of 12 to 48 hours, the single cell multiplies until it has formed a colony of millions or billions of cells, all alike. It may then be transferred to another growth medium forming a stock culture for use in subsequent experiments. A typical pour plate is shown in Fig. 10-1.

In using streak plates for isolating pure cultures, the sterile agar nutrient medium is first poured into a sterile petri plate and allowed to solidify. With an inoculating needle that has been sterilized by flaming, a loopful of the suspension is streaked over about a fourth of the plate surface. The needle is then flamed and another portion of the plate is streaked at right angles to the original streaking. This drags bacteria from the first streak, diluting the number present. With a sterile needle, a third streak is made in the same manner at right angles to the

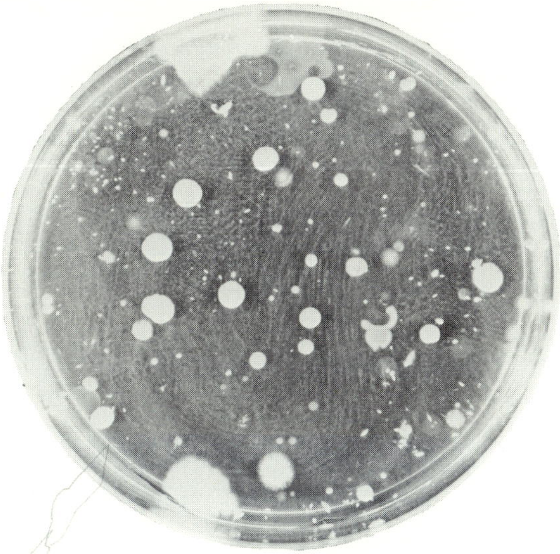

Figure 10-1
Typical pour plate made from dilution of soil. Note large surface colonies, small submerged colonies, and variation in colony shape.

second streak, further diluting the bacteria. A fourth streak is made with a sterile needle at right angles to and across the third streak. When incubated, the plate will show isolated pure colonies at points along the streaks. The scheme is shown in Fig. 10-2, *A*, and a typical streak plate in Fig. 10-2, *B-C*.

In the spread-plate technique, a small amount of the suspension, usually about 0.1 ml, is placed on the agar plate and spread over the surface with a sterile bent glass rod. A typical spread plate is shown in Fig. 10-3, *A* and *B*, and a device for rotating the plate in Fig. 10-3, *C*.

FACTORS INVOLVED IN GROWING MICROORGANISMS

Successful isolation of microorganisms in pure culture involves a variety of factors including proper nutrients, satisfactory growth temperatures, proper amount of oxygen (or lack of it), desirable osmotic properties in the medium, the proper range of acidity or alkalinity (pH), and, with some bacteria at least, the necessary hydrostatic pressure.

Nutrients for the growth of microbes

Just as the gardener grows his flowers in the soil, the microbiologist prepares a suitable mixture of nutrients that becomes a microbial soil. This nutrient mixture is known as a nutrient or culture medium.

To reproduce, microbial cells must be supplied with the appropriate compounds of nitrogen, phosphorus, and sulfur for the synthesis of proteins, deoxyribonucleic acid (DNA), ribonucleic acid (RNA), and the host of other complex molecules necessary for the processes of life. Energy and cell-building materials are derived from carbohydrates, amino acids, and similar compounds.

Small amounts of compounds of magnesium, calcium, iron, and potassium are required, plus trace amounts of salts of manganese, zinc, molybdenum, and similar elements. Certain of the mineral elements constitute a minute but vital part of certain of the metal-enzyme complexes.

Carbon is the building material for the skeletons of all organic—proteins, carbohydrates, fats, and other—molecules. This may be supplied by preformed organic matter, carbon dioxide, or bicarbonates in solution. Microbes growing in a simple inorganic medium without an organic carbon source are classed as *autotrophs*. These autotrophs in turn have been divided into two classes: photosynthetic autotrophs *(phototrophs)*, which includes those forms that derive energy from light and carbon from carbon dioxide or carbonates (higher green plants are in this class), and chemosynthetic autotrophs *(chemotrophs)*, which derive the necessary energy for growth from the oxidation of such compounds as ammonia, nitrites, hydrogen sulfide, sulfur, hydrogen, or iron, and carbon from carbon dioxide or bicarbonates in solution. Forms of this latter type are peculiar to the bacteria.

Borderline forms that require simple organic compounds as hydrogen donors for reducing

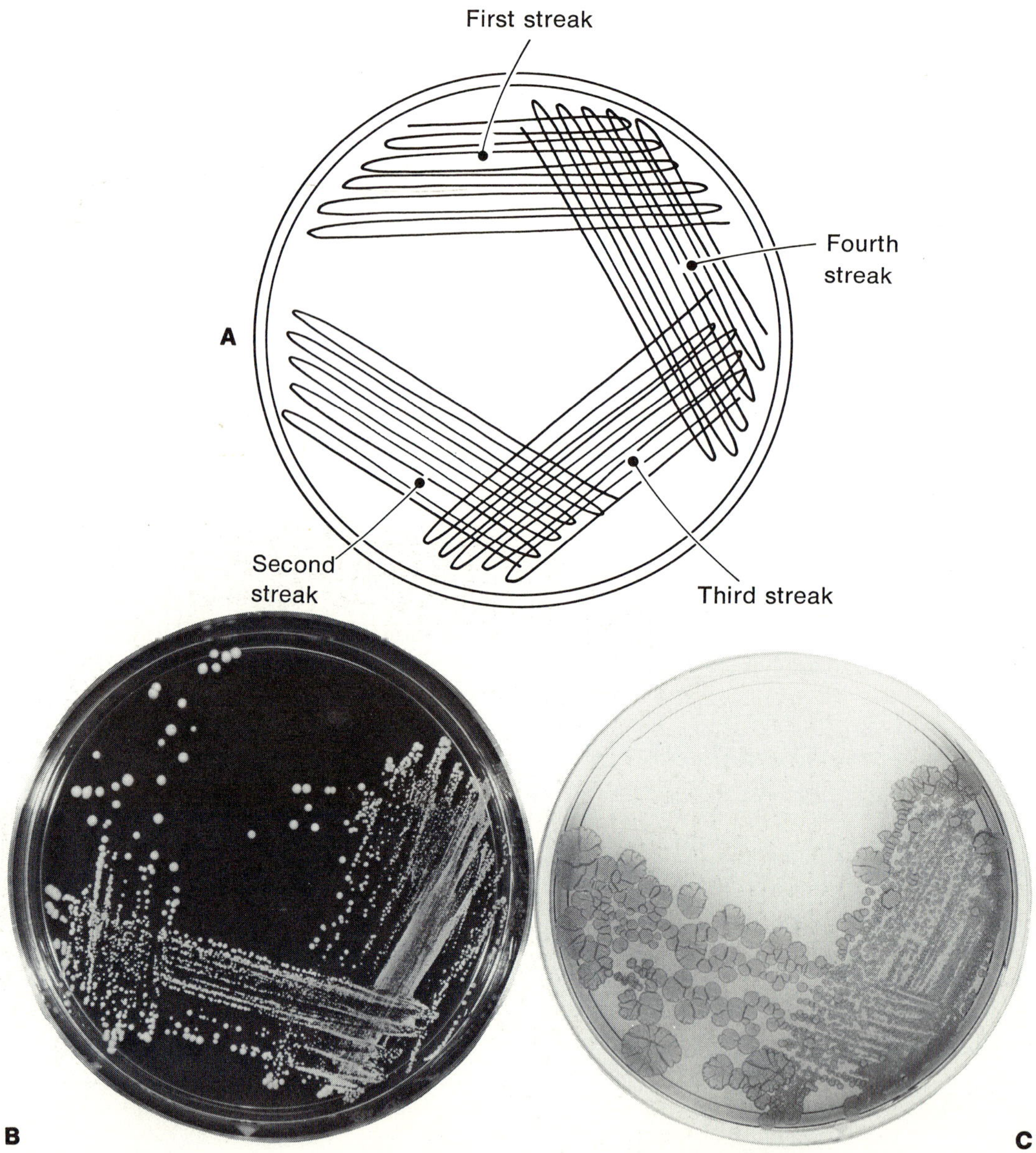

Figure 10-2
A, Procedure for preparing streak plate; **B-C,** typical streak plates with well separated colonies that would permit isolation of pure culture. **B,** Streaked with *Staphylococcus aureus;* **C,** streaked with *Bacillus subtilis* var. *niger* that shows unusual colony appearance and blackening of medium characteristic of strain. (**A,** From Anderson, D.: Laboratory instructions in microbiology, St. Louis, 1969, The C. V. Mosby Co.)

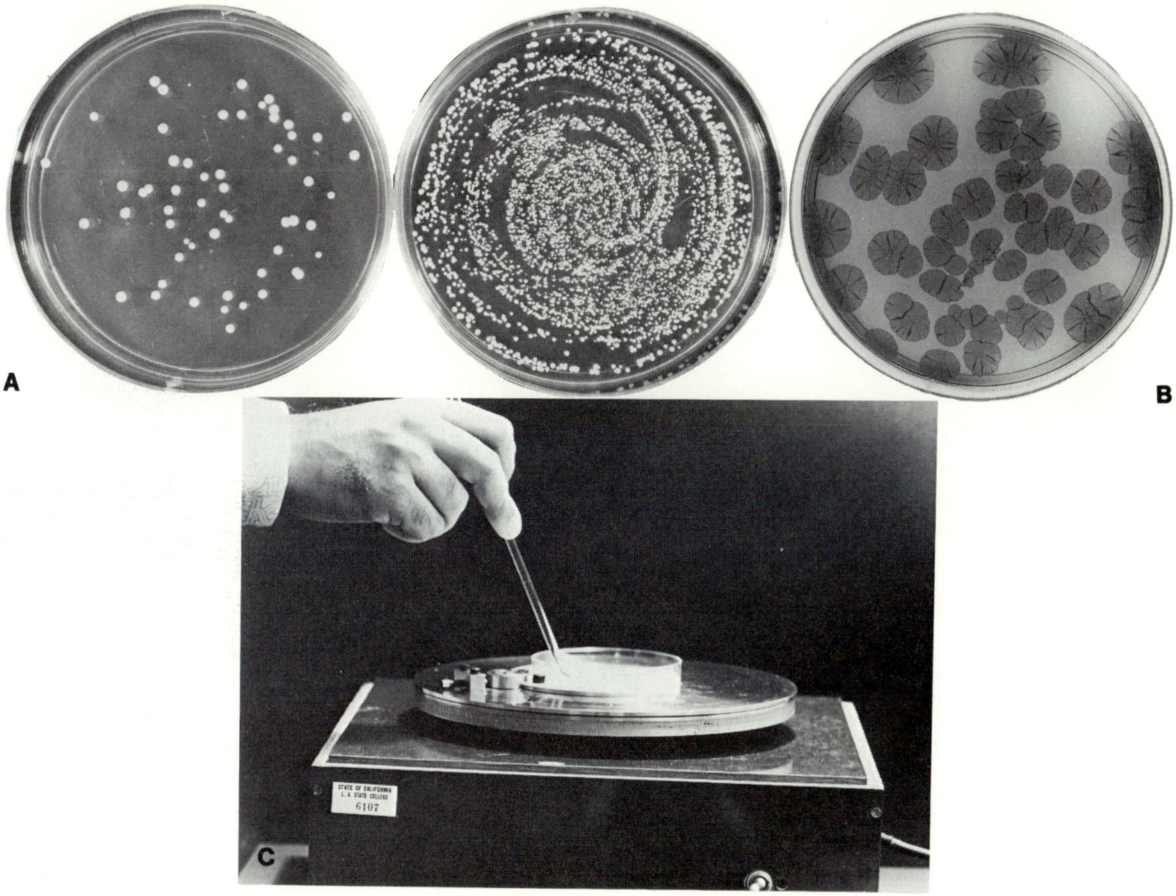

Figure 10-3
Typical spread plates of **A,** *Escherichia coli,* and **B,** *Bacillus subtilis.* Note even distribution and relatively uniform size of surface colonies. **C,** Motor operated rotary table for spreading bacteria on plate by means of sterile glass rod.

CO_2 occur in both groups. As an example, the nonsulfur purple bacteria grow photosynthetically in the absence of oxygen and in the presence of light in a mineral salts photosynthetic medium with a small amount of ethyl alcohol or propionates. In the presence of air and in the dark they grow vigorously on nutrient agar. These intermediate forms are termed *facultative autotrophs.*

The majority of microorganisms, however, are *heterotrophs* (Gr. *heteros* mixed) requiring preformed organic compounds for energy and as electron donors. Man, together with practically all animals except the few photosynthetic protozoa, falls in this category. The nutritional needs of the heterotrophs vary greatly. For example, most strains of *Escherichia coli* and related coliforms, and practically all of the pseudomonads grow well if supplied an appropriate mineral salts mixture plus an organic compound such as glucose to serve as a combined carbon and energy source. Forms that grow on mineral salts plus sugar are termed *chemoorganotrophic*.

Many disease-producing bacteria must be supplied with complex nutrients such as blood, ascitic fluid, or similar compounds, since they have lost the ability to synthesize the necessary vitamins, amino acids, and related growth factors. These bacteria are said to be nutritionally fastidious. The gonorrhea organism is a classic example.

Physical states of media

Microbial media may be used in the liquid, liquefiable solid, or solid states. A liquid medium may be a natural liquid such as milk, a suspension of beef broth, or a solution of pure chemicals (synthetic medium).

If agar or gelatin is added (usually 1.5% agar or 12% gelatin), the medium becomes a liquefiable solid medium. When heated to near the boiling point, the suspension of solid agar is converted to a liquid and remains so until it reaches the solidification point. When cooled to a little below 45° C agar solidifies. Gelatin must be cooled to almost 20° C. Unfortunately, the liquefying temperature of gelatin is only a few degrees above the point of solidification. Many of the pathogens that must be incubated at or near body temperature cannot be grown successfully in gelatin plates. Also, certain bacteria proteolyze (digest) gelatin, quickly turning a plate of solid medium into a foul-smelling liquid.

If agar is added to a liquid medium prior to sterilization, it becomes an agar medium when autoclaved. Thus milk plus 1.5% agar becomes a milk agar.

Solid media consist of materials such as potatoes that are a natural solid state, or those, such as a suspension of raw eggs or egg white, that can be solidified by heating. Because the presence of organic matter in the medium may inhibit the growth of certain autotrophic bacteria, agar or solid organic media cannot be used. Therefore, to provide a solid surface for isolating pure cultures, an inorganic silica gel can be prepared by adding hydrochloric acid to a solution of sodium silicate (water glass). This is immediately poured into a petri dish where it quickly forms a firm, clear gel. This is leached with water to remove the excess acid, and the surface is flooded with a suitable inorganic solution to form a solid isolating medium.

Kinds of media

Over the years, the microbiologists have been pitting their ingenuity against the microbes in attempts to secure pure cultures of the more elusive ones. As a result, an enormous number of formulations of media have been devised. In the early 1930's a survey of the scientific literature yielded some 3,000 formulas. Since that time this list may have been doubled or trebled.

In 1881 Loeffler, who worked in Koch's laboratory, devised a bacterial medium consisting of meat infusion, 1% peptone, and 0.6% salt. Two years later, Miquel suggested that Liebig's beef extract be substituted for the meat infusion, and the familiar nutrient broth was born. This simple formulation, beef extract (0.3%) and peptone (0.5%) (with or without added salt) has become "the microbiology laboratory workhorse." If 0.5% sodium chloride and 1% to 2% whole blood (such as human, sheep, or rabbit), are added it becomes blood agar, which now supports the growth of many of the fastidious pathogens that would not grow on nutrient agar. By dilution of nutrient broth tenfold, the growth of soil actinomycetes is favored. If 1% lactose (milk sugar) is added to nutrient broth, it becomes a standard medium for the presumptive coliform test for water pollution.

Temperature and microbial growth

Microbes, particularly the bacteria, are able to grow at a wide range of temperatures. Psychrophilic (cryophilic) microorganisms grow in the temperature range from −4° to 15° C. Mesophiles (Gr. *mesos* middle) grow from 10° to 15° C, up to about 45° C. The vast majority of microorganisms fall in this class. Thermophiles grow in a temperature range from about 50° to 80° C. Some forms, facultative thermophiles, grow fairly well between 37° C and 55° C, while obligate thermophiles grow only at higher temperatures. A few microorganisms have been found growing at temperatures higher than 90° C.

Psychrophiles are able to grow at low temperatures presumably because they possess enzyme systems able to function at temperatures low enough to inhibit the activity of vital enzymes in most other forms. Psychrophilic microorganisms include the whole gamut of bacteria—gram positive, gram negative, spore formers and non-spore formers, including cocci, vibrios, and long or short rods. They may be aerobes, facultative anaerobes, or obligate anaerobes.

Fortunately psychrophiles, being objectionable in foods chiefly because they may impair the appearance or flavor, have little or no public health significance. They do, however, cause fresh hamburger stored in the refrigerator to spoil in a few days and stored milk to become bitter or off flavor in a relatively short period of time. Fruits stored in the refrigerator often develop mold colonies in a week or so. Foods stored below −10° C are free from attack by psychrophiles.

Mesophilic organisms are all around and in us. Most forms can be placed conveniently into one of two groups: those that grow best at from 10° to 30° C and, those that may grow well in the range from 15° to 45° C but grow best near the body temperature of 37° C. Most of this latter group of mesophiles are normal inhabitants of, or pathogenic to, the body of man and other warm-blooded animals.

The thermophiles are found in hot springs or in heating, decaying organic matter. As well as being thermophilic, many of these forms are markedly thermogenic (heat producing).

The growth temperature ranges of various microorganisms are sometimes termed the cardinal temperatures. The lowest temperature at which a given microbe will grow is called its minimal growth temperature and the highest is called the maximum growth temperature.

Oxygen requirements

Microorganisms vary considerably in their oxygen requirements for growth. Forms that grow only when air is present (obligate aerobes), when grown in a liquid medium tend to form a mat of growth on the surface. Bacteria that can grow only in the absence of air are termed obligate anaerobes. Facultative anaerobes grow either in the presence or absence of air. When these are grown in a tube of liquid medium, the entire tube becomes turbid with growth. Some bacteria will not grow in air but also fail to grow when oxygen is completely absent. These are known as microaerophilic forms. Some so-called microaerophiles actually require an increased proportion of carbon dioxide rather than reduced oxygen.

Osmotic properties of media

The bacterial cell is essentially an osmotic machine. Most types of bacteria, whether spheres, rods, or spirals, have a rigid cell wall perforated with numerous holes or openings giving the cell the appearance of a collander or wire screen. Inside the cell wall is the semi-permeable membrane, which acts as a sac that houses the bacterial cytoplasm, chromosome, and organelles. The bacterial cell with its cell wall resembles a cylinder of wire screen with an inflated balloon on the inside pressing firmly against the enclosing wire. (This simile applies to most bacteria, but some resemble animal cells in lacking the plantlike restricting cell walls.)

If bacteria that have grown in nutrient broth are placed in a solution of 15% sugar, a marked difference in the osmotic pressure in the sugar solution and the components on the inside of the bacterial cell occurs. As a result, water in the cell diffuses preferentially through the cell membrane, reducing the pressure inside the cell wall and causing the membrane-enclosed mass to shrink away from the cell wall. This process is termed plasmolysis. This could be visualized as the change occurring if the balloon inside the screen cylinder is partially deflated.

On the other hand, if the same bacteria are

taken from the sugar solution and placed directly in distilled water, they immediately imbibe water in such quantities that osmotic pressure causes the cells to burst like over-inflated balloons. This over-imbibing of water has been termed plasmoptysis.

The effect of osmotic pressure on growth of microorganisms must be considered in determining the concentrations of nutrients used in bacterial media. For example, few bacteria can withstand salt concentrations higher than 12% sodium chloride. This is why high salt concentrations are used for pickling meats. Jams and jellies do not spoil readily because of their high sugar content. However, certain yeasts and molds are able to grow slowly at sugar concentrations of 60% or more.* They are termed osmophilic forms or lovers of high osmotic pressure. The cells of many bacteria, however, are likely to burst when placed in distilled water. This method (osmotic shock) is often used to release components from inside the cell for biochemical analyses.

Figure 10-4
Highly mucoid colony of *Streptococcus salivarius* produced when grown on nutrient agar containing 5% sucrose, illustrating unusual growth effect of high concentration of carbohydrate. (×20.)

Acidity or alkalinity of media

A vital factor in the growth of microorganisms is the concentration of H^+ ions (hydrogen atom with a positive charge) and negatively charged OH^- ions. Even in pure water these ions exist in extremely low concentrations. This may be illustrated by the following formula:

$$\underset{\textbf{Water molecule}}{HOH} \rightleftharpoons \underset{\textbf{Hydrogen ion}}{H^+} + \underset{\textbf{Hydroxyl ion}}{OH^-}$$

The frequency with which ionization occurs is extremely low and is a dynamic action; that is, while some molecules are undergoing ionization, other ions are reuniting to form the original un-ionized state (HOH). At any given time at room temperature, the concentration of H^+ and OH^- ions will *each* be approximately 1.0×10^{-7} mols per liter. (Thus an average of 1 out of 10,000,000 molecules will be in the ionized state.) This state is considered the neutral point in terms of acidity or alkalinity.

When acids are added to water they undergo ionization. In a strong acid like hydrochloric acid (HCl) practically all of the molecules ionize, while with weak acids only 1% or so undergo ionization. The actual proportion ionizing depends on temperature and concentration.

In sodium hydroxide (NaOH), a large proportion of molecules are ionized to $Na^+ + OH^-$ making it a strong base, while with ammonium hydroxide only a small proportion of the molecules undergo ionization, making it a weak base.

For most living things, life processes go on in a very narrow range of hydrogen ion concentrations. Taking the H^+ of water as 1×10^{-7} grams

*Fig. 10-4 shows an unusual growth response of a bacterium induced by the addition of a moderately high concentration of sugar to a bacterial medium. (Usual sugar additions are 0.1% to 1.0%.) Apparently this combines both a nutritional and an osmotic effect.

per liter (1 in 10,000,000 parts), many forms are limited in their life processes to a tenfold higher or lower concentration, that is, 1.0×10^{-6} (1/1,000,000) grams per liter to 1.0×10^{-8} (1/100,000,000) grams per liter.

Manifestly, dealing with such minute concentrations of hydrogen or hydroxyl ions is a difficult mathematical operation. To change these horrendous decimal fractions into positive values, Sörenson, a Danish worker, developed the pH or potential hydrogen scale.

The definition of pH is "logarithm of the reciprocal of the hydrogen ion concentration expressed as grams per liter." This clever mathematical operation changes very minute decimal fractions into very large positive numbers that in turn are expressed as logarithms.

By way of illustration:

The H^+ concentration at pH 7.0 $= 1 \times 10^{-7}$.

Thus $\frac{1}{1.0 \times 10^{-7}} = 10{,}000{,}000$, the logarithm of which is 7.0. Therefore, the pH at neutral point is 7.0. At pH 6.0 it would be: $\frac{1}{1.0 \times 10^{-6}} =$ 1,000,000.

Two difficulties present themselves in attempting to understand pH: (1) all pH changes are by steps of 10, being exponential rather than arithmetic (thus pH 6.0 represents 10 times as many H^+ as pH 7.0), and (2) as the pH values become greater than 7.0 the proportion of OH^- to H^+ increases exponentially. Practically all life processes cease at about pH 1.0 on the acid side and pH 9 to 10 on the alkaline side.

The pH scale is illustrated in Table 10-1.

Forms that grow only in the acidity range of about pH 5.0 to 1.0 are termed acidophiles, those growing best between about 5.5 or 6.0 to 8.0 are sometimes termed neutrophiles, while those that grow only on the alkaline side at the fairly high (OH^-) concentrations of pH 7.5 to about 10.0 are termed basophiles. (See Table 10-2.)

In general, most bacteria grow well in the pH range 6.0 to 8.0, and most molds and yeast on the acid side pH 2.5 to 5.5. A few will grow in very alkaline media. For example, *Vibrio comma*, the cause of Asiatic cholera, is often initially isolated on a medium having a pH of 9.0. This high alkalinity effectively controls the growth of other intestinal forms.

Table 10-1. The pH scale of hydrogen ion concentrations

	pH VALUE	H^+ IN GRAMS/LITER	OH^- IN GRAMS/LITER
Concentration of OH^- increases by tenfold steps; H^+ decreases at same proportional rate	0	1.0	0.00000000000001
	1	0.1	0.0000000000001
	2	0.01	0.000000000001
	3	0.001	0.00000000001
	4	0.0001	0.0000000001
	5	0.00001	0.000000001
	6	0.000001	0.00000001
Neutral point (H^+ and OH^- concentrations equal)	7	*0.0000001*	*0.0000001*
	8	0.00000001	0.000001
	9	0.000000001	0.00001
	10	0.0000000001	0.0001
	11	0.00000000001	0.001
	12	0.000000000001	0.01
	13	0.0000000000001	0.1
	14	0.00000000000001	1.0

Table 10-2. Hydrogen ion concentrations (pH) in common terms

pH	MATERIALS OR PROCESSES
1	Acidity (as H_2SO_4) tolerated by some sulfur-oxidizing bacteria when initially isolated in a medium at pH 2.8
2	Hydrogen ion concentration of an "acid" stomach
3 4	Juices of lemons, oranges, rhubarb, pineapples, peaches, and tomatoes fall generally in range pH 3.0-4.5
5	Tomato juice agar for lactobacilli pH 5.0
6	Sour milk about 6.0 down to 4.5 (sourness is first apparent to the taste at about pH 6.0 or slightly lower)
7	pH of blood—about 7.35 (most media for bacteria pH 6.8 to 7.4)
8	pH 7.5-8.5—most fresh waters and arid soils
9	pH of barren "black alkali" soils

Buffers

Bacteria that ferment sugars with the production of acid increase the hydrogen ion concentration sufficiently to stop growth or even kill themselves. Alkaline reactions, also inhibitory, arise from the deamination of amino acids, and so forth, which release quantities of ammonia into the medium.

Protection against such metabolic suicide is afforded by compounds known as buffers that are often incorporated in the medium. These are compounds that resist changes in pH of a medium when acids or bases are added. Those most commonly employed are primary and secondary potassium phosphates (KH_2PO_4, K_2HPO_4), although carbonates of calcium and other compounds may be used. Peptones themselves, although added for nutritional purposes, also have some buffering capacity.

Hydrostatic pressure

Bacteria, able to grow and carry on metabolic activity, have been isolated from the deepest part of the ocean. These barophilic organisms (see Chapter 3) will die if they are brought to the surface. However, they have been grown successfully by using a heavy steel cylinder, about 4 inches in diameter and with a hollow center about 2 inches in diameter. This is filled with water along with compressible tubes containing cultures of the bacteria. The chamber is securely sealed and the hydrostatic pressure raised to from 1,000 to 10,000 pounds per square inch by means of a hydraulic jack. Bacteria such as *Escherichia coli* soon succumb in such an environment, but the deep-sea bacteria thrive. They resemble closely the pseudomonads and related forms usually found in seawater.

Elective media or enrichment cultures

Since bacterial populations in natural environments generally represent a vast array of different forms, isolating a certain organism for study in pure culture becomes a major problem, particularly when the desired organism represents only a small fraction of the entire microbial population. Winogradsky and Beijerinck, working independently, were able to isolate unusual bacteria that generally constitute only a minute fraction of the soil flora by employing the elective or enrichment culture method. Basically, this involved the use of a liquid medium that favored the rapid growth of the desired organism allowing it to become the major organism in the medium in a few hours or days.

By this method Beijerinck was able to isolate *Azotobacter*, a free-living, nonsymbiotic, nitrogen fixer from the soil. He prepared a mineral salts mixture without any nitrogen compounds, and added sugar as a carbon source. *Azotobacter*, which could "fix" the nitrogen gas from the air for building its vital body proteins, grew rapidly and soon became the major organism in the mixture. Bacteria that required a nitrogen source grew poorly if at all, depending on the nitrogen compounds that leaked out of the *Azotobacter* cells. When Beijerinck streaked the liquid culture on the surface of a solid nitrogen-free agar or silica gel medium, colonies developed that represented a pure culture of *Azotobacter.* By a similar procedure Winogradsky isolated the nitrate-producing bacteria.

Enriched, differential, and selective media

Many pathogenic bacteria will not grow on ordinary nutrient agar. However, if whole blood, serum, plasma, ascitic fluid, or yeast extract is added, the medium is said to be enriched for the growth of fastidious bacteria. (Do not confuse an enrichment medium with an enriched medium.)

In preparing a differential medium, components are added that permit rapid growth of the desired organism and produce colonies having a distinctive, easily recognizable appearance. A classic example is Levine's EMB (eosin–methylene blue) agar. This is nutrient agar to which has been added 1% lactose (milk sugar) and small amounts of the dyes eosin and methylene blue. On this medium *Escherichia coli* grows rapidly because it can utilize the lactose. Also, in the presence of the two dyes the colonies assume a distinct smooth, blackish appearance with a greenish metallic sheen. Bacteria related to *E. coli* such as *Enterobacter aerogenes* will grow vigorously, but the colonies will be large, brown, and slimy.

A selective medium is primarily an agar medium, which, although it partially inhibits the growth of the desired organisms, will almost or completely inhibit other forms. S-S *(Salmonella-Shigella)* agar is an excellent example. It is made inhibitory or selective because bile salts and dyes are added as inhibitory agents. When a suspension of feces or feces itself is spread or streaked on a plate, the *Salmonella* or *Shigella,* which cause typhoidlike diseases or bacillary dysentery, will produce distinctive colorless colonies. *Escherichia coli,* present in greater numbers in the feces, is largely or completely inhibited. The few strains that grow produce distinctive red colonies.

Stock cultures

Pure cultures of bacteria are often maintained for long periods for experimental use in the laboratory. Such a group of cultures constitutes a stock culture collection. Often cultures must be transferred every few days or weeks. Cultures may also be stabbed into tubes of soft agar and hermetically sealed. Cultures may be held in a relatively unchanged state over a long period by lyophilization or freeze drying. Bacteria are not killed by this rapid desiccation and will retain their viability for long periods, especially when held in a low temperature refrigerator.

SUGGESTED READINGS

Aaronson, S.: Experimental microbial ecology, New York, 1970, Academic Press, Inc.

Blair, J. E., Lennette, E. H., and Truant, J. P.: Manual of clinical microbiology, Baltimore, 1970, The Williams & Wilkins Co. (for American Society for Microbiology).

Cowan, S. T., and Steel, K. J.: Manual for the identification of medical bacteria, New York, 1965, Cambridge University Press.

Gibbs, B. M., and Shapton, D. A., editors: Identification methods for microbiologists, pt. B, New York, 1968, Academic Press, Inc.

Gibbs, B. M., and Skinner, F. A., editors: Identification methods for microbiologists, pt. A, New York, 1966, Academic Press, Inc.

Skerman, V. B. D.: A guide to the identification of the genera of bacteria, ed. 2, Baltimore, 1967, The Williams & Wilkins Co.

11 Identification of microorganisms

ISOLATION AND PERPETUATION OF PURE CULTURES

OUTLINE OF PROCEDURES AND SPECIFIC TECHNIQUES

The microbiologist who isolates an unknown culture of bacteria is often faced with three major questions: to what order, family, genus, or species does it belong, is it identical with any previously described forms, and what does it do, that is, what are its biochemical characteristics, its pathogenicity, and so forth?

To secure the answers he must first be sure that his isolate is a pure culture. He must then establish its distinctive morphological, cultural, and physiological characteristics, and compare them with an accepted source of descriptions of known bacteria, such as *Bergey's Manual of Determinative Bacteriology*, some of the advanced medical textbooks, or published monographs on related forms.

The microbiologist is frequently required to combine microscopic examinations, growth on culture media, environmental responses, nutritional requirements, and physiological reactions in order to establish unequivocally the true identity of an unknown isolate.

The molecular biologist often works with bacteria or other forms without being even remotely concerned with what the microorganism is or its antecedents, so long as it has the enzymes or organelles that he desires to study or is a useful genetic or biochemical tool. On the other hand, a medical microbiologist may be vitally concerned with a quick identification of an organism that has been consistently found in a case or numerous cases of a disease outbreak. He must determine as rapidly as possible whether or not it is a form never before isolated and what antibiotics or other treatments are required for its control.

The microbial ecologist wants to know if the bacterium he has isolated is a new species or if it is closely related if not identical to forms previously isolated and described.

ISOLATION AND PERPETUATION OF PURE CULTURES

Once a bacterial strain has been isolated and its purity established, the experimenter has available a great array of cultural and other identifying tests. In routine practice, in a hospital laboratory where immediate identification of a common organism is imperative, a half-dozen selected procedures may suffice.

On the other hand, if an investigator is attempting to identify a new organism, he may employ fifty or a hundred different tests. Such a pure culture is generally given an identifying code number and placed in the permanent stock culture collection of the laboratory. In some cases, it may be deposited in the American Type Culture Collection.

More than 850 differential tests or observable characteristics have been listed in *Bergey's Manual*. However, only a relatively small number of these tests are used routinely for describing any given organism.

For a test to be applicable it must be easy to perform, give clear-cut positive or negative reactions, be dependable, and be based on some genetically, rather than environmentally, determined feature. Such tests usually involve the production by a bacterium of a specific, easily demonstrable enzyme system, the ability of a bacterium to utilize some unusual or definitive carbon source for cell reproduction (synthesis), or the ability to catabolize carbohydrates fermentatively or by aerobic respiration for the release of energy.

OUTLINE OF PROCEDURES AND SPECIFIC TECHNIQUES

Microscopic examination

Microscopic examination serves to establish the purity of the culture and to characterize the cell morphology. The Gram stain is always used, although such definitive stains as the acid-fast, spore, capsule, or flagella stains may be added.

Examination of a wet mount or hanging drop may determine motility or lack of it. Motility may be determined by stabbing the organism deep into a tube of soft agar nutrient medium using a straight needle. The nonmotile bacterium will grow as a narrow line along the stab. Motile organisms will move out to form clouds

Figure 11-1
Differences in growth of *Staphylococcus epidermidis*, **A**, and *S. aureus*, **B**, on mannitol salt agar. Indicator, phenol red, imparts red color to medium. **A** produces no acid; hence no color change. **B** ferments mannitol to acid changing indicator to yellow as shown by lighter zone in picture.

adjacent to the stab. Motility is not a fixed characteristic because some bacteria that exhibit vigorous motility when grown at 25° C may be nonmotile if grown at 37° C. Also, motile strains may mutate to form nonmotile mutants. Motility is characteristic of the rod-shaped forms, especially the gram-negative types. On the other hand, it is rare indeed among the cocci; only one or two strains of *Sarcina* and *Streptococcus* are known to be typically motile. In addition to using presumptive observations for motility, flagella formation can be demonstrated by flagella stains or electron microscopy. The latter permits more accurate measurement of the number and magnitude of the flagellar waves.

The Gram stain is perhaps the most useful staining procedure. This quickly designates unknown bacteria as gram positive or gram negative. It is immediately helpful if the organism is found to be a gram-positive nonsporing rod, because bacteria of this type are relatively uncommon. If the unknown bacterium is found to be a coccus, the search is further narrowed. When coupled with the characteristic grouping of the cell into clusters, chains, pairs, or packets, the final identification may be relatively easy. With a few forms the Gram stain plus the appearance of a distinctive colony on the original isolating plate may be sufficient to identify both the genus and species of the unknown isolate. Classic examples are *Staphylococcus aureus*, when isolated on mannitol salt agar (Fig. 11-1), or *Streptococcus pyogenes*, growing on blood sugar agar.

Culture methods

Cultural characteristics are indicated by the type of colony produced on agar plates, the appearance of growth on agar slants, the type of growth in broth cultures, and pigment formation in various media. To this may be added such characteristics as the ability of organisms to grow in the presence or absence of air, the minimum, optimum, and maximum growth temperatures, and the limiting pH. Also, some bacteria produce exoenzymes that exude into the agar in a plate culture and hydrolyze certain complex nutrients in the medium to produce a clear, or otherwise visible, zone around the colony. Plate culture media are prepared by adding to a standard agar medium small amounts of a hydrolyzable product such as blood agar, gelatin agar, milk agar, or DNA agar.

In broths, identifying features may be the amount of turbidity produced; growth only on the surface, only at the bottom of the tube, or only below the surface; and rings of growth at the surface. Pigment formation in the medium such as the red color of *Serratia marcescens* or the blue-green, water-soluble pigments produced by the blue pus organism, *Pseudomonas aeruginosa,* also may be a useful characteristic.

On streaked agar slants, either a simple, smooth, flat, or raised streak may develop, the streak may be very slimy and distinctly raised, or it may have a treelike or some other unusual appearance (Fig. 11-2). Here again the development of distinctive red, yellow, orange, blue, or other nonwater-soluble pigments may be significant. Unlike the pigments that are confined to the streak or colony, water-soluble blue, green, or yellow pigments that diffuse out into the surrounding medium are developed by certain pseudomonads.

Colony appearance on agar plates also may be useful in differentiating various bacteria. The colonies may be pinpoint, medium, large, spreading, smooth, rough, slimy, raised, or flat, with entire, lobed, serrated, or other distinctive edges.

Physiological tests

When bacteria are grown on certain special media they may produce specific growth products. These reflect the production by the growing bacteria of enzymes mediated by genetic messages inherent in the bacterial nucleoid. If these products can easily be demonstrated by chemical or other means, they provide a useful method for identifying and differentiating unknown bacteria.

The *fermentation test,* which uses an outer and an inner test tube combination known as a Durham fermentation tube, is perhaps the most widely used of the differential enzyme reactions. It is applicable to those bacteria, principally the Eubacteriales, capable of the anaerobic fermentation of carbohydrates. The test is especially useful with the Enterobacteriaceae.

Tubes of standard broth are prepared that contain 0.1% to 1.0% of the carbohydrate being tested and a pH indicator. To detect the forma-

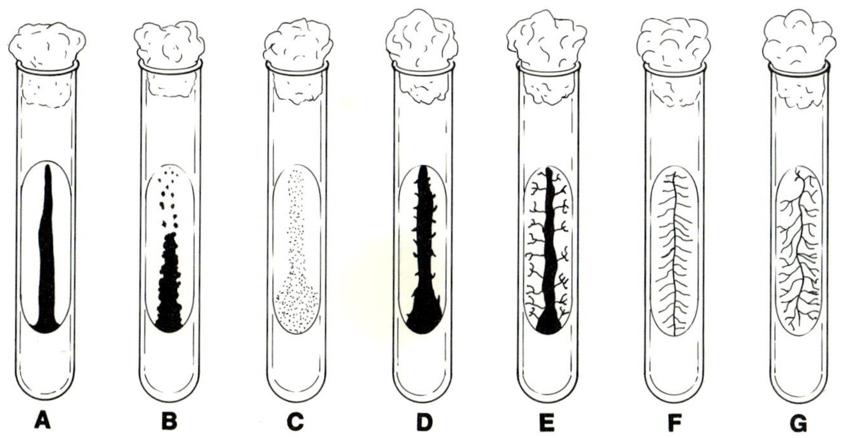

Figure 11-2
Typical appearance of growth of streak cultures of bacteria on agar slants. **A,** Filiform (thread-form), **B,** beaded, **C,** effuse (thin, widely spread out), **D,** echinulate (Latin diminutive for hedgehog), **E,** arborescent (treelike), **F,** plumose (feathery), **G,** rhizoid (rootlike).

tion of gas, the smaller tube is inverted and dropped into the sugar medium prior to sterilization. While the medium is being sterilized, the air inside the tube is heated sufficiently to be driven out so that when the inverted tube is cool it is filled completely with the medium. When bacteria inoculated into the medium begin to grow, they may attack the sugar and, since the outer tube is fairly deep, anaerobic conditions will prevail. As a result, acid may develop, causing the indicator color to change to indicate acidity. If the organism also produces gas from the carbohydrate, gas quickly accumulates in the inner tube. With active gas producers the inner tube may be nearly filled within 24 hours.

The primary sugar used is glucose. A number of other 6-carbon simple sugars such as galactose, and fructose also are used together with 5-carbon sugars. The disaccharides (double sugars) most commonly used are lactose (glucose + galactose), sucrose (glucose + fructose), and maltose (glucose + glucose). *Escherichia coli,* for example, is easily separated in the clinical laboratory from other intestinal bacteria such as *Salmonella* and *Shigella* strains on the basis of its ability to rapidly ferment lactose with the production of acid and gas.

Fermentative organisms fall into two groups, those that produce only acid (A) and those that produce both acid and gas (AG). Some bacteria produce little or no visible change in the medium other than slight turbidity. Hence the test is recorded as no change (NC). Other bacteria attack protein nutrients such as peptone, releasing ammonia that changes the color of the indicator to the alkaline side. These tests are reported as giving an alkaline (Alk) reaction. This is found especially with *Pseudomonas* and other nonfermentative, oxidative bacteria.

Trisaccharides, and polyhydroxy alcohols, such as glycerol (glycerine), and other compounds, also are used for differential tests particularly with bacteria belonging to the Enterobacteriaceae.

Certain bacteria produce decarboxylase enzymes that are capable of splitting off the carboxyl group (—COOH) of certain amino acid molecules to carbon dioxide. The residue remaining is an amine, usually vile smelling, that produces an alkaline reaction in the medium. In the *decarboxylase test,* to a broth medium containing peptone-yeast, glucose, and an indicator such as brom-cresol-purple, is added 0.5% of the amino acid being tested. After seeding with the test bacteria, the cultures are incubated for 1 to 4 days. The tube is tightly closed with a screw cap to insure anaerobic conditions. A positive decarboxylase test is indicated by the development of an alkaline reaction. Although the test works well with several amino acids, lysine is the one most commonly used.

The chemical changes involved are shown at the bottom of the page.

The diamine, cadaverine, which is produced, is of interest because it is the major cause of the noisome odor resulting from the putrefactive decomposition of the human body by bacteria.

The *nitrate reduction test* is also widely used to separate certain types of bacteria. Many bacteria can utilize the nitrate molecule as an oxygen source, reducing it by varying degrees. Nitrate reduction, therefore, may become a valuable test for separating certain bacteria. The products of nitrate reductions may be:

NO_3	NO_2	N_2	NH_3
Nitrate ion	**Nitrite ion**	**Nitrogen gas**	**Ammonia**

The test tubes of nutrient broth plus 1% po-

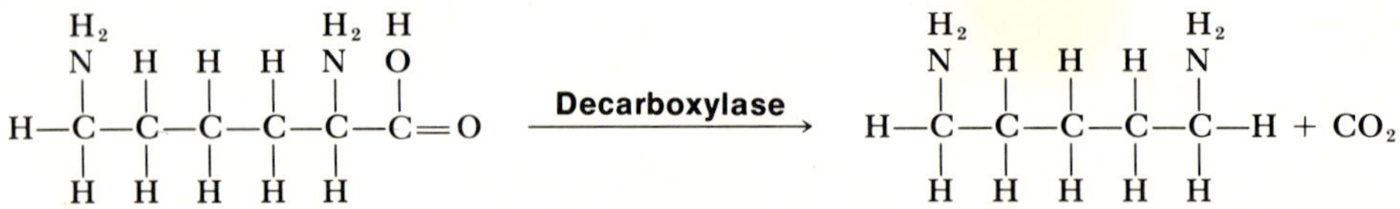

Lysine **Cadaverine** **Carbon dioxide**

tassium nitrate are seeded with bacteria. These are then incubated 12 to 24 hours at temperatures appropriate for the organism. Certain bacteria reduce nitrates only to nitrites, others to nitrogen (N_2) or even ammonia (NH_3). Some aerobes when grown anaerobically derive their oxygen from NO_3 and reduce it completely to N_2. In most cases, only the test for nitrite formation is used. This is performed by adding a few drops each of 0.8% sulfanilic acid and 0.5% α-naphthylamine to the medium. The presence of nitrites is indicated by a distinct pink or red color. If incubation is carried on for a longer period, the nitrites may disappear as a result of being converted into N_2 or even NH_3. The presence of ammonia may be detected by adding a few drops of Nessler's ammonia reagent ($KI + HgCl_2 + NaOH$). A pink-red color indicates the presence of NH_3.

For detecting the reduction of NO_2 to N_2 (gas), to the liquid nitrate medium is added about 0.5% agar to form a soft agar. This is distributed in tubes, sterilized, and inoculated. After inoculation 1 to 2 ml of a melted sterile 2% plain water-agar is pipetted over the surface of the tube culture to insure an air-tight seal for anaerobic conditions. If the organisms reduce NO_3 to N_2, bubbles of gas form in the medium or are trapped beneath the agar seal.

The *hydrogen sulfide test* makes use of the fact that hydrogen sulfide is produced by some bacteria from organic sulfur compounds such as cystine, glutathione or similar organic compounds, or from sodium thiosulfate. The presence of H_2S is detected by placing moistened lead acetate paper at the mouth of a tube of suitable broth culture. In other media, ferrous sulfate may be added as an indicator. In the presence of H_2S this forms a black iron sulfide. The H_2S test is especially useful for identifying certain species of the enteric bacteria.

The *urease test* is another common procedure. Some bacteria can readily hydrolyze urea to ammonia by the following reaction:

$$\underset{\textbf{Urea}}{H_2N{-}\overset{\displaystyle NH_2}{\overset{|}{C}}{=}O} + \text{Urease} + \underset{\textbf{Water}}{H_2O} \longrightarrow \underset{\textbf{Carbon dioxide}}{CO_2} + \underset{\textbf{Ammonia}}{2NH_3}$$

To tubes of a weakly buffered broth medium that contains an indicator for alkalinity, 2% filter-sterilized urea is added. The cultures are then seeded and incubated for 12 to 48 hours at 37° C. An alkaline reaction will indicate urea hydrolysis. This test is used to separate *Proteus* (urease positive) from the urease-negative coliform bacteria.

The *indole test* is a useful procedure. In a medium fairly rich in the amino acid, tryptophan, certain bacteria such as *Escherichia coli* transform the amino acid into the simpler compound, *indole*, by the reaction given below.

The test medium is a 1% tryptone broth high in tryptophan. Tubes of this medium are seeded, then incubated for 24 hours. Kovac's reagent (*p*-dimethylaminobenzaldehyde dissolved in amyl alcohol and hydrochloric acid) is added to the tube. The broth develops a dark red color if indole is present but is yellow if it is absent. Indole is the primary cause of the unpleasant odor of *E. coli* broth cultures.

The *oxidase test* involves the change in the color of an indicator compound induced by the production of the enzyme oxidase by some bacteria. This reaction probably represents an indirect test for cytochrome C.

The procedure consists of flooding a 24-hour

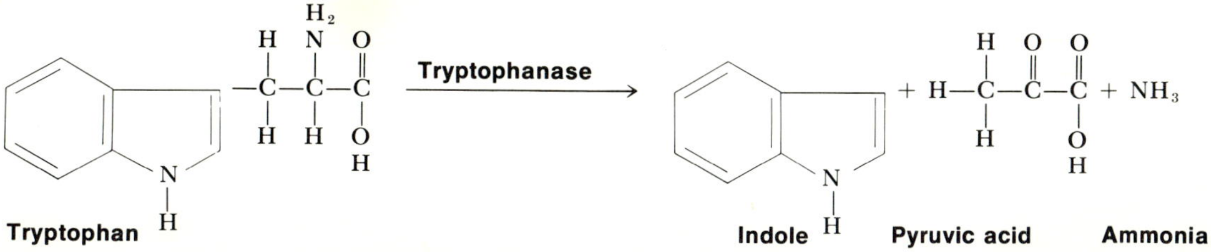

plate culture of bacteria from some clinical source such as a case of gonorrhea with the test reagent, 1% *p*-aminodimethylaniline monohydrochloride. Colonies of oxidase-producing bacteria will turn pink then purple, and finally black. This reaction is important in the diagnosis of gonorrhea because plate cultures of pus from suspected gonorrhea patients may contain colonies of several different kinds of bacteria, making it difficult to pick out the gonorrhea organism. However, with the oxidase reagent the oxidase-positive *Neisseria gonorrhoeae* colonies will stand out.

A rapid filter paper test for oxidase-positive bacteria was devised by Kovac. This consists of placing a few drops of reagent on filter paper, then with an inoculating needle picking up a loopful of growth from a colony or agar slant. The growth is then rubbed into the drop of reagent. If the organism is an oxidase producer, the rubbed-in spot will quickly develop a deep red color, which becomes purple and finally black.

The test is useful for the identification of a number of bacteria such as certain strains of *Pseudomonas*. Moreover, the filter paper can be retained as a permanent record.

In the *litmus milk test*, growth reactions of litmus milk (skim milk and litmus) provide useful observable end reactions. Since milk contains both the sugar, lactose, and the protein, casein, bacteria may enzymatically attack one or both of them. When lactose is attacked by the bacteria one of two things may occur: the sugar may be converted into acid (usually lactic acid), or it may be converted into acid plus gas.

If acid is produced the amount may be sufficient to cause the curdling or clabbering of the milk. When this occurs, the protein, casein, which is in the form of the salt, calcium caseinate (a colloidal sol), is converted to the acid form of casein plus calcium lactate. In this state the casein exists as a colloidal gel. (A colloidal sol is the colloid suspended in water, while the gel represents water held in a colloidal matrix.) If acid is formed the litmus turns red.

With certain bacteria such as some species of *Bacillus*, the lactose is not directly attacked but the growing organism secretes trypsinlike enzymes, which causes the casein to coagulate into a firm, smooth curd without acid being present. This is termed sweet curdling.

After a curd has formed, either by souring or sweet curdling, the growing organism may produce proteolytic or peptonizing enzymes that break down the curd into water-soluble proteoses, peptones or amino acids. This results in the formation of the wheylike, yellowish-tan liquid. If this breakdown continues ammonia may be produced by the deamination (removal of amino groups) of the amino acid. The medium then becomes alkaline and the litmus becomes blue. If the bacteria uses the litmus as a hydrogen acceptor for certain reducing reactions it is changed into the *leuco* or colorless form. The litmus is then said to be reduced.

Nutritional tests

With some bacteria, especially oxidative forms such as *Pseudomonas*, a study of nutritional versatility may be useful in differentiating and identifying unknown bacteria. The technique used consists of adding to a mineral salts solution low concentrations of the test carbon compounds as the sole carbon source. Utilization of a given compound for growth indicates that the bacterium possesses the enzymes necessary to convert the compound into cell components.

Serological tests

The rapid differentiation of strains or even species of gram-negative rods belonging to the Eubacteriales can often be accomplished by means of serological tests. For this purpose, known antisera (see Chapter 24) are produced by injecting known bacteria, such as members of the *Salmonella* or *Shigella* groups, into rabbits. After a series of properly spaced injections, a high level of antibodies will have developed. Blood is then drawn and allowed to clot. The serum that separates from the clot now contains a high concentration of agglutinating antibodies specific for the bacterium that had been injected. Therefore, when the serum is mixed with a saline suspension of the bacterium, the cells will agglutinate or clump. When unrelated bacteria are mixed with the same antiserum, no agglutination occurs.

Antisera produced against various bacteria vary somewhat in their specificity. In addition to clumping the original bacterium they may also clump closely or sometimes remotely related bacteria that share a common antigen. However, when the limitations of the method are recognized and proper compensations are made for the common antigen effect, serological tests often provide useful and rapid methods of identifying certain strains or species of bacteria.

A great many other procedures are also used. Special types of differential or selective media are available that contain inhibitory substances, special nutrients, or indicators. When these are used, various organisms can be identified by the distinctive color reactions or colonies they produce.

Hundreds of such media in the dehydrated form are now available on the market. Some concerns produce the complete medium in a tablet that can be added to a tube of sterile hot water to produce a desired selective liquid medium. Some diagnostic laboratories have developed punch cards for recording tests, which are then fed into a computer providing an instant identification of the unknown organism.

SUGGESTED READINGS

Breed, R. S., Murray, E. G. D., and Smith, N. R.: Bergey's Manual of determinative bacteriology, ed. 7, Baltimore, 1957, The Williams & Wilkins Co.

Skerman, V. B. D.: A guide to the identification of the genera of bacteria, ed. 2, Baltimore, 1967, The Williams & Wilkins Co.

Stanier, R. Y., Palleroni, N. J., and Doudoroff, M.: The aerobic pseudomonads: a taxonomic study, J. Gen. Microbiol. 43:159-271, 1966.

The following commercial publications provide a useful, encyclopedic compendium of the differential media available in dehydrated form for the identification of bacteria, together with information on the developmental background of the various media. They may be obtained by writing the company:

BBL manual of products and laboratory procedures, P. A. Rohde, editor, ed. 5, 1968, Baltimore Biological Laboratory, Division of BioQuest, Cockeysville, Md. 21030

Difco manual of dehydrated culture media and reagents, ed. 9, 1953; Difco Supplementary Literature, 1968, Difco Laboratories, Inc., Detroit, Mich. 48201

12 Classification of bacteria

EVOLUTION OF A CLASSIFICATION OF THE SCHIZOMYCETES

Manifestly, without some formal arrangement of names and descriptions of bacteria, our accumulated knowledge of these forms would be in a state of utter chaos. As an illustration of the confusion that may occur, in 1885 Escherich first described the common intestinal form, *Escherichia coli*, which he called *Bacterium coli commune* (the common rod from the colon). Since that time at least nine other names plus six apparent synonyms have been applied. Thus a single organism has at various times borne fifteen or more published names.

The first formal scheme for classifying living things was devised by the famous Swedish botanist, Carl von Linné (Carolus Linnaeus). He first published his binomial (two name) system of nomenclature in 1753. The basic feature was that he applied two names, analogous to our christian and surnames, to various plants and animals. These are termed the genus and species names. With names of bacteria the first letter of the genus name should be capitalized, the

The genera of interest are:

1. *Mycobacterium* is a form widely distributed in the soil but also responsible for certain diseases. Unlike most of the order it rarely forms filaments but grows as slender acid-fast rods. The most important species of *Mycobacterium* are:
   a. *Mycobacterium tuberculosis* (causes tuberculosis in man)
   b. *Mycobacterium bovis* (causes tuberculosis in cattle)
   c. *Mycobacterium leprae* (Hansen's bacillus, cause of leprosy)
2. *Nocardia* are obligately aerobic, usually slender filaments, often swollen, occasionally branched. They are slightly acid-fast. Mostly soil forms, a few species are responsible for tuberculosis-like infections of the lungs, and other chronic infections.
3. *Actinomyces* species are anaerobic to microaerophilic, some producing actinomycosis, a chronic tissue infection in mouths or other organs of man or animals.
   a. *Actinomyces bovis* (causes lumpy jaw in cattle)
   b. *Actinomyces israelii* (causes actinomycosis in man)
4. *Streptomyces* represents a very large genus of soil forms, producing branching vegetative filaments that do not break up into bacteria-like rods but often produce spore-bearing structures. They form a great proportion of the microbial population in some soils.

ORDER VI—Caryophanales. The Caryophanales (bright nucleus order) are inadequately studied forms found mostly as parasites in the intestines of herbivores or insects. Individual cells are cylindrical or discoid with a conspicuous nucleus.

ORDER VII—Beggiatoales. The Beggiatoales (named for Beggiatoa, an Italian physician) are fresh or marine water, or soil forms occurring mostly in trichomes that bend, flex, or oscillate. They move over a surface with a gliding, jerking motion, resembling some filamentous blue-green algae *(Oscillatoria),* but are devoid of photosynthetic pigments.

ORDER VIII—Myxobacterales. The Myxobacterales (slime rodlet order) are unusual, flexible rods without a rigid cell wall, primarily found in soil or feces of herbivorous animals. They may be isolated by placing particles of soil (or bits of feces) on a plain water-agar to which bacteria such as *Enterobacter aerogenes* has been added. The bacteria furnish nutrients for the myxobacters, which grow out from the colony as thin, slimy, tonguelike protusions filled with actively motile cells. As cultures become older, fruiting bodies resembling minute mushrooms may develop. Resistant cysts may be formed. The group has not been extensively studied.

ORDER IX—Spirochaetales. The Spirochaetales (spiral hair order) are made up of a group of forms without rigid cell walls, occurring as slender, flexuous spirals. They are often difficult to stain with aniline dyes except Giemsa stain. Many can be seen only with the dark-field microscope. Unlike most bacteria they have no flagella; however, they are actively motile due to a rapid lengthwise spinning motion plus, in some forms, a violent lashing motion of the filament. They are difficult to grow on anything except specially enriched media and under carefully controlled conditions of temperature and oxygen tension. Some have never been grown. They are procaryotic, reproduce by transverse fission, and in terms of cell diameter resemble bacteria. In other respects, the largest forms resemble some of the spiral protozoa. They fall in to two major groups:

Spirochetes of the first group are very large (ranging in length from 30 to 500 μm and in width from 0.5 to 2.5 μm), saprophytic forms found in fresh- or saltwater, especially in or around organic matter or in the guts of oysters and other shellfish. The genera are *Spirochaeta*, *Saprospira*, and *Cristispira.*

Members of the second group are much smaller and parasitic or pathogenic in vertebrates. They are slender spirals, ranging from 4 to 16 μm in length and 0.24 to 0.5 μm in diameter. Some can be cultivated on special media. Others can be detected by Giemsa stains or dark-field examination. Of the approximately 40 species, those of special interest are:

1. *Treponema pallidum* (causes syphilis)
2. *Treponema pertenue* (causes yaws or tropical syphilis)
3. *Borellia vincentii* (associated with *Fusobacterium fusiforme* as the cause of trench mouth or Vincent's angina)
4. *Leptospira icterohaemorrhagiae* (causes Weil's disease, a type of jaundice)
5. *Leptospira canicola* (an increasingly important disease of dogs)

Order X—Mycoplasmatales. Mycoplasmatales (fungus, molded order; pleuropneumonia-like organisms) represent an unusual group of forms, still not well defined, that have no rigid cell structure. They grow only on highly enriched media producing very small "fried egg" (thick-center) colonies. The growing elements may be highly pleomorphic bodies, filamentous, myceloid, and sometimes with branching or large, irregular, round cells. Occasionally, almost rod-like forms 2 to 5 μm in length may be found. Very small filter-passing stages are known. The filterable elementary bodies are about the size of viruses but, unlike viruses, are capable of growing on artificial but highly enriched media. Knowledge of these forms is rapidly increasing and indicates that they are involved in far more disease conditions than we formerly suspected. A number of plant and insect diseases are now known to be caused by *Mycoplasma*. One or more species is saprophytic. Isolation in most cases involves filtration techniques or use of antibiotics in the medium to separate the *Mycoplasma* from the bacterial contaminants. Important species are:

1. *Mycoplasma mycoides* (cause of bovine pleuropneumonia)
2. *Mycoplasma pneumoniae* (Eaton agent, cause of primary atypical pneumonia)
3. *Mycoplasma laidlawii* (saprophytic form that can be isolated from sewage, soil, manure, and humus)

NUMERICAL TAXONOMY

Numerical taxonomy is a recently revived classification method providing a procedure for the mathematical determination of the similarity between large numbers of bacterial strains and the expression of this alikeness or similarity.

To determine similarity for a considerable number of strains of bacteria many simple arithmetic calculations are required. Consequently, until computer methods became available, extensive numerical taxonomic studies were not feasible. However, once the data has been accumulated and the program developed, the computer can rapidly perform the necessary operations and print out the similarity information to as many decimal places as desired.

The idea was conceived more than two centuries ago by Michel Adanson, a French zoologist, in an attempt to compare and classify a large number of mollusc specimens. By applying his method, which he first described in 1757, to the accumulated data, he was able to express the similarity between specimens as a simple percentage. After further refinement, details of the method were presented in a paper in 1763 as *la methode naturelle*. This involved the use of a large number of different characters or features, which were either present or absent, that could be expressed by a simple plus or minus. Each feature was given the same value.

The method gained little or no acceptance until, exactly two centuries later (1957), Sneath, an English physician, mathematician, and microbiologist, resurrected Adanson's principles. These he applied to the classification of bacteria under the name numerical taxonomy. He recognized the great value of the computer as the tool that would make this method both feasible and useful. It was based on the following points:

1. Tests (features) were to be selected that could readily be performed and yield results

expressed as simple plus or minus values. A number of such tests (50 to 100 to be performed for each strain of bacteria) was termed as an operational taxonomic unit (OTU).

2. None of these tests was to be redundant. Thus, two different methods that tested for the same character and gave identical results were to be avoided.
3. Each feature (test result) was to be given the same weight.
4. Similarity or affinity between various OTU's was to be expressed as a percentage, calculated by the simple mathematical formula shown below:

$$S = \frac{Ns}{Nf} \times 100$$

S = similarity or affinity
Ns = number of features (test results) that were the same for the two strains of bacteria being compared
Nd = Number of features that were positive for one strain (OTU) but negative for the other or vice versa
Nf = Ns + Nd

In cases where test results were negative for both strains, the negative results were ignored. Likewise if a test was not performed or not possible with the given strain, it was listed as NC (no comparison). If one strain was NC for a given character or feature but the other strain was positive the comparison was not made and the results were not included.

Examples of some of the tests or features used in numerical taxonomy are:

1. Ability or inability to produce indole from tryptophan
2. Motile or nonmotile
3. Motile by peritrichous flagella
4. Motile by polar flagella
5. Gram positive, gram negative
6. Starch hydrolysis
7. Deoxyribonuclease secretion
8. Ability to decarboxylate lysine or other amino acids

In making the comparisons, results are arranged in table form similar to the Roman square used in genetics. If 100 to 200 strains are compared for 50 to 100 tests, manifestly an almost astronomic number of simple mathematical operations are involved. It is in such cases that the computer becomes invaluable.

To illustrate the method of determining similarity, results of IMViC tests, plus H_2S production and urea hydrolysis tests for five strains of bacteria are recorded in Table 12-1.

As a reminder, IMViC is a mnemonic for the indole, methyl red, Voges-Proskauer, and citrate tests. The indole reaction tests for formation of indole from tryptophan. The methyl red test measures the fermentation of glucose required to produce a high enough concentration of hydrogen ions to result in a pH of below 4.5. Some bacteria will produce acid but not in high enough concentration to give a positive MR test. The Voges-Proskauer test tests for the production of acetylmethylcarbinol. The "i" in IMViC is added to make the mnemonic easy to pronounce. The citrate test demonstrates the ability to utilize citrates as a sole carbon source.

The H_2S and urea hydrolysis tests are added simply to provide more tests or features.

The results of this hypothetical study are shown in Table 12-1:

Table 12-1. Strains compared

	STRAINS				
	A	B	C	D	E
Indole	+	−	+	−	−
Methyl red	+	+	−	+	−
V-P	−	−	+	+	+
Citrate	−	+	−	+	−
H_2S	−	+	+	−	+
Urea	−	+	+	+	+

In examining the comparison chart (Table 12-2), keep in mind that when a strain is compared with itself only the positive tests are considered. Thus, when A is compared with A, only 2 positive tests are involved. It is therefore

Table 12-2. Comparisons of similarity

	A	B	C	D	E
A	$\frac{2}{2}$ 100%				
B	$\frac{1}{5}$ 20%	$\frac{4}{4}$ 100%			
C	$\frac{1}{5}$ 20%	$\frac{2}{6}$ 33%	$\frac{4}{4}$ 100%		
D	$\frac{1}{5}$ 20%	$\frac{3}{5}$ 60%	$\frac{2}{6}$ 33%	$\frac{4}{4}$ 100%	
E	$\frac{0}{5}$ 0%	$\frac{3}{5}$ 60%	$\frac{2}{4}$ 50%	$\frac{2}{5}$ 40%	$\frac{3}{3}$ 100%

expressed as 2/2 = 100%. With *B*, four tests were positive hence the expression 4/4 = 100%.

In comparing *A* and *B*, only one test was positive for both strains but with four additional tests, one strain was positive and the other negative, hence the expression:

$$\frac{1}{1+4} = \frac{1}{5} = 20\% \text{ similarity}$$

The observed similarity can be expressed as follows:

60% = B-D, B-E, C-E, D-E
33% = C-D
20% = A-B, AC, AD
0% = A-E

The above results suggest the following interpretations:

1. B-C-D-E apparently belong to the same genus, perhaps the same species.
2. C-D may be separate species within the same genus or separate genera.
3. A is a genus distinct from the remaining forms.
4. Results obtained from only six tests are of little value for a definitive numerical taxonomic study.

In the near future numerical taxonomic data will be stored in computer centers. The microbiologist need only to perform a prescribed battery of tests, submit the information, and receive a print-out statement giving the percentage similarity of the unknown organism to known forms.

GUANINE-CYTOSINE RATIOS

A decade or so ago the discovery was made that the DNA of the nucleoids of bacterial cells that carry the genetic code could be isolated rather easily. The DNA contains four purine or pyrimidine bases: guanine, cytosine, adenine, and thymine. When a purified, denatured DNA preparation was overlaid on the surface of a cesium chloride solution and centrifuged at high speed in a preparatory centrifuge, the fractions isolated after centrifugation gave a method of estimating the relative proportion of guanine + cytosine (G-C) to the other two bases, hence, the G-C ratio.

$$\text{G-C ratio} = \frac{\text{G-C concentration}}{\text{G + C + A + T concentration}} \times 100$$

Thus the G-C ratio expresses the percentage of the bases in the DNA that are guanine and cytosine.

In addition to the relative rate of sedimentation in cesium chloride solution, the G-C ratio can also be calculated on the basis of the melting temperature of the extracted DNA.

It was soon discovered that the G-C ratios of closely related bacterial strains generally agreed within a percent or two, while entirely different strains might have markedly different G-C ratios. Hailed at first as the answer to all problems of bacteria classification, it soon became apparent that a great deal of overlapping occurred between various species and genera. Its great usefulness now seems to be in separating out aberrant species that have crept into a given genus. Up to 1970 the G-C values for more than 450 species of bacteria had been determined. The values vary widely, ranging from 30% to 75%. Some authorities are of the opinion that the greatest value of G-C ratios in classification will be in establishing the family groupings within the orders of bacteria.

The information thus far accumulated indicates that the G-C ratios of the vascular plants fall between 35% and 45%. The higher animals show similar values.

SUGGESTED READINGS

Breed, R. S., Murray, E. G. D., and Smith, N. R.: Bergey's Manual of determinative bacteriology, ed. 7, Baltimore, 1957, The Williams & Wilkins Co., pp. 1-28.

Marmur, J.: A procedure for the isolation of deoxyribonucleic acid from micro-organisms, J. Mol. Biol. 3:208-218, 1961.

Marmur, J., Falkow, S., and Mandel, M.: New approaches to bacterial taxonomy, Annu. Rev. Microbiol. 17:329-372, 1963.

Skerman, V. B. D.: A guide to the identification of the genera of bacteria, ed. 2, Baltimore, 1967, The Williams & Wilkins Co., pp. 4-17.

Sneath, P. H. A.: Some thoughts on bacterial classification, J. Gen. Microbiol. 17:184-200, 1957.

Sneath, P. H. A.: The application of computers to taxonomy, J. Gen. Microbiol. 17:201-226, 1957.

13 Sterilization procedures

Methods of sterilization designed to destroy or remove microbes have made possible the isolation of pure cultures from the mixed array of microbes found in nature, protection against those forms that produce disease or death, preservation of foods, and aseptic performance of surgical operations.

DEFINITIONS

Terms used to describe the sterile or near sterile state have been coined over the years and have often been used very loosely or even completely misused. Consequently, it is necessary to define clearly the terms we employ. The following definitions are derived from Webster's International Dictionary or from official or semi-official sources such as the federal Food and Drug Administration or the American Public Health Association:

sterile The state of being rendered free from all living microorganisms including bacterial spores and mold spores. If viruses are present they are inactivated when sterility is achieved. The sterile object or material may still have the hulks of the dead microorganisms or the inactivated virus particles. The term sterile, like the word pregnant, is an absolute term. An object or material is either sterile or nonsterile. Sterilization not only involves killing of the microbial cells, it also can be achieved by complete removal of microbes by filtration. The presence of a single detectable viable organism on the surface of an object or in a liquid or semiliquid medium renders it nonsterile.

sterilization The process of achieving sterility. This may be achieved by heat, radiant energy, supersonic waves, filtration, or other physical means. Chemicals, usually termed disinfectants or germicides, may also produce a sterile state. Most chemicals, however, leave a toxic residue that renders a sterile container incapable of supporting the subsequent growth of microorganisms when a nontoxic sterile medium is added and the medium is inoculated. Chemical sterilization is, therefore, of little value in pure culture studies.

sanitary Used to describe an object or material freed from all dangerous microbes but not necessarily sterile. Agents that produce this state are often chemical rather than physical in nature and are termed sanitizers or sanitizing agents. They are generally used on surfaces.

pasteurization Essentially a sanitization process. Milk, beer, wines, and similar liquids may be heated to temperatures below boiling, which kills off most of the microorganisms, including the pathogenic forms, without sterilizing the liquid. In most cases 90% to 99% of the microbes may be killed. The reason for using a lower than boiling temperature is that higher temperatures may impair the taste, color, or some other desirable quality of the liquid.

STERILIZATION BY DRY HEAT

Flaming or incineration

Any metal object or surface that can be heated to red heat can be sterilized by flaming or incineration. It will remain sterile only until contaminated from the air or by other contact with the environment. By this process bacteria and other cells are quickly cremated leaving only the cell ash. The process is most commonly used to sterilize inoculating needles. It carries

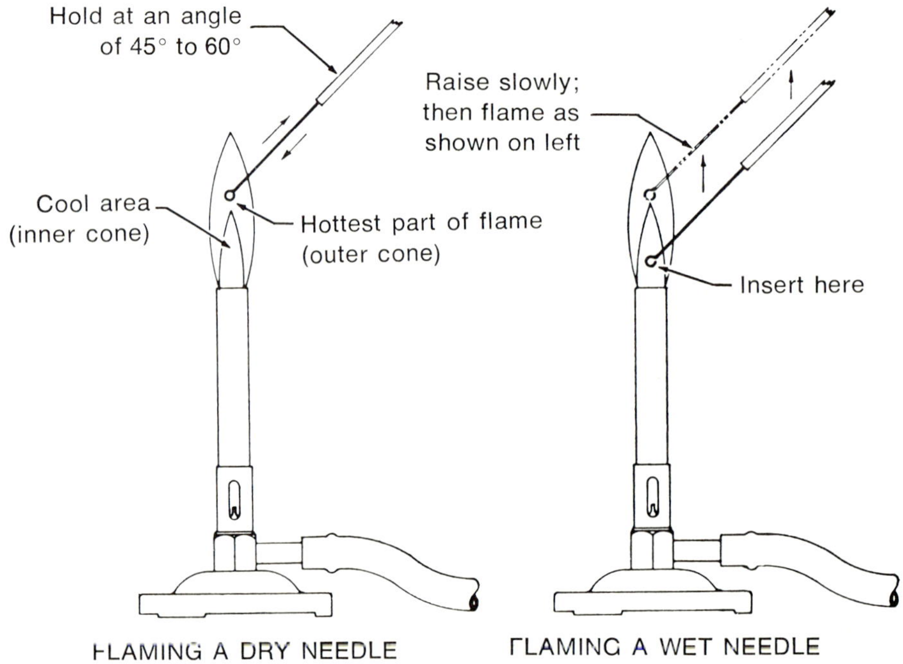

Figure 13-1
Proper methods for flaming. (From Anderson, D.: Laboratory instructions in microbiology, St. Louis, 1969, The C. V. Mosby Co.)

some hazard because if an inoculation loop filled with bacteria is placed directly in the hottest portion of the flame, steam may form causing the mass to sputter and pop out onto the table surface. Most bacteria inside the mass are still alive and are capable of growing. A considerable number of accidental laboratory infections have occurred from contact with viable bacteria in this manner. A bacteria-filled loop can be safely sterilized if it is first thrust into the cool inner cone, then slowly raised up into the hottest part of the flame (Fig. 13-1).

In laboratories where virulent viruses are being studied, the exhaust air from the air conditioning system may be incinerated before being exhausted to prevent accidental carriage of airborne infective agents into the outside air.

Dry sterilization

Glass or metal objects that can be heated in an oven for long periods at relatively high temperatures can be sterilized by using dry heat. This is a relatively slow method because as bacteria are heated the proteins dry out and their resistance to denaturation becomes greater. Koch, nearly a century ago, found that anthrax spores were killed only after 3 hours heating at 140° C. Similar spores were killed by boiling for 10 minutes at 100° C.

In general, spores are much more resistant to dry heat than are vegetative cells. However, at 160° C the resistance of spores is abruptly reduced, greatly enhancing the rapidity of killing.

The *U. S. Pharmacopeia* recommends a temperature of 170° C for 2 hours; the *British Phar-*

macopoeia, 150° C for 1 hour, and the *Russian State Pharmacopeia,* 160° to 170° C for 1 hour. A practical range and time is 150° to 180° C for 1 to 3 hours depending on how full the oven is. With petrolatum, mineral oils, and similar substances, sterilization in the autoclave is not feasible because of the slow rate of heat penetration. They can, however, be successfully sterilized in an oven if exposure periods of 3 to 4 hours at 160° C are used. Glassware should not normally be wrapped in paper because this chars at 180° C and bursts into flame not much above this temperature. Cotton plugs also char. Most plastic stoppers disintegrate at even lower temperatures so they cannot be dry-heat sterilized. A housewife can sterilize dry glassware in an oven heated to about 325° F for 2 hours or so.

STERILIZATION BY MOIST HEAT

Moist heat is generally considerably more effective than dry heat. Mold spores are about 2 to 10 times as resistant to dry as to moist heat. Moist heat may be applied by boiling, by flowing steam, by steam under pressure, by intermittent boiling, by heating intermittently at lower than boiling temperature, or by pasteurization.

Boiling

The use of boiling, while generally an excellent method of sanitization, cannot properly be called sterilization. Instrument sanitizers for boiling surgical and similar instruments cannot correctly be called sterilizers. The reason is that the spores of many soil bacilli are killed slowly in boiling water; a few will resist boiling for 2 to 3 hours. Viruses and mold spores are almost as susceptible to boiling temperatures as vegetative bacteria such as *Escherichia coli.* If, in an emergency, surgical instruments must be sanitized by this method, they should be washed until meticulously clean before boiling begins. A minimum boiling time of 30 minutes is desirable. Addition of 2% sodium carbonate to the water greatly enhances the possibility of the instruments' becoming sterile rather than merely sanitary.

Flowing steam at atmospheric pressure

Flowing steam can produce about the same surface temperature as boiling water after an object has been heated to a steam temperature. However, it may be lower if a cool surface condenses much of the steam.

Fractional or intermittent sterilization

This procedure, first developed by Tyndall in 1877, is sometimes called tyndallization. Initially it was widely used, but today it is employed only for the sterilization of organic compounds that disintegrate rapidly when heated to the above boiling temperatures used in the autoclave. Fractional sterilization can be accomplished by means of the Arnold sterilizer. This is a metal cupboard with a chamber at the bottom in which water is boiled. The free-flowing steam passes upward through perforated shelves and out the top. In place of the Arnold sterilizer, the autoclave can be employed by leaving the vents open and the door ajar to allow the steam to flow free without pressure. The medium can also be placed in a boiling-water bath. Sterilization is accomplished by heating to boiling temperatures for 1 to 30 minutes a day on 3 successive days, with the medium being incubated between heatings. The theory conceived by Tyndall is that during the first exposure all vegetative cells and the most sensitive spores will be killed. During the subsequent incubation period most of the remaining spores will germinate. As vegetative cells, they will then be killed. The third heating destroys the remaining slow germinators. Today sterilization by filtration has largely replaced tyndallization.

Lower than boiling temperatures

Some material such as blood serum will coagulate at boiling but not at lower temperatures. Lower-than-boiling temperatures may therefore be used with some success. This consists of heating the material being sterilized in a ther-

Table 13-1. Effect of air trapped in autoclave on the temperature of sterilization

PRESSURE IN LBS/SQ IN	TEMPERATURE WITH ALL AIR REMOVED °C	°F	TEMPERATURE IF NO AIR IS RELEASED °C	°F
5	108	226	72	162
10	116	240	90	193
15	121	250	100	212
20	127	260	109	228
25	131	267	115	240
30	134	274	121	250

mostatically controlled water bath at 60° to 80° C for 30 to 60 minutes a day on 4 to 7 successive days. It is used only where filtration is not feasible.

Steam under pressure—autoclave sterilization

In 1681 Papin described a "digester or engine for softening bones" that involved steam under pressure. Chamberland, in Pasteur's laboratory in the early 1880's, applied the same principle in the design of his "autoclave" (translated freely this means self-locking), to be used for sterilizing bacterial media. His original device was much like our present-day pressure cooker. Modern autoclaves have become elaborate devices with automatic air exhausts, thermostatically controlled temperatures, temperature recorders, and automatic timers, but the principle remains the same. It consists of boiling water in a closed container with an open valve that allows all entrapped air to escape. When the point is reached that a plume of pure steam pours out, the escape valve is closed and the pressure begins to rise. (The importance of having the air completely exhausted is illustrated in Table 13-1.) At a pressure of 15 pounds per square inch, the temperature (at sea level) will have risen from 100° C (212° F) to 121° C (250° F). At this temperature, even spores will be killed after a 5 to 10 minute exposure. In practice, sterilization is carried on at 15 pounds (121° C) for 15 to 20 minutes with test tubes or small flasks of media or 30 minutes or longer with very large containers. After the desired time, the heat (or steam) is turned off, the steam is allowed to escape slowly, and the autoclave returns to atmospheric pressure. (Rapid escape of steam may result in media boiling out of containers or plugs popping out of tubes.) The sterile media are removed to avoid continued heating that may cause deterioration. With wrapped glassware, bandages, and packaged instruments, the sterilization may be prolonged to ensure that all air entrapped between packages is replaced with live steam. Also, when loaded with such packages, the autoclave may safely be rapidly exhausted to atmospheric pressure after sterilization. Some autoclaves may be exhausted further. The resultant vacuum causes a rapid drying of the material. This is known as the rapid exhaust and dry cycle.

The autoclave is a remarkably effective sterilizing device. Dry heat at 121° C (the temperature generally used with the autoclave) in a dry oven requires 2,000 times as long to kill the spores of *Bacillus subtilis* as the same temperature in the autoclave.

It must be remembered that the temperature at any given pressure depends on the initial boiling point of water. The final total pressure used must be equivalent to about two atmospheres. Thus at sea level (pressure already one atmosphere, roughly 15 lbs/sq in) the boil-

Table 13-2. Effect of altitude on autoclave sterilization temperature

ALTITUDE	BOILING POINT OF WATER °F	°C	APPROXIMATE ATMOSPHERIC PRESSURE LBS/SQ IN	*STEAM PRESSURE REQUIRED ON AUTOCLAVE GAUGE
Sea level	212	100	14.7	16.0 ± 1
0.2 mile	210	98.4	14.1	16.6 ± 1
0.4 mile	208	97.8	13.5	17.1 ± 1
0.6 mile	206	96.6	13.0	17.7 ± 1
0.8 mile	204	95.5	12.5	18.2 ± 1
1.0 mile	202	94.4	12.0	18.7 ± 1
2.0 miles	192	88.9	9.7	21.1 ± 1

*To give 121° C (250° F)—standard temperature for sterilization

ing point of water will be 100° C. However, in Denver, the mile high city, the atmospheric pressure is only about 12.0 lbs/sq in and the boiling point only 94.4° C. Therefore, the autoclave must be operated at about 18.7 lbs pressure to produce a temperature of 121° C (see Table 13-2).

The principles of autoclave operation also apply to the operation of the household pressure cooker since the autoclave, with all its elaborate controls, is still essentially a glorified pressure cooker.

To ensure the sterility of materials and media that have been autoclaved, some sort of testing device is necessary. This is of special importance in hospitals where such things as surgical instruments and bandages in wrapped packages are being sterilized and sterilization must be achieved in all cases to assure the safety of the patients.

A number of methods have been devised for testing for sterility. The direct method consists of unwrapping a package of presumably sterile bandages and with sterile scissors cutting off a piece of the cloth. This is then placed in a suitable medium such as thioglycollate broth. If after 24 to 72 hours of incubation no growth has developed, the sample (and the autoclave load) is assumed to be sterile.

Another method consists of impregnating a strip of filter paper about microscope slide size with 10^6 spores of *Bacillus stearothermophilus*, *B. subtilis* or *Clostridium sporogenes*, all of which are resistant to sterilization. After drying, each strip is sealed hermetically in paper or aluminum foil envelopes. The test strip is then placed inside a package of gauze, wrapped instruments, or similar material. After sterilization the package is unwrapped, the spore strip removed aseptically and dropped into a suitable liquid medium. If no growth occurs, complete sterility is indicated.

Instant recorders made by printing heat-sensitive dyes on paper that is placed in or on a package or flask are available. The dye is one that changes color when heated at 121° C for 15 minutes. Other rapid indicators consist of tiny cylinders of plastic or similar material sealed inside a glass vial. When an adequate temperature has been maintained for the desired period of time, the cylinder melts and often changes color.

FILTRATION

Liquids are often sterilized by passing them through filters with pores small enough to retain bacteria and other microorganisms but in most cases allowing viruses, if present, to pass through. Such liquids are sterile in terms of bacteria, yeasts, and molds but not necessarily in terms of viruses. Finer pores will remove the viruses but the rate of flow of the liquids may be slow. In many cases, the presence of viruses is not a

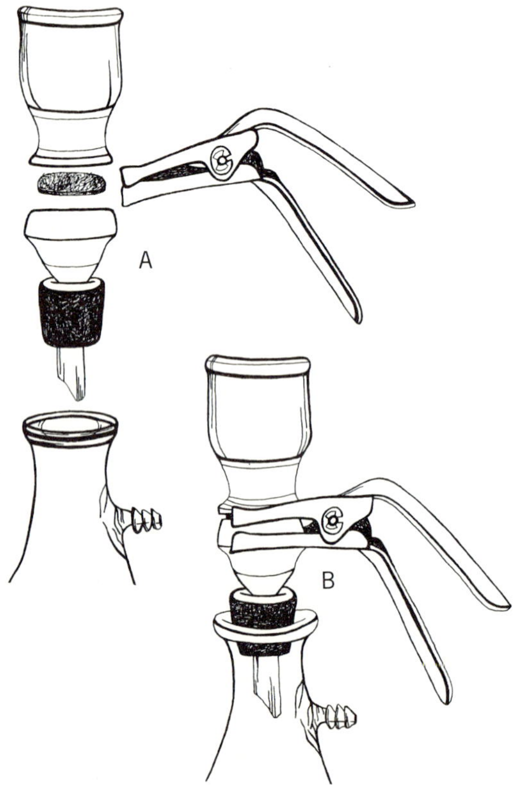

Figure 13-2
Millipore filter. **A,** Disassembled; **B,** assembled. (From Swatek, F. E.: Textbook of microbiology, St. Louis, 1967, The C. V. Mosby Co.)

significant factor unless they happen to be phages to which a given test organism is sensitive.

Chamberland, of autoclave fame, in 1884 developed an unglazed porcelain filter using quartz sand and kaolin to make a hollow candle. This was widely used, especially in the earlier days of virus study, to separate viruses from bacteria or host-cell components. It was also used extensively for filter-sterilizing bacterial media.

Thick asbestos pads (Sietz filters) have also been widely used. While generally quite effective they possess some disadvantages. The pore sizes are irregular, a positive charge is carried, bacteria may grow through, and tiny bits of particulate asbestos may be carried into the filtered medium. Filters have also been made of a plaster of Paris and diatomaceous earth mixture (Berkefeld and Mandler type filters), or sintered glass.

All of these types of filters are still used to some extent for certain purposes. However, the membrane filter (MF) has replaced them for most applications. The first membrane filters were made by Zsigmondy and Buchman in 1922 in Germany. Although the filters were available before World War II, microbiologists in the United States evinced little interest in them until about 1945 when they were brought to the United States from Germany as a scientific prize of war. They were at first classed as secret because of their apparent value in biological warfare but by 1950 they came into public use for water testing (see Chapter 17). They were sold under trade names, such as the Micropore or Millipore filters. The latter has now come into widespread commercial use in the United States.

The filters are made of thin (150 μm) disks of collodion or certain plastics supported for filtration by sintered glass, sintered iron, or similar flat supports (Fig. 13-2). Although the disks are very thin, when adequately supported they will withstand pressures up to 10,000 psi (pounds per square inch). Pressures or suction levels commonly used range from 5 to 100 psi. These thin membranes have millions of pores per square centimeter. Only about 20% of the volume is solid material. The pores are also uniform in size for a given type of membrane. In the Millipore filters some 14 pore sizes of membranes are available ranging from 14 μm down to 0.025 μm. For removal of bacteria the pore size used is 0.45 μm. Sizes finer than this will also remove viruses.

Because of the thinness and the toughness of the filters, they filter rapidly, producing absolute filtration; that is, all particulates larger than the pore size are retained. Thus with an appropriate pore size, yeast cells can be retained while bacteria pass through. These filters are now finding a great many uses. Not only can bacterial media be filter-sterilized but cold pasteurization of beers and wines is practiced for removing bacteria and yeasts from the fermented products. A great variety of other uses have developed such as purifying water for various uses and sterilization of solutions of antibiotics, insulin, and tissue culture media. Air may be sterilized by passage through membrane filters.

Another important use is the isolation of bacteria from water. Bacteria that have been retained on the surface of the filters may be grown by placing the filter on the surface of an appropriate solid medium in a petri plate. Small filters (Swinney type) may be attached to hypodermic syringes and used to filter very small quantities of liquid.

ULTRAVIOLET LIGHT AND IONIZING RADIATION

Ultraviolet light has long been known to have a distinct germicidal effect. The wavelengths most destructive to microorganisms fall between 2000 and 3000 Å (200 to 300 nm). The lethal effect rises rapidly from both extremes, reaching its greatest effectiveness at 2650 Å (265 nm). Neither the suntanning wavelength or black light is especially germicidal. Until the development of the cold cathode mercury vapor lamps in the 1930's, it was not feasible to use ultraviolet as a sterilizing or sanitizing agent. The so-called germicidal lamp produces about 80% of its rays at 2537 Å, near the peak of germicidal efficiency. Ultraviolet light exhibits its germicidal effect primarily because it is strongly absorbed by the DNA molecule, especially by the pyrimidine bases. Its major effect seems to be the formation of thymine dimers (double molecules), thereby disrupting the genetic message carried by the DNA.

Ultraviolet damage is not always permanent. Injury at times may be reversed if the cells are exposed to bright visible light with wavelengths between 3300 and 4800 Å. The repair mechanism appears to be one of splitting the lethal thymine (and cytosine) dimers.

Although ultraviolet light has definite germicidal power it has certain marked limitations. It has little penetrating power through glass, water, or liquids, and is not easily reflected. Consequently, its chief value is as an air-sterilizing or sanitizing agent, and as a sanitizer for flat exposed surfaces. It has proved to be valuable for the control of airborne infections in hospital corridors, operating rooms, air conditioning systems, food processing plants where surface contamination of foods causes a problem, and in the cosmetic industry where contamination of the cosmetics must be avoided during the packaging process. Ultraviolet light has also been used for sterilizing the surfaces of space vehicles.

Germicidal ultraviolet lamps must be used carefully. Exposure for any period of time causes marked reddening (erythema) of the skin and conjunctiva of the eyes. (See Chapter 22.)

The sun is a powerful emitter of germicidal radiation. However, the layer of ozone in the atmosphere absorbs most of the radiation below 2900 Å. Were it not for ozone, sunlight would damage or kill every exposed cell on earth. Visible sunlight does have some germicidal effect. However, it is only about 1/30,000 as great as that of a germicidal lamp.

In addition to ultraviolet light, X-rays and gamma rays such as those from cobalt 60 and beta rays from strontium 90 all exhibit germicidal effects. X-rays are used against microorganisms primarily to induce mutations for use in genetic experiments. Gamma rays, with their substantial penetrating power, have been used experimentally to sterilize food in cans. However, this is limited by some technical difficulties. Beta rays have only slight penetrating power, so their usefulness is limited largely to surface sterilization or sanitization. Both types of radiation have been employed to some extent for sterilizing such things as plastic items, animal fibers, and sutures.

ULTRASONICS—USE OF SUPERSONIC VIBRATIONS

Sound waves above the upper frequencies heard by the human ear (faster than 16 kilocycles per second) set up waves or cavitations in water or solutions that quickly disrupt bacterial cells.

While of little or no practical value as a method of sterilization or sanitization, an ultrasonic vibrator immersed in a liquid culture of bacteria will quickly tear the bacteria apart, releasing their internal contents. This provides a very useful method for the release and study of intracellular enzymes, nuclear material, or organelles.

OTHER LETHAL PHYSICAL METHODS

Desiccation or ordinary air drying is sufficient to kill many kinds of bacteria, at least nonsporing forms. However, many, after complete drying, still retain their viability for long periods, especially if protected by organic matter. For example, *Mycobacterium tuberculosis* dried in sputum has been found to retain its viability for months. On the other hand, such sensitive bacteria as *Neisseria gonorrhoeae* or *N. meningitidis* will die quickly when the temperature is lowered and drying occurs.

Freezing, that is, slow, natural freezing, causes the fairly rapid death of some bacteria, while others retain their viability for long periods when frozen in ice or other liquids.

Quick freezing, as used in the preservation of foods, is much less injurious to microorganisms than slow freezing. When quick-frozen foods melt, microbial growth ordinarily begins immediately. Consequently this warning is placed on most frozen foods—if thawed do not refreeze.

Freeze drying, the latest method of food preservation, essentially combines quick freezing and rapid desiccation. When used with bacterial cultures, enzyme preparations, and so forth, it is known as lyophilization. The process is essentially one in which the material is kept in a vacuum, and as the moisture is drawn off it is quickly condensed by supercooling at extremely low temperatures using such things as Dry Ice and liquid nitrogen. Under these conditions practically all of the bacteria or other microorganisms in food retain their viability. When such food is reconstituted, it should be eaten immediately or cooked.

ETHYLENE OXIDE

Ethylene oxide, $(CH_2)_2O$, is a chemical compound belonging to the epoxide group. Its sporicidal power and the fact that it vaporizes readily from objects, leaving no toxic residue, makes it a practical agent for producing sterility. For this reason it is included with the physical means of sterilization. While not as efficient as autoclaving, it can be used as a special purpose method for sterilizing heat-sensitive materials such as plastic petri dishes, pipettes, cystoscopes, plastic tubing, fabrics, paper, leather, wood, packaged items, electric cords, catheters, mouthpieces, nebulizers, and blankets. It has also been used for sterilizing dried fruits and spices. However, there is some indication that nutritional value may be impaired.

Ethylene oxide (ETO) is a liquid at low temperature, and boils at 10.7° C. It vaporizes quickly and has remarkable penetrating power. It is also highly explosive. To overcome this danger a mixture of 10% ETO and 90% CO_2 is available in pressure cylinders under the trade name, Carboxide.

ETO is toxic to man. The maximum safe concentration is about 100 parts per million. For short periods, 30 minutes or less, much higher concentrations seem to be tolerated safely.

The great penetrating power is illustrated by the observation that at moderate pressure and after a long enough time ETO will penetrate and sterilize a whole barrel of spices.

In actual use, a special type of gas-tight autoclave is employed, capable of being hermetically sealed and able to stand pressures of several pounds. The material is first dehumidified. The air in the autoclave is evacuated, the autoclave heated to about 60° C, and gas at several pounds pressure introduced. The gas concentration used is usually between 450 to 800 parts per million. Exposure times of 8 to 20 hours are required to kill spores. When removed from the autoclave, the gas quickly evaporates from the now sterile items. Material packaged in airtight coverings will remain sterile indefinitely. Thus a carton of 500 petri dishes (packed 20 per plastic sleeve) can furnish sterile petri dishes to a small laboratory for months.

The sporicidal effect of ETO is much greater than that of formaldehyde. Also, ETO is effective against viruses.

The sterilizing ability of ETO has provided the microbiology laboratory, hospital, and so forth with a wide variety of sterile, single-use, disposable articles such as hypodermic syringes, pipettes, test tubes, petri dishes, and a constantly increasing array of surgical and medical items.

STERILIZATION OF PLANET-IMPACTING SPACE VEHICLES

The sterilization, which implies complete absence of living microorganisms or their spores, of space vehicles designed to impact a planet such as the moon, or Mars is a complex problem. Not only must surfaces be sterilized, but the interiors of porous materials, liquids, and so forth, must also be sterile.

The problem of proving that an object is sterile is also replete with difficulties. Test organisms are usually spores of *Bacillus subtilis*, the thermophile, *B. stearothermophilus*, and the tough anaerobe, *Clostridium sporogenes*.

Cold sterilization (processes requiring and producing very little heat) is generally used. Applicable sterilants for this purpose are ethylene gas or ethylene gas dissolved in methanol. Ethylene gas is not a liquid at room temperature. However, when mixed with methanol (50:50) it boils at room temperature so that it may be applied as a liquid but will evaporate without leaving any residue. A fresh solution of formaldehyde in methanol has also proved effective.

Sterilants used must not damage the delicate electronic circuits, corrode metal surfaces, or react with the various plastics used, especially in the electronic system. Even the solid propellant might require sterilization. Liquids must be sterilized internally. This is often accomplished by adding a sterilizing agent that after a period of time will decay into a nontoxic, nonreactive compound.

SUGGESTED READINGS

Jay, J. M.: Modern food microbiology, New York, 1970, Van Nostrand Reinhold Co., Chap. 9.

Lawrence, C. A., and Block, S., editors: Disinfection, sterilization and preservation, Philadelphia, 1968, Lea & Febiger.

Norris, J. R., and Ribbons, D. W., editors: Methods in microbiology, vol. 1, New York, 1969, Academic Press, Inc., Chap. 3.

Perkins, J. J.: Principles and methods of sterilization in health sciences, ed. 2, Springfield, Ill., 1969, Charles C Thomas, Publisher.

14 Disinfectants, antibiotics, and chemotherapeutic agents

CHEMICAL AGENTS FOR DISINFECTION

MODE OF ACTION OF DISINFECTANTS AND RELATED CHEMICAL AGENTS

KINDS OF DISINFECTING AGENTS

ANTIBIOTICS

THE NEW PENICILLINS AND OTHER ANTIBIOTICS

SENSITIVITY TESTING

NONMEDICAL USES OF ANTIBIOTICS

THE SULFONAMIDE DRUGS

OTHER NEW CHEMOTHERAPEUTIC AGENTS

In Chapter 13 methods of sterilization were discussed. There the emphasis was upon complete destruction of all bacteria, primarily by mechanical means. This chapter, on the other hand, focuses on chemical means of killing bacteria or stopping their growth, either outside or inside the body. The chemical methods tend to be more selective in their destruction of microbes.

CHEMICAL AGENTS FOR DISINFECTION

Terms relating to chemical disinfection are defined below:

disinfection Literally, to rid of infection, the process of rendering an object free from disease-producing organisms. British, and to a lesser extent American, workers include both physical and chemical means. More commonly, however, disinfection implies the use of chemicals for this purpose.

disinfectant Any chemical agent that destroys pathogens on the surface of inert objects or in inert materials but is not applied directly to the body. Although disinfectants kill disease producers, they do not necessarily destroy spores or viruses.

antiseptic (against sepsis) A compound that can be applied to and remain on the surface of the skin or mucosa and kill or inhibit the growth of infection-causing microorganisms. It is usually in the form of a salve or ointment. Mouth washes or gargles cannot correctly be called antiseptics unless in the concentration used and during the brief period they are in contact with the tissues they are able to kill or inhibit bacteria.

bactericide A compound that kills bacteria. Generally this definition is limited to those that kill vegetative bacteria. Spores may or may not be killed.

bacteriostat A compound that inhibits the growth of bacteria without killing them. When the toxic substance is removed, the bacteria may resume growth.

germicide In practice, a term essentially synonymous to bactericide or disinfectant. It implies that the organism kills germs (usually vegetative bacteria).

sporicide An agent that kills spores and so represents the ultimate in germicidal effectiveness.

viricide A compound or agent that permanently inactivates viruses.

fungicide An agent that specifically kills fungi or their spores.

fungistat A compound that merely inhibits the growth of fungi.

algicide An agent, usually added to water in a storage reservoir or swimming pool, that kills algae and their cysts.

sanitizing agent An agent that sanitizes (renders free from pathogens) any surface. The term is generally used with eating utensils, dairy equipment, and food processing surfaces, and may be chemical or physical.

Factors influencing action of chemical agents are as follows:

1. Concentration—at a high concentration a given compound may be an active bactericide, at a lower concentration a bacteriostat, and at a very low concentration may stimulate bacteria growth.
2. Temperature—in general the higher the temperature at which a disinfectant is used the greater will be its effectiveness.
3. Organic matter—if organic matter such as milk, or blood serum is present on or in the object or material being disinfected, the disinfectant may quickly lose its germicidal power.
4. Germicidal spectrum—some chemical agents may be active against a wide variety of microbes while others have a limited range of effectiveness.

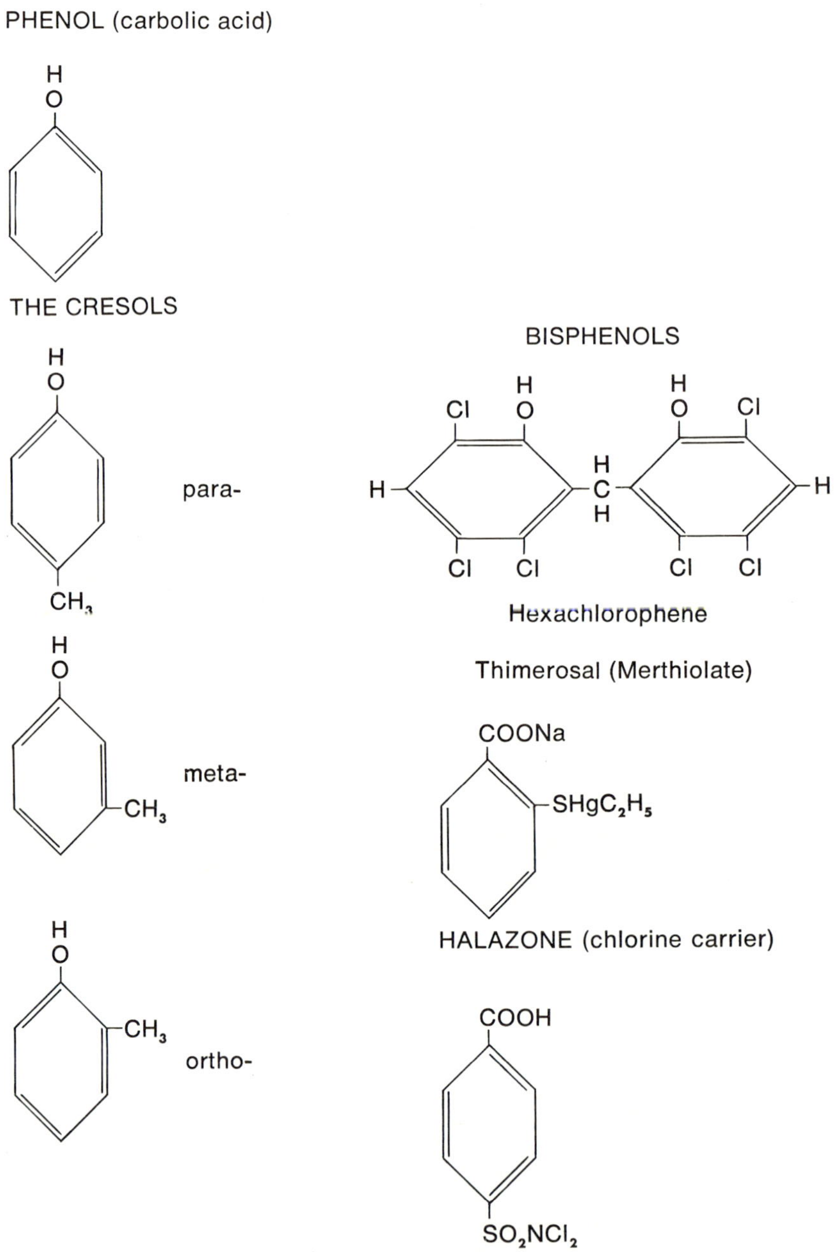

Figure 14-1
Molecular structures of representative disinfectants.

MODE OF ACTION OF DISINFECTANTS AND RELATED CHEMICAL AGENTS

The mechanism of antimicrobial action is still not completely understood. However, the actions involved include the following:

1. Interference with the integrity of the cell wall—in mature bacteria the cell wall may be selectively removed exposing the fragile cytoplasmic membrane, or in the growing cell the agent may interfere with the formation of the cell wall itself.
2. Disruption or reduction in the integrity of the cytoplasmic membrane—in either disruption or reduction, cytoplasmic materials leak out of the cell rendering it incapable of continued metabolic activity. Generally the cell dies.
3. Action upon cell protein—some agents act as nonspecific protein coagulants and as such denature any or all proteins of the cell to the extent that the cell is irreversibly damaged and death ensues. Alcohol exerts such an effect.
4. Interference with enzyme activity—the metabolism of the cell is an orderly chain of enzyme reactions. The entire cell metabolism may suddenly halt if a single (or several) critical enzyme reactions are blocked. However, if the blocking compound can be removed, the cell metabolism is revived and the cell resumes its normal proliferation. The best known agents of this type are the compounds of mercury.
5. Interference with polymerases—compounds that interfere with the polymerases responsible for the synthesis of DNA or RNA strike directly at the basis of cell reproduction and metabolism thus blocking all future proliferation of cells.
6. Action against covalent bonds—another type of reaction involves the blocking or breaking of covalent bonds, thus flattening or distorting the three-dimensional structure of certain enzymes so that they can no longer function.

KINDS OF DISINFECTING AGENTS

Because more than 8,000 compounds have been listed with the Food and Drug Administration as disinfecting or antiseptic agents, it is impossible to critically compare all the various agents now on the market.

The major groups of disinfectants or antiseptic compounds are as follows:

Phenol or phenol derivatives

Phenol (or carbolic acid), a coal-tar product, was discovered in 1834. It soon came into use as a means of destroying or masking the odors of sewage or similar odoriferous materials. LeMaire (1860) was the first to recommend its use on wounds. Lister (1865) found that concentrations of 2% to 5% were effective in preventing the highly fatal infections following compound fractures. In 1868 he published his famous work on antiseptic surgery in which he described spraying the operative area with phenol.

Phenol has three serious drawbacks. It must be used in relatively high concentrations, it is injurious to tissues, and it has a pronounced odor.

The addition of a methyl (CH_3) group to phenol yields cresol. The germicidal power of cresol is considerably greater than that of phenol. Phenol may be further modified by attaching either an ethyl (CH_3CH_2) group instead of the methyl group, or both. Also chlorine (Cl) or bromine (Br) groups may be attached. These variations provide a wide assortment of compounds of varying disinfecting power and toxicity.

Notable among these derivatives is hexachlorophene (or bis 3,4,6 trichlorophenol). (The prefix *bis* means two or twice.) The molecule is made by combining two phenol molecules through a $-\overset{H}{\underset{H}{C}}-$ bond and adding three Cl groups to each phenol (Fig. 14-1).

Hexachlorophene is insoluble in water but has the rare property of remaining stable when combined with soaps or detergents. It is used in

3% concentrations in various germicidal or bath soaps. Because of its insolubility, it is presumed to remain in germicidal concentrations on the skin. A question has been raised as to its possible cumulative toxic effects.

Mercury compounds and their derivatives

Mercuric chloride, bichloride of mercury or corrosive sublimate ($HgCl_2$), was first introduced as an antibacterial agent by Koch in 1881. Geppart (1889) found that it was not sporicidal. Since that time it has been shown that it is a bacteriostat rather than a bactericide. Because of extreme toxicity it is no longer used to any extent.

A good many synthetic organic disinfectants have been developed that provide the bacteriostatic action of mercury without the high toxicity of the bichloride. However, they are not consistently active against both gram-positive and gram-negative bacteria. These compounds carry such trade or chemical names as Merthiolate, Mercresin, Mercurochrome, and Phenylmercuric Acetate. They vary in sporicidal effect.

Chlorine and hypochlorites

Chlorine gas or chlorine gas combined with ammonia has as its primary application the disinfecting of water supplies. The hypochlorites, calcium hypochlorite (chloride of lime) or sodium hypochlorite (chlorinated soda), while extremely useful in water disinfection, are also active germicides. The laundry bleaches sold under such trade names as Clorox or Purex contain 5% sodium hypochlorite and in relatively low concentrations (10 to 100 ppm) are active germicides for use on glassware. However, these concentrations may corrode metal objects.

Chlorine water and hypochlorites have two distinct disadvantages for general use: disagreeable odor, and marked instability in the presence of organic matter. Hypochlorites act primarily as oxidizing agents. Hydrogen peroxide likewise owes its germicidal power to the release of nascent oxygen.

Iodine and its derivatives

Tincture of iodine (5% I_2 in about 50% alcohol) was first listed in the U. S. Pharmacopeia in 1830. Davies (1839) was the first to recommend tincture of iodine for the treatment of wounds. It was used with considerable success on wounds during the Civil War (1862) and was shown to be sporicidal as early as 1880. The germicidal effect does not vary greatly with gram-positive and gram-negative bacteria, making it a broad-spectrum germicide.

Its effectiveness appears to be greater than that of chlorine or bromine. Tissue toxicity of iodine is lower than that of phenol. It also has considerable tissue penetrating ability.

In the past, tincture of iodine often contained 5% to 7% I_2 in 87% ethyl alcohol. At this concentration the skin was sometimes burned. Today's 2% solution, while still highly germicidal, rarely causes burning of the skin.

Iodine preparations of various sorts are available that liberate free iodine in the presence of water. These are termed iodophors (iodine carriers).

Iodine solutions are also useful in emergencies for the disinfection of drinking water and swimming pools, and the sterilization of sutures or surgical instruments. Also, when added to alcohol (70% ethanol + 0.25% I_2), iodine provides an effective mixture for sterilizing clinical thermometers.

It has been used for the control of skin infections such as ringworm and athlete's foot.

Alcohols

While ethyl alcohol (ethanol) and isopropyl alcohol (isopropanol) are fairly useful as disinfectants, they do have certain limitations. Although they act quickly against vegetative bacteria (including *Mycobacterium tuberculosis*), if no organic matter is present, they are only slightly sporicidal and their action ceases as soon as they evaporate.

Because they are not sporicidal, they should not be used for cold sterilization of surgical

A. Ammonium chloride molecule

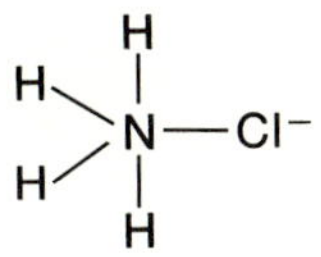

B. General formula for the quaternary ammonium chloride halides, based on the ammonium chloride molecule. A great number of varieties have been synthesized.

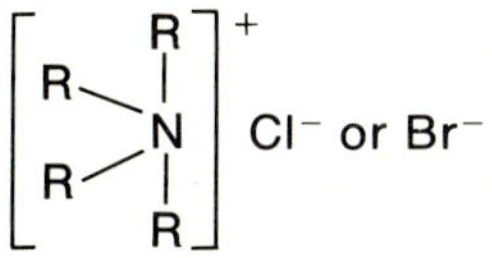

C. Molecular structure of the original member of this group, known by the trade names Zephiran Chloride or Roccal.

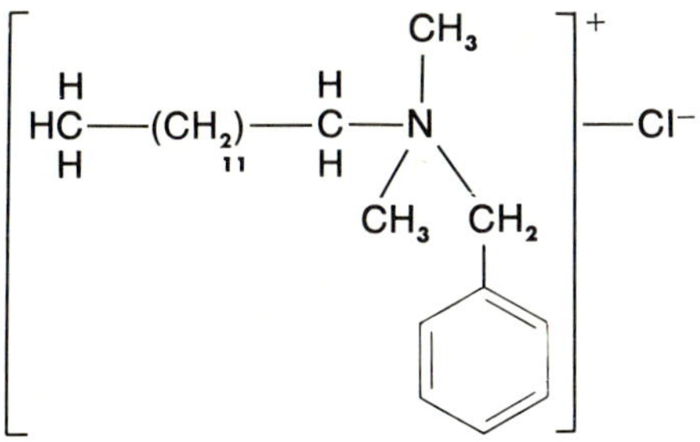

Ionizes in solution to form a positively charged complex active molecule or cation and a negatively charged chloride ion.

Figure 14-2
Structural formulas for representative cationic detergents.

instruments, sutures, or hypodermic syringes or needles. Alcohol "sterilized" syringes and needles have also been involved in the patient-to-patient transfer of the virus of infectious hepatitis.

With ethanol, water must be present as a carrier. Thus the most effective concentrations are about 70% to 80%. Rubbing alcohol is commonly in this range. Isopropanol, unlike ethanol, shows greater germicidal effect with increased concentration, that is, 99% is better than 70%. When used as a skin wash, one advantage of alcohols is that they defat the skin. While the skin is far from sterilized, an alcohol wash enhances its relative cleanliness.

Cationic and other detergents

The term detergent is derived from a Latin term meaning to cleanse. It can be properly applied to ordinary soaps as well as to synthetic detergents.

Ordinary soaps, (primarily sodium salts of the

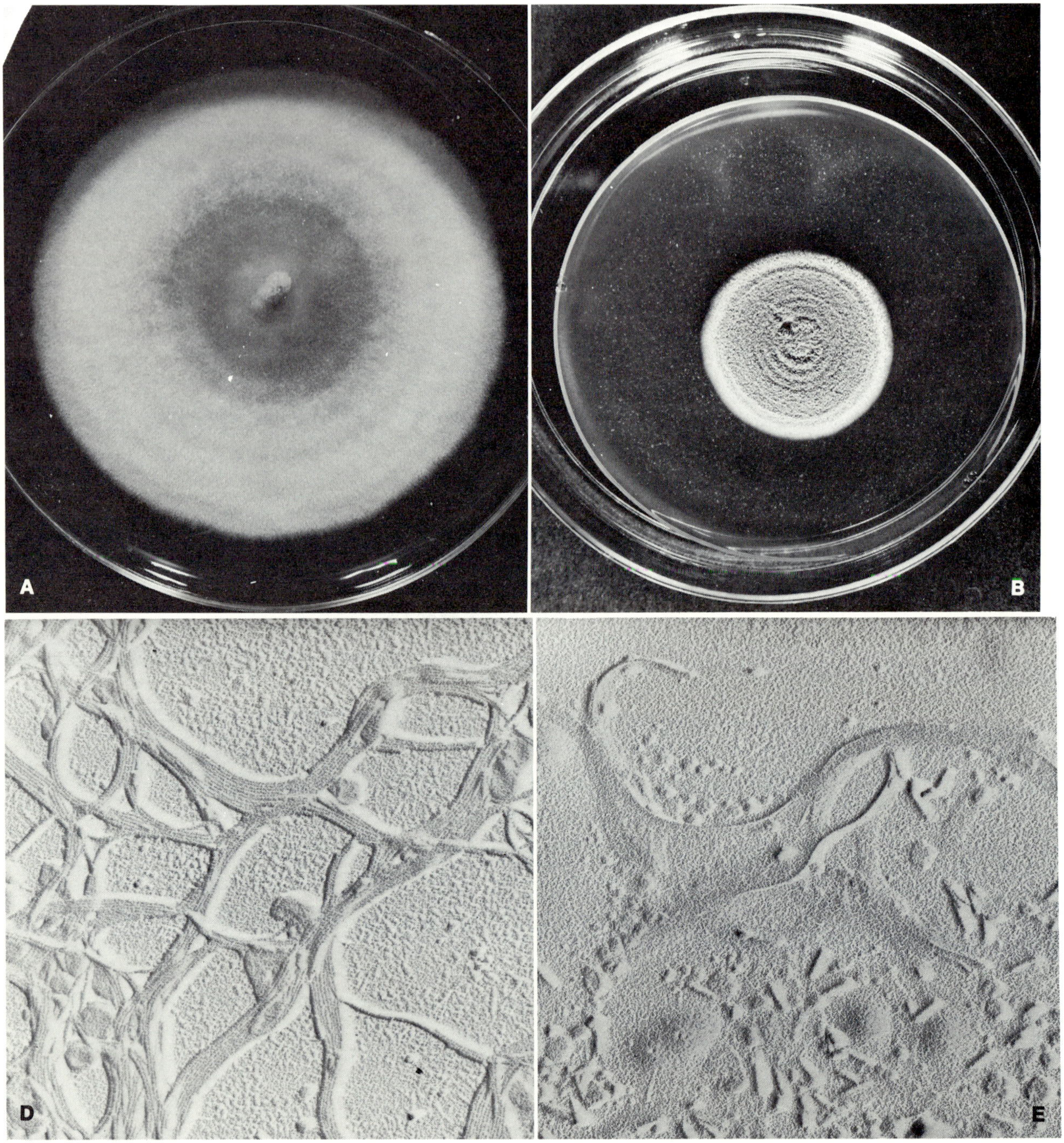
A
B
D
E

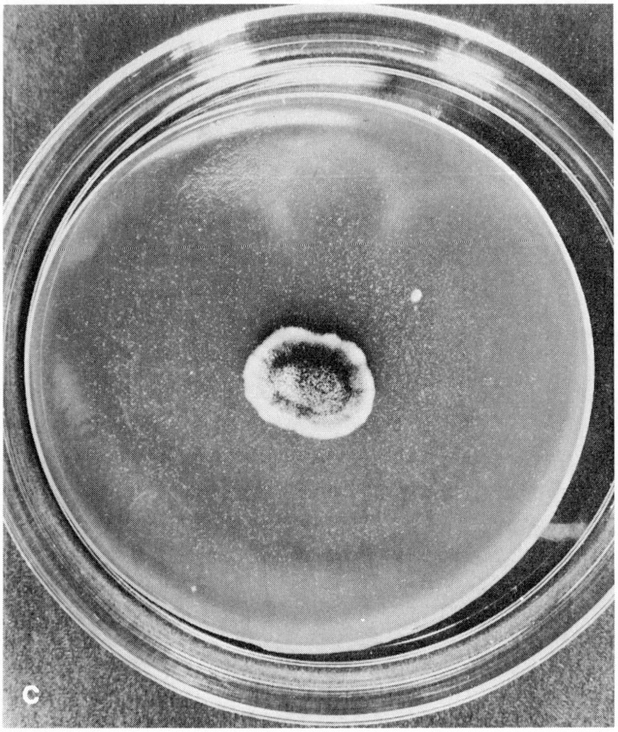

Figure 14-3
A-C, Effect of anionic detergent (LAS) on growth of mold, *Curvularia* sp., on Sabouraud's agar. **A,** 7-day giant colony without detergent; **B,** growth with 0.05% detergent added; **C,** growth in presence of 0.2% detergent; **D-E,** electron micrographs showing flagella-dissolving power of LAS on flagella harvested from *Pseudomonas aeruginosa.* **D,** Original flagella, metal shadowed with uranium. **E,** Similar preparation after 5 minutes in 0.1% LAS. (Courtesy Foster Eubank.)

higher fatty acids) have limited value as disinfectants. Their chief value arises from their ability to reduce the surface tension of the skin surface. This defatting action serves to float away many of the bacteria. The synthetic detergents have one property not possessed by the sodium soaps. They do not form insoluble calcium soaps and thus can be used successfully in hard water.

Based on their mode of molecular alignment in water, synthetic detergents have been divided into three categories: cationic, anionic, and nonionic.

The cationic detergents are used primarily as disinfectants. The most widely used molecular variety includes derivatives of ammonium chloride. These are termed quaternary ammonium chloride compounds or "quats." They are so called because the four hydrogens surrounding the nitrogen atom are replaced with hydrocarbons or substituted hydrocarbons with one to twelve carbons in the chain or with benzyl or similar cyclic molecules. In water, the molecule is ionized into active complex portions that carry a positive charge, while the single chloride ion carries a negative charge. This explains the designation, cationic detergent. The comparison between the ammonium chloride and cationic detergent molecules is shown in Fig. 14-2.

The germicidal effect of a cationic detergent was first demonstrated in 1936 by Domagk, the Nobel laureate who also discovered the sulfonamides. The molecular structure of the compound he used is shown in Fig. 14-2. It is called a dodecyldimethyl benzylammonium chloride and is probably more familiarly known under the trade names, Roccal and Zephiran Chloride.

Today, many different cationic formulations are available, not all of them derivatives of ammonium chloride. They show considerable variation in germicidal effectiveness and other properties. However, they tend to be uniformly active against most gram-positive and gram-negative bacteria.

The anionic detergents are primarily sodium salts of organic acids such as alkyl benzene sulfonic acid or long-chained sulfates such as sodium lauryl sulfate. They are termed anionic detergents because the positive charge is carried by the sodium while the balance of the molecule has a negative or anionic charge. On a soiled dish in dish water, the molecules are arranged with the hydrophilic, positively-charged sodium in the water phase while the lipophilic negatively-charged complex ion is immersed in the fatty surface of the soiled area. As a result of the greatly reduced surface tension the detergent lifts away and solubilizes the surface soil.

In a bacterial culture the molecules are aligned with the sodium ion in the water side of the cell surface while the cell wall attracts the lipophilic, negatively-charged portion of the surfactant molecule. Gram-positive bacteria are much more susceptible to injury than gram negatives (Fig. 14-3).

The third type of detergents, the nonionics, tends to show slightly less detergency and also only limited antibacterial action. However, they tend to be rather readily biodegradable.

ANTIBIOTICS

Nearly a century ago (1877) Pasteur and Joubert observed an example of an antibiotic action that they considered an "antagonism." Vuillemin in 1889 first applied the terms "antibiose" and "l'antibiotic" to a situation where one growing microbe destroyed or prevented the growth of another. The term was soon limited to the effect as seen in vitro, while "antagonism" was used to describe the effect in vivo. Waksman revived and extended the definition in 1942 as follows: "An antibiotic is a chemical substance, produced by microorganisms, which in *low concentrations* has the capacity to inhibit growth or even destroy bacteria and other microorganisms."

The first antibiotic employed commercially was called pyocyanin, the water-soluble blue pigment produced by *Bacillus pyocyaneus (Pseudomonas aeruginosa)*, the blue pus organism. Isolated in 1890 by Emmerich and Loew, it was incorporated in a preparation for topical application to surface infections in 1899. It was still available commercially until 1930. Two properties limited its value: it was too toxic to be injected, and it was relatively unstable.

In 1928 Alexander Fleming, in the laboratory of St. Mary's hospital in London, noticed that a colony of *Penicillium* had developed at the edge of a petri plate heavily seeded with *Staphylococcus aureus*. In a zone around the mold, the colonies of bacteria appeared to have been partially lysed by some compound that had diffused from the mold into the agar. Accordingly he grew the mold in a broth culture, then filtered it. When a drop of the filtrate was placed on the surface of a plate seeded with the *Staphylococcus aureus*, a clear zone again formed. The following year, he reported this observation at the International Microbiological Congress and suggested that the filtrate might be added to blood agar for use of isolation of the whooping cough bacteria. The filtrate would inhibit the gram-positive cocci and allow the gram-negative whooping cough bacteria to grow without competition.

He gave the name penicillin to this unidentified antibacterial agent in the filtrate. The mold was sent to the American mycologist, Dr. Charles Thom, who identified it as *Penicillium notatum*. Although this new antagonistic agent showed great promise, two English mycologists, Clutterbuck and Raistrick, who attempted to isolate the compound, warned that it was far too unstable to be of any value.

In 1938 Dr. Howard W. Florey, Director of the Sir William Dunn School of Pathology at Oxford, and his research team, which included Dr. Ernst Boris Chain, a biochemist, and Professors Heatley and Abraham, began a study of penicillin. By 1939 they had isolated a crude extract from the mold cultures. The extract was found to have remarkable germicidal action on *Staphylococcus aureus*. By 1941, the British Ministry of War recognized that it possessed tremendous military value. However, continual bombings made production in Great Britain hazardous. By a joint agreement with the United States, Florey and Chain left England and set up a mold fermentation on a large scale using the huge fermentation vats of a whiskey distillery that had been closed by the government to conserve wheat. Out of this arrangement grew the industrially important submerged aeration culture method. Using the equipment available, the mold, *Penicillium notatum,* an aerobic form, was grown in the 30,000 gallon stainless steel closed vats by bubbling sterile air through them. By 1943 about 29 pounds of the relatively pure penicillin had been produced, and large scale production of the antibiotic became feasible.

The great advantage of penicillin was its low toxicity for the patients. This was explained when its bactericidal effect was finally shown to be due to interference with the formation of the bacterial cell walls, particularly those of gram-positive bacteria, which were so fragile they soon succumbed to osmotic shock.

Mycoplasma that lacked a cell wall were found to be unaffected by penicillin. (Our own cells lack a cell wall, hence are not injured by penicillin.)

An exception is *Treponema pallidum.* Although it has no distinctive cell wall it is still susceptible to penicillin. However, it differs materially from the fragile mycoplasma in other respects.

Prompted by the discovery of penicillin, Selman A. Waksman and his group in the United States became interested in the search for antibacterial agents. One of his former graduate students, René Dubos, in 1939 isolated two antibiotics, gramicidin and tyrocidin, produced from *Bacillus brevis.* (A mixture of the two is called tyrothricin.) Although markedly inhibitory toward certain gram-positive bacteria, they were far too toxic for other than topical applications.

Waksman turned his energies toward the isolation of antibiotics from the the actinomycetes. He had long known that actinomycetes often inhibited the growth of soil bacteria. In 1940, he and his student, Woodruff, isolated and crystalized a highly potent antibiotic. This they called actinomycin. Appropriately the recently isolated actinomycete that produced it was named *Streptomyces antibioticus.*

In 1944, Schatz, Bugie, and Waksman isolated streptomycin from *Streptomyces griseus.* This had unusual properties not possessed by penicillin. It proved to be active against gram-negative bacteria and to some extent against *Mycobacterium tuberculosis.* This discovery has been called the next major advance after penicillin.

On the basis of their various contributions, Fleming, Florey, Chain, and Waksman were made Nobel laureates, and Fleming and Florey were both knighted by the British Government.

Immediately after World War II the pharmaceutical companies began an almost frantic search for new antibiotics. The molds yielded only one additional significant antibiotic, Cephalosporin, produced by a species of *Cephalosporium* and isolated in 1952. It resembled penicillin in several respects but was less potent. However, it could be used against penicillin-resistant strains of staphylococci, with patients who had become allergic to penicillin, and against a few gram-negative forms.

The bacilli yielded only two additional groups of antibiotics. These were the bacitracins produced by *Bacillus subtilis* and *B. licheniformis,* and the polymyxins produced by *B. polymyxa.*

However, the actinomycetes proved to be amazingly fruitful. In one survey, out of a total of 769 strains of streptomyces tested, 370 were inactive but 399 produced antibiotic-like compounds. A recent survey indicates that more than 500 antibiotics have been isolated from actinomycetes. Most of them have been far too toxic or unstable to have any great value. Indeed within the last decade no new outstanding antibiotics have been found.

The search for new antibiotics has given way largely to attempts to chemically modify the present biologically produced antibiotic or to develop a procedure for producing the antibiotics by chemical synthesis rather than by conventional microbiological methods.

If we include antibiotics that can be used only by topical application and those applicable for only nonmedical uses, there are currently about 50 different basic kinds of antibiotics on the market. To this list can be added about ten natural or chemically modified varieties of penicillin. A number of these antibiotics, such as actinomycin D, are produced only in small quantities for research purposes. The picture is further confused by the practice in the pharmaceutical industry of using several trade names for a given antibiotic rather than the single generic name.

One or more than one of the following types of antimicrobial activity will account for the inhibitory action of most of the antibiotics. Antibiotics may:

1. Interfere with cell wall production or damage existing cell walls
2. Damage the cell membrane causing lysis or leakage of cell contents
3. Interfere with protein synthesis including synthesis of certain enzymes
4. Directly damage the nucleic acids of the cell or interfere with DNA or RNA synthesis

THE NEW PENICILLINS AND OTHER ANTIBIOTICS

The recently achieved chemical modifications of the nucleus of the penicillin molecule are a remarkable example of biochemical wizardry.

When penicillin first became available it was hailed as the control agent for all staphylococcus infections. However, within a short time a few strains were found to be resistant to the antibiotic, and soon the penicillin-resistant staphylococcus strains became so widespread that they posed a major health problem.

The discovery was made that the penicillin-resistant strains produced the enzyme penicillinase, which split the penicillin molecule rendering it ineffective.

In 1960, through a fortuitous circumstance, one of the pharmaceutical laboratories discovered that by omitting an acyl side chain donor from the formula for the penicillin medium, the particular *Penicillium* strain used produced 6-amino-penicillanic acid. This was the precursor to penicillin G and other penicillins and proved to be amenable to chemical modifications. Out of this discovery grew several new kinds of penicillin produced by chemically attaching a new kind of side chain to the nucleus. These new compounds still possess relatively high antibiotic activity but are nearly all immune to attack by penicillinase. From this discovery also came methicillin, oxacillin, nafcillin, diphenicillin, and ampicillin. Table 14-1 shows the relative molecular structures of each. The last of these, ampicillin, proved to be a broad-spectrum antibiotic capable of inhibitory or lethal action against several more varieties of bacteria than the original penicillin, including some gram-negative forms.

Cephalosporin C, which resembles penicillin, also has a molecular nucleus that can be chemically modified. This modified agent has proved useful against several gram-negative bacteria as well as against penicillin-resistant staphylococci. It can also be used successfully with patients who have become allergic to penicillin.

The tetracyclines, as their name indicates, have an unusual molecular structure, four fused 6-carbon rings. Their primary value is their broad antibacterial spectrum; they are active against both gram-positive and gram-negative bacteria. The tetracyclines apparently inhibit cell synthesis by interfering with ribosome production and function. They are effective in the treatment of rickettsial and chlamydial diseases, and also have several practical, nonmedical applications.

The most significant of the tetracyclines are chlortetracycline (Aureomycin), discovered in 1948, a growth product of *Streptomyces aureofaciens*, and oxytetracycline (Terramycin), produced by *S. rimosus*, isolated in 1950.

The polymyxins B and E (Coli-Mycin) are unique because they damage the bacterial membranes in a manner resembling that of the cationic detergents. Unlike many of the antibiotics, they are bactericidal to nongrowing as well as growing cells. They normally must be injected and must be used with caution since they may produce kidney and nerve sheath damage because of their detergent-like properties.

Novobiocin, produced by *S. niveus*, has a spec-

Table 14-1. Comparative molecular structures of natural and semisynthetic penicillins

PENICILLIN TYPE OR COMPOUND	TYPE STRUCTURAL FORMULA	R-GROUP
Penicillin: basic structure	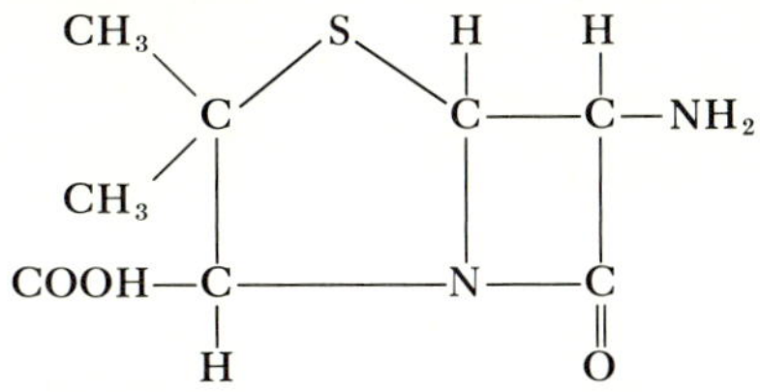	–
6-Aminopenicillanic acid (6-APA)	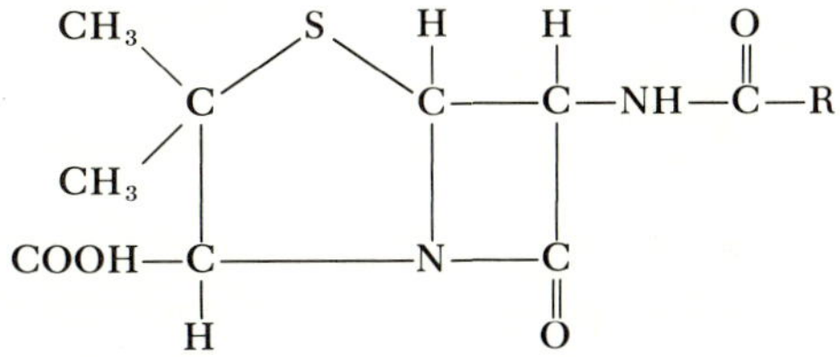	–
Penicillin G (benzyl penicillin)	–	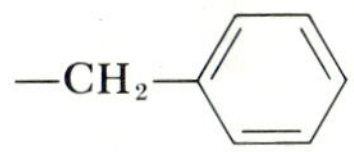
Phenethicillin (phenoxyethyl penicillin)	–	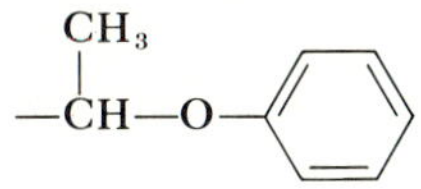
Propicillin (phenoxypropyl penicillin)	–	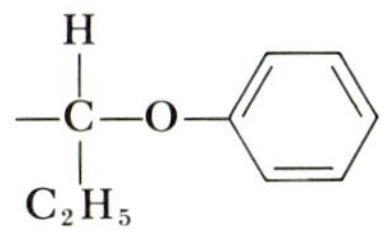
Phenbenicillin (phenoxybenzyl penicillin)	–	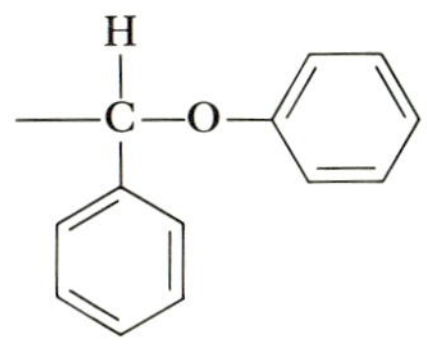
Methicillin (dimethoxybenzyl penicillin)	–	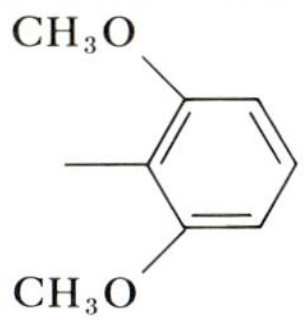
Ampicillin (aminophenylacetamido penicillin)	–	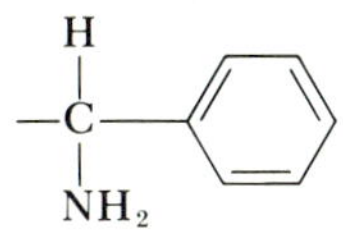
Oxacillin (cloxacillin)*	–	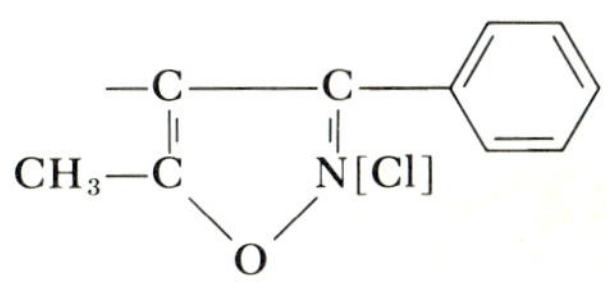

*When Cl atom is added to oxacillin molecule

trum similar to penicillin but in addition is active against some gram-negative bacteria, especially *Proteus* strains. Because of this, it is extremely useful for the control of some bladder infections. It damages the bacterial membranes, causing the cells to become leaky. This action is primarily bacteriostatic but may prove slowly bactericidal.

Chloramphenicol is of special interest for two reasons. It is now produced synthetically and it possesses a very broad antibacterial spectrum. However, it must be used with care because it may produce toxic reactions.

SENSITIVITY TESTING

Antibiotics often are effective against only certain bacterial strains. In addition, a considerable number of strains of bacteria resistant to certain antibiotics have appeared in the last two decades. As a result, before antibiotics are administered to control a given bacterial strain its sensitivity to several antibiotics should be tested.

This is done primarily by isolating the causative organism in pure culture. The organism is heavily streaked on a petri plate containing a suitable agar medium. Small paper disks impregnated with known amounts of several antibiotics are then dropped at intervals on the plate. Incubation for 12 to 24 hours is usually sufficient for the bacterial lawn of bacteria to develop. If the organism is sensitive to a given antibiotic, a clear zone appears around the disk. If an appreciable number of resistant strains are present, they develop small isolated colonies inside the clear zone or inhibited growth. (Fig. 14-4, *A-E*).

Many physicians now merely prescribe a

Table 14-2. Antibiotic activity spectrum of several common antibiotics

ANTIBIOTIC	SOURCE ORGANISM	ACTIVE AGAINST
Bacitracin	*Bacillus licheniformis, B. subtilis*	Chiefly gram-positive bacteria
Cephalosporin	*Cephalosporium* species	Chiefly gram-positive bacteria
Chloramphenicol	*Streptomyces venezuelae* (now produced synthetically)	Gram-positive and gram-negative bacteria, rickettsiae, and chlamydia
Chlortetracycline	*Streptomyces aureofaciens*	Gram-positive and gram-negative bacteria, rickettsiae, and chlamydia
Coli-Mycin B	*Bacillus colistinus*	Gram-negative bacteria
Cycloserine	*Streptomyces garyphalus, S. orchidaceus*	*Mycobacterium tuberculosis*
Fumagillin	*Aspergillus fumigatus*	*Entamoeba histolytica*
Griseofulvin	*Penicillium patulum*	Some pathogenic fungi
Neomycin	*Streptomyces fradiae*	Certain gram-positive and gram-negative bacteria, and *Mycobacterium tuberculosis*
Nystatin	*Streptomyces noursei*	Some pathogenic fungi
Penicillin	*Penicillium chrysogenum, P. notatum*	Chiefly gram-positive bacteria
Polymyxin B	*Bacillus polymyxa*	Gram-negative bacteria
Streptomycin	*Streptomyces griseus*	Chiefly gram-negative bacteria
Tyrothricin	*Bacillus brevis*	Gram-positive bacteria
Viomycin	*Streptomyces puniceus*	*Mycobacterium tuberculosis*

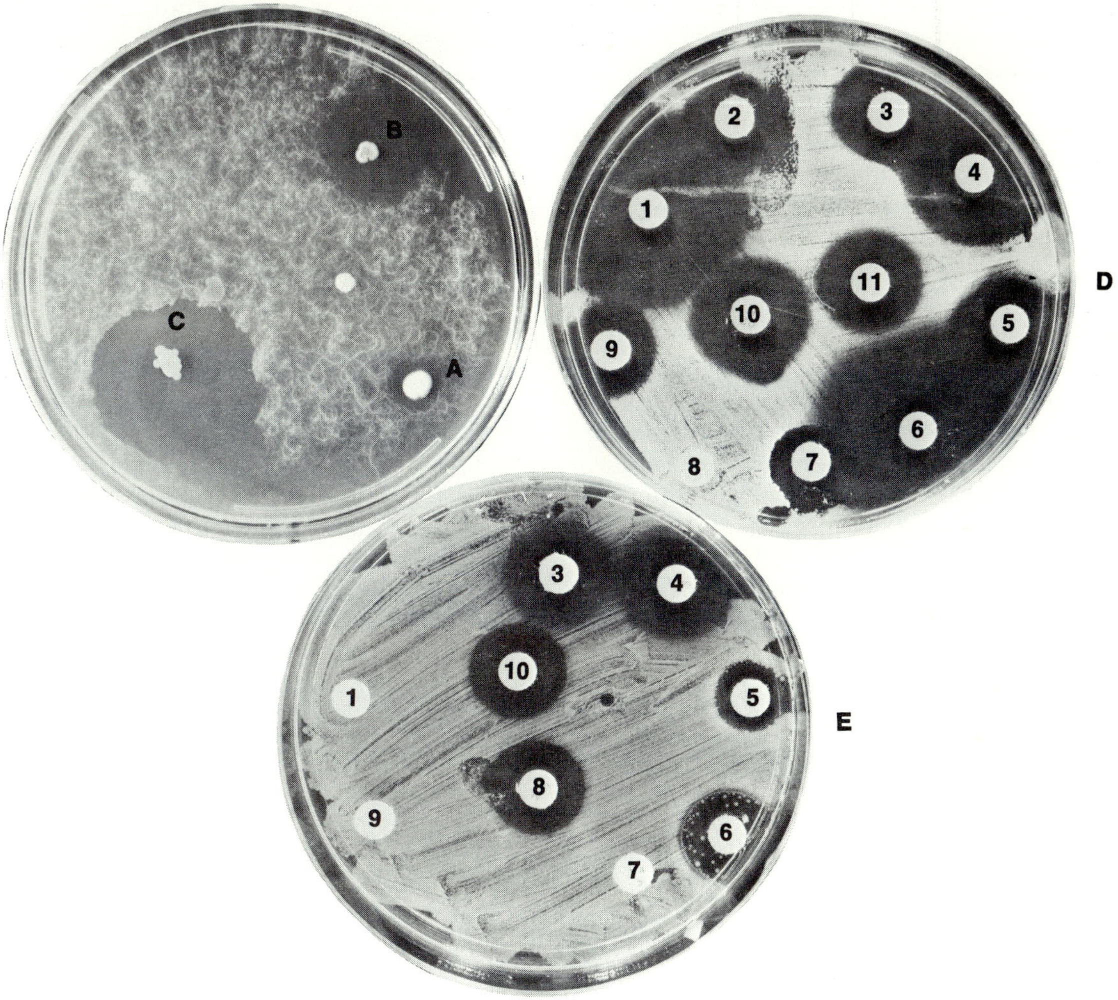

Figure 14-4
Antibiosis in nature. Soil bacteria count plate showing growth-inhibiting (antibiotic) action of three actinomycetes colonies, **A, B, C,** against spreading bacilli that have covered surface of plate. **D-E,** Antibiotic sensitivity test. Disks placed on surface of seeded plates showing differences in sensitivity of *Staphylococcus aureus,* **D,** and *Escherichia coli,* **E.** Antibiotic code: **1,** penicillin; **2,** erythromycin; **3,** Chloromycetin; **4,** Terramycin; **5,** Mandelamine; **6,** ampicillin; **7,** Vancomycin; **8,** nystatin, **9,** oleandomycin; **10,** dihydrostreptomycin; **11,** polymyxin B. (Note growth of resistant strains of *E. coli* in clear zones around **6.**)

broad-spectrum antibiotic without sensitivity testing, in the hope that it may control the offending organism. While this is sometimes a matter of expediency it is not always the best or most successful way of treating the disease.

Promiscuous use of antibiotics is a questionable practice for several compelling reasons. The patient may develop toxic reactions to the antibiotic itself. This is especially true with penicillin. Also, while the initial administration of the antibiotic may quickly kill the sensitive strains of the offending organism, too small a dose or too short a period of administration may permit the uncontrolled growth of antibiotic-resistant strains.

Oral antibiotics may be destructive to the normal bacteria in the intestine that hold in check certain yeasts and other undesirable forms. With these controls removed, the patient may suffer from a massive growth of the undesirable forms in the intestinal tract. With infants and the aged, this sometimes produces serious or even fatal consequences.

NONMEDICAL USES OF ANTIBIOTICS

Antibiotics have been used successfully, especially in greenhouses, for the control of certain bacterial diseases of plants. These agents may be applied as sprays or be added to the irrigation water or to the soil itself.

Broad-spectrum antibiotics added to feeds of poultry, pigs, calves, and so forth, in many cases increase the rate of growth. This suggests that the observed growth increases may be due to control of undesirable bacteria, especially intestinal forms.

In 1958 the Food and Drug Administration gave permission to poultry dressing plants to add up to 10 ppm of chlortetracycline or oxytetracycline to the slush ice used to cool the eviscerated birds. Initially, this greatly lengthened the shelf life of the packaged poultry due to the control of psychrophiles responsible for spoilage. However, the effect later became less apparent because antibiotic-resistant strains developed.

In 1959 the Food and Drug Administration approved the use of the addition of the tetracyclines to wash waters used in preparing cuts of fresh fish for packaging. The observed effects were similar to those noted with poultry.

In growing tissue and single-cell cultures in virus laboratories for the proliferation of viruses, antibiotics are added to control bacterial contamination. The viruses are unaffected by the antibiotics.

Penicillin, widely used by early bacterial geneticists for the isolation of mutant strains of bacteria, is still employed to some extent. Actinomycin D, because of its effect on DNA synthesis, has also provided a valuable experimental tool for the bacterial geneticist.

On November 4, 1966, when the storm-swollen Arno river poured over its banks and submerged many of the art treasures in Florence, Italy, it appeared that many of these treasures would be destroyed. Even though they were rescued as soon as the water subsided and taken to higher ground to dry, green, black, white, and pink molds began to appear on many of the priceless paintings. Although the masterpieces had been saved from the flood, it now appeared that they were to become food for the molds. However, this threat disappeared when they were sprayed with the fungicidal antibiotic, nystatin.

THE SULFONAMIDE DRUGS

Until eclipsed by the antibiotics, the sulfonamides were hailed as the panacea for the control through chemotherapy of all bacterial infections. These were unique chemical agents, more toxic to the bacterial cells than to the tissues of the patient.

In 1932, Gerhard Domagk tested some azo dyes, synthesized by his co-workers in a German industrial laboratory, for their antibacterial action. One of them, Prontosil, showed, along with a low toxicity, a marked ability to cure experimental streptococcus infections in laboratory animals. This laid the foundation for the synthesis of a variety of sulfonamides. On the basis of this discovery Domagk was awarded the Nobel Prize in 1939.

When first developed, the various molecular varieties of sulfa compounds proved remarkably effective against many gram-positive bacteria. Initially *Neisseria, Shigella, Proteus, Klebsiella,* and *Escherichia* were also susceptible to the sulfa drugs. However, the drugs had two fatal weaknesses. Drug-resistant strains of bacteria easily evolved and patients often developed sensitivities or other undesirable reactions.

A classic example of the development of a sulfa-resistant bacterial population occurred with the gonorrhea organism, *Neisseria gonorrhoeae.* Today a high percent of the gonorrhea organisms isolated in this country are sulfa-resistant.

In spite of the weaknesses, sulfa compounds are still used extensively, especially for the control of urinary tract infections. Often they are combined with antibiotics in the hope of securing a synergistic therapeutic response.

OTHER NEW CHEMOTHERAPEUTIC AGENTS

In addition to antibiotics and the sulfonamides, a number of chemicals molecularly unlike either group have been synthesized. All have one characteristic in common; they possess high toxicity (in low concentrations) for microorganisms, and relatively low toxicity for living animal tissue.

Significant among these new antimicrobial compounds are the nitrofurans (nitrofurazone) and the completely unrelated nalidixic acid. The first is a broad-spectrum compound, while the latter is especially effective in the treatment of infections caused by *Escherichia coli, Enterobacter aerogenes, Klebsiella pneumoniae, Shigella flexneri,* and *Salmonella typhimurium.*

SUGGESTED READINGS

Harris, M.: Pharmaceutical microbiology, London, 1964, Baillière, Tindall & Cox, Ltd.

Lawrence, C. A., and Block, S., editors: Disinfection, sterilization and preservation, Philadelphia, 1968, Lea & Febiger.

Perkins, J. J.: Principles and methods of sterilization in health sciences, ed. 2, Springfield, Ill., 1969, Charles C Thomas, Publisher.

Waksman, S. A.: What is an antibiotic or an antibiotic substance?, Mycologia 39:565-569, 1947.

Waksman, S. A.: The actinomycetes: A summary of current knowledge, New York, 1967, The Ronald Press Co.

15 Metabolic activities of microorganisms

DEFINITION OF TERMS

BACTERIA AS TOOLS FOR THE STUDY OF METABOLISM

NATURE AND KINDS OF ENZYME ACTIONS

METABOLIC PATHWAYS

REGULATION OF ENZYME ACTION

FAMILIAR EXAMPLES OF MICROBIAL ENZYMATIC ACTIVITY

Man had long been concerned with the operation of his body; however, he proved to be a complex, variable, and unpredictable experimental animal. Only a few individuals were willing to participate in experiments and the devotion to such experimentation was not always wholehearted. When rats proved to be useful, the basics of life processes began to be elucidated. However, some workers turned to microorganisms as experimental tools. The animal physiologists were often reluctant to apply these findings to animal physiology, so that although by the late 1920's a great deal of experimental work had been done with microorganisms, its significance was still questioned.

Among those who entered this intellectual arena was A. J. Kluyver, Director of the Laboratorium voor Microbiologie of Delft, Holland (Leeuwenhoek's home town). In his concept of comparative biochemistry he stated his belief that metabolic processes found in bacteria and other microorganisms had their counterparts in the life processes of higher forms. Evidence soon made it apparent that microorganisms, and especially bacteria because of their simplicity, were ideal tools for the study of the basic metabolic processes.

By 1940 the electron microscope had become a useful new tool. Soon after 1950, its value was greatly increased when it became possible to make thin sections of bacteria and begin the intricate and involved process of coupling organelles with metabolic functions. By 1960 many more precise analytical methods, such as highly efficient refrigerated centrifugation, electrophoresis, partition chromatography, ultraviolet spectrophotometry, radioactive labelling techniques, and a host of other unbelievably sensitive devices, were applied.

It has now become possible to carry our knowledge of life down to the very molecules involved. Out of this has come the tremendously revealing study of molecular biology.

DEFINITION OF TERMS

Bacterial metabolism (Gr. *metaballein* to change) includes the whole complex of all the chemical reactions involved in the reproduction of bacterial cultures, whether in the test tube or in nature. This in turn is divided into two types of processes, anabolism and catabolism.

Anabolism (Gr. *anabolē* a raising up) represents the biosynthetic chemical processes involved in the building up of simpler molecules, such as sugars, and amino acids, into the great macromolecules of living tissue, proteins, complex carbohydrates, DNA, RNA, lipids, and so forth—the end purpose being the expansion of one bacterium into two cells, continuing ad infinitum.

Catabolism (Gr. *katabolē* a casting down) represents the many chemical reactions in which organic and inorganic materials are broken down to supply energy and precursor molecules (building blocks) for the biosynthesis of cells. The catabolic processes, therefore, supply both the building materials for biosynthesis and the energy needed to complete the process. In brief, catabolism is an energy-yielding process while anabolism is an energy-requiring process.

Enzymes (Gr. *en* in + *zyme* leaven or ferment), which are organic catalysts, are the chemical agents in the living cells that bring about the unbelievably complex array of life processes. These processes can occur without enzymes but at a slower rate.

Enzymes are not used up in the process and do not change the end point of the reaction. During the operation they attach to a given substrate molecule (the material acted upon by the enzyme) and cause the chemical change that they are programmed to produce because of their molecular configuration. As soon as it is completed, the enzyme itself leaves this changed molecule and goes to the next substrate molecule. All this occurs in microseconds. A proportion, which varies with enzymes and substrates, of about one enzyme molecule to a million substrate molecules appears to be sufficient to carry on a normal enzymatic reaction. Although not used up during the catalytic reaction, enzymes wear out and must continually be replaced in the cell.

Many enzymes are complexes made up of both the basic protein, the apoenzyme, and a second prosthetic group or coenzyme. Some of the coenzymes are members of the vitamin B complex, others are organic-metallic compounds. The enzymes and coenzymes may attach or separate reversibly.

Vitamins of the B group that may serve as coenzymes include nicotinamide, riboflavin, pantothenic acid, folic acid, biotin, thiamine, pyridoxine, and cobalamin.

Enzymes apparently operate by means of a patch of atoms or site on the surface of the complex enzyme molecule that attaches to a corresponding site or atomic group on the substrate molecule in a sort of "lock and key" fashion. The enzyme serves as the "key" that instantaneously unlocks the substrate "lock," then moves to the next molecule. Thus molecules of substrate must not only have the same chemical composition but also the same steric or three-dimensional molecular arrangement if they are to be modified by enzymes.

The so-called transport chain proves an interesting example of the protein-coenzyme molecule in action. In this process, oxidation of a substrate molecule in a sequence is accomplished by the enzyme's removing hydrogen (primary dehydrogenation) from the molecule and attaching the hydrogen to an active site on its own molecule. The substrate molecule is thus oxidized and the enzyme reduced. Examples of enzymes of this type are the yellow enzymes or flavoproteins. These are made up of a protein plus the vitamin riboflavin in the form of flavin mononucleotide (FMN) or flavin adenonucleotide (FAD).

A second type of respiratory transfer enzyme is the cytochrome group, while a third is the nonprotein low molecular weight group, the quinones. As an example:

$$\text{Flavoprotein} + \text{succinic acid} \longrightarrow \text{fumaric acid } (-2H) + \text{reduced flavoprotein } (+2H)$$

At this point the flavoprotein molecule is rendered inactive by the (2H) molecules on the receptor sites. These may be removed by shuttling them to a molecule of atmospheric oxygen or to a cytochrome or quinone molecule. This whole operation requires only microseconds. In the case of atmospheric oxygen the process is as follows:

$$\text{Reduced flavoprotein } (+2H) + O_2 \longrightarrow \text{oxidized flavoprotein } (-2H) + H_2O_2 \text{ (hydrogen peroxide)}$$

Hydrogen peroxide is extremely toxic to cells if appreciable amounts accumulate. Aerobic bacteria are protected against hydrogen peroxide by the enzyme, catalase. This catalyzes the reaction:

$$H_2O_2 + \text{catalase} \longrightarrow H_2O + 1/2\ O_2 \text{ (nascent oxygen)}$$

In rapidly growing cells, the action of catalase goes on continuously at an unbelievably rapid rate.

BACTERIA AS TOOLS FOR THE STUDY OF METABOLISM

Of all living things, bacteria are perhaps best suited for the study of the basic metabolic processes of life. Each cell is relatively simple in its external and internal morphology. Each is a self-sufficient entity obsessed, as it were, with a single great driving force, the reproduction of more cells like itself. Because of the small size of the cells, a bacterial population represents a tremendous total surface area. This is significant because the basic life processes are primarily initiated as surface reactions.

The rapidity of reproduction is also of major importance. Consider that a single cell can in 24 hours undergo enough reproductions and enough cell generations to represent more individuals than the entire population of the earth.

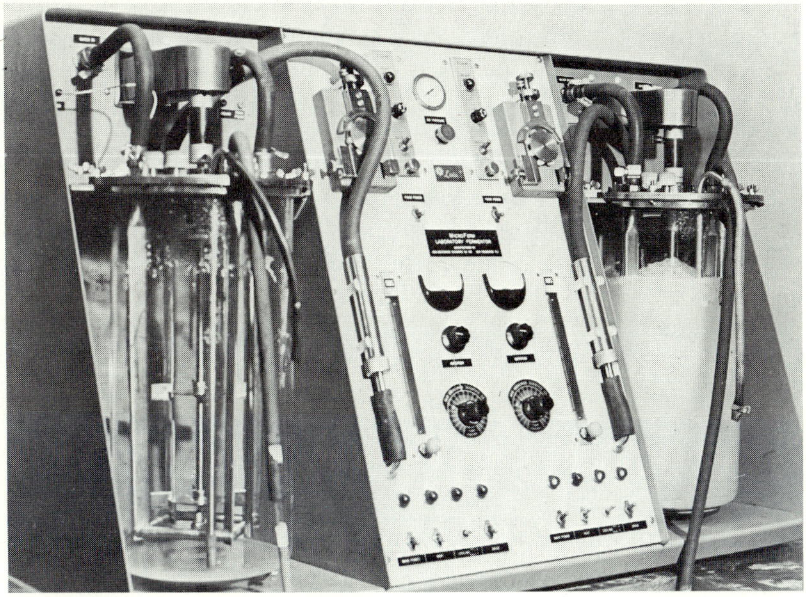

A B

Figure 15-1
Laboratory fermenter used to grow large quantities of bacteria for enzymatic studies. Temperature and aeration are automatically controlled.
A, Empty growth chamber, showing impeller blades for active agitation;
B, chamber containing actively growing culture of bacteria. After suitable growth period several grams (wet weight) of packed cells may be obtained. This is sufficient for most enzymatic studies.

Coupled with these factors is the ease with which bacteria can be grown and their enzymatic versatility.

Methods of estimating growth

In metabolic studies of the bacteria, their responses to such things as temperatures, nutrients, and species differences are often measured in terms of comparative rates of growth. This involves estimates of numbers of individual viable cells or of total numbers of cells usually determined at hourly intervals over a 24-hour period. (A number of ingenious devices have been developed to permit the growth of large enough quantities of bacteria to permit extensive enzyme studies [Fig. 15-1].) A variety of methods may be used, some of which are described below:

Plate counts involve an estimate of the number of viable bacteria by preparing serial dilutions of the bacteria in sterile saline or distilled water. A volume of 1.0 ml of each dilution is then transferred to one or more sterile petri plates. Over the dilution is poured a suitable sterile agar nutrient medium that has been melted and cooled to 45° C. This is well mixed, the agar allowed to solidify, the plates incubated for 24 to 72 hours depending on the species, and the number of colonies counted. This number multiplied by the dilution factor is recorded as the number of bacteria per milliliter. This value is not completely accurate. Although a colony generally may develop from a single cell, occasionally it may represent the offspring of two or three or a small clump of cells. Thus plate counts usually represent somewhat fewer than the actual number of viable cells. However, the plate count provides perhaps the best available method of estimating the number of viable cells in a suspension.

Another method of estimating numbers of viable cells is the most probable number method (see Chapter 17). This is used for bacteria that grow only in liquid cultures. Here again values are only an estimate but a statistically defensible one.

An interesting new proposal for the estimation of bacterial numbers of viable bacteria is based on a determination of the amount of adenosine triphosphate (ATP) produced by actively growing cells. By means of a combined chemical and partition chromatographic method ATP is isolated. This is then measured on the basis of the amount of bioluminescence produced in a firefly luciferin-luciferase system using an electronic luminescence biometer. A direct correlation has been found between the viable cell counts and bioluminescence measurements. The method has also been proposed as a means of estimating the viable bacterial populations in various materials such as dairy products, water, blood, and urine.

Counts of total numbers of individual bacterial cells can be made with the Petroff-Hausser counting chamber. This is a modification of the hemocytometer used for counting red and white blood cells. Both living and dead cells are counted. Counts also can be made electrometrically using a modification of the Coulter blood cell counter. This counts electronically the number of cells in a given volume of fluid passing through a gate. Not only are the cells counted but a computer-like device can measure their size, and record and print out the proportion of cells of different sizes.

Another widely used method involves the spectrophotometer. This estimates the relative cell concentration on the basis of the increase in density, turbidity, or cloudiness of a liquid culture caused by the increase in numbers of bacterial cells. This method does not separate living from dead cells, and results may be influenced by capsule production or variations in the size of individual cells during a period of growth. However, the method is widely used and because of its simplicity replaces the more time consuming plate count if a rough estimate of cell reproduction is sufficient.

The dry weight of samples of cells taken at intervals during an experimental growth period also may be used for purposes of comparing growth responses.

The bacterial growth curve

In order to visualize the differences in growth resulting from variations in incubation temperatures, nutrients, and so forth, arithmetic graphs have been prepared by plotting numbers of cells against time. However, the line rises so rapidly and to such a height that it is often difficult to plot, compare, or interpret such results.

Growth of bacteria is a geometric rather than an arithmetic progression. Therefore, a comparison of results can best be visualized by plotting the logarithm of the number of bacteria against time. The idea of plotting logarithms developed early in the 1900's. Buchanan in 1918 visualized the growth of bacteria in terms of

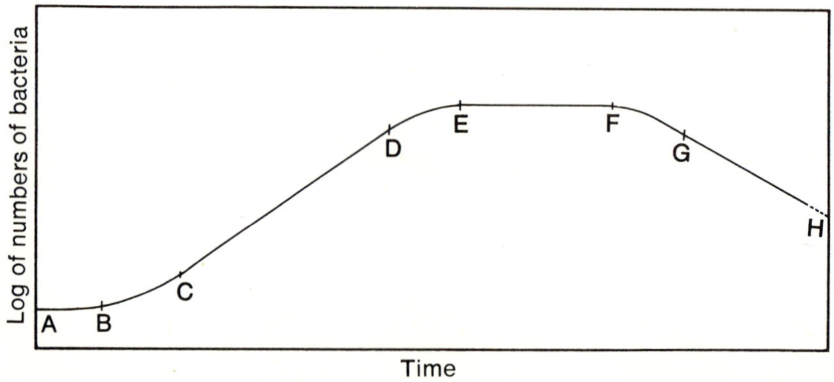

Figure 15-2
An idealized growth curve.

idealized growth curve with eight phases of growth (Fig. 15-2).

The stages as shown on the growth curve may be interpreted as follows:

1. *Latent or initial stationary phase* (Fig. 15-2, A-B). For a short period after the bacteria are inoculated into the new growth medium, they are busily engaged in adjusting to the new environment, and synthesis of the necessary enzymes is underway. However, no appreciable increase in the number of cells occurs.
2. *Lag phase* (Fig. 15-2, B-C). Some of the cells are beginning to reproduce but cell reproduction still lags.
3. *Logarithmic phase* (Fig. 15-2, C-D). All of the cells have completed the adjustment, are in their physiological youth and reproducing at their maximum rate. When the logarithms of counts made at intervals are plotted against time and the points on the graph are joined, a straight line appears. This is a significant value because from it the growth rate or coefficient of the rate of growth and the generation time values can be calculated. These provide a visual comparison of the effect of different factors on the rate of growth.

   With many bacteria, the log phase, as it is called, will last for 6 to 12 hours, but with some bacteria or media it may be 24 hours or longer.
4. *Phase of negative acceleration* (Fig. 15-2, D-E). At this point a portion of the cells begins to reproduce more slowly as nutrients are used or waste products begin to accumulate.
5. *Maximum stationary phase* (Fig. 15-2, E-F). At this stage the growth curve is a straight horizontal line indicating a maximum number of viable cells. The number of cells reproducing are now equal to the number dying off. This phase may persist for a few or many hours depending on the organism and medium used.
6. *The accelerated death phase* (Fig. 15-2, F-G). Begins when the number of cells dying exceeds the number reproducing. This goes on at an increasing rate until it reaches a steady rate.
7. *Logarithmic death phase* (Fig. 15-2, G-H). Represents a steady rate of death in which cells die at a maximum but regular rate. This is shown by a straight line. The slope may not be as steep as the log phase. Theoretically this steady rate should continue until all of the cells have died.

   In actual practice, the point is reached in which a small number of highly resistant bacteria may remain viable for an extended period before the culture is devoid of living cells. This represents the final phase and varies greatly with bacteria and media. (The death phases are not applicable to spore-forming bacteria because the spores may retain viability for an indefinite and prolonged time.)

Plotting of the logarithms of the number of cells enables the investigator to calculate the following several useful values:

1. Generation time can be calculated by the following method:

   $$g = \frac{t \log 2}{\log b - \log B}$$

   g, Generation time (average time for one cell to become two or the initial number of cells to double; this may be expressed in minutes or fractions of hours). B, Number of bacteria at beginning of the log phase or some other predetermined point in logarithmic growth. b, Number of bacteria, usually at the end of the log phase. t, Time (hours or minutes between points B and b).
2. Number of generations may be calculated as follows:

   $$n = \frac{t}{g}$$

   Thus if the elapsed time was 6 hours (360 minutes) and the generation time was 20 minutes, the n (number of generations) would be:

   $$\frac{360}{20} = 18 \text{ generations}$$

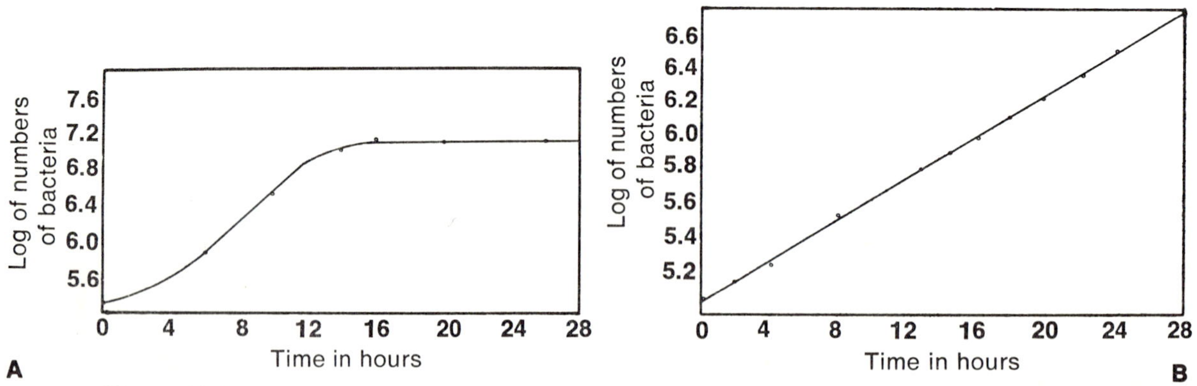

Figure 15-3
A, Growth curve of *Rhizobium meliloti* in mannitol nitrate medium, generation time 2.62 hours. **B,** Growth curve of *R. japonicum* in same medium. Generation time 6.28 hours. Note that organism remains in logarithmic phase throughout the 26-hour period.

3. The rate of increase per cell, coefficient of the rate of growth or K is calculated as below:

$$K = \frac{2.3}{t} \log \frac{b}{B}$$

This value is often used to compare responses of different strains or species of bacteria to growth temperatures, nutrients, and so forth.

Examples of the application of growth rate comparisons made by the author for *Rhizobium meliloti*, the alfalfa root nodule bacterium, and *R. japonicum*, the "slow growing" soybean root nodule bacterium, are shown in Fig. 15-3.

Anatomy of the bacterial cell as related to metabolism

The individual bacterial cell represents the ultimate agent studied by the bacterial physiologist. The structure of the cell bears a direct relation to the physiological changes involved. Therefore, a brief consideration of the cell should be enlightening.

Members of the Schizomycetes, the great group of the simple and higher bacteria, vary in relation to the presence or absence of cell walls as mentioned previously. Of the 10 orders, 7 possess rigid cell walls, 3 do not. However, only a small number of bacterial species have been used as experimental agents in basic studies of bacterial metabolism. Of these, the lion's share of experiments have involved the use of *Escherichia coli*, while a considerable number of investigators have used *Bacillus subtilis*, *B. megaterium*, or related spore-forming rods. Some work has been done with *Pseudomonas* species or the various cocci. In the small remaining number of contributions, a varied group of bacteria has been employed. Very little experimentation has involved the use of the Schizomycetes without distinct cell walls.

The primary function of the cell wall is to protect the cell membrane and maintain the distinctive shape of various bacteria, particularly the rod-shaped forms. The walls perform other functions such as holding back certain large organic molecules, or allowing only a small proportion to enter. In addition, they provide antigenic components useful for identifying various bacteria, especially members of the Enterobacteriaceae. However, they are not completely indispensable, since on suitable media bacteria from which the cell walls are removed are able to function in a limited fashion.

With gram-positive bacteria, the cell walls are thick (15 to 80 nm), dense, and make up 10% to 25% of the dry weight. They represent an alternating network of N-acetylmuramic acid (L. *murus* wall) and N-acetylglucose amine. In addition, a basic layer of glyceropeptide with alternating linkages in a polysaccharide is also generally present.

The cell wall of the gram-negative bacteria is a thin, more complex layer of lipopolysaccharides plus an outer layer that protects the cell from the action of lysozyme. This must be removed before lysozyme can directly attack the cell wall.

To study the internal physiological activities of the bacterial cell, the internal complex of organelles and enzymes must be released and isolated in the cell-free state.

The initial method was to grind the bacteria with a mortar and pestle with sharp sand particles or fine, crushed glass. This ruptured most of the cells but often disrupted many of the enzymes and their organelles.

Cells also can be ruptured using the so-called French press in which frozen bacteria are placed in a steel cylinder that has a very small aperture; a tight-fitting cylinder is inserted, and the bacteria are forced through the aperture with great pressure or by sharp driving blows on the top of the inner cylinder.

Supersonic devices, in which ultrasonic vibrations set up violent cavitations in a bacterial suspension literally tearing apart many of the cells, also may be used. All of these methods may disrupt the cell contents.

Objections to these violent methods of disruption have been met by the relatively gentle method of selectively removing cell walls by means of the enzyme, lysozyme, to release enzymes and organelles in an essentially undisturbed state. Lysozyme is a unique enzyme widely distributed in nature, which may perform an important function in protecting us against many infectious bacteria.

In the early 1920's, Alexander Fleming, who later discovered penicillin, found in nasal washings, tears, and certain animal tissues, an enzyme that quickly dissolved the cell wall of the gram-positive coccus, *Micrococcus lysodeikticus*. Other gram positives were also found to be readily lysed by this enzyme, which he called lysozyme (literally, a lysing enzyme). These bacterial cell walls apparently serve as a natural substrate for lysozyme. The type of cross linkage in the sugar-peptide substance of the cell wall determines the degree of susceptibility to lysis.

Most gram-negative bacteria tend to resist the action of lysozyme. However, gentle treatment with certain detergents changes the molecular structure of the cell-wall substance enough to permit lysis to occur. Lysozyme also has been crystallized as a pure enzyme from egg white. This preparation is frequently used in the laboratory for lytic disruption of bacterial cells.

After gram-positive bacteria are treated with lysozyme, the cell wall is completely removed, leaving an osmotically fragile structure. With the rigidity of the cell wall gone, a rod assumes a spherical shape. It can easily be ruptured by the addition of distilled water or by rough handling. In this state it is known as a protoplast. If placed on or in a hypertonic medium, such as nutrient broth or agar with 20% sucrose, the protoplasts remain intact, respiration and protein synthesis continue, protoplast weight increases, and perhaps division occurs. However, no cell processes continue for more than a limited time.

With gram-negative bacteria, cells must be treated with a mild alkali or detergent before lysozyme can act. Even so, removal of the cell wall is not complete. The resultant bodies with fragments of cell wall attached are known as spheroplasts and are fragile enough to be gently ruptured for the study of the cell components.

After protoplasts or spheroplasts are ruptured, large fragments of ruptured membrane still remain. These can be washed and concentrated by centrifugation for electron microscope examination or subsequent biochemical studies.

Cell membrane as a regulatory mechanism

The cell membrane is a remarkable structure, a thin osmotic barrier that determines what enters or leaves the cell.

Our knowledge of these membranes has come largely from rate of uptake by protoplasted cells of different sugars and other compounds measured by means of radioactive labelled molecules of these compounds.

Cell membranes apparently selectively allow some molecules to enter but turn others away. Molecularly speaking, the cell membranes must be busy areas. This movement of molecules through the membranes has been called active transport. Because this movement resembles enzyme action in many respects, these receptors or regulatory agents have been termed permeases. It is known that most active transport requires energy from ATP to operate.

The membranes are poorly permeable to highly hydrophilic substances, for example sugars, organic acids, and amino acids, except as they are selectively passed through. A number of permeases for various molecules or molecular varieties have been identified for *Escherichia coli*.

Use of radioactively labelled sugars and measurement of the amount of labelled sugar inside washed whole cells after varying periods of exposure affords a method of estimating the rate of passage.

In the case of amino acids, a single transport system appears to be able to bring in several related amino acids. For example, leucine, isoleucine, and valine seem to be transported through the membrane by the same permease. In some cases, auxotrophs deficient for one amino acid (unable to synthesize it) will suffer growth inhibition if supplied the desired amino acid plus another closely related type.

In addition to bringing molecules in, the permeases must also transport molecules out. These include waste products such as lactic acid, and antibiotics, and also exoenzymes (extracellular enzymes) that dissolve complex molecules in the surrounding medium. Examples are enzymes that digest or hydrolyze starch, gelatin, casein, and so forth (Fig. 15-4). Some of these enzymes seem to be attached to the membrane lipids even when extended outside the cell. Those still attached have been called ectoenzymes and those released outside the cell, exoenzymes. The function of these enzymes may be digestive in nature, that is, hydrolyzing the complex protein molecules into component amino acids, and the starch molecules into maltose or glucose. These small molecules may then be pushed or pulled by the permeases through the cell membrane. Other exoenzymes or ectoenzymes apparently function outside the cell, synthesizing capsules, slime, or similar materials from components in the medium.

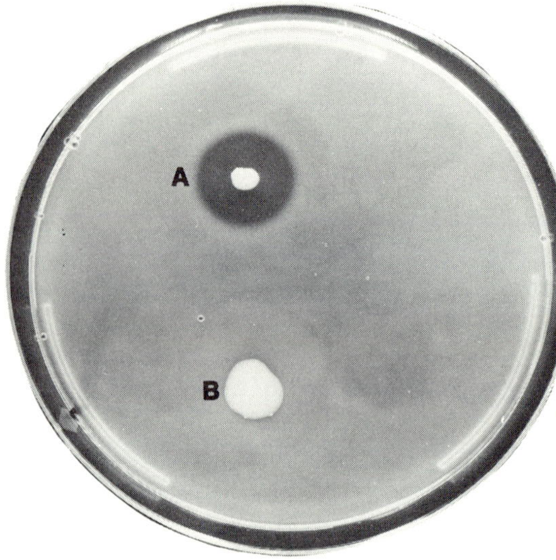

Figure 15-4
Growth of two strains of *Pseudomonas* sp. on milk agar. Strain **A** proteolyses the casein while **B** lacks the proteolytic enzymes.

Internal organelles

The internal organs, or organelles, of bacteria have been difficult to delineate because often only thin sections of the bacteria can be studied and three-dimensional shapes are hard to visualize. However, the cell membrane often folds in to form structures known as mesosomes. These frequently appear as structures separate from the cell membrane itself. They can sometimes be visualized in negatively-stained whole cells. Such stains reveal them to be associated with the cell membrane and to be "much folded." The mesosome function is still not completely understood. The location is often near the septum or dividing point suggesting a relation to cell division (Fig. 15-5). A possible function in protein synthesis has also been considered. On the other hand, in some bacteria it seems to be associated with the cell nucleoid or chromosome. Some bacteria, especially the photosynthetic bacteria, show many of these cell membrane infoldings; others relative few.

NATURE AND KINDS OF ENZYME ACTIONS

The basic actions performed by enzymes are to provide the materials and energy required to synthesize the bacterial cell, and to synthesize the new cell itself.

The names of enzymes end in -ase and the word itself may describe either the substrate acted upon, the mode of action, or the product. Some common enzymes, such as pepsin, may carry trite names with little or no descriptive value. As a result, different investigators in different countries have, until the last few years, applied different names to a given enzyme. For this reason earlier papers dealing with enzymology were sometimes difficult to fully comprehend.

Because of this confusion, in 1964 a commission appointed by the International Union of Biochemistry proposed a classification made up of six major divisions of the enzymes based on the total reaction that they produce. The various enzymes in turn were given an identifying code number that identified the nature of the action, specification, and so on, followed by a short descriptive trivial name such as "fructose diphosphate aldolase." These major division names, together with some representative types of action, are listed below.

Oxidoreductases. Enzymes of the oxidoreductase group catalyze oxidation-reduction actions. This may be accomplished by oxidation through the removal of hydrogen by the dehydrogenases, or direct action of atmospheric oxygen as a hydrogen acceptor. The primary example is cytochrome oxidase. The basic reaction involves:

$$4\ \text{protein-}Fe^{++} + O_2 \rightleftarrows 4\ \text{protein-}Fe^{+++} + H_2O$$

The cytochrome oxidase is oxidized in the process; oxygen is reduced to H_2O. The enzyme may be quickly reduced by accepting hydrogens and the operation repeated.

Transferases. Transferase action involves the transfer of different molecular groups not free during the transfer process. These enzymes catalyze the transfer of a given group from donor to acceptor. As an example, a methane (CH_3) group may be transferred from *S*-adenosyl methionine to guanidoacetic acid, transforming it into creatine.

Hydrolases. Hydrolases are enzymes that catalyze the addition of a chemical bond in a complex molecule by the introduction of a water molecule causing it to cleave into two or more simpler molecules. Examples are the amylases that cleave the starch molecule into maltose and glucose, the various proteinases that digest gelatin, casein, and so forth, into peptides and amino acids, or the ureases that hydrolyze the urea molecule into carbon dioxide and ammonia.

Lyases. Lyases are similar to the transferases except that the transferred groups are free during the process. Examples are the taking-up or giving-off of CO_2, NH_3, H_2O, or HCHO in the free state.

Carbonic anhydrase, a lyase, splits carbonic acid (H_2CO_3) into $H_2O + CO_2$ or reverses the operation. Decarboxylases for pyruvic acid ($CH_3COCOOH$) can split off CO_2 and leave acetaldehyde (CH_3CHO).

Isomerases. Isomerases actuate a rearrangement of atoms or groups within the molecule. The number of atoms of each type remains the same, but the resultant molecule will have different properties than the first.

For example, through the action of DPNH the molecular structure of diacetoacetic acid is rearranged to $\alpha\beta$dihydroxyisovaleric acid.

Ligases. Ligases (literally enzymes that tie) link two molecules together by synthesizing new C-C or C-N bonds. They are active in pro-

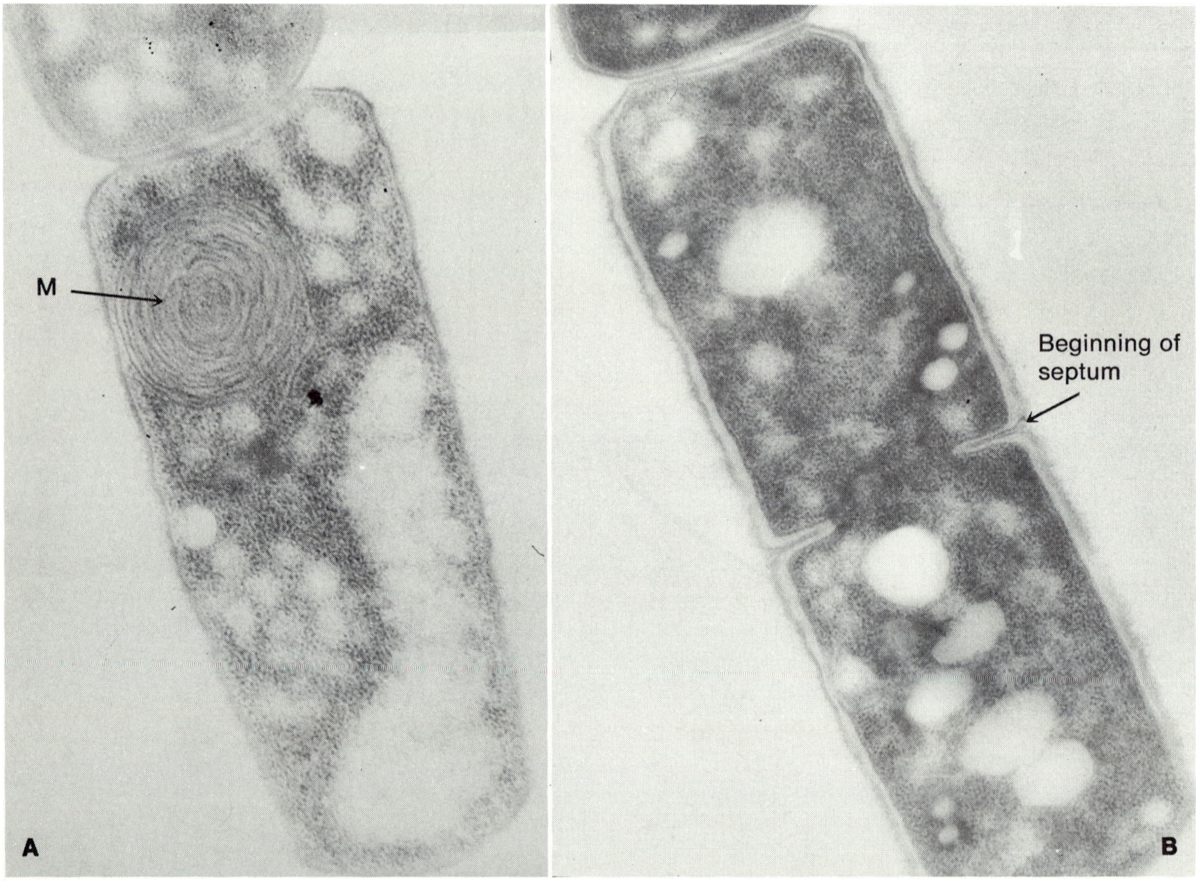

Figure 15-5
A, Thin section of *Bacillus megaterium* showing the organelle, mesosome, M, or middle body. The position suggests that it may be associated with the formation of the cross wall or septum. (×135,000.) **B,** Illustrating first step in cell division, an electron

tein formation, linking amino acids together to form various types of proteins including the enzymes themselves. These are amino-acid–activating enzymes of which 20 are known (one for each protein amino acid). Polysaccharides that constitute the bacterial capsules are synthesized from glucose or other sugar molecules through the action of ligases.

Enzyme action in living cells represents a beautifully integrated and interrelated series of catalytic actions. The processes of supplying energy, building materials, and various special molecules are tied directly into the process of synthesizing the cell components. The speed and effectiveness of the whole series of interactions is amazing. Its operation involves the mediation of probably hundreds of different enzymes. Some systems seem to be identical in many kinds of living forms, while others seem to be almost species specific.

Species of bacteria can be differentiated on the basis of the enzyme systems that provide

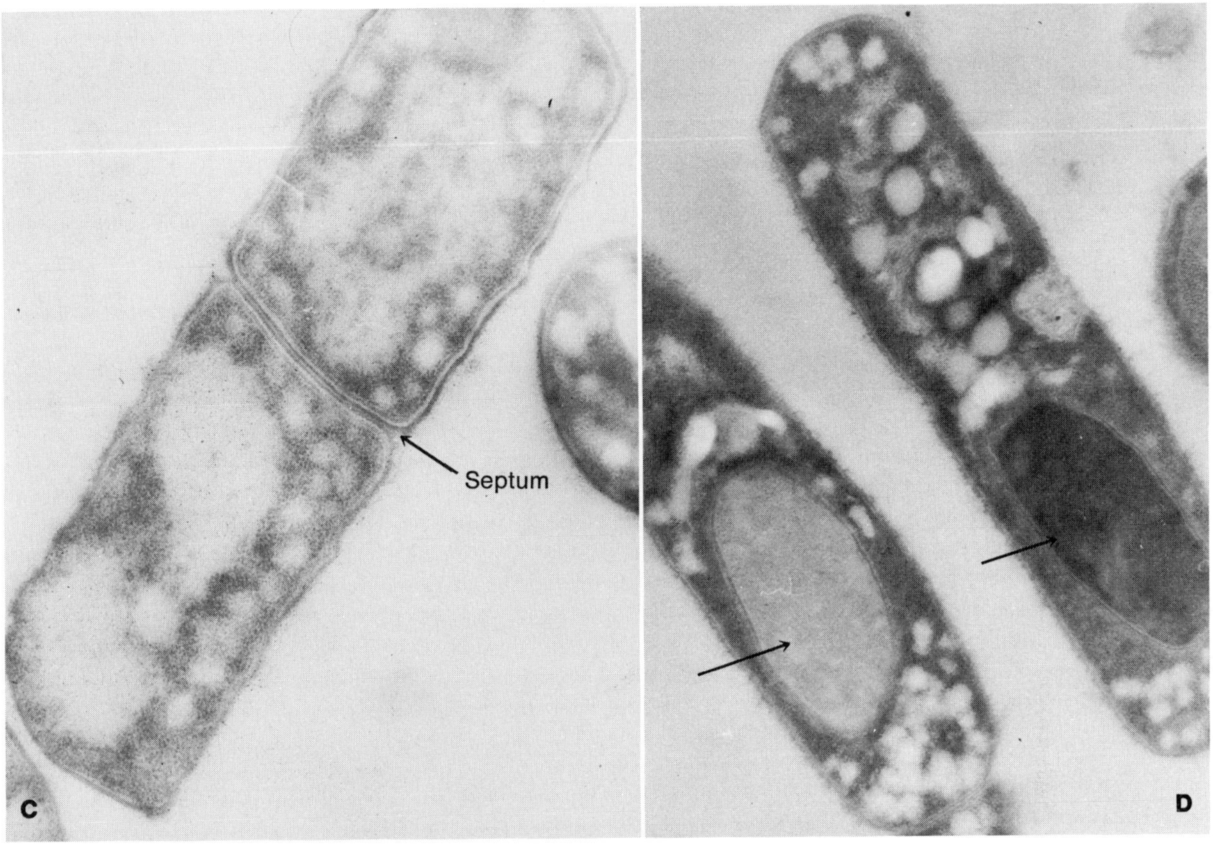

micrograph of thin section of *Bacillus megaterium* shows beginning of septum formation. (×135,000.) **C,** Septum completely formed and cells ready to divide. (×135,000.) **D,** Thin section of *B. megaterium* showing two stages in spore formation. (Courtesy Dr. J. C. Lara.)

them with the pattern of distinctive enzyme responses with which we identify them in the laboratory. For example, *Escherichia coli* is separated from the *Salmonella* species on the basis of its ability to ferment lactose with the production of acid and gas.

METABOLIC PATHWAYS

In elementary biology textbooks, enzymatic reactions are often written as though they were simple, one-step processes. For example it is made to appear that:

Under anaerobic conditions, glucose + yeasts → ethyl alcohol + carbon dioxide, or under aerobic conditions, glucose + yeasts → carbon dioxide + water. This, however, is not the case. Either process proceeds through a series of steps. Some may require energy, others yield energy. Each step is characterized by a specific enzyme and an intermediate product.

The terms metabolic pathway or enzymatic

scheme are used to describe the series of steps necessary to complete a given process.

The EMP and similar pathways

Vitally important among the many known pathways is the glycolytic pathway that results in the conversion of glucose to pyruvic acid. This may be considered an anaerobic fermentation in that it occurs in the absence of oxygen and yields only partially oxidized end products.

Three different pathways or schemes have been identified. These are briefly described in the following paragraphs.

Using *Saccharomyces cerevisiae* as the experimental organism, Embden, Meyerhof, and Parnas discovered the first pathway. It is now identified by their initials as the EMP pathway. It has since been found in fermentative bacteria and many other living forms. Counting all of the changes in molecular configuration and intermediate compounds involved, a total of ten enzymes are required to mediate the change from glucose to pyruvic acid. This amounts to splitting glucose, a 6-carbon polyhydroxy alcohol, into two 3-carbon highly oxidized pyruvic acid molecules.

In the initial steps, energy must be supplied at two points by introducing two high-energy, adenosine triphosphate (ATP) molecules that release a high energy phosphate and are reduced to adenosine diphosphate (ADP). This is an important method of supplying energy for biological processes. With this initial supply of energy, the remainder of the enzymatic process continues to operate. In the end it finally yields four ATP molecules. Counting the two introduced initially, the whole process yields a net energy gain of two ATP molecules. A major value of the pathway is to provide energy for life processes without the complete breakdown of the glucose molecule.

The two pyruvic acid molecules that remain may be used fermentatively by the homofermentative lactic acid bacteria to produce D or L lactic acid by a simple expedient of the enzymatic addition of two hydrogens. Or, it may be converted into ethanol and carbon dioxide by the yeast cell enzymes. The enteric bacteria (*Escherichia*, *Salmonella*, and *Enterobacter*) form mixtures of acids from pyruvic acid through various steps.

A simplified picture of the steps involved in the EMP pathway follows (keep in mind the fact that each step is mediated by a special enzyme):

1. Activation of hexose
   - a. *Glucose* converted to *glucose-6-phosphate* (1 high-energy phosphate group added from ATP, forming ADP)
   - b. *Glucose-6-phosphate* changed to *fructose-6-phosphate* (intramolecular rearrangement)
   - c. *Fructose-6-phosphate* converted to *fructose-1,6-diphosphate* (another high-energy phosphate added from ATP)
2. Oxidation and breakdown of 6-carbon sugar to two 3-carbon molecules
   - d. *Fructose-1,6-diphosphate* split into two *glyceraldehyde-phosphate* molecules
   - e. Two *glyceraldehyde-phosphates* oxidized to two *1,3-diphosphoglyceric acids* (inorganic phosphate added to each, pair of hydrogen atoms removed from each, creating unstable molecules)
   - f. Two *1,3-diphosphoglyceric acids* converted to two *3-phosphoglyceric acids* (2 high-energy phosphate groups transferred to 2 ADP's, forming 2 ATP's)
   - g. Two *3-phosphoglyceric acids* changed to two *2-phosphoglyceric acids* (intramolecular rearrangement)
   - h. Two *2-phosphoglyceric acids* converted to two *phosphoenol pyruvic acids* (water molecule removed from each, creating unstable molecules)
   - i. Two *phosphoenol pyruvic acids* changed to two *pyruvic acids* (2 high-energy phosphate groups transferred to 2 ADP's, forming 2 ATP's)

Overall reaction:

Glucose (C_6) → 2 pyruvic acid ($2C_3$)
ATP molecules added – 2

ATP molecules produced—4

Net energy gain—2ATP

A second pathway, the Entner-Doudoroff (ED) scheme, found only in bacteria, differs somewhat from the EMP pathway and yields only one ATP molecule.

The third, or hexose monophosphate (HMP), pathway differs from the other two. It is widely distributed among living forms. Like the Entner-Doudoroff pathway, in terms of energy it yields only one ATP.

The Krebs cycle

Of major importance, particularly to aerobic forms, is another pathway, the tricarboxylic acid or Krebs cycle (Krebs and Johnson, 1932), so called because it is cyclic in operation.

This is often meshed into EMP and similar cycles and provides a means of securing the energy still held in the released pyruvic acid. It also provides an indirect means of securing energy from carbohydrates and sugars, and from fats after their hydrolysis by lipases to fatty acids and glycerol. Proteins can become energy sources after hydrolysis to amino acids and deamination of the amino acids. With carbohydrates, sugars, fats, and proteins, the conversion of each compound to a form permitting its introduction into the Krebs cycle requires a series of enzymatic actions. Most of these represent supplementary metabolic pathways, each with a special function. Each step uses or yields energy or building materials, and all contribute to the primary function of the bacterium—synthesizing more cells.

The Krebs cycle is the primary energy-yielding method of aerobic organisms, while the glycolytic cycle is the primary pathway for the anaerobes, although facultative anaerobes take advantage of both. In the interoperation of the Krebs cycle and the EMP pathway, the pyruvic acid molecule, still rich in energy, must first be reduced to a 2-carbon molecule. This is accomplished by the removal of a CO_2 molecule through the intermediation of acetyl-coenzyme A.

The 2-carbon acetate thus formed is now drawn into the cycle by union with the 4-carbon oxaloacetate. This forms the 6-carbon citric acid with its three carboxyl groups. The molecule now passes through a series of eight additional steps during which 2 carbons are converted to CO_2 by union with O_2, yielding their maximum amount of energy. Thus all three of the carbons of the pyruvate will have been converted to CO_2. The amount of energy produced is surprisingly high. Whereas the anaerobes had to be satisfied with 2 ATP molecules from a molecule of glucose, the Krebs cycle itself yields 15 for each pyruvate molecule oxidized in each turn of the cycle, or a total of 30 for the original glucose molecule. In addition, reoxidation of the two NADH molecules built up in the EMP cycle yields 6 more ATP molecules. Thus the aerobe secures a total of 38 ATP units of energy from a glucose molecule while the anaerobe is limited to two.

The entire system of interacting metabolic pathways is amazingly intricate, involving as it does an array of compounds that must be oxidized to secure the energy they contain, while others are split into simpler molecules to provide useful building blocks to be built into complete cells. This implicates a complex of carbohydrates, proteins, fats, DNA, RNA, and a host of other large molecules together with their component parts. The picture the bacteria has yielded to the bacterial physiologist has given us a remarkably new insight into this almost incomprehensible problem.

The biosynthetic enzyme complex

Although the discussion thus far has emphasized the two major energy-yielding processes, it should not be implied that these are the primary enzyme operations in the metabolism of the cell. In fact, the biosynthetic or anabolic process might well be considered the most important of all of the enzymatic activities because it seems to be the ultimate objective of each bacterial cell.

However, anabolic and catabolic activities cannot be separated except in the mind of man as he tries to visualize the amazing phenomenon of cell reproduction. Synthetic action would be impossible without energy and building material. On the other hand these are of only transitory value to the cell without biosynthesis.

The whole combined operation of anabolism and catabolism involves a myriad of interrelated cycles, of cycles within cycles, all intermeshed into a smoothly operating sequence, at least during the logarithmic phase of growth.

Some appreciation of the intricacies of the enzymatic machinery of the bacterial cell can be gained from a unique, well-studied property of *Escherichia coli.* This bacterium possesses the ability to grow in a mineral salts medium with glucose as the sole carbon source. The medium furnishes nitrogen in the form of the ammonium ion, phosphorus as phosphates, sulfur as sulfates, plus small amounts of magnesium and iron. In addition, some formulations include traces of rare inorganic compounds.

The raw materials available must be combined into the complex macromolecules of the cytoplasm, enzyme proteins, special proteins for the flagella, the cell membranes and organelles, the cell wall, the DNA, RNA, and the host of other vital molecules, large and small, required for the synthesis of new bacterial cells.

The truly remarkable feat is that in such a minimal medium, *E. coli* is able to synthesize enough cells to double its numbers once or twice each hour for 12 hours or longer, then continue the operation at a diminished rate for many more hours. If the spent medium is continually replaced with fresh medium the high rate of growth can be maintained for longer periods.

REGULATION OF ENZYME ACTION

One of the marvels of nature is the balanced, steady fashion in which a normal, wild-type bacterial cell reproduces itself. This is doubly remarkable when you appreciate the great complexity of the enzyme operations with their multiplicity of major and minor metabolic pathways. Each is directly or indirectly interrelated and the proper operation of one contributes to the success of the next. Coupled with this is the fact that two opposite types of reactions are going on in this metabolic factory—the degrading of organic compounds to supply energy or smaller molecules as needed, and the reverse type of action, biosynthesis, which is essentially the stitching together of the small molecules to form the huge proteins, fat, carbohydrate, DNA, RNA, and other molecules involved in building up a new cell. Many of the biosynthetic operations appear to be reverse operations of the degradative process. However, nature has avoided a monumental, physiological traffic jam by providing a separate set of enzyme systems for each.

In addition, a precise system of enzymatic controls must be constantly in operation if growth is to go on in a smooth, balanced, steady fashion. This is complicated by the dynamic nature of the environment in which the bacterium is growing.

In a fresh medium with ample nutrients and no waste products, the amount and kind of enzymes needed will be different than at a later stage in the growth cycle when nutrients begin to be depleted, waste products accumulate, and perhaps enzymes wear out.

In addition to this, enzymes are of two types, constitutive and induced or inducible. Constitutive enzymes are always present while inducible enzymes are produced only when the inducer, or product to be attacked, is present in the medium (environment). With the induced enzyme a special type of regulatory action is required to turn on the production of the desired enzyme. We now recognize that the genes needed to produce an inducible enzyme are present in the DNA of the cell. However, a repressor gene blocks the operation of this set of genes until the presence of the inducer overrides the repressor action.

The best studied example of induction is the enzyme system responsible for lactose fermentation by *Escherichia coli.* If a strain of *E. coli* is placed in glucose broth, the glucose is immediately moved into the glycolytic (EMP) pathway and transformed into pyruvic acid, which in turn is subsequently changed to acid and gas.

When the same *E. coli* strain is inoculated into lactose broth, no initial reaction occurs. Enzymes for lactose fermentation are absent because the lactose-utilizing enzyme system is inducible. Therefore, the first step is that the lactose, as an inducer, overcomes the action of the repressor and the production of lactose-utilizing enzymes begins. The message must be transcribed on the messenger RNA and transported to the ribosomes, which begin to make the special proteins that constitute the lactose-utilizing system involving several enzymes.

The cell membrane prevents lactose in the medium from passing through until a galactoside permease is synthesized. This pushes or pulls the lactose molecule inside the cell. Once inside, the lactose molecule is still of no use to the cell until the freshly synthesized β-galactosidase hydrolyzes the disaccharide into a molecule of glucose and one of galactose. The glucose is immediately caught up in the EMP pathway and converted to pyruvic acid. The galactose has the wrong molecular configuration and it must, therefore, be converted into glucose. Three enzymes, galactokinase, transferase, and epimerase, rearrange the molecule to glucose-l-phosphate. It is then fed into the EMP pathway and converted into pyruvic acid.

All of these operations require time. Supported by the nutrients in the nutrient broth, the cells grow at a fairly rapid rate but two or more cell generations may elapse before active utilization of lactose occurs and growth is accelerated by the added supply of available energy.

A considerable number of inducible enzymes are known but all show the same type of lag period before enzyme activity reaches its maximum. Among the other regulatory actions, accumulation of an excess of end products at the terminus or pool of a pathway may slow down the synthesis of one or more involved enzymes. This again is a slow-reacting process becoming evident only after one or more cell generations. To turn it off completely requires the additional time for the enzymes already present to wear out.

On the other hand, the direct regulation of the action of the enzymes themselves brings an almost immediate slowdown or speedup of the production of enzyme end products.

Enzymes subject to such immediate control are known as allosteric enzymes. These are usually large protein molecules with several duplicator units and with more than one kind of attachment site. These sites include the catalytic site to which the substrate is attached and the allosteric site to which small molecules of a different shape, known as allosteric effectors, are attached.

Allosteric enzyme molecules are both large and flexible. By changes in molecular shape, more or fewer sites may become active as substrate catylizers. The activators also speed up the action. However, if inhibitors "complex" with the enzyme, the molecule may be sufficiently warped that one or all of the substrate sites becomes nonreceptive. Since they can be activated or inhibited by small molecules, unrelated in shape to the substrate molecule, the enzyme action can be immediately increased or decreased.

Thus enzymatic actions may immediately be influenced quantitatively. Allosteric enzymes are usually located strategically at a certain point near the beginning of a pathway. When products begin to pile up at the end of the pathway, the action of the allosteric enzyme may be slowed down, which slows down the operation of the entire pathway. Allosteric enzymes seem to be found in many pathways. By initiating or suppressing enzyme synthesis the cell is able to exercise a sluggish type of control. However, feed-back inhibition operating through allosteric enzymes, may exercise an immediate control of enzyme action. Other mechanisms, not yet discovered, also may function. In any event, effective control mechanisms make the growth of bacteria the smooth balanced operation that it generally is. Only when mutations alter the normal genetic messages, or toxic inhibitory substances are introduced, are the catabolic and anabolic enzyme actions sufficiently disrupted to interfere with the normal pattern of growth.

FAMILIAR EXAMPLES OF MICROBIAL ENZYMATIC ACTIVITY

This brief introduction to microbial metabolism may aid in translating some commonly seen evidences of microbial activity in terms of the major metabolic pathways involved.

Rising of bread, for example, is due to the conversion of the glucose molecule by the yeast, *Saccharomyces cerevisiae,* into carbon dioxide and ethyl alcohol. The dough rises because of entrapment of the great amount of carbon dioxide produced while the ethyl alcohol is driven off by the process of baking. The alcohol is responsible for the pleasant odor of baking bread and the eye-stinging sensation when the oven is opened.

In making bread, flour and a suspension of yeast cells are mixed into dough. However, the complex starch molecules are useless to the yeast because they cannot pass through the membranes of the yeast cell. In the wet dough, naturally occurring amylases (starch-hydrolyzing enzymes) become active and rapidly cleave the starch molecule into its component maltose and glucose molecules.

The glucose and maltose molecules thus released from the starch, with the aid of the proper permeases, are quickly transported through the cell membrane. The glucose enters directly into the EMP pathway, and the fermentative process begins. Maltose, however, must first be hydrolyzed into its two component glucose molecules before it can enter the glycolytic pathway.

Glucose is quickly converted to pyruvic acid. The 3-carbon molecule is then enzymatically split into a molecule of carbon dioxide plus a molecule of acetaldehyde. The formula of the reaction is as follows:

$$\underset{\textbf{Pyruvic acid}}{CH_3COCOOH} \longrightarrow CO_2 + \underset{\textbf{Acetaldehyde}}{CH_3CHO}$$

The relatively unstable acetaldehyde is then enzymatically reduced to ethyl alcohol (ethanol) by the addition of two hydrogens as shown below:

$$\underset{\textbf{Acetaldehyde}}{CH_3CHO} + 2H \longrightarrow \underset{\textbf{Ethyl alcohol}}{CH_3CH_2OH}$$

Although the baker is primarily interested in the carbon dioxide produced, the main concern of the producer of wine or spirits is the ethyl alcohol. The brewer is interested in both fermentative products.

It should be remembered that while the fermentative process is in vigorous operation the whole gamut of biosynthetic pathways are also synthesizing a multitude of new yeast cells.

The homofermentative production of lactic acid is seen in the spontaneous souring of raw milk, and in commercial products resulting from the lactic acid fermentation. In yogurt, acidophilus milk, and many of the ripened cheeses, *Streptococcus lactis* may play a part, but various species of *Lactobacillus* play the major role. The lactic acid fermentation involved in all of these processes is termed homofermentative because there is only one major fermentation product.

The enzymatic pattern in the production of lactic acid is essentially the same with both the streptococci and lactobacilli. Glucose, or galactose that has been converted to glucose-6-phosphate, enters the EMP pathway and emerges as pyruvic acid. Enzymatic hydrogenation adds 2 hydrogens and the pyruvic acid is converted to lactic acid according to the formula below:

$$CH_3COCOOH + 2H \longrightarrow CH_3CHOHCOOH$$

Pyruvic acid **Lactic acid**

Large numbers of cells are needed before any appreciable accumulation of lactic acid occurs. Milk, for example, does not begin to taste sour until the population of lactic acid bacteria has reached 20,000,000 to 40,000,000 bacteria per milliliter.

The activities discussed above represent primarily anaerobic fermentations. On the other hand, the Krebs cycle probably furnishes most of the energy required for the vigorous bacterial growth represented by the formation of a slimy layer on the surface of spoiling meat or the yeast layer that sometimes develops on the surface of spoiled fruit juices.

SUGGESTED READINGS

Doelle, H. W.: Bacterial metabolism, New York, 1969, Academic Press, Inc.

Kluyver, A. J., and van Niel, C. B.: The microbe's contribution to biology, Cambridge, Mass., 1956, Harvard University Press.

Sharon, N.: The bacterial cell wall, Sci. Am. **220**:92-98, May 1969.

Sokatch, J. R.: Bacterial physiology, New York, 1969, Academic Press, Inc.

Stanier, R. Y., Doudoroff, M., and Adelberg, E. A.: The microbial world, ed. 3, Englewood Cliffs, N. J., 1970, Prentice-Hall, Inc.

16 Genetics of microorganisms

As a distinct field of science, microbial genetics is scarcely three decades old, although it has expanded rapidly in the last decade.

The door to genetic knowledge was first pushed slightly ajar by Mendel (1861) with his classic sweet pea experiments. Miescher, without recognizing that his findings might have any relation to heredity, in 1869 isolated from white cells in pus a substance that he called nuclein. We now know that what he had found was impure DNA. Further information came in 1882, when Flemming, using stains on sections of tissue, observed deeply staining bodies that he called chromatin (color). Waldeyer (1888) applied the name chromosomes (color + body) to these deeply stained structures in cells. Altmann (1899) recognized the relation of nuclein to the nucleus and called it nucleic acid.

By this time a number of scientists had begun to suspect a relationship between chromosomes and heredity. However, the door to the science of genetics was not fully opened until Mendel's work was rediscovered, more or less independently, in papers published in 1900-1902 by DeVries, Tschermak, Correns, and Bateson, which provided the solid foundation necessary for genetics to become a science.

However, the plant and animal geneticists were so engrossed in harvesting the rich fields of new knowledge in their own disciplines that they paid no attention to microorganisms as tools for the study of basic genetics. In fact, many of them questioned that the laws of genetics could apply to the microbes.

DEVELOPMENT OF MICROBIAL GENETICS

The first step toward the use of microbes as genetic tools came in the early 1930's when Lindegren, utilizing the discovery by Shear and Dodge (1927) of sexuality in the mold, *Neurospora*, found that he could separate pure haploid strains that could then be genetically crossed. *Neurospora*, the common pink bread mold, was found sometimes to have two types of growth equivalent to male and female strains. When spores or mycelia of the haploid male (A) and haploid female (a) strains were seeded into opposite sides of a sterile petri plate containing a suitable sterile medium, mycelia quickly grew toward the center of the plate. At the point where the colonies came together, a series of tiny black dots developed resembling tiny grains of pepper. These were the result of a brief period of diploidy in which the 7 chromosomes of the male (A) strain intermingled with the 7 chromosomes of the female (a) strain. In the realignment of chromosomes, occasional crossovers occurred that gave rise to new gene combinations. The reduction-division involved in the formation of ascospores sometimes showed an alignment of new haploid spores that were genetically unlike either parent. Microscopic examination at the juncture of the (A) and (a) mycelia revealed structures known as perithecia, which have been described as drawstring purses filled with cigars. The "cigars" are long asci (sacs) housing 2 to 8 tiny ascospores arranged in a neat line. Lindegren, with a forceps, popped open the perithecia and squeezed out the asci onto the moist surface of a petri dish filled with firm water-agar. Then, using a tiny flattened needle (while working under a dissecting microscope), he hoed out the 8 individual ascopores and arranged them in a straight line on the surface of the agar. An agar block was then cut out around each spore and the single haploid spore transferred to a suitable medium. The single spore grew into a pure haploid culture. In the 8-spored ascus of *Neuro-*

spora crassa, 4 of the spore cultures were generally found to resemble one parent, and 4 the other. Some, however, showed combinations of characters not seen in either parent. By isolating a large number of such haploid cultures and crossing them with opposite strains, genetic data began to be accumulated. Initially, the characters observed were easily seen phenotypic ones, such as pink, red, and albino-colored growth. Later it was found that *Neurospora crassa* could be grown on a simple medium containing mineral salts, glucose (as an energy and carbon source), plus a minute quantity of biotin since the mold lacked the ability to synthesize this vitamin. This was called a minimal medium, one on which the naturally occurring wild-type strains would grow. These forms were able to synthesize their own amino acids, B-complex vitamins (except biotin), purine and pyrimidine bases, and so forth.

Occasionally strains were found that were unable to grow on a minimal medium. They would, however, grow well on an enriched medium such as one containing casein digest (amino acid source) and yeast extract (B-vitamin source). This was called a complete medium. Forms that required auxiliary nutrients were termed auxotrophs. Such forms were found only on rare occasions, that is, about once in 15,000 isolations. To increase the frequency of such mutations, cultures were exposed to X-rays. The incidence was now increased a hundredfold or more.

It was also found that if two unlike auxotrophs lacking the ability to synthesize different growth compounds were crossed (A and a), a small number of wild-typelike forms developed that were able to grow on a minimal medium. Ryan and Lederberg (1946) coined the term, prototroph, to describe such new recombinant strains in bacteria. It was reasoned that the difference in auxotrophs represented differences in enzyme systems that were gene mediated. This led to the concept of "one gene—one enzyme" advanced by Beadle and Tatum, which eventually won them the Nobel prize. (This has been modified to one gene—one polypeptide.) *Neurospora* immediately was accepted as a powerful genetic tool. Lindegren later applied a similar technique to *Saccharomyces cerevisiae* (the beer yeast), which also became a tool for genetic studies.

BACTERIA AS GENETIC AGENTS

In the early 1940's bacteria were still looked upon as being outside the genetic pale. Most of them were smaller than the individual eucaryotic chromosomes; no one had proved that they possessed a nucleus; and any bacterial variations were considered merely environmental responses.

However, as various findings of genetic significance were published, a good many investigators became interested in bacteria as agents for the study of genetics. Two major problems still remained to be answered: does true spontaneous mutation occur in bacteria, and what kind of nuclear structures, if any, do bacteria have?

Spontaneous mutations in bacterial populations

Prior to 1940, mutations (rare events occurring in 1 in 10^6 to 10^9 cells) were being encountered in bacterial cultures. These, however, were thought by many microbiologists to be induced by contact with a mutagenic agent such as a phage or antibiotic. The idea that such mutants were either preexisting forms in a bacterial population, or the result of continuous spontaneous mutations and not merely induced by the mutagen was not widely accepted.

The fluctuation test

The first proof of the presence of preexisting mutations, or continuing spontaneous mutation in a bacterial population came from the fluctuation test of Luria and Delbrück (1943). They observed that when a culture of *Escherichia coli* B was lysed by a phage, occasionally colonies developed in the center of the clear phage plaque. When these bacteria were grown separately in pure culture, they were found to be phage resistant. Here then were mutant strains of bacteria. Luria and Delbrück were prompted to ask, "Are these mutants induced by contact with the phage, or are they spontaneous mutations already present in the population or continually arising independent of the phage?"

Their approach was primarily a mathematical one. They seeded a 20 ml broth culture with bacteria, then divided it into two 10 ml portions. The first 10 ml portion was incubated for 24 hours, then pipetted and spread in 0.2 ml portions on the surface of 50 phage agar plates. The second 10 ml portion was initially pipetted in 0.2 ml amounts in sterile tubes, *then* incubated 24 hours. Next, the contents of each tiny tube was spread on a series of 50 phage agar plates. Each 0.2 ml of inoculum represented 10,000,000 to 20,000,000 cells. The number of phage-resistant colonies was counted on the plates of both series and the results compared.

The experiment was based on the following assumptions: If phage-resistant colonies developed only as a result of contact with the phage, then the numbers would be approximately the same on all of the plates, or if phage-resistant colonies represented a preexisting mutation or mutations arising spontaneously in a rapidly growing culture, then the numbers of phage-resistant colonies would be the same on all plates of the first series but would fluctuate or vary greatly in the second series. The latter proved to be the case (see Table 16-1).

Table 16-1. Fluctuation in numbers of phage-resistant bacteria

PLATE NO.	0.2 ML UNDIVIDED 24 HR CULTURE	0.2 ML DIVIDED 24 HR CULTURE
1	12	1
2	14	0
3	17	0
4	13	6
5	19	0
6	14	0
7	20	101
8	21	0
9	17	0
10	18	3
11	13	48
12	16	0
13	15	1
14	20	0
15	12	0
16	14	4
17	15	0
18	19	0
19	16	27
20	13	0

From these experimental results, Luria and Delbrück concluded that bacteria mutate to phage resistance without previous contact with the phage and that such mutation goes on continuously at a very low rate. They calculated a mutation rate of 2.45×10^{-8} mutations per bacterium per time unit. This same type of fluctuation was also demonstrated using streptomycin.

Newcombe's spreading experiment

Many microbiologists and geneticists were still not convinced that spontaneous mutation occurred in bacteria. In 1949, Newcombe, an English worker, added further proof of bacterial mutability with his simple but ingenious spreading experiment. He seeded the surface of 12 nutrient plates, each with about 50,000 *Escherichia coli* cells. After incubation of 6 hours, a myriad of minute colonies had developed, and the total number of cells had increased to nearly 3,000,000,000. At this point he sprayed half the plates with a phage suspension and to the second series of plates he added a small amount of sterile saline, and with a sterile bent glass rod spread the tiny colonies over the surface of the agar. After spreading, he sprayed the second or spread series of plates with phage. All of the plates were then incubated for 24 hours and the number of phage-resistant colonies counted. Newcombe reasoned logically that if mutants developed only after contact with the phage, the number of phage-resistant colonies on the undisturbed and spread plates would be the same. On the other hand, if phage-resistant mutants were present in the initial inoculum or muta-

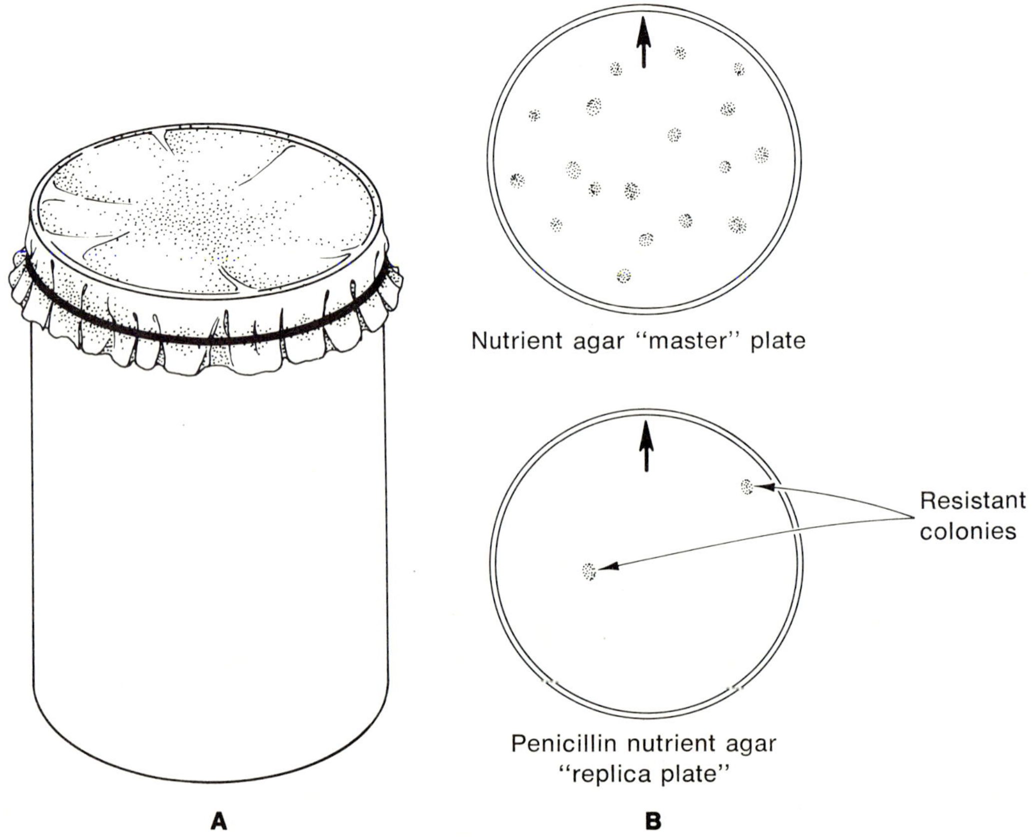

Figure 16-1
Replica plating technique. **A,** Replicating cylinder with velveteen in place; **B,** master plate and replica plate. (From Anderson, D.: Laboratory instruction in microbiology, St. Louis, 1969, The C. V. Mosby Co.)

tions occurred during the 6-hour incubation period, the number of resistant colonies would be materially greater on the spread plates. This proved to be the case. The 6 undisturbed plates showed a total of 240 colonies while the total for the 6 spread plates was 12,638. In spite of this proof, a few die-hards still insisted that the experiment was not clear cut because the bacteria had been in contact with the phage.

Replica plating technique

This objection was not answered until Lederberg and Lederberg (1952) came up with the beautifully simple, but highly effective, replica plating technique (Fig. 16-1). This method, besides furnishing futher proof of spontaneous mutation in bacteria, is also an excellent tool to simplify the task of transferring great numbers of mutant bacteria to a variety of media. A cylinder of wood, slightly less in diameter than the inside of a petri dish, is used. On this is placed a piece of sterile velveteen, pile side up, held firmly in place with a metal collar or rubber band. A petri plate, which has been surface seeded (to yield only surface colonies) is now inverted gently over the surface of the velveteen bringing the colonies in contact. Upright pile threads of the cloth serve as myriad tiny inoculating needles, retaining bacteria from the imprinted colonies. The bottom half of a pour agar plate of test medium is now inverted gently over the velveteen bringing the agar in contact with the colony imprints. Thus in a single operation a hundred or more colonies can be transferred from a master-plate impression. The initial contact with the seeded plate results in about 10% to 30% of each colony's adhering to the velveteen. This is sufficient to permit the subsequent effective inoculation of 6 to 8 plates from one replication master plate.

This device was ideal for the mutation experiment for which it was designed, which consisted of heavily seeding a nutrient agar plate with bacteria to provide about 10^8 bacteria. The edge of the plate was marked for reference purposes, and the plate was then inverted over the sterile velveteen leaving the imprint of thousands of small colonies. A phage plate was inverted over the cloth and touched gently to seed with bacteria. Only a few phage-resistant colonies developed because all of the sensitive bacteria were immediately lysed. Using the position of one of the resistant colonies on the phage plate as a reference point, a small block of colonies were cut out from the master plate to include the resistant colony. This block was shaken in sterile saline and the suspension used to seed a second plate. The proportion of mutants could thus be increased perhaps a hundredfold without contact with the phage. This series of transfers was repeated several times until many of the colonies would then be resistant mutants and easily isolated. The same results were also secured with nutrient agar containing streptomycin. In fact, after several replications it is possible to secure a pure culture of phage- or streptomycin-resistant mutants.

The point is that by this method of indirect selection a pure culture of phage- or streptomycin-resistant mutants that had never been in contact with phage or streptomycin could be isolated.

Using nutrient broth tubes and a simple dilution technique, Cavalli-Sforza and Lederberg (1956) were also able to isolate a pure culture of phage-resistant *Escherichia coli* that had never been in contact with the phage. This was called the indirect or "sib" selection method. It has also been used successfully for selecting out in pure culture pre-adaptive (preexisting) strains of antibiotic-resistant mutants.

The nuclear structure of bacteria

Bacteria, as procaryotic forms, are now recognized as possessing a simple, primitive chromosomal structure unlike that of the higher eucaryotic plants and animals. Bacteria lack the nuclear membrane, chromosomes, and nucleolus of higher forms. Instead, their nuclear structure consists of an enormously long circular strand of DNA resembling a string of beads. This is made up of a double strand of DNA molecules arranged in a helix. When straightened out this would have the appearance of an infinitely thin but extremely long ladder. Until the early 1940's bacteria were variously thought to be without a nucleus, to be all nucleus, or to have nuclear material spread evenly throughout the cell. When bacteria were stained, the entire cell took up the stain. (We now know that both the DNA and RNA take up the stain with equal avidity.) Robinow in 1942, by ingeniously treating cells with surfactants and RNase to remove the RNA, was able to demonstrate deeply staining structures, sometimes dumbbell shaped, at both ends of the cell or extending through the center of the cell. In subsequent experiments during the next few years he was able to identify this material as DNA.

The real picture of the bacterial DNA chromosome or nucleoid did not materialize until Cairns in 1963 published pictures of a radioactive-labelled chromosome of *Escherichia coli.* He grew a strain of *E. coli* K12 for about two cell generations in a medium containing H_3 thymidine, which imparts a radioactive label to the DNA of each cell. The bacteria, after treatment with Duponol (a detergent), were then lysed by the enzyme, lysozyme. The cell contents were collected on a membrane filter and dialyzed to remove extraneous material. The membrane was placed in contact with a photographic film and held in the dark for two months. The radiation emitted by the tritiated thymidine left a photographic image (as grains of precipitated metallic silver) of the bacterial chromosome. These images appeared as somewhat tangled circles that had a circumference of 1,100 to 1,400 μm. (This is remarkable when you consider that *E. coli* cells are generally 1 to 2 $\times$ 0.8 μm in size.) Thus the strand that makes up the bacterial chromosome averages at least 1,000 times as long as the bacterial cell itself. The circularity of the bacterial chromosome has since been repeatedly established with electron micrographs. The circularity of the genetic material (DNA or RNA) of phages is now also recognized.

Unlike the cells of higher forms, including the mold, *Neurospora,* and yeasts, bacteria lack typical chromosomes and the familiar mitotic apparatus. Consequently cell reproduction in bacteria is primarily asexual in nature, and so it was thought for a long time that bacteria would continue to go from cell generation to cell generation without any possibility of genetic changes. Now however, a number of unique ways in which genetic material can be transferred from one bacterial cell to another have been observed. These include transformation, conjugation, transduction, and sexduction.

TRANSFORMATION

The discovery of transformation as a method of genetic transfer in bacteria has definitely established the genetic role of DNA, and has pointed a major way by which we may soon achieve artificial genetic control in plants, animals, and perhaps even man.

In 1928 Griffith, an English worker, while experimenting with various capsular types of pneumococcus *(Diplococcus pneumoniae)* injected a small number of living Type II *(R)* (rough, noncapsulated, nonpathogenic) bacteria mixed with a large number of killed Type III *(S)* (smooth, capsulated, highly pathogenic) into mice. Many of the mice died, and from their heart blood Griffith obtained pure cultures of Type III *(S)* bacteria. Something had passed from the dead Type III *(S)* bacteria that changed the nonvirulent Type II *(R)* strain into a virulent Type III *(S)* strain, which killed the mice. This process was called transformation. Immediately, other workers investigated the phenomenon and were able to duplicate the results. In addition, they found that cell-free extracts of Type III *(S)* cells were able to transform living Type II *(R)* cells into Type III *(S)* pathogens.

Not until 1944, however, were Avery, MacLeod, and McCarty able to prove that the transforming principle (TP) was actually deoxyribonucleic acid (DNA). For their experiment they grew many liters of Type III (S) pneumococcus cultures, disrupted the cells, and finally extracted a minute amount of the nucleic acid fraction or DNA. They then added this to living Type II *(R)* cells and were able to transform them to Type III *(S)* pathogenic strains. Even dilutions as low as 1:600,000,000 were able to bring about this change. This discovery represented a major scientific breakthrough and established the whole new field of DNA genetics. Someone has said that this discovery was as important to biology as the discovery of nuclear fission was to the physical sciences.

If DNA extracted from a streptomycin-resistant strain of pneumococci is added to a rapidly growing culture of streptomycin-sensitive pneumococci, most of the cells quickly become streptomycin resistant. After the DNA fragment is incorporated into the bacterial chromosome, the progeny of each transformed cell will continue to show the transformed character. Only about 5 minutes elapses between the entrance of the DNA into the cell and the incorporation of DNA into the bacterial chromosome. A transforming DNA fragment represents about 1/200 of a bacterial chromosome. Thus a DNA fragment of 7 μm would carry about 21,000 base pairs. This is enough to supply the code for about 10 to 14 proteins.

If only a few billionths of a gram of DNA from a streptomycin-resistant strain is added to a million or so streptomycin-sensitive cells and allowed to remain in contact for about 20 minutes, several hundred thousand of these cells will be transformed genetically to streptomycin-resistant cells. Without the added DNA only a cell or two, or perhaps none of them, would show a spontaneous mutation to streptomycin resistance.

Some transformation does appear to take place under natural conditions. This can be demonstrated by the simple expedient of growing two genetically distinguishable strains of pneumococci together. New strains that are intermediates between the two strains will appear in small numbers. This indicates a bilateral exchange of genetic material between the two strains. The source of the DNA involved in such natural transformation still is not clear. It may represent excretion or leakage from living cells, or may come from the disintegrating hulks of the few dead cells.

Experiments have shown that in transformation the passage of DNA fragments occur only when the cell is in a competent state. The mechanism of competence is still not entirely clear.

Competence or transformability may appear suddenly and explosively in a culture, then disappear almost as rapidly after an hour or two. It does not appear until near the beginning of the stationary phase when the number of bacteria has increased to 1,000,000 or 10,000,000 cells per milliliter. Competence appears in a synchronous manner. The number of competent cells may increase a millionfold in ten minutes. This suggests the presence of an activator substance. Such a substance has been recently isolated and has been shown to be a protein or proteinlike compound with a molecular weight of about 10,000. It is functional only in a slightly alkaline medium, and an infinitely small amount (a few hundred micrograms) is sufficient to produce competence in several hundred thousand cells in as little as ten minutes.

According to a recent survey, in addition to the widely studied *Diplococcus pneumoniae* and *Haemophilus influenziae,* transformations have been demonstrated in the following genera of the Schizomycetes: *Acinetobacter, Agrobacterium, Bacillus, Escherichia, Micrococcus, Mima, Moraxella, Mycoplasma, Neisseria, Pseudomonas, Rhizobium, Sarcina, Staphylococcus, Streptococcus,* and *Xanthomonas.* Transformations have been observed both within and between species.

Among the eucaryotic forms, those in which transformation has been found include: *Aspergillus niger, Drosophila melanogaster, Mus* species (mice), human bone marrow, and leukemic and lymphocytic cell cultures.

CONJUGATION OR RECOMBINATION IN BACTERIA

Using the minimal and complete medium technique, Gray and Tatum (1944) isolated a considerable number of *Escherichia coli* strains deficient in the ability to synthesize certain growth factors, and requiring the addition of these growth factors to a minimal medium in order to grow. Lederberg and Tatum (1946) grew in yeast extract medium two triply deficient mutants, Y-10 (requiring threonine, leucine, and thiamine) and Y-24 (requiring biotin, phenylalanine, and cystine). When a heavy growth was obtained, the cells were harvested, washed, and seeded on a minimal medium. About one out of every million cells produced a visible colony on the minimal agar. Filtrates failed to produce the effect, hence it was not transformation or crossfeeding. The explanation advanced was that a genetic recombination had occurred in which genes were shared to produce nondeficient or prototrophic cells able to grow on the minimal agar.

The following schematic representation of the

crossing of the two triply deficient strains illustrates the reaction (designations A, B, C, and so forth, used in place of names of deficient genes).

Genetic make-up of auxotrophs and resultant prototroph

Strain Y-10		*Strain Y-24*
abcDEF (auxotroph deficient for genes abc)	×	ABCdef (auxotroph deficient for genes def)

(prototroph)
ABCDEF

(Production of prototroph—rare event occurring once out of each 1,000,000 to 10,000,000 auxotroph cells)

The importance of this discovery was that it demonstrated for the first time that, even though it was a rare event, a cell-to-cell exchange of genetic material was possible. As this was studied more intensely, the following facts became apparent:

1. For such a recombination, cells of both parent cultures have to be living.
2. Live cells of the parents have to be in contact.
3. Genes or determiners are arranged in a linear fashion conforming with the idea of a long circular chromosome.
4. Some cells are donors, some recipients. This suggested maleness and femaleness. Later a fertility (F) factor was demonstrated. This was a small circular strand of DNA that could be transferred independently from the male (F^+) to the female (F^-) strain. This then could grow independent of the nucleoid as an episome. When this occurred the F^+ cell became F^- and the F^- cell became F^+.

Genetic transfers were very rare events ($1:10^6$ or 10^7). However, independently two mutant strains were soon discovered in which such genetic recombinations were about 1000 times more frequent. These bacteria were termed Hfr (high frequency of recombination) strains. It was then found that in the F^+ strain, the F factor existed as an episome; in the Hfr strains the factor was incorporated as part of the bacterial chromosome.

On conjugation of Hfr strains, the circular bacterial chromosome apparently breaks and begins to feed into the female call at the origin or end adjacent to the F factor DNA. About eight minutes elapses before the first evidence of genetic transfer is noted. After this time more and more genes are fed into the recipient cell. When conjugation is interrupted by spinning the cells in an electric blender, the transfer of genetic markers stops. Under natural conditions spontaneous separation of cells generally occurs before conjugation is complete, leaving only an incomplete diploid. (Bacteria are normally haploid.) This is termed a merozygote (partial egg). Only in rare instances are the complete bacterial chromosome and the F factor (at the final end of the chromosome) transferred to produce a true diploid state. The transfer of chromosomal materials requires as much as $1\frac{1}{2}$ hr or five or so cell generations.

Later it was discovered that male cells possessed a very thin flagella-like structure that apparently served as a channel through which the DNA strand from the male passed into the female or receptor cell. These structures were termed fertility fimbriae (Duguid and co-workers) or pili (Brinton). In *Escherichia coli,* male cells usually possess two to four F pili. Conjugating cells observed under the phase microscope have been described as resembling a very active dog (the male cell) on the end of an almost invisible leash (the pilus) held by the dog's mistress, the female cell. The frequency with which such transfers occur in nature is not known but conjugating strains, especially of *E. coli,* have become magnificent tools for the microbial geneticist. Some microbiologists maintain that direct cell surface contact rather than pili is responsible for genetic transfer in conjugation.

TRANSDUCTION

Another unique means of transfer of genetic material between bacterial cells is known as transduction (literally leading across).

It was discovered by Zinder (a doctoral candidate of Lederberg's) and his mentor (1952) while testing for transformation, and was a truly serendipitous discovery. They made use of the "bacterial bundling board" developed by Davis in testing for transformation. The Davis device consisted of a large U-shaped culture tube with a bacteria-tight sintered glass filter fused in the bottom of the U. Different cultures could be grown in each arm, and with pressure and suction the liquid medium could pass from one side to the other. However, no bacterial cells would pass through. Thus, if transformation occurred, bits of DNA from one culture could be transferred to the other.

For Zinder and Lederberg's experiment, illustrated in Fig. 16-2, two different strains of *Salmonella typhimurium* were used. The first, 22A, was tryptophan requiring; the second auxotroph, 2A, was histidine requiring. Three Davis tubes were set up. The first was inoculated into both side arms with 10^8 22A cells. Pressure was applied to both side arms every 20 minutes for four hours. The populations from both sides were then harvested and streaked on minimal agar plates, but no growth occurred.

The second tube was inoculated in both arms with 10^8 2A cells. After mixing and four-hours' growth, no growth on minimal agar occurred from either arm.

The third tube was inoculated on one side with 10^8 22A cells, the other with 10^8 2A cells. After mixing and growth as above, inocula were streaked on minimal agar plates. The plate inoculated with strain 22A showed 10 prototroph colonies per 1×10^6 cells, while no prototrophs developed on the plate inoculated with strain 2A. They called the agent that passed through the filter FA (filterable agent). It was not destroyed by heating to 56° C (hence not bacteria) nor by DNase (hence not naked DNA). This pointed to DNA from host but within some sort of a protective coating.

The FA was soon identified as a lysogenic strain of phage, P22. It was found that 1 in 10^5 to 10^6 phage particles released from lysis of the donor bacterium carried pieces of host chromosome into the recipient bacteria that they infected.

We now know that the transducing phage apparently smuggles a fragment of DNA from its previous host cell inside its own protein coat. When the phage DNA is injected into the susceptible host this other DNA fragment is also injected. Cells that are lysogenic now have incorporated into their chromosome not only the phage DNA, but the fragment of new DNA from the previous strain. Thus the new genetic information becomes a part of the bacterial genome.

As with transformation, the piece of genetic DNA is small. Although about the same size as the phage genome itself, it represents only about 1/100, or occasionally 1/50, of the bacterial DNA.

Another type of genetic exchange, which is similar to transduction, involves the transfer of a fragment of host DNA attached to a male F factor that breaks away from the host chromosome in an Hfr strain. This is a rare event that has been termed sexduction.

MOLECULAR BIOLOGY AND BREAKING OF THE GENETIC CODE

With the establishment by Avery, and coworkers of DNA as the genetic substance, the repository of the genetic messages that determine the nature of the mature individual, a tremendous stride forward was made in genetic knowledge. Emphasis on control of disease, two world wars, and a lack of the necessary sophisticated equipment, methods, and concepts, had stopped the early molecular biologists from breaking the genetic code, if indeed they suspected that a universal one existed.

When Watson and Crick came up with their helical model of the DNA molecule (1953), the stage was set. With the newfound understanding of bacterial mutability, the electron microscope visualization of the bacterial chromosome or nucleoid in the early 1960's, such procedures as electrophoresis, chromatography, and differential centrifugation, the necessary tools were at hand.

Today, we have a picture of the action of DNA and the translation of its messages into enzymes,

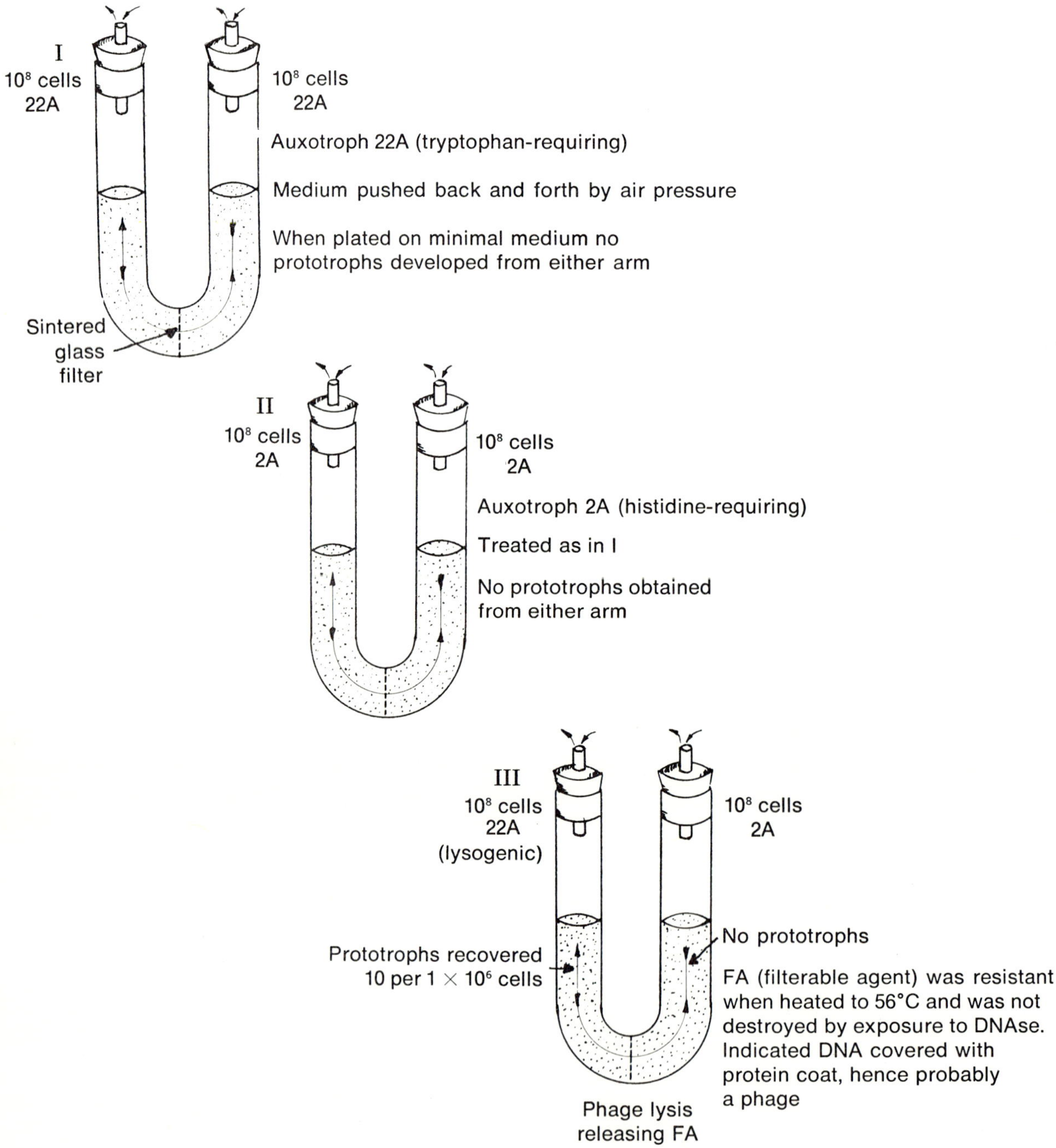

Figure 16-2
Transduction experiment of Zinder and Lederberg.

Table 16-2. The genetic code

AMINO ACID	CODE FOR PRODUCTION OF THE AMINO ACID
Glycine	GGU, GGC, GGA, GGG
Alanine	GCU, GCC, GCA, GCG
Phenylalanine	UUU, UUC
Valine	GUU, GUC, GUA, GUG
Leucine	UUA, UUG, CUU, CUC, CUA, CUG
Isoleucine	AUU, AUC, AUA
Cystine	UGU, UGC
Methionine	AUG
Serine	UCU, UCC, UCA, UCG, AGU, AGC
Threonine	ACU, ACC, ACA, ACG
Aspartic acid	GAU, GAC
Asparagine	AAU, AAC, GAU, GAC
Glutamic acid	GAA, GAG
Glutamine	CAA, CAG
Arginine	CGU, CGC, CGA, CGG, AGA, AGC
Lysine	AAA, AAG
Tyrosine	UAU, UAC
Proline	CCU, CCC, CCA, CCG
Histidine	CAU, CAC
Tryptophan	UGG
None	*Nonsense codes* UAA, UAG, UGA

their products, and cell growth, even though knowledge is increasing so rapidly that only an approximate picture is available. However, these points are now apparent:

DNA serves two purposes: it functions as the pattern, or template, for its own replication, and it is the bearer of the code that determines the kind of messenger RNA (mRNA) involved in the production of the entire armamentarium of enzymes used for the proliferation of the parent cell. In the bacterial cell, a code for the production of a thousand or more different enzymes has been postulated.

The code or language of the genetic message is now recognized as being made up of a triplet of three nucleic acid bases: Included are three of the four acid base components—adenine (A), guanine (G), cytosine (C), and uracil (U)—of the RNA nucleotide. Thus the message U-U-U (uracil-uracil-uracil) means "make a molecule of the amino acid, phenylalanine." (See Table 16-2.)

The three-base code for a given amino acid is termed a codon. Since there are four DNA bases, which serve as templates for mRNA molecules, the language has a possibility of 64 three-letter words. There are only 20 amino acids isolated from proteins, so some combinations would be ambiguous or degenerate. A few of these triplets will not code for any amino acid. However, some combinations code for more than one amino acid.

A given enzyme involves a combination of a considerable number of amino acids arranged in a complicated three-dimensional pattern, which is termed a polypeptide. The function of these various enzymes involves the multitude of operations necessary to the vital functions of the cell. This great variety of operations moves at an amazing rate. The machinery of a given cell may turn out another completely equipped and functional cell in as little as 20 minutes, while at the same time discharging all the waste materials into the surrounding medium. Manifestly with such an amazing activity of molecular machinery, some operations might produce too much of a given product.

A whole array of control agents functions. Many enzyme codes include operon genes that control a given process that in turn are controlled by an operator gene. The operator gene, with the aid of repressor genes that turn off the "machine" when too much product has been produced, turns off or on a given operation. (Repressor genes operate only when functional operon genes are present.) The stopped process may be turned on again by an operator gene that functions essentially as a derepressor. In other cases the enzyme process is turned off as a result of the excess accumulation of end product. This is termed feedback or feedback control.

The present picture of the role of DNA is that it serves as a template for the synthesis of mRNA. This short-lived mRNA moves into the organelle, ribosome, where it serves as the attachment point for individual amino acids. These amino acids are picked up and maneuvered into place by transfer RNA (tRNA), small soluble molecules, which are assisted in the attachment by special enzymes. *Escherichia coli* may have as many as 15,000 ribosomes per cell.

INDUCTION OF BACTERIAL MUTATIONS

A mutation is recognized as a sudden change in the enzymatic or growth pattern of a strain of bacteria (or other organisms) that continues as a fixed character through subsequent cell generations. In bacteria this is most often seen as the loss of the ability to ferment a sugar, hydrolyze a complex molecule, or synthesize some growth product. It may also represent the loss (rarely the gain) of some easily seen phenotypic character, for example, the loss of pigment production in *Serratia marcescens* (Fig. 16-3).

Mutation production is especially easy to visualize when you consider that the bacterial nucleotide is a great long double strand, and a great many changes can occur in one or more of the nucleotide base pairs. One or more of the base pairs, or a base pair in one strand but not the other, may be lost. A short section may break away, then attach at another point, changing the whole linear code of the genetic message (translation). A short section of one strand may be turned around in relation to the other (inversion).

If a small section of the DNA chain is deleted, a particular enzyme system may be lost either partially or completely, and the resultant cell then becomes a deficient mutant. If the enzyme system (or systems) lost is vital to cell growth, the cell fails to reproduce. This then becomes a lethal mutation. Sometimes the message is so garbled that no amino acids are coded. These then are functional "nonsense" codons. Such nonsense codons may cause the polypeptide chain to terminate prematurely and enzyme function is lost. Some mutations simply reduce the effectiveness of a given protein quantitatively, but not qualitatively. Production of most growth products requires a series of steps involving several enzymes. As previously mentioned, this series is called an enzyme pathway. If a mutation blocks the pathway at a given point, the reaction may proceed along another pathway that is sometimes less efficient than the first. This is known as a shunt.

In nature mutations generally are rare events. In *Escherichia coli*, for example, mutations occur once in every 1,000,000 to 10,000,000 cell divisions. This is of interest when you consider that the nucleoid of *E. coli* is about 3×10^6 base pairs long and that in a mutation some change has occurred in one or more of these base pairs.

Mutagenic agents

Any agent that causes mutations in bacteria or other test forms becomes a most useful tool for the geneticist by increasing the number of mutant strains available for experimentation. The mutagenic action of X-ray has been known for nearly 40 years. Various types of semilethal damage, such as DNA breakage or deletions, have been produced that result in various genetic abnormalities. Ultraviolet light also produces demonstrable genetic changes.

Keep in mind that the DNA molecule, made up

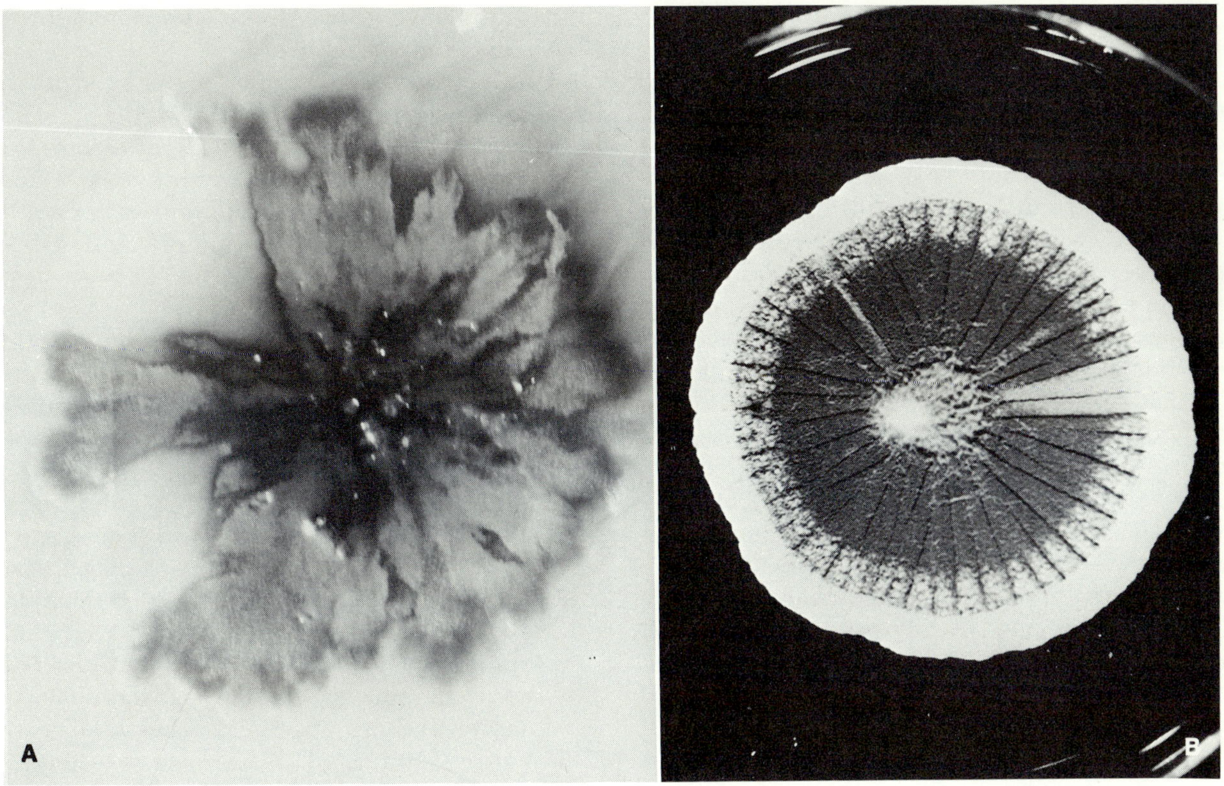

Figure 16-3
A, Sectored giant colony of *Serratia marcescens,* 14-day growth, following exposure to ultraviolet light. Note different colors developing from differently effected chromosomes.
B, Light-colored segment in a colony of *Penicillium* sp. probably representing mutational change in localized portion of mycelium from which it arose.

as it is of two base pairs, is held together by hydrogen bonds between the pairs. The complementary pairs are: adenine (purine) always opposite thymine (pyrimidine) (A-T), and guanine (purine) is always opposite cytosine (pyrimidine) (G-C). In other words the molecule is internally complementary. However, the RNA triplets are coded from the DNA, coding shows the same complementary except that *uracil* (methylthymine) replaces thymine, A-U is opposite G-C.

Any physical or chemical agents that can induce mutations by changing the genetic message is termed a mutagen. Although X-ray and ultraviolet light have been recognized as mutagens for many years, it was only about 1945 when the first chemical mutagen was discovered. Since that time many chemical mutagens have been found.

During World War II the mutagenic effect of mustard gas was first demonstrated in Edinburgh. In Germany urethane was shown to produce similar effects. Other chemical mutagens include formaldehyde, pyrogallol, and hydrogen peroxide. Caffeine, a purine derivative, has been shown to produce chromosomal breaks in plant, human, and bacterial cells.

Base analogues, similar but not identical to purine or pyrimidine bases, may produce mutations by replacing a similar base. Thus a pyrimidine base may be replaced by 5-bromouracil. Formaldehyde, hydroxylamine, ethyl amines, and so forth, are all fairly active mutagens.

In 1960 a new and very powerful chemical mutagen, 1-methyl-3-nitro-nitroso-guanidine (NG or NTG) was discovered quite by accident in the course of testing the cancer cell inhibiting power of a variety of compounds. At moderately high concentrations, it killed tissue and bacteria cells, but at lower concentrations, many of the cells that survived were found to be genetically modified.

Starting with about 1,000,000,000 cells, about 10,000 survived exposure to a medium containing 2 mg NG/ml. Of the survivors about 10% showed some mutational change. About 1 in 20 mutants were mutated at more than one site. This mutagen is now widely used in the microbial genetic laboratory. However, it must be used with care because of its danger as a human carcinogenic agent.

In addition to the rather drastic mutation-inducing changes previously mentioned, less drastic ones may be transitions (replacement in which the purine, adenine, is replaced by the purine, guanine, or the pyrimidines, thymine and cytosine, may replace each other). More drastic than transitions are the transversions in which a purine replaces a pyrimidine.

One of the effects of transitions or transversions is that the triplet codon may be modified to produce a different amino acid. This may have little effect on the resultant enzyme or may profoundly effect its activity.

When nonsense triplets are formed, formation of the enzyme polypeptide may be terminated resulting in the loss of enzyme action.

PHAGE GENETICS

Phages (bacterial viruses), because of their small size and complete host dependence, would appear to be unlikely candidates as experimental agents for genetic studies. However, since about 1950 they have proved to be useful in studies of basic aspects of genetics.

Until recently most of the phage work has revolved around *Escherichia coli* and the T-odd (T-1,3,5,7) and T-even (T-2,4,6) phages.

Studies have now been extended to other *E. coli* phages, phages of *Pseudomonas, Bacillus,* and so on, as well as the single-stranded RNA.

Several properties of phages make them useful for genetic studies:

1. Their genetic material is in the form of a circular strand of DNA (or RNA) similar to the nucleoid of bacteria. Measurements by the Cairns method show it to be about 100 μm in length compared with the 1,000 to 1,400 μm length of the DNA strand of the host cell. In terms of base pairs, which carry the genetic message, phages have about 150,000 compared with 3,000,000 for the host cell.
2. Spontaneous and induced mutations occur with the phages, similar in frequency and nature to those observed in bacteria.
3. Various strains of phage may be identified on the basis of rather easily recognizable genetic markers. One of the most commonly used means of identification by genetic markers is by the nature of the plaques formed when individual plaques develop on a plate heavily seeded with susceptible bacteria. Wherever a single phage particle is present a plaque or cleared area will form in the "lawn" of growing bacteria. The phage suspension must be diluted enough to produce only a hundred or so plaques per plate. The plaques vary sufficiently in their characteristics to make them useful as genetic markers. These characteristics may be size (large, small, minute); appearance (clear-cut, fuzzy, haloed, turbid, star-shaped, and so forth); or combinations of size and shape.
4. Host range is also a useful genetic marker. Some produce plaques on *E. coli* B (the original strain used for phage genetic studies);

others may produce plaques on *E. coli* K12 but not on *E. coli* B.

5. Some strains are temperature sensitive. These produce plaques on *E. coli* at 25° C but not at 42° C.
6. Osmotic-resistant strains have also been identified. Normally, when a phage is placed for a period in a solution high in salt, and then rapidly diluted with distilled water, the protein coat of the phage will burst. Some mutants, however, are resistant to osmotic shock and do not rupture. It is thought that the protein coat of these mutants is sufficiently pervious so that the salt inside the head leaks out before the protein coat takes up enough water to cause the coat to burst.

If two strains showing genetic marker differences infect a single cell at the same time, the DNA may intermingle. As a result when the phages are released by the cell burst, three types of plaques may be formed: Those characteristic of phage I, those characteristic of phage II, and those unlike or intermediate between I and II.

The location of genes, both of bacteria and phages, have been mapped on the circular nucleoid in relation to other known genes.

UNIVERSALITY OF THE GENETIC CODE

Remarkable as it may seem, the genetic code is apparently the same for all forms of life—from virus to elephant. Thus a given triplet code for the production of a particular amino acid is the same, regardless of the living form in which it occurs. Luria has likened it to the words "of a book printed in a universal language."

Because of this, genetic code information derived for phages or viruses is applicable to all higher plants or animals. This explains the true value of bacterial or phage genetic studies and the significance of microorganisms as tools for the modern molecular biologist.

SUGGESTED READINGS

Braun, W.: Bacterial genetics, ed. 2, Philadelphia, 1965, W. B. Saunders Co.

Fraser, D.: Viruses and molecular biology, New York, 1967, The Macmillan Co.

Mortimer, R. K., and Hawthorne, D. C.: Yeast genetics, Annu. Rev. Microbiol. **20**:151-166, 1966.

Norris, J. R., and Ribbons, D. W., editors: Methods in microbiology, vol. 3A, New York, 1970, Academic Press, Inc., Chaps. 6 and 7.

Stent, G. S.: Molecular biology of bacterial viruses, San Francisco, 1963, W. H. Freeman and Co. Publishers.

Watson, J. D.: The double helix, New York, 1968, Atheneum Publishers.

PART 2

Practical aspects of microbiology

17 Microbiology of water and sewage

Our very lives are dependent upon the availability of safe and adequate water. The continued existence and quality of our cities and industries depend on a pure, unpolluted supply of fresh water. Great civilizations and populous cities could never have developed without such water supplies. Not only must water be adequate for domestic needs, but our industries require amazing quantities for washing, cooling, and other types of nonconsumptive uses. Little wonder then that ecologists are so concerned with the current widespread pollution of our water supplies.

The total supply of water presently available cannot be considered a fixed amount, because as pollution increases, less and less water is available for use.

Water pollution, except for a few cases of purely chemical pollution, is primarily a microbiological problem. This may involve water as a carrier of disease, or the need for active microbial decomposition of organic pollutants.

KINDS OF WATER

Water supplies may be divided conveniently into two categories:

1. Surface waters include streams, lakes, and even shallow wells. Surface waters passing through or in contact with human habitation are almost invariably polluted.

When undisturbed by man, surface waters generally possess a natural microbial flora. Algae of various types are present as sessile inhabitants of a stream bed or shoreline or as free-living forms. Periodically, when temperatures and nutrients are right, the algae population may burgeon until it assumes the proportion of a bloom and the waters may teem with free-living algae. The algae contribute oxygen through photosynthesis and become an important food source for fish and other aquatic forms. As the algae die, they may yield great amounts of organic matter that stimulates the growth of the normal bacterial flora, which in turn provides food for a free-living protozoan population. In natural, unpolluted waters an ecological balance normally exists that waxes and wanes in cycles from season to season. The bacterial flora in such undisturbed waters will be quite different from that of polluted waters. Many of the bacteria will be pseudomonads, while polluted water carries in addition, intestinal bacteria, soil forms, or forms peculiar to rotting organic matter.

2. True ground waters, in sharp contrast with surface waters, are waters from underground sources free from any contact with surface waters. They are found in springs and deep wells. In most cases they will contain very few bacteria. The bacterial counts of true ground water are often 1 to 2 bacteria per milliliter. In contrast,

surface waters sometimes carry thousands of bacteria per milliliter.

With true ground water, the temperature varies little from summer to winter, often rising inexplicably a degree or so during the winter and dropping slightly during the summer.

Ground waters are not invariably free from contamination. In areas underlain with limestone, channels may develop through which surface water may enter, then be carried to rise again to the surface as springs or to be pumped out in wells. Very coarse gravel strata may also allow contamination of underground waters by polluted surface waters.

DANGERS OF POLLUTION

The dangers associated with pollution are transmission of intestinal diseases, impairment of quality and usefulness of water supplies by loading them with putrescible organic matter, and chemical poisoning of water supplies. The end result is that not only is the water unsafe to drink, but the clear stream or lake loses its aesthetic and recreational values.

NONDOMESTIC WATER USES

Nondomestic waters, those used for recreation, irrigation, industrial cooling waters, dairy and food processing, manufacturing of paper, plastics, and cosmetics, pose problems quite different from those problems associated with water for household use only.

Often water entirely acceptable for drinking may be undesirable for reconstituting milk, or washing fish or poultry. While it may be free from coliforms, it may contain considerable numbers of psychrophiles. Although the maximum growth temperature of these bacteria is around 20° C, psychrophiles may grow fairly rapidly even at 5° C. As a result, refrigerated, reconstituted milk soon may develop an off-flavor, or packaged fresh poultry in refrigerated display cases may develop slime, unpleasant odors, or become discolored. Most of the common offenders are members of the genera *Pseudomonas*, *Flavobacterium*, or *Achromobacter*. Generally, these organisms have little public health significance. However, *Pseudomonas aeruginosa* can produce intestinal infections or enteritis in children.

Processed, potable waters to be used for special purposes are often given additional treatment to make sure psychrophiles or nuisance bacteria are removed or killed. Thus waters used to formulate cosmetics are often pasteurized at 143° F for 30 minutes, or 161° F for 15 seconds.

Irrigation water may become a carrier of plant disease–producing bacteria, and also of animal and human intestinal disease agents.

WATER AS A CARRIER OF DISEASE

Surface waters in streams and freshwater lakes are almost invariably potential carriers of disease. The known waterborne pathogens include the following: Of the bacterial diseases, the typhoid organism, *Salmonella typhosa*, was at one time in the United States the major waterborne pathogen. Today, thanks to water treatment, especially almost universal chlorination, and constant testing of most water supplies, this disease has become rare indeed. Other members of the genus *Salmonella*, such as *S. paratyphi* (paratyphoid A), and *S. schottmuelleri* (paratyphoid B), may still occasionally cause waterborne epidemics. Added to these are the rather rare, waterborne *Shigella* infections, and those caused by pathogenic strains of *Escherichia coli* and *Paracolobactrum* or other enteric intermediates.

Fortunately, since the turn of the century, cholera epidemics, caused by waterborne pathogens, have ceased in the United States and Europe. However, a devastating cholera pandemic appeared in the Middle East in 1971, with major outbreaks in parts of Lebanon, Syria, Egypt, Israel, Libya, Guinea, and Pakistan.

Protozoan diseases, especially amebic dysentery caused by *Entamoeba histolytica*, may often be waterborne.

The fungus disease, histoplasmosis, caused by the yeastlike fungus, *Histoplasma capsulatum* may possibly be transmitted by water supplies. The spores of this form survive for long periods in water. In cities securing their drinking water supplies from the Missouri, Mississippi, and Ohio rivers, the incidence of histoplasma skin-test positives among the population is exceptionally high.

Enteric (intestinal) viruses also pose a serious problem. The diseases include infectious hepatitis, poliomyelitis, Coxsackie virus infections, and those caused by members of the poorly defined ECHO (enteric cytopathogenic human orphans) virus group.

Proof of the role of water as a carrier of disease

Although water may have long been suspected as a carrier of disease, it was not until 1854 that water was proved epidemiologically to have this nefarious role.

In 1854 London had one of its recurrent outbreaks of cholera, which in the St. James parish alone resulted in 700 deaths. A commission, with Dr. John Snow as its secretary, was appointed to investigate the epidemic. His report is a classic of epidemiological detective work. He reported that of 73 persons who died of cholera, all had one contact in common, their drinking water. A break in an adjacent masonry sewer had allowed sewage to seep into the well. When he "broke the handle off the pump," the epidemic ceased.

The first and very dramatic proof of the efficacy of filtration of water came in 1892. A cholera outbreak occurred in Hamburg, Germany, and before it ran its course, 17,000 of the 640,000 inhabitants suffered from cholera, and of these, 8,605 died. The city's domestic water came from the Elbe River, which was diverted into the water mains with no treatment except brief settling. The city of Altona, population 143,000, was contiguous with Hamburg, separated only by a dividing street. It also obtained its water from the Elbe River but *below* the Hamburg sewer outfall. Because of the poor quality of the water at that point, the city installed a treatment plant consisting of a sedimentation basin and slow sand filters.

Robert Koch, fresh from his studies of cholera in India, investigated the epidemic and was astounded to find that on the Hamburg side of the dividing street many cases of cholera were reported, while immediately across the street in Altona no cholera was found. Indeed, no cases of cholera occurred in the entire city of Altona.

In spite of the effectiveness of modern filtration and chlorination in preventing waterborne disease outbreaks, occasional epidemics are still periodically reported as the result of a breakdown of these defenses.

Classic among these is an epidemic that occurred in Seymour, Indiana, in July 1940. Following heavy rains that raised the level of the White River coupled with improper operation of the treatment plant, an explosive epidemic occurred. Before it subsided there were 2,250 cases of severe diarrhea and gastroenteritis. Of these, 884 cases were diagnosed as typhoid and resulted in 27 deaths.

Unless constant vigilance is maintained, any city using a polluted source of domestic water may also face waterborne epidemics due to cross-connections, back pressure, improper chlorination, or floods.

CLASSIFICATION OF DOMESTIC WATER

Domestic water supplies may be classed as:

safe A water is safe when devoid of any evidence of fecal pollution. Such water may, however, have undesirable tastes or odors due to inorganic salts or other materials.

potable A water that is not only safe but is clear, sparkling, and free from any objectionable tastes or odors. This is the ultimate in desirable water quality.

polluted A polluted water is one that is unsafe or undesirable for domestic use. It comes from streams or lakes into which sewage or industrial wastes have been dumped. It may or may not be a carrier of intestinal pathogens, although it generally is. The term, polluted, is used very loosely.

contaminated The term is often applied to water that carries pathogens. Such water may be desirable in terms of clarity and absence of undesirable tastes and odors, but still be definitely unsafe.

METHODS FOR DETERMINING SAFETY OF WATER SUPPLIES

In considering a new municipal water source such as a well or spring, a sanitary survey of the area might reveal potential sources of pollution. Likewise, if a chemical analysis reveals high concentrations of chlorides and nitrates, these might come as leachings from the soil, or they might be indicative of remote sewage pollution. A microscopic examination for algae and protozoa would suggest sewage pollution if typical sewage microorganisms were found. However, the most useful and significant tests are bacteriological. These include estimates of numbers and kinds of bacteria present.

Tests for bacteria in water

As early as 1883, Koch described a method of estimating numbers of bacteria in water. Modifications of this method now constitute the standard plate count (SPC) as described in Chapter 10.

The plate count (number of colonies per milliliter of water) for safe water was once set at 200 (later reduced to 100) per milliliter. Today the public health codes do not specify a plate count for drinking water, although maximum plate counts for swimming pool or bathing water may be specified.

Far more significant are tests that reveal the presence of bacteria common to the intestines. In 1885 Escherich found an organism common to human intestinal discharges that we now call *Escherichia coli*. This organism is almost universally present in the intestines. Therefore, its presence in water might be an excellent indicator of sewage pollution. *E. coli*, when grown in a nutrient broth medium plus glucose, produces acid and gas, fermentatively. Theobald Smith suggested that sewage pollution might be detected by adding measured portions of water to the J-shaped Smith tube and then observing for gas accumulation in the long arm of the tube. (Later the commonly used inverted gas tube was developed by Durham.) It was soon found that bacteria with no sanitary significance would also produce gas from glucose. However, the ability to produce acid and gas from lactose was more nearly differential for *E. coli*, because many other glucose fermenters did not also attack lactose.

By 1917 the standard water test medium became nutrient broth plus 1% lactose. Investigators found some bacteria (often intestinal forms), other than *E. coli*, that were also lactose fermenters. This entire lactose-fermenting group was termed coliforms. The most common other member of the coliform group is *Enterobacter (Aerobacter) aerogenes*. This organism can be found in the intestines, but it also can grow in decaying organic matter.

Accordingly, the lactose broth was modified by reducing the lactose content to 0.5%, and adding inhibitory substances such as ox bile, bile salts, or the surfactant, lauryl sulfate, making the test more selective. Still organisms other than fecal coliforms occasionally gave false-positives.

Therefore, to the presumptive test was added the confirmed test. Several types of agar media were devised that give distinctive colonies with *E. coli*. These include:

1. Levine's eosin–methylene blue agar on which *E. coli* produces a colony with a typical green metallic sheen. *Enterobacter aerogenes* produces a much larger mucoid colony with a brownish center, rarely showing a sheen.
2. *E. coli* colonies grown on Endo agar are reddish to cerise, often with a gold metallic sheen.
3. Other differential agars include brilliant

green bile agar, violet red bile agar, and several other agars formulated to give distinctive color to colonies with *E. coli* or the coliforms in general.

Positive presumptive tests may also be confirmed by inoculating a large loopful (3 mm loop) of growth from a positive lactose broth tube into tubes of brilliant green bile, EC *(E. coli)* medium, boric acid broth, formate-ricinoleate broth, or several other lactose media designed selectively to permit growth and gas formation by *E. coli*. Most of these media contain bile or bile salts, or a surfactant, and often inhibitory dyes.

Incubation of tubes of a lactose broth at 44° to 46° C is also used, because *E. coli* will grow and produce gas at this temperature while other coliforms will be inhibited. This is the basis for the Eijkman and similar differential tests.

Quantitative estimates of numbers of coliforms in water

Coliform numbers in high quality or treated drinking water are difficult to determine by the plate count because they may represent fewer than 1 bacterium per milliliter of water. Accordingly, the MPN (most probable number) test is generally used. This is based on probability tables as originally conceived by Poisson. In its simplest form, five Durham tubes of lauryl tryptose broth are inoculated with 10 ml of water. These are incubated at 35° C and gas production is read at 24 and 48 hours. Using an MPN table, the number of tubes showing gas will be read as coliforms/100 ml of water. For example, a single positive tube (of the five) is reported as 2.2/100 ml with a 95% confidence range of 0.1 to 12.6.

For sewage or ocean water, a commonly employed procedure is to use three tubes in each of three dilutions. The first trio is inoculated with 1.0 ml, the second with 0.1 ml, and the third with 0.01 ml. Gas is read at 48 hours, and the positives confirmed on brilliant green bile broth. If, for example, all of the 1.0 ml tubes and one of the 0.1 ml tubes showed gas, the MPN read from the tables would be reported as 430 coliforms or *E. coli* per 100 ml of sewage or seawater with a 95% confidence range of 70 to 2100.

Fecal streptococci or enterococci

In addition to coliforms, we normally carry in our intestines the enterococci. *Streptococcus faecalis* (first isolated by Escherich in 1886) is a representative form. Members of this group are not found in nature in the absence of fecal pollution, do not grow in water, and are more resistant to chlorination than *E. coli* and related coliforms. They will survive longer in brackish waters than the coliforms but die out quickly in the soil.

If enterococci are found in chlorinated water that has no coliforms it may indicate pollution and the probable presence of viable enteric viruses.

Tests for enterococci are similar to the coliform tests involving the use of liquid or solid glucose-containing media, with sodium azide added to inhibit gram-negative and most gram-positive bacteria. Presence of the test organism is indicated by acid production and the presence of typical pinpoint colonies on the confirmatory agar.

Membrane filter technique

The membrane filter (see Chapter 15), sold under such trade names as Millipore, and Isopore, has one distinct advantage for water testing. With the plate count, only 1 ml of water can be used, while with the MF up to 1,000 ml of water can be tested at one time, by filtering a water sample and then placing the MF on the surface of a differential medium, such as Levine's eosin–methylene blue agar. *E. coli* can be detected easily by the typical green sheen of the colonies. This gives a direct estimate, which is more accurate than the MPN test, of *E. coli* present in water. Confirmed tests are possible within 6 to 24 hours. With the usual 47 mm filter with a pore size of 0.45 μm, colony counts of up to 5,000 are possible if a dissecting microscope is used.

The membrane filter technique is valuable as a field test procedure. It also is useful for rapid testing of water under emergency conditions. Results, however, do not agree fully with MPN results.

PURIFICATION OF DOMESTIC WATER SUPPLIES

Most of the major cities of the United States obtain their domestic water from surface sources such as rivers or lakes, which often are heavily polluted.

Treatment designed to change polluted water to a potable water supply usually involves the following steps:

Sedimentation, with or without the use of a chemical flocculating agent, results in a settling out of the coarse particles that tend to carry down bacteria and similar smaller particles. When aluminum sulfate, ferrous sulfate, and so forth, are added to waters already alkaline, or made so by the addition of calcium hydroxide, jellylike colloidal floccules form that settle out readily.

The water may then be passed through fairly deep sand filters. The floc is caught in the sand, and forms a rather efficient filter removing most of the organic particles and bacteria. Most of the enteric viruses are also caught, but not inactivated, in the filter layer.

After filtration, water is greatly improved in quality, appearance, and sanitary properties, but is generally not entirely free from potentially dangerous intestinal bacteria. Therefore, the water is generally chlorinated.

In terms of concentration of chlorine required to kill bacteria in 15 to 30 seconds, the following values have been determined:

1. Most intestinal and respiratory pathogens require 0.15 to 0.25 ppm.
2. *E. coli* is somewhat more resistant and requires 0.2 to 0.4 ppm.
3. Bacterial spores require 2.5 to 280 ppm.

SEWAGE

Wastes carried by sewage vary greatly. They may come from domestic sources or pour in from industries. In most cases, by the time sewage reaches the great trunk sewers it is a mixture of both.

Domestic sewage carries (1) fecal wastes, which are a constant source of disease-producing bacteria, protozoa, or viruses; (2) large amounts of readily decomposable (putrescible) organic matter contributed largely by garbage grinders; (3) substantial amounts of almost pure cellulose primarily in the form of toilet paper; and (4) a variety of washing compounds, soaps, detergents, and other organic or inorganic compounds.

On the other hand, industrial wastes may be relatively low in fecal matter and putrescible matter, but high in mineral oils and a great variety of inorganic or organic chemicals. These are often toxic, and may resist removal or biodegradation.

In this connection, the problem of biodegradability of detergents has stimulated a great deal of public outcry, part of it completely sound and justifiable but much of it hysterical and often short-lived based on a lack of solid facts.

Since 1965 all detergents are presumably biodegradable. However, this is a relative term, because of the 30 to 40 varieties of detergents some are rather quickly converted to carbon dioxide and water plus bacterial cell substance, while others undergo biodegradation more slowly. Of the detergent molecules, about 60% to 80% will be converted to $CO_2 + H_2O$ while 20% to 40% will be incorporated into the cells of the degrading bacteria. Some of the formulations contain nitrogen and sulfur compounds. On biodegradation, the atoms of these compounds will emerge primarily as nitrates or sulfates, or be tied up in the proteins and nucleic acids of bacteria. Phosphates, if present, may be incorporated into bacteria and/or other microbial cells, or if in excess may remain in the sewage effluent.

SEWAGE DISPOSAL

Sewage treatment or sewage disposal methods have as their primary objectives, (1) the destruction or removal of pathogens of various types, (2) the oxidation of readily decomposable (putrescible) organic matter to carbon dioxide, water, nitrates, sulfates, and so forth, plus microbial cell bodies, and (3) removal of as much as possible of the slowly decomposable residue such as lignins, and waxes that would otherwise form undesirable deposits in stream beds, or along stream channels of streams into which the treated sewage effluent is dumped.

Disposal of sewage may be direct without specific treatment, or after primary, secondary, or tertiary treatments.

Without treatment

In the past it was customary to dump sewage directly into a stream, lake, or the ocean without any sort of treatment. However, within the last few decades public opinion has been aroused. The consensus, now supported by federal and state laws, is that no city has the right to pollute a stream or lake used by another as a source of water.

The theory used in the past to justify such direct disposal was that water contained considerable dissolved oxygen that would enable the aerobic bacteria to completely decompose the putrescible organic matter in the sewage. The end products theoretically would be carbon dioxide, water, nitrates, sulfates, phosphates, and bacterial cells that would ultimately die or settle out. This natural oxidative process has been termed mineralization, and represents the maximum level of oxidation of organic matter.

In practice, however, unless the volume of the stream is many times that of the inflowing sewage, much of the oxygen will be used immediately. The stream then becomes milky, and some odor will be apparent. In this state of partial deoxygenation, the stream is termed "stale." When all of the oxygen has been exhausted, the water becomes blackened and the odor of hydrogen sulfide, indole, and other putrefactive compounds develops, and the stream is termed "septic." As the level of oxygen begins to drop, game fish will be killed. The coarse trash fish, such as carp, will live on somewhat longer but will be killed eventually. In lakes or the ocean, beaches will be fouled and swimming will become unsafe by pollution with raw sewage.

Hydrogen sulfide from sewage-laden water often reaches a level that causes the paint on houses along the stream, lake, or ocean to turn black due to the formation of lead sulfide.

The possibility of untreated sewage transmitting diseases, such as typhoid and typhoid-like infections, amebic dysentery, and the enteric virus diseases, represents a constant health hazard.

Lagoonization, however, presents a promising method of making sewage safe without further direct treatment. Raw sewage is allowed to flow into a large lagoon or pond. When filled, it remains undisturbed for a period, usually 30 to 60 days or even longer. During this period the natural purification process becomes operative. The coarser particles settle to the bottom where anaerobic decomposition occurs. In a pond 5 feet deep, about 75% of the decomposition is anaerobic. The water-soluble putrescible compounds are vigorously attacked by the aerobic or facultative anaerobic bacteria. These in turn may be ingested by the rapidly developing protozoan fauna, which in turn are ingested by the metazoans.

Concurrently, algae, chiefly *Chlorella*, grow rapidly, stimulated by the presence of a supply of nitrates, phosphates, and sulfates. Their rapid photosynthesis adds quantities of oxygen to the water, hastening the growth of aerobic bacteria and the oxidative destruction of putrescible organic matter. In a period of 30 to 60 days, the foul gray sewage is replaced by clear water. A manyfold decrease in bacteria occurs, while the pathogens succumb to the waste products of the aerobic bacteria or are engulfed by protozoa, eviscerated by *Bdellovibrio bacteriovorans*, or exploded by phages.

At the end of the self-purification cycle, the water can be dumped into a stream, pond, or lake without hazard. In some cases, it is allowed to flow into shallow basins where it sinks into the ground and replenishes the underground water supply. If it undergoes further filtration followed by chemical treatment, it may be suitable for reuse.

A number of water-poor towns in the north central part of the United States have utilized lagoonization followed by further treatment to permit reuse of sewage to supplement or replace a dwindling water supply. Such sewage has been successfully recycled several times before its load of mineral salts became too high.

Sewage irrigation has been used as a means of sewage disposal. In practice, raw sewage is allowed to flow alternately over a series of plots of coarse, sandy soil, often seeded with food crops such as grains. The sewage solids are trapped at the surface, and are rapidly decomposed by the aerobic bacteria after the sewage has been diverted to another plot. The sand also serves as a filter allowing the purified water to filter down into underground water basins.

Primary treatment

In primary treatment the sewage may merely be emptied into large settling basins and held for several hours. During this period, enough of the solids settle out to remove about a third of the bacteria and putrescible matter.

To hasten sedimentation, ferrous sulfate, or some other soluble iron or aluminum compound, may be added to the water. If the water is sufficiently alkaline, an insoluble colloid, ferric hydroxide (or aluminum hydroxide), will form. This floc, as it is called, is made up of large, amorphous gelatinous masses that settle out rapidly. The slimy masses sweep much of the coarser organic matter down with them, leaving the sewage relatively clear.

In addition to simple sedimentation or flocculation, the effluent may be chlorinated, primarily to kill the pathogenic bacteria or inactivate the enteric viruses.

Anaerobic methods for the primary treatment of sewage are used for homes, or other small installations such as camps or schools. For home use, the septic tank is employed. Septic tanks are usually divided into two compartments by a baffle. The raw sewage flows into the first compartment where the solids are settled out and undergo slow anaerobic decomposition. When the level of the dividing baffle is reached, the top portion of the settled sewage flows over the baffle into the second or siphon tank. From here the effluent is carried by gravity into underground lines of tile with open joints laid over a bed of gravel. This constitutes the absorption field through which the effluent flows then sinks into the soil.

Secondary treatment

After sewage has undergone primary treatment, the effluent may contain about 50% of its original organic load. It may, therefore, be subjected to secondary treatment to further reduce the putrescible material and bacterial content. This is essentially a means of greatly accelerating the process of natural self-purification of the primary effluent.

Slow sand filters represent the first type of secondary treatment developed. In this process, the effluent from primary sedimentation was allowed to flow over a 5 to 6 feet deep basin filled with coarse sand. The sewage flowed down through the sand, and the filtrate was drawn off at the bottom. When initially charged with sewage, the flow-through was rapid and only a small part of the bacteria and solids were removed. However, in a few days a slimy layer, called the *schmutzdecke* (filter skin) by the Germans, developed on the surface of the sand and extended into the interstices. This consisted primarily of slime bacteria such as *Zoogloea* and *Caulobacter*-like forms. This greatly increased the efficiency of filtration but the rate of flow steadily decreased, greatly diminishing the capacity of the filter.

The modified rapid sand filter is more practical, but less efficient. Another objection to sand filters is the size of filtering area required. An acre of conventional lagoons is required for the treatment of sewage from 100 to 300 persons, an acre of intermittent sand filters can handle the sewage from about 1,500 persons, but the conventional trickling filters treat sewage from 37,500 persons per acre of surface, and an acre of activated sludge tanks is capable of treating sewage from a population of 120,000. Consequently, except for very small towns, the sand filters have been largely replaced by more efficient methods.

The trickling filter method is not a true filter, but it probably came into existence as a result of the observation of the operation of the *schmutzdecke* that developed on sand filters. The trickling filter consists of a basin about 5 to 6 feet deep filled with lemon-sized, crushed stones. The effluent from the primary settling tank is sprayed over the surface of the rocks from a series of fixed pipes with end nozzles producing an umbrella-like spray, or the more efficient self-propelled rotary sprinkler is used.

If trickling filters are well operated, they will remove 75% to 90% of the putrescible material. If operated at a relatively low rate, a filter less than six feet deep will handle about 2,000,000 to 4,000,000 gallons per day (MGD) per acre.

An activated sludge plant in operation. The Los Angeles Hyperion Sewage Disposal Plant is a classic example of a combined activated sludge (secondary treatment) and primary treatment plant, and serves some 3,000,000 persons. Greater Los Angeles is a sprawling megalopolis situated in several valleys; a great metropolitan center with a complex of many smaller urban towns. Sewage from most of these funnel into the huge Hyperion Sewage Disposal Plant situated near El Segundo on the Pacific Ocean. Figs. 17-1 through 17-5 illustrate the operation of this plant. The sewerage system is a great underground network of pipes, running from small subsidiary lines to great trunk lines. Some sewage must travel as far as 45 miles, and the plant load is about 300 million or more gallons per day. When it arrives at the Hyperion Plant, the sewage is septic (completely devoid of oxygen). It is gray-black in color, and has a stale odor often livened by traces of hydrogen sulfide. Contrary to belief, it is almost entirely water, rarely carrying more than 0.1% total solids.

The first step of this treatment system involves the flow of the raw sewage through 1-inch bar screens to remove large objects. The sewage next flows into a huge settling basin where the slowly decomposable fat and oils are skimmed off the surface, and the coarse particulate matter, mostly sand and gravel, settles to the bottom (Fig. 17-2). The settled sewage now flows into a series of large tanks. It stands for a period in the first tank, then flows over the top of a baffle into the second tank. This is repeated through several tanks (Fig. 17-3), permitting the suspended organic matter or sludge to settle out. As the sewage flows over the baffles it picks up dissolved oxygen to hasten aerobic decomposition.

The accumulated sediments are scraped from the bottom of the tanks by moving wooden bars and pumped into the sludge digestion tanks (Fig. 17-4). Here an anaerobic decomposition begins. This continues until the readily decomposable compounds have been changed to simple compounds including a substantial percentage of marsh gas or methane (CH_4), which is trapped in the top of the dome-shaped cover of the digestion tank. This is then "scrubbed" and used to drive the huge Worthington engines that generate electricity or drive the air pumps used to aerate the sewage in the "aeration tanks," and the remainder of the gas goes to an adjacent electric power plant.

After primary treatment, the liquid sewage, now relieved of a portion of its load of organic matter and bacteria, flows into huge aeration holding tanks. Here great quantities of air are

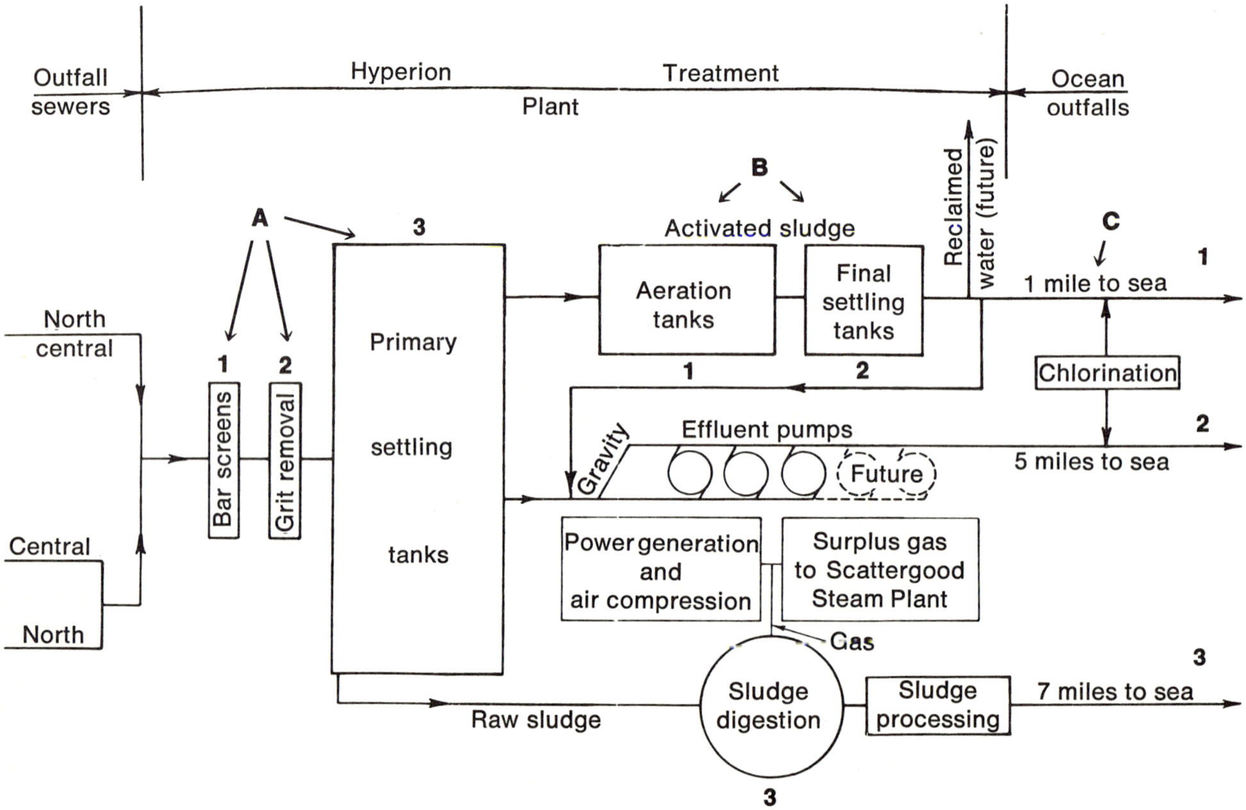

Figure 17-1
Flow diagram, Hyperion activated sludge sewage treatment plant, City of Los Angeles. Treatment steps are as follows: **A,** primary, **1, 2, 3; B,** secondary, **1, 2, 3;** and **C,** final disposal into Pacific Ocean, **1, 2, 3.** (Courtesy Los Angeles City, Bureau of Sanitation.)

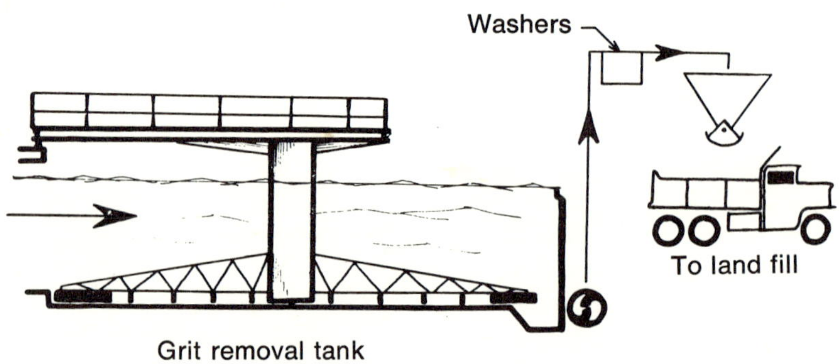

Figure 17-2
Method for removing sand and gravel from inflowing raw sewage by rapid settling (see Fig. 17-1, *A,* 2). (Courtesy Los Angeles City, Bureau of Sanitation.)

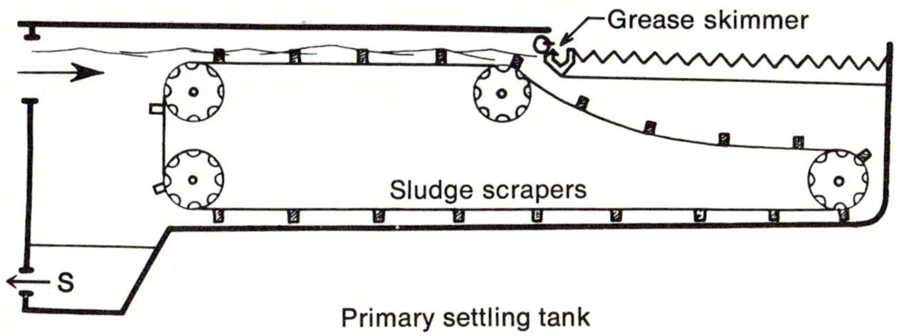

Figure 17-3
Primary removal of readily settleable organic matter and floating grease (see Fig. 17-1, *A, 3*). Settled sludge accumulates at **S** and is pumped into sludge digestion tanks (see Fig. 17-1, *B, 3*). (Courtesy Los Angeles City, Bureau of Sanitation.)

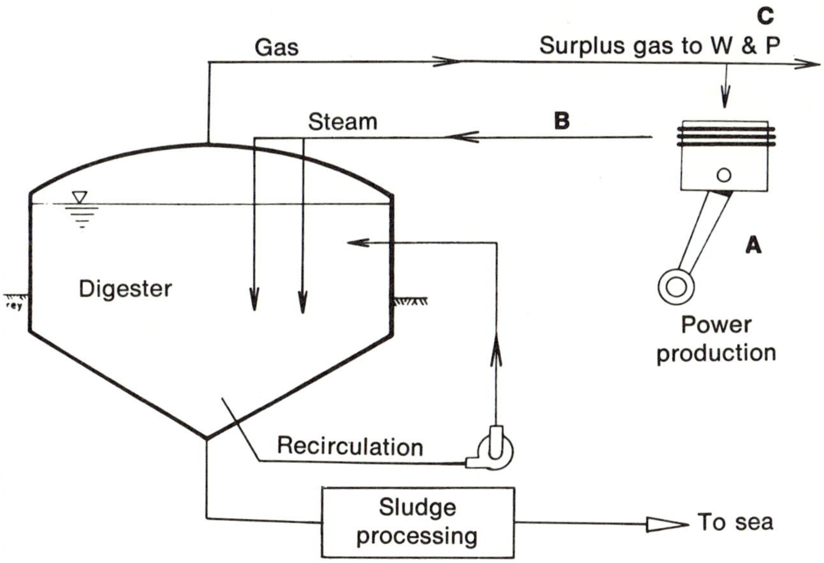

Figure 17-4
Sludge digestion process (see Fig. 17-1, *B, 3, C, 3*). (Courtesy Los Angeles City, Bureau of Sanitation.) Note: Gas from digestor resembles natural gas in many respects. It is used as fuel to operate the huge Worthington internal combustion engines which are used to generate electricity to operate pumps and aerators. It is also burned to produce steam to raise the temperature of the digestor tanks to thermophilic levels. Waste gas is sent to an adjacent city power plant to provide fuel for the steam boilers, to operate the steam turbine-powered generators.

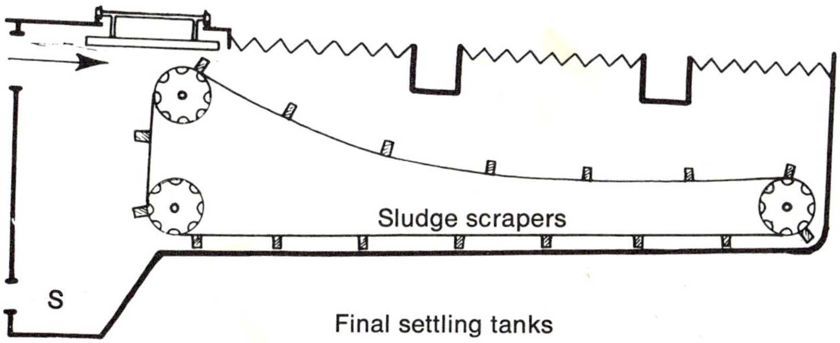

Figure 17-5
Final settling tanks for removal of sludge following aeration in secondary treatment (see Fig. 17-1, *B, 2*). Sludge accumulates at **S** and is pumped into sludge digestors. (Courtesy Los Angeles City, Bureau of Sanitation.)

blown through the sewage (about 1.5 cu ft air/gal). This serves two purposes: the aerobic bacteria are stimulated to decompose much of the organic matter to CO_2, NO_3, and SO_4, while floccules of the slower decomposing organic matter are formed to serve as bacterial niduses. These represent vigorously oxidative living masses. As these slimy floccules grow, they begin to settle out, carrying with them most of the bacteria and remaining organic matter. This process continues for about 6 hours, then the aerated sewage with its flocculated sludge flows into static settling basins (Fig. 17-5). Here the remaining sludge settles to the bottom. Part of this is used to seed the freshly filled aeration tanks, and the balance goes to the sludge-digesting tanks. The settled sewage, now carrying only about 5% of its original load of organic matter,

Figure 17-6
Major operational units, Hyperion Sewage Disposal Plant (see Figs. 17-1 to 17-5). **A,** Battery of aeration tanks in operation; **B,** closeup of aeration tanks; **C,** battery of sludge digestion tanks; **D,** collection (cylindrical) and storage (spherical) tanks for gas from sludge digestion tanks.

flows from the top of the tanks. It is then chlorinated. Initially it was dumped into the ocean through a mile-long pipe, 20 feet in diameter, emptying on the ocean floor into 45 feet of water. (This water is pure enough that a portion of it has been experimentally pumped into the ground to replenish the underground water supply for the wells that supply the neighboring cities.)

Since 1961 the secondary effluent has been discharged through a 5-mile pipeline into the ocean where it is dumped at a depth of 200 feet over the edge of a deep submarine canyon.

During rush periods, the plant does not have the capacity to treat all of the sewage. The excess is given primary treatment, consisting of settling, preliminary aeration, and chlorination. It then flows through the 5-mile pipeline.

The sludge that is settled out from the various processes undergoes anaerobic decomposition in the huge sludge tanks. When decomposition is complete (about 21 to 30 days) the slurry of sludge flows into a 7-mile long pipe and is discharged into the ocean at a depth of 320 feet at the edge of a deep submarine canyon.

To control any possibility of beach pollution, samples are taken daily for coliform tests from some 14 beach monitoring stations along the Santa Monica Bay and from a similar number of points around the outfalls in the bay itself.

Tertiary treatment

With serious consideration being given to the direct reuse of sewage for domestic purposes, many new innovative procedures are being tested.

The purpose is to create from treated sewage a safe water supply essentially devoid of organic matter, bacteria, or viruses. It should be low in nitrates, clear, sparkling, and without any objectionable tastes or odors, and of a quality suitable for direct reuse. Some of these new tertiary treatment methods include:

1. Superaeration using pure oxygen is a recently tried method of sewage treatment. In most cases, pure oxygen gas is blown over the surface of a large reaction tank. This results in greatly increased growth of aerobic bacteria and some degree of direct chemical oxidation. Pilot plants are now in operation that permit the production of pure oxygen gas as a part of the sewage treatment plant operation. The cost is not excessive and the increased efficiency obtained may justify widespread use of this method.
2. The use of higher (thermophilic) temperatures are being studied. If the temperature is increased to the lower thermophilic range, a marked increase in effectiveness of treatment and a shortened holding time is attained. Also, the rate of destruction of pathogens is materially increased.
3. To enhance the rate of decomposition of cellulose fibers, studies have been made in which pure cultures of *Cellulomonas* strains were successfully introduced into secondary and tertiary aeration tanks.

PROBLEMS INVOLVED IN REUSE OF SEWAGE

With the present constant need for greatly increased water supplies, the reuse or recirculating of sewage effluents appears to be inevitable. In fact, it has already been in operation to a limited extent with some small, water-poor cities. However, certain sociological and technical problems must be solved before this practice can become widespread. Some of these problems are public disfavor, high salt levels, high nitrate levels, and the difficulty of complete removal of intestinal viruses.

Public disfavor

The idea of using your own sewage as a water source is repugnant to many people. However, in many cases these same people live in cities where the primary source of water is a large river such as the Mississippi, the Ohio, or the Hudson. In many cases the sewer outfall of one city is not too far from the intake of the next city down the river.

Until the public is convinced that advanced treatment methods, capable of producing a completely potable water from sewage, are available, however, public resistance will constitute a major obstacle.

High salt levels

In some small water-poor cities in the Midwest, forced by drought to recycle their sewage, the buildup of salt concentration has been a major problem. This has limited the number of times the sewage could be reused. Efforts are being made to remove the accumulated salts by electrodialysis. If this can be done economically, a major objection to recycling will have been met.

High nitrate levels

In the treatment of sewage, either by natural or artificial methods, the proteins are first converted to amino acids. These are deaminated yielding considerable amounts of ammonia, which under aerobic conditions is quickly converted to nitrates.

A high nitrate content in domestic waters is objectionable for two reasons:

1. For a number of years we have known that if infants are fed a bottle formula made with water high in nitrate, the nitrates may react with the hemoglobin, changing it into stable methemoglobin that has lost its oxygen carrying capacity. Affected infants are sometimes called "blue babies" (not to be confused with infants suffering from congenital heart defects).
2. Until recently, waters high in nitrates were not considered harmful to adults because their intestinal flora includes bacteria that reduce nitrates to nitrites and ultimately to nitrogen gas. However, suspicions have now been voiced that nitrates in drinking water or certain preserved foods may be the instigators of some types of cancer.

Methods are being studied in which nitrates in the effluent may, under anaerobic conditions, be converted to N_2 by adding pure cultures of certain denitrifying bacteria.

Removal of intestinal viruses

One of the major concerns in water treatment and direct reuse of sewage is the successful removal of the enteric or intestinal viruses. Epidemiological evidence that water plays a major role in transmission of virus infections is still largely presumptive. However, waterborne epidemics of infectious hepatitis have been well authenticated. The most spectacular epidemic was an outbreak that occurred in New Delhi, India, following massive pollution of a part of the water supply; in areas of the city served by another water treatment plant, no cases of the disease occurred.

The problem of satisfactory removal of viruses from treated water has not yet been entirely solved. However, recently developed methods of chemical flocculation have shown great promise. The chief objection is the further increased salt content.

SPECIAL PROBLEMS RELATED TO SEWAGE DISPOSAL

In addition to the basic problems involved in sewage disposal, some related secondary problems have been recognized. These include poisoning due to mercury in fish, eutrophication, recently recognized indirect effects on marine algae, and sources of pollution other than municipal or industrial sewage.

Mercury poisoning from fish

Recently the public has been shocked to find that fish caught in certain freshwaters may carry potentially dangerous amounts of methyl mercury. This highly toxic organic compound of mercury may cause irreparable nerve damage and chromosomal aberrations that could give rise to abnormal offspring.

Bacteria play a major role in the production of this compound. Industrial wastes that carry inorganic or phenyl mercury compounds may accumulate in the mud at the bottom of rivers or lakes. Anaerobic bacteria convert these accumulated compounds into the ionic CH_3Hg^+, or the highly volatile methyl mercury ions. These ions escape into the water above the bottom, where they are taken up by fish and other forms. Evantually they are concentrated in the body fats of fish. For example, the body fat of certain pickerel taken in the Great Lakes has been found to contain 3,000 times more mercury than the surrounding water.

The problem, while spectacular, is apparently not acute in the United States, but in Japan a considerable number of deaths have been found to be caused by eating fish and shellfish taken from mercury-contaminated waters.

Eutrophication

A recently recognized problem that concerns the ecologist is eutrophication (Gr. well-nourished), a condition that may lead to the conversion of former freshwater lakes into bogs. Although this condition is not synonymous with pollution, pollution may be a major contributing factor. Eutrophication may rapidly develop in lakes that are generally shallow and rich in nutrients, with seasonal oxygen deficiencies in the lower depths. When plant nutrients (especially nitrates or phosphates) from sewage, run-off from farmlands, and other sources enter the lake, a tremendous growth of free-floating algae may develop. When these exhaust their nutrients and die, they settle to the bottom where decomposition, primarily by anaerobic bacteria, may occur. This results in a continually deepening rich, black, humic, organic layer. This layer furnishes an ideal soil for shallow water plants that continue to encroach from the shore deeper and deeper into the increasingly shallow lake. Eventually a clear, deep lake may be converted into bog. Such a lake has "died."

The dying of a lake may have many undesirable results. Not only may the aesthetic and recreational appeal of the lake be lost forever, but also the results may be disastrous for cities that derive their domestic water supply from the lake. The algae bloom may clog the filters of the treatment plant, the water may develop undesirable tastes and odors that are difficult to remove, while the increased organic content resulting from anaerobic decomposition makes proper chlorination difficult if not impossible.

Two plant nutrients have been indicated as major culprits in eutrophication: the soluble compounds of nitrogen and phosphorus. Sewage is relatively rich in such compounds, although run-off from adjacent, well-fertilized farm lands may also be a major contributor to the overnourishment of the lake.

Recently some ecologists have characterized the high phosphate detergents poured into sewage as the major cause of eutrophication. Many public health workers, however, question this as an oversimplification of a many-faceted problem. Certainly the problem is a real one beginning to be seen in a number of major lakes. The most spectacular threat is apparent in the Great Lakes, where overnourished Lake Erie is rapidly becoming a dying lake. It is akin to the corpulent man who continues to overeat as he waits for a fatal coronary. Likewise, there may still be time to prevent it.

Indirect effects of sewage on marine algae

Sewage pollution may produce unexpected undesirable results. Along the Southern California coast a large forest of valuable kelp is currently being depleted. The cause was not apparent at first. A careful study revealed that sewage dumped into the ocean stimulates the growth of sea urchins, which feed on the baby kelp plants, hence resulting in a significant loss of this valuable natural resource.

MICROBIOLOGICAL PROBLEMS OF SWIMMING POOLS AND BATHING BEACHES

Swimming constitutes a tertiary use of water. The microbiological, physical, and chemical problems differ from those inherent in water supplies and in sewage disposal.

Today's swimming pools, both private and public, are largely of the recirculating type. The pool is initially filled with water of drinking quality, and then regularly filtered and chlorinated. In a well-designed and operated pool, the entire contents is filtered three or four times in each 24 hours, and chlorine additions are controlled to provide a reasonably uniform level. As the water is depleted it is replenished with the same high quality drinking water. As a result, the major concern for the health of the swimmer is not with intestinal pathogens for they are not initially present and are not likely to be added to any great extent; moreover, good swimmers rarely ingest any appreciable amount of water. In addition, in a properly chlorinated pool the available chlorine content is high enough to serve as a mild disinfectant in which the bacteria are not likely to survive longer than a few seconds or at most minutes.

Any pathogens that may be present come from the swimmers themselves. Infections, if they occur, are primarily of the skin, nose, throat, or eyes. If an individual swimmer suffering from a cold or other nasal infection dives into the pool, the pressure created may carry bacteria-laden water into the frontal, ethmodial, maxillary, or the sphenoidal sinuses. In rare instances the middle ear may become infected. The bacteria involved are primarily staphylococci and streptococci, which vary in pathogenicity and the severity of the infections they evoke.

Any individual with nasal, throat, or eye infections may expel clouds of the pathogens. Many of these are quickly killed by the residual chlorine, but in a crowded pool an adjacent swimmer may acquire the still viable bacteria.

In addition to the ubiquitous staphylococci and streptococci, a number of other pathogens may find their way into the nasal area, eyes, and so forth, often as a mixed flora. The adenoviruses have been shown to be transmitted by swimming pool contact. It is logical to suppose that other respiratory viruses such as those causing the common cold, influenza, and so on, also may be so transmitted. When poliomyelitis was prevalent, several instances were reported in which the poliovirus had apparently infected susceptible swimmers through infected swimming pool water.

Impetigo has reportedly been spread from child to child by swimming pool contact. It is possible, however, that instead of being waterborne it may have been transmitted by true direct contact.

Inclusion conjunctivitis, known also as swimming pool conjunctivitis, is caused by an infective agent belonging to the chlamydial group, and closely related to the cause of trachoma or granulated eyelids. It apparently can be carried through swimming pool water from a person with an acute eye infection. This organism can also produce an active infection of the genital tract; the infection is transmissible through sexual intercourse or is carried through swimming pool water, especially if the water is not chlorinated. Proper chlorination largely prevents the spread of this pathogen.

Fresh cuts or skin abrasions may occasionally become infected, usually with streptococci or staphylococci. When a knee or elbow is abraded against the side of the pool, a chronic infection with an atypical mycobacterium can develop, and may persist for some time unless treated.

Occasionally ringworm infections of the feet occur. However, these infections are not usually acquired in the swimming pool itself but from such places as walk-ways or shower stalls.

Swimming pool waters are customarily disinfected by treating with chlorine gas or hypochlorites to produce a level of 0.4 to 0.6 ppm of residual chlorine. If chloramines are used (not generally approved by public health agencies), the residuals must be at least 2.0 ppm.

Recent studies indicate that chlorine or hypochlorite residuals of 1.0 ppm are desirable. Contrary to public opinion, this level does not cause eye irritation. Such irritation is the result of an acid reaction (when chlorine gas is used) or an alkaline reaction that develops from hypochlorites. It is, therefore, important to maintain not only an adequate chlorine residual level, but also an essentially neutral pH.

In addition to chlorine, iodine and bromine are sometimes used to disinfect swimming pools.

The tests for microbial safety usually involve MPN tests for coliforms or plate counts. However, the presence or absence of enterococci may be more significant than the presence or absence of coliforms.

As a precaution it is important to make sure that a pool is not overloaded with swimmers. Every pool has a capacity beyond which the safety of swimming is greatly reduced. Also, persons with colds or other respiratory infections, or skin infections should be barred from swimming pools.

Growth of algae can be controlled by the use of copper sulfate, increasing the chlorine, iodine, or bromine level, or using one of the organic algicides.

Swimming beaches present a different and somewhat more complicated problem. Unlike the swimming pool, the danger of sewage pollution is very real. Also it is generally most difficult to chlorinate the waters of open beaches. In some heavily used freshwater beaches underwater pipes are laid in a grid fashion to distribute hypochlorites into the water. In some instances, so-called chloroboats are used.

The primary safety measures are the constant testing of water for the presence of coliforms, and, where possible, controlling the overloading of the beach, especially in the case of freshwaters. Coliform levels above a safe point justify closing of the beaches.

In addition to the possibility of infections with intestinal bacteria and viruses when beaches are thronged with bathers, the possibility of person-to-person transmission of pathogens is always present. In fact, it poses a more serious threat than with swimming pools because the waters are generally not chlorinated.

Well-controlled beaches, frequently tested for fecal contamination, are reasonably safe for swimming. However, they are probably not as safe as the well-controlled and well-operated conventional swimming pool.

SUGGESTED READINGS

Berg, G., editor: Transmission of viruses by the water route, New York, 1966, Interscience Publishers.

Gainey, P. L., and Lord, T. H.: Microbiology of water and sewage, Englewood Cliffs, N. J., 1952, Prentice-Hall, Inc.

Lawrence, C. A., and Block, S. S., editors: Disinfection, sterilization, and preservation, Philadelphia, 1968, Lea & Febiger.

McKinney, R. E.: Microbiology for sanitary engineers, New York, 1962, McGraw-Hill Book Co.

Salvato, J. A.: Environmental sanitation, New York, 1958, John Wiley & Sons, Inc.

Standard methods for the examination of water and waste water, ed. 13, New York, 1971, American Public Health Association.

18 Microorganisms and foods

An important factor in food quality is often the nature and kind of the microorganisms present. Certain measures to control the microflora are in the hands of public health and other agencies, but most control rests with the housewife, the restaurant cook, or the consumer himself, and it can be done effectively only if that person has some concept of microorganisms in relation to food.

The nutritional needs of microorganisms are often similar to those of men or animals. Thus microbes and men are continually competing for foods and foodstuffs.

FOOD SPOILAGE—GENERAL CONSIDERATIONS

If the appearance, odor, taste, or texture of a food is sufficiently changed from the accepted norm by microbial action, it is considered spoiled and inedible.

Table 18-1. Microbes involved in spoilage of representative foods

FOOD	TYPE OF SPOILAGE	MICROORGANISMS COMMONLY RESPONSIBLE	CONTROL MEASURES
Bread	Mold	*Rhizopus nigricans* (black bread mold) *Penicillium expansum* (blue bread mold) *Neurospora sitophila* or *N. crassa* (pink bread mold)	Calcium propionate added to flour; refrigeration of bread in summer
	Rope (strings of slime)	Mucoid strains, usually of *Bacillus subtilis*—spores not killed during baking	As above; spore counts of flour to eliminate flours high in spores
	Souring	Lactic acid bacteria or coliforms	Cleanliness; refrigeration in summer
Carrots, celery, lettuce, cabbage, etc.	Soft rots; soft, mushy, almost slimy spots; unpleasant odor	Chiefly *Erwinia (Pectobacterium) carotovora*	Avoid bruising and contact with infected containers
	Soft black rots	Molds such as *Alternaria, Rhizopus,* occasionally *Penicillium*	
Fruit juices	Mold (surface)	*Penicillium* sp. most common cause	
	Clouding	Nonfermenting wild yeasts	
	Clouding; alcoholic fermentation	Fermenting wild yeasts	Cover containers to avoid air exposure
	Clouding; unpleasant odor; slime in tomato juice	*Lactobacillus* sp., *Leuconostoc* sp. (favored by acidity, pH 5.0; lack of sugar)	
Meats (fresh beef, etc.)	Slime (surface); odor	Complex group of bacteria—pseudomonads, *Achromobacter, Proteus,* etc.	Low refrigeration temperatures inhibit growth
	Rancid odor	Lipolytic (fat-splitting bacteria)—mostly pseudomonads	Same
	Musty odor	Saprophytic actinomycetes	Avoid surface contamination with soil
	Surface spots—red, yellow, etc.	*Serratia marcescens, Flavobacterium* sp., micrococci, some pigmented yeasts	Refrigeration; avoid surface contamination
	Discoloration—green, grey, brown	Mostly pseudomonads; *Achromobacter* sp. attack myoglobin	Same
Ground meats	Sour or unpleasant odor; slime	Chiefly mixed surface flora incorporated by grinding; *Proteus,* coliforms, pseudomonads (psychrophiles), etc.	Refrigerate properly; short holding period
Processed meats (frankfurters, etc.)	Mold	Any of several–most commonly *Penicillium* sp.	Same
	Slime (surface)	Various bacteria, yeasts	
Eggs (in shell)	Rotten	Pseudomonads, *Proteus, Alcaligenes,* etc.; lysozyme in egg inhibits initial growth	Proper refrigeration; avoid cracked eggs
	Mold filaments; rotting	Occasional molds on surface penetrate cracks in shell	

The distinctions between spoiled and unspoiled foods are sometimes illogical, controlled largely by the mores of a particular social group. Thus, if milk sold as "sweet" milk is soured by *Streptococcus lactis*, we will reject it; yet cultured buttermilk, soured by the same organism, is completely acceptable.

Foods vary greatly in their susceptibility to microbial deterioration. This depends on the moisture level, pH and protein or sugar content, and similar factors. Flour (13% to 15% moisture), dried milk (8%), dry beans (15%), wheat (10% to 15%), dehydrated vegetables (14% to 20%), and dried eggs (10% to 11%) keep indefinitely as long as they remain dry. The lower limit of available moisture required for microbial growth is about 18% for bacteria, at least 20% for yeasts, and as low as 13% to 16% for molds.

Because dry foods can be kept for long periods, they are termed nonperishable (a relative rather than absolute term). Other foods, with medium moisture content and a protective coating, for example, potatoes, apples, and nutmeats, are termed semiperishable. Foods high in moisture that must be stored under refrigeration are considered perishable. This group includes meats, fish, poultry, eggs, milk, and many fruits and vegetables.

The availability of oxygen is an important factor. If foods are exposed to the open air, the molds (which are essentially aerobic), those yeasts that grow best aerobically, and the obligately aerobic bacteria can all grow. If oxygen is limited or absent, growth will be confined to certain yeasts and the facultative or obligately anaerobic bacteria.

Most of the spoilage organisms are present naturally in the food environment. Spoilage may be controlled to some extent by washing the products, disinfecting the surfaces of tables, conveyers, and so forth, or adding antibiotics or certain disinfectants to wash waters, especially in the case of fish or poultry. The use of germicidal ultraviolet lights in the food-handling areas is often helpful in reducing surface contamination.

If the protein content of a food is high and the pH near neutral, growth of bacteria is favored.

In prepared meats with a high salt content, particularly those held in pickling brines, halophilic bacteria may grow rather rapidly. On the other hand, surface molding of meats high in salt may occur more commonly than the formation of surface slimes by halophilic bacteria.

In concentrated sugar products such as honey, when growth does occur it is most commonly due to osmophilic yeasts. Honey often contains substances that are fungicidal; this may prevent molding. High sugar content, coupled with an acid pH, favors yeast and mold growth. This explains the molding of fruits, and the spontaneous alcoholic fermentation by naturally occurring yeasts of expressed fruit juices such as apple cider or grape juice. The concentration of sugars in fruits and vegetables is not high enough to inhibit bacterial growth and the inhibitory factor for bacteria appears to be acidity rather than sugar content.

Another nutritional aspect leading to spoilage may be the concentration of B-vitamins in a given food. High concentrations generally stimulate the growth of spoilage organisms, especially bacteria. Spoilage of eggs appears to be inhibited by the presence of the antibacterial enzyme, lysozyme, in the egg white.

Types of spoilage of representative foods, microorganisms involved, and control measures are summarized in Table 18-1.

METHODS OF FOOD PRESERVATION

Canning

Prompted by the offer of a prize of 12,000 francs for the development of a practical method for preserving food to feed Napoleon's armies, Nicolas Appert, a French confectioner, in 1795 began a series of experiments on food preservation. His procedure was successful, and in 1809 he received the coveted prize. The following year he published a book describing his method, which involved placing food in wide-mouthed glass jars, closing them with cork stoppers, and heating for several hours in a boiling water bath. In 1810 Peter Durand, an Englishman, patented a method for preserving foods in "tin canisters" — soon shortened to "cans."

Until about 1850 the industry used only boiling water temperatures, but attempts were made with some success to raise temperatures by adding calcium chloride or salt to the heating baths, or by substituting oil for water.

In 1874, a retort, pressure cooker or autoclave, was patented by Schriver, a canner in Philadelphia. With this device it was now possible by increased steam pressure to raise the holding temperature well above boiling. The retort was a real breakthrough and gave great impetus to canning. About 1920 several sporadic outbreaks of botulism traced to canned olives and other nonacid foods prompted the National Canner's Association to develop a series of recommended procedures based on careful experiments. Application of these methods and standards has essentially eliminated canned goods as a source of botulism, and has greatly reduced spoilage.

The problems facing the canner are highly complex. Killing botulism and spoilage bacteria in nonacid foods requires a much longer period of heating and higher temperatures than for acid foods. Also, a narrow temperature range exists between preserving and overcooking food products.

Theoretically, the canner seeks to achieve a sterile product. This does not always occur, and the state of the canned product is described as "practical sterility." This implies that *Clostridium botulinum* spores have been eliminated, together with the even tougher spores of the thermophilic spoilage anaerobes or facultative anaerobes. Spores of innocuous aerobic bacteria sometimes persist but may be unable to grow.

Spoilage of canned goods may involve "swells" in which the spoilage bacilli produce enough gas to cause the top or bottom of the can to bulge. However, with high acid fruits, a nonmicrobial swell may result from the reaction of the fruit acid with the metal of the can releasing hydrogen gas (Fig. 18-1).

Another type of spoilage is the "flat sour" in which the spoilage organism produces quantities of lactic acid and other acids but no gas. If hydrogen sulfide is produced, the spoilage is described as a "stinker," and the inside of the can will be blackened due to the formation of metallic oxides.

After removal of cans from the retort they must be quickly cooled to prevent the growth of the thermophilic spore formers that might otherwise grow during a prolonged cooling period.

Although spoilage does occasionally occur, the success of canning as a method is an amazing monument to applied microbiological knowledge.

Refrigeration

Refrigeration is effective in extending the keeping qualities of foods because many of the mesophilic forms have a minimal growth temperature above that of a properly operating refrigerator, consequently, they fail to grow. However, obligate or facultative psychrophilic bacteria, and molds and yeasts will grow slowly and spoilage will eventually result. (See Chapter 10.)

Quick freezing

Slow freezing, in which 3 hours to 3 days may be required for freezing food solid, preserves foods. However, when thawed, the foods are soft, juices exude, and the quality is impaired. This is due to the formation during freezing of large ice crystals that rupture cells of the plant or animal tissue.

On the other hand, quick freezing of food, usually in small containers at temperatures as low as −50° F, achieves a frozen state in 30 minutes or less. This rapid freezing forms only very small ice crystals that do not rupture the cells. Such foods keep for relatively long periods if stored at very low temperatures. Before quick freezing, fruits or vegetables may be blanched

Figure 18-1
Spoilage action in canned goods. **A,** Can with leaky seam, subject to microbial spoilage; **B,** normal can; **C,** "swell" due to growth of gas-producing bacteria in can or chemical production of hydrogen from acid in fruit; **D,** spoilage of canned fruit due to gas-producing yeasts (pressure has forced fruit juice through seal of lid).

or scalded to reduce surface organisms and inactive enzymes that might damage quality during storage. After thawing, bacteria that survive in the frozen foods may begin to grow, impairing the quality or giving rise to food poisonings. Frozen foods that have thawed should *never be refrozen.*

Ultralow temperatures

Packaged foods may be frozen with liquid nitrogen at about −150° F. Placed in an insulated container, they may be shipped considerable distances without requiring holding in refrigerated compartments.

Drying, dehydration, or desiccation

From time immemorial, man has preserved foods by sun drying them. At an early time he found that meat dipped in hot brine, then dried in the sun, would keep indefinitely. Likewise, fruits or other materials spread on trays or flat rocks would be sun dried successfully. In the case of fruits, not only is the water removed by drying, but the sugar and acid concentration is sufficiently increased to exert considerable preservative action except against osmophilic yeasts or molds.

To hasten the drying process, artificial heat came to be used. Large evaporators were developed that would quickly dry fruits or vegetables on trays or moving belts in a stream of hot air. With liquids such as milk, various juices, soups, coffee, or other drinks, heated drums were revolved through the liquid, and the dehydrated film scraped off when drying was complete. The spray-dry method is now widely used. Milk or other liquids are blown as a fine mist into a huge barrel-like container or "cyclone" through which hot, dry air is continually blown. As the moisture is removed, the solids fall to the bottom as a fine, dry powder. Large quantities of dehydrated milk, eggs, and soup are now on the market.

Freeze drying or lyophilization

Dehydrating foods requires heating at relatively high temperatures. Recently a method of dehydration at low temperatures in a vacuum, with the water being trapped in supercooled containers has been developed. This is known as freeze drying or lyophilization. Not only are the food components themselves largely unchanged, but even the microorganisms are still alive. Such foods can be reconstituted practically in the original state merely by replacing the initial water content. However, it should be remembered that once food is reconstituted it is equally as perishable as the initial product since it still has its original microflora.

Irradiation

Two types of irradiation have been used in food preservation, ultraviolet light and gamma radiation. If large meat refrigerators are equipped with germicidal ultraviolet lamps the meat can be held at higher temperatures than usually required without development of surface molding or slime. This allows the natural autolytic enzymes to break down meat structures, making the meat more "tender." Meat so stored is often advertised as "tenderized."

When gamma radiation first became available from such sources as cobalt-60 it was hailed as the ultimate food preservation method. The rays, which can penetrate up to 5.5 inches and have considerable germicidal effect, were seen as ideal methods for the cold sterilization of foods in cans. However, several practical difficulties became evident. When the product was irradiated long enough for sterilization, it was often discolored and the oxidation produced undesirable flavors. Also, the killing power varied with the bacteria present and *Clostridium botulinum* spores were not consistently destroyed. The cost of the method raised another practical problem.

Low dosage exposure has been found to increase the shelf-life, especially of meat products, without any impairment of appearance or flavor. Research still continues on gamma and other radiation as a food-preserving method.

Chemical preservatives

The origin of salt as a food preservative is lost in the distant past. Today a number of foods, especially beef, pork, and fish, are preserved with salt. For example, in preserving hams, a 15% brine is used with sugar, sodium nitrate, and sodium nitrite added. These last two preserve the red color of the meat by converting the myoglobin to a stable form. The sugar serves a dual purpose; not only does it improve the flavor, but it initially stimulates the growth of bacteria that reduce the nitrates to nitrites, which are, themselves, growth inhibitors. The hams are cured at 36° to 38° F. If spoilage does occur, it is usually due to halophilic bacteria introduced in the salt itself.

Smoking with wood smoke has long been used as an adjunct to salt preservation of meats. The smoke yields a mixture of compounds known as pyroligneus acids, which adhere to the surface. These include such preservatives as phenols, cresols, and formaldehyde in small quantities. Smoking adds flavor, reduces surface molding, provides a gloss, and tenderizes the underlying meat. In the past, heavily pickled, heavily smoked hams would keep without refrigeration. Today hams are much less heavily preserved and must be refrigerated.

Acetic acid is the active agent in vinegar. The highly acid pH achieved with the 2% to 4% acetic acid solution found in commercial vinegar accounts largely for its preservative action, although the acetate ion itself is inhibitory to certain bacteria.

Spices and flavoring agents such as horseradish, garlic, and onions act as bacteriostatic agents as long as their highly volatile oils are still present. When these oils evaporate microbial growth may be resumed (Fig. 18-2).

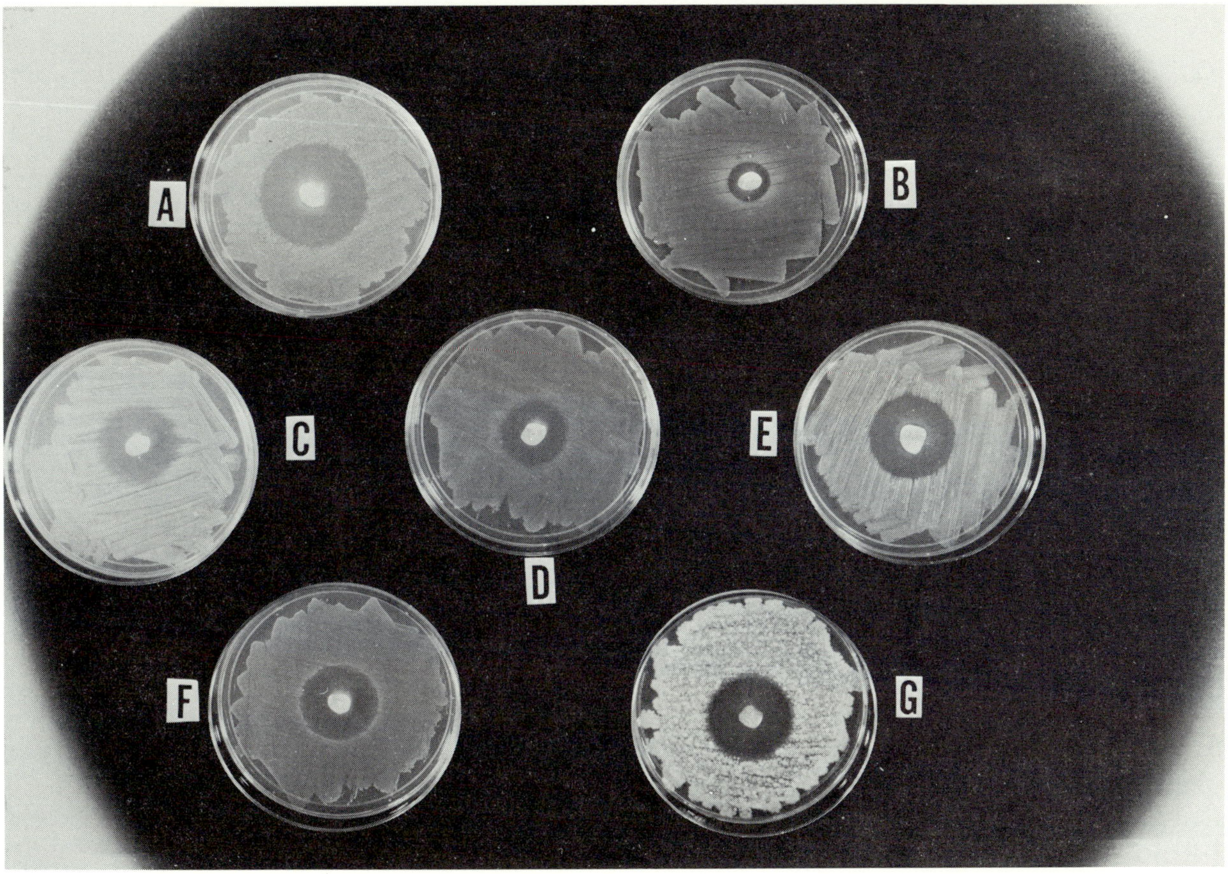

Figure 18-2
Inhibitory action exhibited when small slice of garlic is placed on surface of seeded nutrient agar plate. **A,** *Escherichia coli;* **B,** *Serratia marcescens;* **C,** *Sarcina flava;* **D,** *Pseudomonas* sp.; **E,** *Sarcina lutea;* **F,** *Staphylococcus aureus;* **G,** *Bacillus polymyxa.* (Note: Action is temporary. When pungent garlic oil evaporates, clear zones are filled with bacterial growth.)

Many spices, especially oil of cloves (eugenol) and oil of cinnamon (cinnamic aldehyde), have substantial germicidal power.

FERMENTED FOODS

A number of common foods are produced by fermentation processes, especially those involving the lactic-acid-producing bacteria. These include sauerkraut, fermented (dill-type) pickles, and green and ripe olives. The basic procedure consists of packing the plant material in deep containers and adding salt (NaCl) directly or as a brine to prevent the growth of proteolytic spore formers, aid in securing anaerobic conditions, and permit the unimpeded growth of various lactobacilli. This results in vigorous lactic acid formation. Flavoring materials such as dill or vinegar may be added. Green or ripe olives are first immersed in lye (1.25% to 1.5%

Table 18-2. Microflora involved in producing representative fermented foods

FOODS	SALT (NaCl) CONCENTRATION	INITIAL BACTERIA	INTERMEDIATE FLORA	FINAL FLORA	FINAL ACIDITY
Sauerkraut	About 2.5%	Coliforms	*Leuconostoc mesenteroides*	*Lactobacillus plantarum*	pH 2.5*
Fermented pickles (dill type)	8% increased to 15%	*Pseudomonas* sp. *Flavobacterium* sp. *Achromabacterium* sp.	*L. mesenteroides* *Streptococcus faecalis* *Pediococcus cerevisiae*	*Lactobacillus brevis* followed by *L. plantarum*	1.0% to 1.5% lactic acid
Green or ripe olives	Similar to fermented pickles				

*1.25% lactic acid, 0.3% acetic acid, 0.6% ethyl alcohol.

NaOH) to remove the bitter glucoside, oleuropein. Bacteria involved and other data are shown in Table 18-2.

FOOD INFECTIONS AND FOOD POISONINGS

Outbreaks of food-borne diseases may develop in two different ways:

1. The foods may merely serve as mechanical carriers of the pathogenic bacteria, protozoa, or viruses in numbers large enough to enable the pathogens to establish themselves in the face of the protective action of the body. Once established, the pathogens continue to reproduce during a period of incubation until their numbers and metabolic products are sufficient to produce the symptoms characteristic of the disease.
2. In the true food poisonings, great numbers of bacteria grow in a food before it is eaten, producing toxic products that result in the violent fulminating reactions that occur a few hours after the food has been consumed. In some cases, the massive numbers of bacteria alone may be responsible for the distressing symptoms of food poisoning.

Bacterial food poisoning represents unpleasant episodes in the lives of practically everyone. The food poisoning usually runs its brief stormy course without its etiology being determined. Only in rare outbreaks are public health agencies called in to determine the cause. As an example, the California State Health Department summarized its findings on 705 outbreaks, totaling 14,831 cases, investigated between 1959-1964. Of the total outbreaks, 48% were of unknown etiology, 20.8% were due to staphylococcus enterotoxin, 8.1% involved *Salmonella* strains, 4.6% were due to *Clostridium perfringens*, while 1.4% involved other bacteria such as *Bacillus cereus*, *Streptococcus* species, and *Enterobacter aerogenes*. The remaining cases were due to chemicals or plant poisons.

Staphylococcus enterotoxins

Certain strains of *Staphylococcus aureus* produce substantial quantities of enterotoxin, so called because of its specificity for the intestinal tract. Dack and co-workers in 1930 first demonstrated the toxic nature of *S. aureus* in an outbreak of food poisoning.

Staphylococci food poisoning is a true toxic poisoning rather than an infection. Outbreaks are usually due to foods in which the bacteria grow rapidly when held at room temperatures for a few hours. A variety of foods may be involved, such as custards, cream puffs, sliced meats, certain salads, sandwiches made with salad dressing, and various meat products, including ham. The foods are usually infected by food handlers suffering from boils or other staphylococcus infections on their hands, or by droplets from sinus or throat infections.

The period between ingestion of foods and initial symptoms may be as short as 1 hour or as long as 6 hours, ordinarily 2 to 4 hours. The onset is usually abrupt and violent with severe nausea, cramps, vomiting, diarrhea, and prostration. Temperature and blood pressure may drop dramatically. The disease is self-limiting, generally subsiding within 24 hours or so. Deaths rarely occur.

The toxin can withstand heating, so that cooking infected food does not render it nontoxic. Control involves proper refrigeration and eliminating food handlers with known staphylococcus infections.

Salmonella food poisoning

Food-borne salmonelloses (other than typhoid and paratyphoid), may grade imperceptibly from simple intestinal infections preceded by a prolonged incubation period to the dramatic fulminating food poisonings. Only this latter aspect will be considered.

The genus, *Salmonella*, with its more than 1,000 serological types, is widely distributed. The organisms frequent the intestines of cattle, sheep, and especially swine and poultry. The family cat or dog may sometimes be a carrier. A fraction of a percent of humans are thought to be chronic carriers. Thus, various meats may become infected during the slaughtering process, powdered eggs may carry salmonellae, and food handlers themselves may infect the food they handle.

Proper cooking and refrigeration aborts most potential infections, but occasionally foods may be prepared and allowed to remain at room temperature for a long period before being served. Inadequately cooked meats, especially poultry or pork, may also be responsible for food poisonings.

The serotypes most commonly involved are *S. typhimurium* (for which mice and rats are the natural reservoir), *S. heidelberg, S. enteritidis* (frequently encountered in cattle), *S. panama, S. newport*, and *S. oranienburg*.

Large outbreaks may occur in hospitals, institutions for children, among military personnel, and following consumption of food at church suppers, or other situations where large numbers eat a common meal. However, salmonella food poisoning can and does occur in the home.

The incubation period is usually 12 to 24 hours, although it may be as short as 6, or as long as 48 hours. The symptoms are acute and distressing, characterized by nausea, explosive vomiting and diarrhea, abdominal pain, moderate fever, and prostration. The stools are usually greenish and foul smelling. The symptoms generally persist for 2 to 3 days, or occasionally as long as 6 days, before they subside. Bacteria may be discharged in the feces for 3 days to 3 weeks, but chronic carriers are rare. The carrier state sometimes persists for months among young babies. Mortality is low, confined largely to infants or older debilitated persons.

Botulism

Botulism is the classic example of true toxin food poisoning and is caused by the most potent of all the bacterial toxins. (As little as 1/1,000,000 of a gram will kill a 20-gram mouse.)

It was named botulism by Mueller in 1870, and van Ermengen (1896) applied the name *Bacillus botulinus* to the organism isolated from an outbreak similar to that described by Mueller. Two species are now recognized as causing botulism: *Clostridium botulinum* (types A, B, C, D, E, and F) and *C. parabotulinum*.

The organisms are widely distributed in nature, and consequently spores may often be present in foods. However, certain factors limit growth, such as acidity (below pH 5.0), growth of other bacteria such as the lactic acid producers, refrigeration temperature (below 10° C for types A and B_1, 3° C for types E and F), and lack of anaerobic conditions. Nitrites will inhibit growth, and if growth has occurred boiling the food for 15 to 30 minutes will destroy the toxin.

Food products most commonly involved are home-canned, nonacid vegetables, lightly salted cured meats, pickled fish less acid than the limiting pH, and inadequately smoked or salted dried fish.

Human outbreaks are most commonly due to types A and B, while type C is encountered in poisonings of ducks and other wild birds. Type E toxin is most commonly associated with fish or fish products.

When food containing the toxin is ingested, the toxin is absorbed through the intestines into the blood stream. Symptoms may appear in 18 hours or may be deferred for as long as 4 days. An outstanding symptom is double vision, followed by difficulty in swallowing and slurred speech. Although vomiting may sometimes occur, constipation rather than diarrhea is common. By the third to seventh day death may come as a result of paralysis of the respiratory muscles.

Fortunately, the disease is now extremely rare, with only a score or so of cases per year in the United States.

The toxin is a true neurotoxin, and once it has been ingested little can be done to neutralize its effect. Its action is to interfere in some manner with acetylcholine release at the myoneural junctions. Although antisera are available, they exert little or no effect after the symptoms develop. The fatality rate averages about 50%.

Clostridium perfringens food poisoning

Clostridium perfringens is widely distributed in nature. Most of us carry this organism in our intestines, and it, or its spores, is found in soil, sewage, natural fertilizers, and water.

The organism, being a spore former, is not killed in cooking. In certain foods such as rolled roasts, that are high in protein, B-vitamins, and lecithin, the spores germinate readily. If the cooked meat is held at about 40° to 45° C (104° to 118° F) for a few hours, the numbers may increase dramatically. No specific toxin has been identified, but ingestion of several hundred million bacteria apparently produces the symptoms. The incubation period is usually 8 to 20 hours. The symptoms include great quantities of intestinal gas, nausea, cramps, and diarrhea. Vomiting and fever rarely occur. Symptoms usually subside after 21 to 24 hours. Control of poisoning involves the immediate consumption or rapid cooling of meats after cooking.

Bacillus cereus food poisoning

The *Bacillus cereus* organism is responsible for a type of food poisoning similar in most respects to *Clostridium perfringens* outbreaks. The organism is widely distributed in the soil, and, when present in very large numbers in food, may cause violent but brief food poisoning symptoms. Although rarely encountered in the United States, frequent outbreaks have been reported in Scandinavia and Hungary.

OTHER MICROBIAL TOXINS

In addition to the well-known types of bacterial food poisonings, severe and sometimes fatal symptoms that follow soon after toxin-containing food is ingested may be caused by other microorganisms or rare bacteria.

Shellfish poisoning

Shellfish poisonings have been known for more than 200 years. A series of outbreaks along the California coast (1928-1937) led finally to the conclusion that shellfish developed their toxicity as a result of ingesting the red tide dinoflagellate, *Gonyaulax catenella*. The California mussel, *Mytilus californianus*, and the bay mussel, *Mytilus edulis*, both were found to accumulate a toxin from *Gonyaulax* when the algae concentration exceeded 200 cells/ml. Below 200 cells/ml the toxin was excreted or destroyed without accumulation. At the height of a red tide, the concentration has often reached

20,000 to 30,000/ml. The blooms usually last about two weeks, after which time the mussels soon lose their toxicity.

The Alaskan butter clam binds the poison for a year or more. As a result, a very low concentration of *Gonyaulax* may be sufficient to make the mussel highly toxic.

The symptoms of poison strike quickly. About 30 minutes after the clams are eaten, the lips and tongue begin to tingle and a progressive paralysis ensues. If death has not occured within 24 hours, the prognosis is good.

Toxic metabolites from molds

During the last decade, a great deal of interest has developed in a group of powerful toxins produced by molds and known as aflatoxins.

These were first recognized in 1960 in Great Britain when a number of young turkeys died as a result of eating moldy peanut meal. A careful study showed that the toxin was produced by the growth of *Aspergillus flavus*. The name, aflatoxins was coined from the genus initial and the species name. *A. parasiticus*, *A. fumigatus*, and *Penicillium puberulum* strains also have been found to produce aflatoxins.

Initially it was thought that moldy peanut and cottonseed meals were primary sources of the toxin; however, cornmeal, wheat, oats, buckwheat, rice, and soybeans also have been found to support growth and aflatoxin production by *A. flavus* and other active molds.

When moldy grains were fed to rats, liver tumors developed; moldy cottonseed meal caused liver tumors in young trout.

Emphasis thus far has been on aflatoxin poisoning of domestic animals fed moldy grains. However, the suspicion is developing that aflatoxin might have an undesirable effect, at least indirectly, on humans. Milk from dairy cattle fed aflatoxin-contaminated grain was shown to contain the toxin. The compound is extremely heat stable, and will withstand autoclaving.

SUGGESTED READINGS

Frazier, W. C.: Food microbiology, ed. 2, New York, 1967, McGraw-Hill Book Co.

Hammerstrom, R. J., and Hill, W. F., Jr., editors: Proceedings of the Gulf and South Atlantic States shellfish research conference, Cincinnati, Ohio, 1969, USPHS Bureau of Water Hygiene.

Lawrence, C. A., and Block, S. S., editors: Disinfection, sterilization, and preservation, Philadelphia, 1968, Lea & Febiger.

Recommended methods for the microbiological examination of foods, ed. 2, New York, 1966, American Public Health Association.

Recommended procedures for the examination of sea water and shellfish, ed. 4, New York, 1970, American Public Health Association.

Slanetz, L. W., Chichester, C. O., Gaufin, A. R., and Ordal, Z. J., editors: Microbiological quality of foods, New York, 1963, Academic Press, Inc.

Thatcher, F. S., and Clark, D. S., editors: Microorganisms in food: their significance and methods of enumeration, Toronto, 1968, University of Toronto Press.

19 Microbiology of milk and dairy products

In America, approximately one-fifth of our average food intake is made up of milk and its products. Practically everyone consumes some milk or milk derivative, knowingly or unknowingly. Thus an interest in the microbiology of this important food is justified.

Not only does milk or its various derivatives represent a vital human food, but it also provides an excellent medium for the growth of many kinds of microorganisms.

COMPOSITION OF MILK

In spite of its nutritional value, milk is mostly water (87% to 87.5%). Its nutrients are as follows:

1. The protein content averages about 3.2%. Of this 80% is casein; the balance is present in the whey—the portion remaining after the casein is removed. The white color of milk is due primarily to the calcium caseinate. Milk proteins are ideal in that they are complete proteins and high in the essential amino acids.

   Four varieties of casein have been identified: alpha, 50%; beta, 30%; gamma, 5%; and kappa, 15%. This last fraction contains some sialic acid, which aids in keeping milk proteins in a colloidal suspension. Another, but small fraction, is lactalbumin, said to be responsible for the skin that develops when milk is boiled.
2. Butterfat content averages about 3.8% but varies considerably with the breed of cattle producing the milk.
3. Milk contains about 4.8% carbohydrates made up primarily of lactose or milk sugar. Lactose is remarkable in that it only is known to occur in milk. It is a disaccharide made up of a molecule of glucose and a molecule of galactose. Its limited occurrence in nature, and its unusual molecular structure may explain why so many bacteria are unable to attack lactose fermentatively.

Although most of the butterfat is a product of the body metabolism of the cow, recent studies indicate that as much as 10% may be contributed by the bacteria and protozoa in the rumen. Likewise, many of the amino acids in the milk protein come from the same source.

When the casein is removed from milk for cheese making, the watery part or whey remains. This contains serum components, much of the sugar, lactalbumin, minerals, and water. The serum components contain antibodies and in diseases such as *Brucella* infection, their presence in whey aids in detecting the disease. Also, the antibodies inhibit bacterial growth in raw milk for the first few hours after the milk is drawn.

MICROORGANISMS IN MILK AND ITS PRODUCTS

Microorganisms, especially bacteria, bear a direct relationship to milk as a carrier of pathogens, as microbes that spoil milk or impair its quality, or as fermenting forms that convert milk and its products into desirable foods such as buttermilk, yogurt, and cheese.

Milk-borne diseases

A few decades ago, milk was an important carrier of a variety of diseases, but today Grade A pasteurized milk and dairy products of similar quality are among the safest of all foods. Milk-borne epidemics have become rare indeed. The application of microbiological principles to the industry has wrought this amazing transformation. These measures include inspection of dairy herds and elimination of disease-carrying animals, immunization of the cattle, meticulous control of the sanitation of milk handling and milk handlers, pasteurization, and refrigeration.

Milk-borne epidemics still occur in hygienically less advanced countries, or in our own country in areas where milk is carelessly produced from untested cows, carelessly handled, and consumed raw.

Milk-borne diseases fall into three categories: diseases from infected animals, diseases from human carriers, and diseases transmitted to milk from the environment, for example, infection of milk by polluted wash waters.

Infected dairy cattle may produce milk that carries the bacteria responsible for tuberculosis, brucellosis, septic sore throat, and staphylococcus food poisoning. In addition, the Q fever rickettsia can be transmitted from infected dairy cattle.

Typhoid and septic sore throat epidemics arising from human carriers are now rare. Milk-borne diphtheria epidemics have disappeared, because diphtheria, like typhoid, has practically vanished from the United States.

Toxigenic staphylococci from infected cattle or human carriers still occasionally give rise to explosive food poisonings. These are traced on rare occasions to unrefrigerated milk, but more commonly to improperly refrigerated foods made with infected milk.

The pathogenic strains of *Escherichia coli* have recently replaced in importance the *Salmonella* pathogenic types, especially among children who have been involved in outbreaks of summer diarrhea. Likewise, nonspecific types of enteritides may arise from the consumption of milk or milk products containing large numbers of killed or living *E. coli, Proteus, Pseudomonas,* or *Clostridium perfringens.* These may find their way into otherwise completely safe milk through polluted water used for washing the equipment, and may multiply as a result of inadequate refrigeration.

A few authenticated cases of milk-borne protozoan infections have been described. Amebic dysentery is the most common of the protozoan infections, although in rare instances the protozoa of balantidiasis and giardiasis may be involved.

Even high quality, safe, Grade A pasteurized milk may become contaminated if handled carelessly in the home or restaurant.

Bacteria and impairment of quality

Defects in the quality of milk, such as impaired flavor, gas formation, bitterness, objectionable odors, and sliminess, may result from the growth of a variety of bacteria. However, such defects are not often encountered today because high quality, properly pasteurized milk rarely contains spoilage organisms, and adequate refrigeration delays or prevents growth of such forms long enough for the milk to be consumed before the defects become apparent.

Raw milk, if held at or near room temperature for 6 to 48 hours, often undergoes spontaneous or normal souring due to the production of lactic acid by *Streptococcus lactis* that soon causes the milk to curdle or clabber. This type of fermentation is not objectionable because the organism is homofermentative; that is, only one kind of acid, lactic acid, is produced. This gives the "clean" sour taste found in buttermilk. Curdled sour milk or clabbered milk is considered a delicacy by many people. Significantly, souring does not become apparent until the numbers of *Streptococcus lactis* reach 10,000,000 to 40,000,000 per milliliter, a population many times greater than encountered in raw high grade milk held under refrigeration.

Pasteurized milk will sometimes sour if held in a warm place, due to certain thermoduric strains of *Streptococcus lactis, S. cremoris,* or *S. thermophilus* that have survived pasteurization.

With all undesirable milk spoilages, huge numbers of bacteria are necessary before impaired quality is apparent. Some of the types of impairment and the organisms involved are as follows:

1. Sliminess or ropiness in milk is most often due to *Alcaligenes viscolactis,* although some strains of *Escherichia* and *Enterobacter,* and *S. cremoris* var. *hollandicus* also may produce a similar reaction. Occasionally these bacteria survive pasteurization.

2. Sweet curdling, caused by a renninlike enzyme instead of a high level of acidity, may be produced by some strains of *Bacillus subtilis* and *B. cereus*, by strains of *Proteus*, some micrococci, and some strains of *S. faecalis* var. *liquefaciens*. With this last organism, sweet curdling may precede acid production.
3. *Escherichia coli* produces souring, curdling, and marked gas formation, while some strains of *Enterobacter aerogenes* produce quantities of gas that cause the milk to become frothy without producing enough acid to result in curdling. *Clostridium perfringens* may produce a "stormy" fermentation in which quantities of gas are combined with curdling. This may result in the curd being blown out of the tube or bottle.
4. Bitterness may result from proteolysis of the milk proteins. A variety of organisms, particularly pyschrophiles, may produce this effect.
5. Fat splitting or lipolysis is mostly encountered in butter, and is characterized by development of a markedly rancid odor. *Pseudomonas fluorescens*, *P. fragi*, and *Achromobacter lipolyticum* are the forms most commonly involved.
6. Odors (other than those absorbed from feed eaten by the cow) may be produced by heavy bacterial growth. These include musty odors caused by pseudomonads, and medicinal odors involving *Enterobacter aerogenes*.
7. Milk is sometimes turned red by a massive population of *Serratia marcescens*. "Blue milk" results from the combined growth of *Pseudomonas syncyanea* and *Streptococcus lactis*. These were often encountered in the past, but such spoilage is rarely seen today. Pink yeasts may grow noticeably in cottage cheese or sour cream.

METHODS FOR INSURING SAFETY OF MILK

In addition to the inspection and testing of dairy cows, and sanitary inspection of dairy barns and processing plant equipment, pasteurization is a prime factor in the production of today's remarkably safe milk.

The principle of heat pasteurization was developed by Louis Pasteur in 1868 to cure the "wine disease" in which spoilage organisms so impaired the quality of wine that the whole wine industry of France faced economic ruin. This method was applied to milk about 1895 and proved most effective in the control of milk-borne diseases.

Pasteurization has two objectives: the killing of disease-producing bacteria and viruses, and destruction of bacteria responsible for the impairment of the quality of milk and its products.

Pasteurization is not sterilization. It involves the heating of milk to a temperature that kills pathogenic bacteria and apparently viruses, together with most of the spoilage bacteria. It results in the destruction of 90% to 99% of the microorganisms. Spores of bacteria usually withstand pasteurization along with certain heat-resistant nonsporing bacteria.

Vegetative bacteria that resist pasteurization but do not grow at pasteurization temperatures are termed thermoduric or heat-enduring forms.

The thermoduric bacteria include various strains of *Micrococcus thermophilus*, strains of *Streptococcus faecalis*, species of *Bacillus*, and possibly a few strains of *Escherichia coli* and *Enterobacter aerogenes*. *Microbacterium lacticum* is frequently found in pasteurized milk and is one of the most heat resistant of the nonsporing bacteria. Thermoduric forms are generally not pathogenic, and do not impair the keeping qualities of properly refrigerated milk. However, if milk is held at temperatures above 10° C (59° F), growth and quality impairment may occur.

On the other hand, thermophilic bacteria may actually grow during pasteurization by the holding method. In general, thermophiles have growth temperature optima around 50° C. Some will not grow below 45° C (obligate thermophiles), but others (facultative thermophiles) grow down to 25° C (room temperature), and consequently are easily detected by the standard plate count. Most of the thermophiles are spore-forming bacilli, although *Lactobacillus thermophilis* is the one nonsporing form frequently encountered.

Two methods of pasteurization are commonly used.

1. The holding, vat, or low-temperature long-time (LTLT) method. This consists of heating milk, cream, ice cream mix, or similar products in a large vat. For high quality raw milk, the temperature generally specified by public health regulations is 145° F (62.8° C) for 30 minutes. With cream, ice cream mixes, and so forth, higher temperatures (155° F) may be used. The temperature used is one that will destroy pathogens and most spoilage bacteria without imparting any detectable burnt flavor.
2. The high-temperature short-time (HTST) method for pasteurizing high quality, Grade A raw milk. This involves quickly heating the milk to 161° F (71.6° C) for 15 to 17 seconds, then cooling it immediately. The HTST method is effective only when used with a high quality, low bacteria-count raw milk.

Freshly drawn raw milk is not sterile. It usually contains 200 to 500 bacteria per milliliter. However, most of these bacteria and those picked up from the milking equipment and containers are eliminated during pasteurization.

In modern dairies producing high quality raw milk, extreme precautions are taken to prevent bacterial contamination. The udder and the adjacent parts of the cow are washed with a spray of warm water, then usually wiped with towels moistened in disinfectant before the cups of the milking machine are attached to the teats. The milk is drawn directly into a large refrigerated vat where it is quickly cooled to near the freezing point. When milking is finished, the milk is pumped into large insulated tank trucks, which are essentially massive thermos bottles. These transport the milk to control-processing plants where it is pumped directly into refrigerated holding vats. Almost immediately the milk is passed through an HTST pasteurizer, cooled in seconds, and immediately bottled in sanitized bottles or cartons. Placed in racks, these are moved immediately into refrigerators to await distribution. Thus the milk may never have human contact or be exposed to daylight until poured into the consumer's glass.

A few food faddists have objected that pasteurization of milk reduces its food value, but these objections are not borne out by sound scientific experiments.

PROCEDURES FOR TESTING MILK FOR QUALITY AND SAFETY

Standard plate count

In the 1890's, public health workers became concerned with milk as a possible transmitter of disease. The agar plate count method developed by Robert Koch in 1883 offered a means of assessing and comparing the quality of milk supplies. Within a decade it was being used for this purpose, and by 1910, preliminary steps were taken to standardize the procedure. In 1916 the first standard plate count method (SPC) was published by the American Public Health Association. Since that time improvements and modifications of the method have appeared in the various editions of *Standard Methods for the Examination of Dairy Products*.

Basically the SPC method consists of diluting 1.0 ml milk in 99 ml of sterile water, which is then shaken 25 times. Using a sterile 1.1 ml pipette, 0.1 ml of the diluted milk is transferred to a first plate, and 1.0 ml to a second plate. Sterile plate count agar (tryptone glucose extract agar or milk protein hydrolysate agar), which has been melted and cooled to 45° C, is poured into the plate, mixed well with the milk dilution by rotation of the plate, and allowed to solidify. It is then incubated at 35° C for 48 ± 3 hours. The number of colonies developing is then counted. Numbers are calculated by multiplying the number of colonies by the dilution. Plates showing 30 to 300 colonies are selected. Fewer than 30 colonies introduces a great deal of sampling error, and numbers greater than 300 are difficult to count because of colony crowding.

Manifestly, the SPC gives only comparative, rather than absolute, results because psychrophiles, thermophiles, anaerobes, and other specialized groups are not counted.

Estimates of numbers of psychrophiles are made by seeding plates with 1 ml and 0.1 ml of milk and incubating the plates at refrigerator temperature (5° C) for 7 days. For thermophiles, the incubation temperature used is 55° C, and the incubation period, 48 hours.

Detection of the presence of *Escherichia coli* and related bacteria as an indication of fecal contamination, especially of pasteurized milk, may be of major concern. For this test, 1 ml portions of milk are plated without dilution using a special *E. coli* medium, violet red bile (VRB) agar. After the poured plate has solidified, a thin layer of the sterile melted medium is poured over the surface to ensure subsurface colonies and a semi-anaerobic condition. *E. coli* develops as a large (1 mm) colony, red in color, and generally surrounded by a zone of precipitated bile.

In addition to VRB agar, the equally selective desoxycholate lactose agar may be used. Raw milk, even of high quality, generally shows some typical *E. coli* colonies, but properly pasteurized milk rarely shows any appreciable number of coliforms.

Colony counts of more than 10 coliforms per milliliter in pasteurized milk, therefore, indicate improper handling of raw milk, improper pasteurization temperatures, or leakage of raw milk during pasteurization.

Milk is sometimes tested for coliforms by using the modification of the MPN test as applied to water, using selective media.

Direct count of bacteria in milk

The plate count detects only those bacteria that grow on the SPC medium at the temperature and oxygen conditions used. Moreover, the results are only educated approximations because a colony may develop from a single bacterium or a clump of bacteria. However, the plate count is the only satisfactory method for estimating numbers of bacteria in high quality milk because of the relatively small numbers present.

To provide a better indication of kinds of bacteria in milk, Breed and Brew developed the direct microscopic method for estimating numbers of bacteria. It consists of spreading 0.01 ml of milk on a slide as a thin film over an area of 1 sq cm. After the film is air dried, it is fixed over boiling water, flooded with xylol to remove the unstainable fat globules, then with alcohol to remove the xylol, and then stained with methylene blue.* Thirty to 50 fields are examined, the bacteria counted either as clumps or individual cells, and the average number multiplied by the microscopic factor determined for the microscope used. With a 6.4× eyepiece and an oil immersion objective it will be 300,000 and with a 10× eyepiece, 500,000. In other words a single field represents 1/300,000 or 1/500,000 ml of milk.

This method is applicable only to milk relatively high in bacteria, such as process milk used for making cheese or butter, and is not applicable to high quality Grade A raw or pasteurized milk. Counts are generally two to eight times higher than obtained by the plate method, depending on whether clumps or single cells are counted. It has the advantage that udder infections can be readily detected, and improperly cooled milk gives a different picture than milk produced under unclean conditions.

Standards for the quality of milk based on SPC results have been set up by the United States Public Health Service, state, county and city agencies.

For example, in Los Angeles County milk must meet the following standards based on the SPC:

1. Grade A raw milk for direct consumption—not more than 15,000 bacteria per ml; no standard for coliforms
2. Certified raw milk—not more than 5,000 bacteria per ml; coliform count not more

*Newman-Lampert stain, a stain and solvent mixture, may also be used. This dissolves the fat globules and stains the bacteria in one operation.

than 10 per ml (less than 0.3% of the milk sold in populous Los Angeles County is raw milk; most of it certified, although the practice of certification soon may be discontinued)

3. Grade A raw milk to be processed as Grade A pasteurized milk—not more than 75,000 bacteria per ml; no coliform standard
4. Grade A pasteurized milk—not more than 15,000 bacteria per ml by the SPC; not more than 10 coliforms per ml

The above standards are similar to, or more restrictive than, the requirements of most states or major cities.

BUTTERMAKING

Commercial butter contains about 80.5% butterfat (federal regulations specify 80%), 16.5% water, 2% salt (if added), and 1% curd. It may be sold as sweet cream butter, made from unsoured cream, or sour cream butter made from spontaneously soured cream with the acidity neutralized. Butter cultures may or may not be added. Butter is tested for butterfat, salt, and often for certain special groups of bacteria.

Microorganisms are important in butter because they impart flavor or bouquet to butter, or they may act as spoilage organisms that produce rancidity, other off-odors, off-flavors, or color defects. Butter spoilage may be due to molds, lactose fermenting yeasts, and fat-splitting bacteria such as *Achromobacter lipolyticum*. These are rarely encountered today because of sanitary precautions exercised in butter production, and refrigeration. The primary organism in commercial butter cultures is *Streptococcus lactis*, which is mixed with smaller proportions of *Leuconostoc citrovorum*, *L. dextranicum*, or *S. diacetilactis*.

DESIRABLE FERMENTATION PRODUCTS

Milk fermented by the naturally occurring lactic acid bacteria, such as *Streptococcus lactis*, *S. thermophilus*, the various homofermentative lactobacilli, and certain of the *Propionibacterium* species, yields pleasant tasting products long considered desirable foods. The natural habitat of the fermentative microorganisms is the grains and grasses used to feed the dairy cattle. The microflora of the common commercial fermented milk products is described in Table 19-1.

Table 19-1. The microflora of the common commercial fermented milk products

PRODUCT	NATURE OF PRODUCT	MAJOR FERMENTER
Buttermilk (churned)	Remains after cream is churned; contains flecks of butter; organisms added in butter culture	*Streptococcus lactis* (Fig. 19-1)
Buttermilk (cultured)	Culture added to milk; enriched with powdered milk	*S. lactis*
Hampshire sour cream	20% homogenized cream; 0.5% to 1.0% butter culture added; powdered milk may be added	*S. lactis* (Fig. 19-2)
Yogurt	Powdered milk solids added to milk to give custardlike consistency; organisms in culture not established in intestines	*Lactobacillus bulgaricus*, plus *S. lactis*, *S. thermophilus*, occasionally yeasts (Fig. 19-3)
Acidophilus milk	Milk sterilized or boiled; pure culture of *Lactobacillus acidophilus* added; culture can establish itself in intestines	*L. acidophilus* (dies out in older cultures) (Fig. 19-4)

MICROBIOLOGY OF CHEESE

Cheeses fall into two classes, soft unripened cheeses and ripened cheeses.

Soft unripened cheeses

Soft, unripened cheeses are represented by low-fat cottage or pot cheese, and baker's cheese, and by high-fat cream and Neufchatel cheeses. These cheeses keep under ordinary refrigeration for only 1 to 2 weeks, although the high-fat cheeses, being lower in moisture, have better keeping qualities.

Cottage cheese was originally made by allowing raw milk to curdle or clabber spontaneously, and then heating slightly to hasten separation of curd and whey. After stirring to break up the curd, the whey was drained off through a screen or cloth bag, leaving the cottage cheese.

Commercially, cottage cheese is now made from skim milk or powdered skim milk to which buttermilk culture is added until a firm curd is formed. Sometimes rennet is also added. Milk held at 29° to 32° C (84° to 90° F) with 5% starter added will curdle in 4 to 6 hours. The curd is then cut into cubes, the whey drained, and the curds usually washed with water. Depending

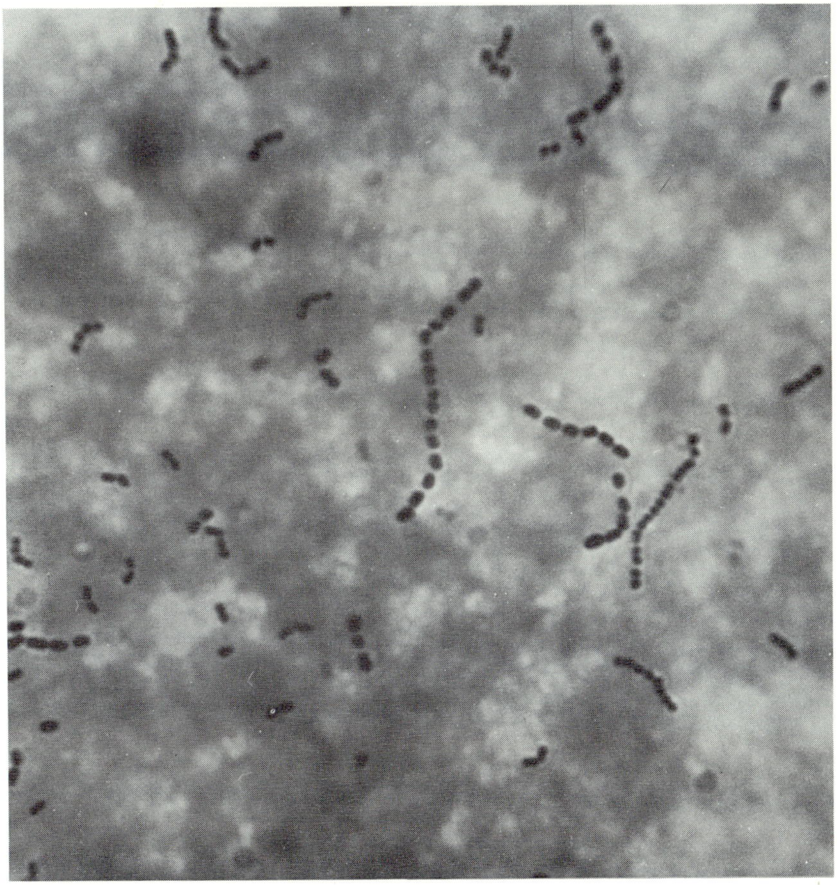

Figure 19-1
Stained mount of *Streptococcus lactis* in buttermilk (×3,000).

on the variety of cottage cheese desired, cream or salt may or may not be added. Water content is about 80%. (Fig. 19-4.)

Baker's cheese, for use in baking, is similar to cottage cheese except that it is almost fat free, has a smooth consistency, and contains about 65% to 75% water.

Spoilage of cottage cheese generally involves molding or production of a bacterial slime coupled with an unpleasant putrefactive odor. The mold most commonly encountered is *Oidium lactis* (Fig. 19-5), while sliminess may be due to *Alcaligenes*, *Pseudomonas*, *Proteus*, *Enterobacter*, or *Achromobacter* species.

Cream cheese is made with cream (16% to 20% butter fat) that has been pasteurized and homogenized. Culture or starter and rennet are added. After a satisfactory curd is formed, it is heated to 46° to 54° C (115° to 129° F), the cut curd is placed in a muslin bag and the whey drained off. When the moisture content is reduced to about 55%, bags are emptied and salt (0.8% to 1.0%) is blended into the curd, which is then packaged. Neufchatel cheese is similar

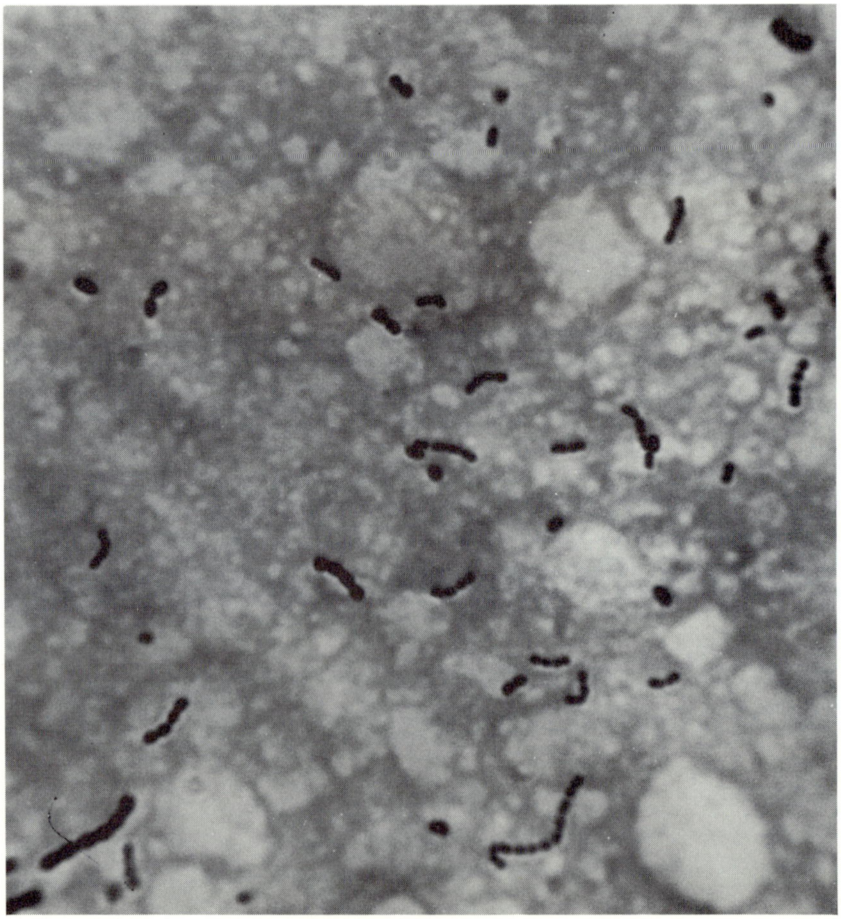

Figure 19-2
Stained mount of Hampshire sour cream showing *Streptococcus lactis* or *S. cremoris* (×3,000).

but has about 65% water content. Although the somewhat lower moisture content enhances keeping qualities, these two cheeses are subject to the same types of spoilage as cottage cheese.

Ripened cheeses

All ripened cheeses are produced by essentially the same initial process although the distinctive characteristics of various cheeses develop during ripening.

Preparation involves the following steps:

1. Milk, either raw or pasteurized, is placed in large stainless steel holding vats, and a culture of lactic acid bacteria is added. These cultures contain selected strains of *Streptococcus lactis*, *S. cremoris*, and homofermentative lactobacilli such as *Lactobacillus bulgaricus*, *L. casei*, *L. lactis*, and *L. helveticus*. For cheeses "cooked" at 49° to 54° C, *S. thermophilus* also may be added.

The holding temperature depends on the type of cheeses desired, but it is usually 38° C or may be as high as 42° to 46° C. Lactic acid is developed quickly, together with other products that

Figure 19-3
Stained preparation of yogurt showing *Lactobacillus bulgaricus* and *Streptococcus lactis* (×3,000).

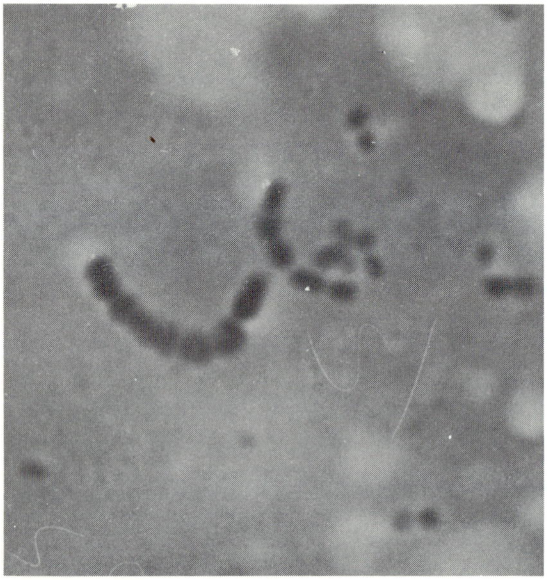

Figure 19-4
Stained mount of cottage cheese with *Streptococcus lactis* (×3,000).

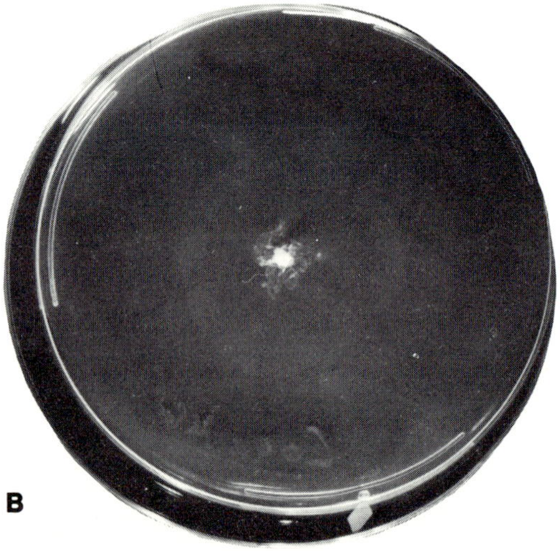

B

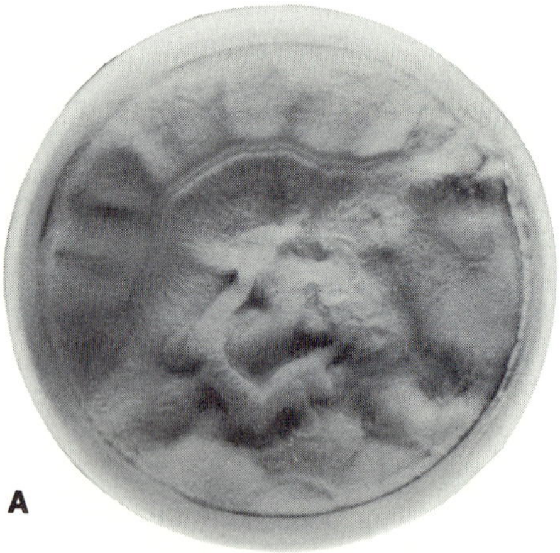

A

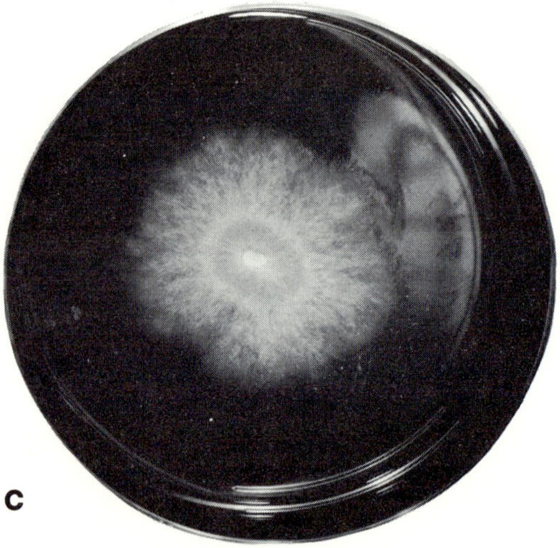

C

Figure 19-5
Common spoilage organism of cottage cheese, *Oidium lactis,* growing as a solid mat on the surface of carton of cottage cheese, **A,** Same organisms growing on plate cultures of **B,** corn meal agar; **C,** Sabouraud's dextrose agar, **D,** tryptone glucose yeast extract agar; **E,** Sabouraud's maltose agar; **F,** mode of spore formation in mycelium (×1,200); **G,** filaments of mycelia and spores (×1,200).

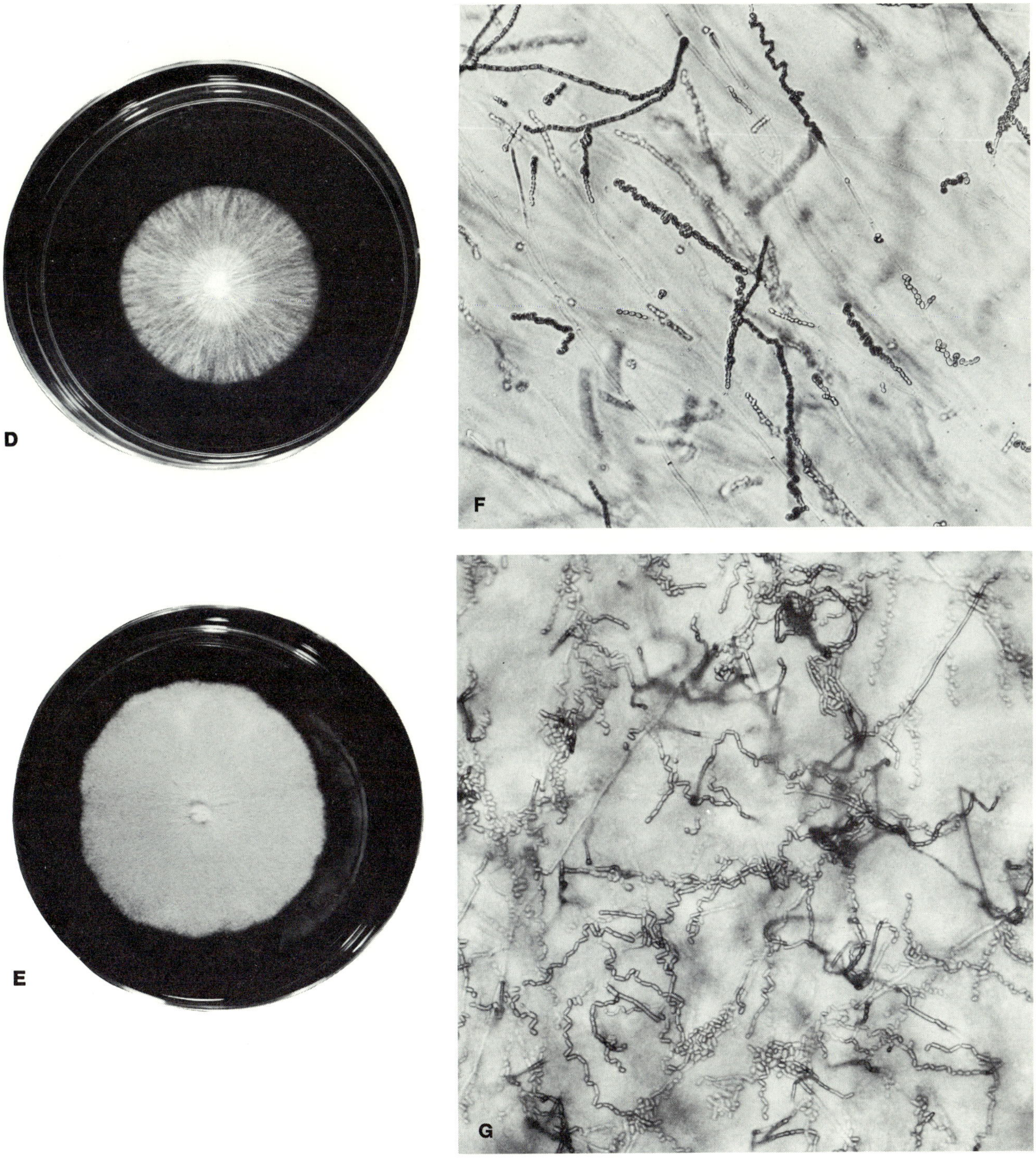
D
E
F
G

Table 19-2. Microbiology of ripened cheeses

CHEESES	MOISTURE %	BUTTERFAT° %	MAJOR RIPENING ORGANISMS	LENGTH RIPENING PERIOD (MONTHS)
Hard (grating) Parmesan Romano Sapsago	32 to 34	–	*Streptococcus liquefaciens, S. zymogenes* (milk boiled prior to adding culture)	6 to 14
Hard (slicing) Cheddar Colby	39	50	*S. lactis, S. cremoris* later replaced by *Lactobacillus casei* or *L. plantarum*	3 to 12 at 0° to 5° C; 3 to 4½ at 10° to 15° C
Swiss (Emmenthal and Gruyère similar in many respects)	40	44	*S. thermophilus*, lactobacilli; final ripening, *Propionibacterium* sp. (lactic acid converted to propionic and acetic acid, and CO_2, which impart sweetish aroma and "eyes")	6 to 9
Semisoft Provolone Edam Gouda	45	40 to 46	Similar to Cheddar	3 or more
Mold ripened (outside) Camembert (French)	–	–	Initial ripening involves *S. lactis;* primary ripening organism *Penicillium camemberti* (outside surface) plus surface smear of yeast and *Brevibacterium linens*†	
Mold ripened (inside or "veined") Gorgonzola (Italian), Roquefort or blue, (American) Danish blue	41 to 43	30 to 32	Spores of *P. roqueforti* seeded into milk or curd; *S. lactis* imparts initial high acidity; pressed cheese may be punctured with needle to allow air to enter	Foil wrapped after 3 months; ripening continues.
Smear ripened Brick Muenster (Limburger similar but 50% moisture)	44 to 49	50	*S. lactis, S. thermophilus* added as starter to pasteurized milk; molded bricks placed in brine (2 days), then in ripening room at 15° C in humidity of 90% to 95%; soon covered with reddish brown, slimy growth of yeasts, micrococci, and *B. linens*†	2
Unripened Monterey Jack	44 to 49	50	Semisoft, unripened, lactic acid curd cheeses	Eaten within 30 days

° Expressed as percent of dry solids
† *B. linens* (L. *linens* spreading over, smearing) natural flora of dairies

in small quantities impart aroma or bouquet to the ripening curd.

2. Rennet, a product containing the milk-curdling enzyme, rennin, which is obtained primarily from calves' stomachs, is now generally added to hasten and improve curd formation. Rennet acts directly on the kappa casein, and because of its ability to stabilize the casein colloids, curdling occurs in about 30 minutes.

3. When curds have the desired firmness, the curd is "cut" with gridlike "cheese-harps." It is then placed in circular or rectangular molds and pressed to remove the whey.

4. Salt is then added, either by rubbing it on the surface of the pressed cheese, or floating the cheese in brine. The primary purpose is to prevent surface spoilage by proteolytic bacteria.

5. Ripening now begins. Temperature and humidity are controlled to produce the desired product.

During ripening, which normally must proceed for at least 2 months, three microbiological processes operate: (1) The casein curd undergoes proteolysis in which the firm protein is converted into the more water soluble polypeptides, peptides, and to some extent, amino acids. The "rubbery" cheese now becomes softer, crumbly, and brittle. (2) Bacteria attack the residual lactose, producing in addition to lactic acid, mixtures of acetic, propionic, butyric, and other volatile acids that give the cheese its distinctive odor and flavor. (3) Lipolytic bacteria may also hydrolyze some of the butterfats to yield butyric, caproic, and other fatty acids that also contribute to the flavor and aroma of the cheese.

Soft cheeses, such as Limburger, are ripened primarily by a surface smear of bacteria. For this reason, the cheeses must be small, and are often protected with a hard paraffin covering.

With hard cheeses, ripening is largely due to activity of bacteria within the cheese. This explains why it is possible to make the huge "cartwheels" of some cheddar cheeses.

The microbiology of ripened cheeses is summarized in Table 19-2.

SUGGESTED READINGS

Davis, J. G.: Cheese, basic technology, vol. I, New York, 1965, American Elsevier Publishing Co., Inc.

Foster, E. M., Nelson, F. E., Speck, M. L., Doetsch, R. N., and Olson, J. C.: Dairy microbiology, Englewood Cliffs, N. J., 1957, Prentice-Hall, Inc.

Hammer, B. W., and Babel, F. J.: Dairy bacteriology, ed. 4, New York, 1957, John Wiley & Sons, Inc.

Patton, S.: Milk, Sci. Am. 221:59-68, July 1969.

Standard methods for the examination of dairy products, ed. 12, New York, 1967, American Public Health Association.

20 Industrial applications of microorganisms

The industrial microbiologist is essentially a pragmatist who uses his ingenuity to convert the enzyme systems of microbes into productive industrial machines.

The very uniqueness of microorganisms in the world of living things makes them peculiarly adapted to meet the demands of the industrial microbiologist. Microorganisms possess an unbelieveable array of enzymes. Coupled with this is their enormous metabolic activity. Someone has calculated that a single cell of the common yeast, *Saccharomyces cerevisiae*, can metabolize 3,000 times its own weight of glucose in an hour. This unbelievable activity is made possible largely by the enormous relative surface area of the billions or more cells often found in each milliliter of a vigorously growing liquid culture of bacteria or yeasts.

The third factor is the almost unbelievable rapidity with which one cell can become two. With many bacteria, the generation time may be an hour or more, but with some strains of *Escherichia coli*, under ideal conditions of nutrients, temperature, oxygen, and pH, it may be as short as 20 minutes. In such a culture, growing unimpeded for 24 hours a single cell could become theoretically more than 10^{18}, that is, 1,000,000,000,000,000,000 cells. Fortunately, this enormous cell accumulation is only a mathematic artifact. In actual practice, cell division may go on at the maximum or logarithmic rate for only a short time of about 9 to 12 hours, depending on nutrients, temperature, aeration, and other factors. At the maximum population level or stationary phase, the number of cells dividing becomes equal to the number of cells dying, producing a plateau or sort of bacterial "population zero" state.

SCOPE OF INDUSTRIAL MICROBIOLOGY

Industrial microbiology is a vast field, and in order to provide an overview, the topics discussed will be limited to the following:

1. The alcoholic fermentations, including wines, beers, and industrial alcohol and other organic solvents
2. The organic acids
3. Foods and food additives including proteins and amino acids, vitamins, and use of whole microbial cells as foods
4. Enzyme preparations including amylases, and proteinases
5. Pharmaceutical products including antibiotics, sterol and other molecular conversions, and the ergotin alkaloids
6. Miscellaneous industrial applications of microorganisms such as the use of crude cultures, and bacterial insecticides

ALCOHOLIC FERMENTATIONS

Beers and wines might well have been included in the discussion of food microbiology, but since they relate so closely to the production of industrial ethanol they are inserted at this point.

Beer

The process of brewing consists of a fermentation of beer wort by a selected yeast strain to yield beer that contains about 5% to 8% alcohol plus noncarbohydrate grain solids, and about 91% water. The process, while essentially continuous, is a relatively slow one requiring about 90 days from wort to finished beer. In modern breweries every step is meticulously controlled by means of electronic measuring and recording devices.

The initial raw materials consist of malted barley, coarsely ground rice, and corn grits. For flavoring, a "tea" is made by boiling hop blossoms for a short time to obtain certain desirable water-soluble compounds.

Steps in the production of beer wort are as follows: A suspension of rice and corn grits is boiled for a 30-minute period in the so-called rice cooker. This prepares the starch for later saccharification. Malted barley is next ground and placed in a mashing tub. Here, enough water is added at about 100° F to make a slurry or mash. (Malting of barley is usually not done at the brewery but in malting establishments. The process consists of moistening selected varieties of brewing barley and allowing the grains to germinate. When the plumule or embryonic leaf of each seed is about $\frac{3}{4}$ inch in length, the grain is placed in a tumbler and the plumules knocked off. The grain then is dried and sold to the breweries as malted barley. During the germination process a considerable amount of amylase or starch-splitting enzyme develops within the seed. This converts much of the starch to maltose or malt sugar, a disaccharide containing two molecules of glucose.) Steeping the ground barley malt extracts the water-soluble maltose and the amylase. The suspension of rice and corn grits is mixed with the malted barley extract and much of the starch is then saccharified to dextrin or the simpler maltose by the amylase. The liquid portion is pumped into the brew kettle and boiled for $1\frac{1}{2}$ hours. Near the end of this period the hop tea is added for flavor. The strained liquid is called wort and is essentially the growth medium for the yeast.

The wort is quickly cooled to 45° F, and a precise quantity of yeast suspension is added. The yeast used is a carefully selected strain of *Saccharomyces cerevisiae*, usually variety *carlsbergensis.*

The fermenting process now begins and continues for 7 days or longer, depending on the kind of beer. During this period the yeast cells reproduce rapidly and the sugars are converted to CO_2 plus ethanol (ethyl alcohol). The cells tend to settle to the bottom of the fermenting tanks.

The young beer is piped into the main fermenting tank for an additional 7 days, during which time most of the older yeast cells will have settled out. From the fermenting tank the young beer goes into the aging vats where it stays for about 2 weeks of primary aging. At this point the beer may be pumped into a second series of aging tanks for secondary aging or krausening, that is, adding a small amount of young beer from the initial fermentation process. The new yeast cells yield added carbon dioxide to increase the effervescence. The length of the holding period varies.

At the end of aging, the beer is forced through asbestos (or membrane) filters. This removes nearly all of the yeast cells and enhances the brilliance and clarity of the beer. The finished beer, cooled to 34° F, is now dispensed into kegs, bottles, or cans.

The beer in the keg still contains a small number of yeast cells and must be kept under refrigeration until sold. However, the bottled or canned beer is pasteurized in the container at 140° F for 13 minutes, or filtered through membrane filters to remove the yeast cells, then quickly cooled to room temperature. Bottled and canned beer will keep indefinitely.

The removed yeast cells represent a valuable by-product. For every pound of yeast added as seed, three pounds of yeast cells are produced. These are checked microscopically for condition of the cells. The most vigorous batches are used to seed the young beer. The less desirable batches are dried and sold for use as feed in fish hatcheries, added to poultry feed, and so forth. However, some concerns debitter the yeast and press it into tablets that are sold for human consumption as brewer's yeast.

In modern well-operated breweries meticulous sanitation prevents spoilage of the product. Careful laboratory tests also mean an early detection of contaminants. The contaminants most commonly found are the slime producers, especially *Leuconostoc*, and the acid producers, *Acetobacter*, *Lactobacillus*, and *Streptococcus lactis* or related streptococci.

Wine

Neolithic man probably domesticated grapes, discovered the wine fermentation, and developed some sort of wine industry. According to available records, wine making changed little between 2500 B.C. and A.D. 1800. Wines were prone to spoilage and rarely kept longer than a year, until Pasteur introduced the heating of wines or pasteurization. About 1890 the use of pure cultures of wine yeasts was first introduced.

Although wine can be produced from many fruits other than grapes, essentially all commercial wine is made from grapes. Many varieties of wines are currently available but three major types will account for most of these varieties:

1. Table wines primarily result from spontaneous fermentation of the pressed grapes by the yeasts normally on the surface of the fruit. They contain 6% to 10% ethanol.

Before the fermentation of the crushed grapes begins, sulfur dioxide gas in a predetermined amount is bubbled through the must, or sodium bisulfite is added. This kills most of the microorganisms introduced from the surface of the grapes. However, the *Saccharomyces cerevisiae* var. *ellipsoideus* survives the so-called sulfuring treatment and carries on the fermentation. In some cases selected wine yeast cultures are also added.

When wine spoils, *Leuconostoc* may produce slimes, *Lactobacillus* and *Acetobacter*, excessive acidity. Heterofermentative yeast strains may cause the wine to become cloudy and develop undesirable flavors.

2. Sparkling or effervescent wines are clear, bright wines containing carbon dioxide under pressure that causes the wine to effervesce when the bottle is opened. The high concentration of carbon dioxide may be achieved by carbonating the wine. However, the superior quality wines are carbonated by using a special strain of carbon dioxide–producing yeasts, and adding sugar to increase the gas production. Champagnes are representative of this class of wines and can be made from grape or apple juice.

3. Fortified wines are natural wines to which "spirits" or brandy has been added to raise the alcohol content to about 20%. They are primarily aperitif or dessert wines, and include sherry, port, Madeira, and Malaga.

In natural fermentation, various yeast strains will tolerate from about 5.0% to 11.5% ethanol before their growth is stopped by their own waste product.

Industrial alcohol production

Ethanol is the most widely used of the organic solvents. At one time, much of it was produced by the yeast fermentation of such raw materials as black strap molasses, surplus corn, potatoes, sulfite liquors from the paper industry, or even wood chips or sawdust hydrolyzed with an acidic solution at high temperature and pressure in an autoclave. Today, chemical synthesis, using ethylene from the oil refineries or natural gas, has largely replaced alcoholic fermentation as a source of ethanol.

Acetone and butanol production

The critical need for both acetone and butanol during World War I gave great impetus to the microbial production of these products. Until the end of World War II, about two-thirds of the butanol and one-tenth of the acetone used in the United States came from the fermentation industry. By this time butanol had become important not only as a paint and lacquer solvent, but also in the production of certain plastics. However, for economic reasons, chiefly the cost of the source raw material such as molasses, the synthetically produced butanol has now largely replaced the fermentation product. In the microbiological process the primary organism used is the spore-forming anaerobe, *Clostridium acetobutylicum*. This organism will ferment sugars with the production of several end products including acetone, N-butanol, isopropanol, formic and acetic acid, and other compounds.

ORGANIC ACID PRODUCTION

Acetic acid production

Acetic acid, which imparts sourness to vinegar, is produced by a variety of bacteria in the course of their fermentative processes. *Acetobacter* or *Acetomonas*, when grown in the presence of wine, cider, or malt liquors high in ethanol, oxidizes the alcohol almost quantitatively into acetic acid. This homofermentative reaction was probably first recognized when men began to produce wines. It is an effective way of producing vinegar as a useful food additive.

Although a tremendous demand exists for chemically pure acetic acid in the chemical and pharmaceutical industries, the vast quantities required are now produced by direct chemical synthesis. For the food industry, a strictly synthetic vinegar, lacking as it does the bouquet and flavor nuances of natural vinegar, has never been popular. Consequently food vinegars are still supplied by the relatively more expensive, time-honored method of vinegar making.

In Europe, wine vinegars are widely used; in the British Isles, malt vinegar is the most popular. Malt vinegar is essentially beer made without hops, then acetified. In the United States, cider vinegars are perhaps the most popular.

The acetic acid bacteria, originally grouped in the genus *Acetobacter*, produce acetic acid from ethanol and some strains produce gluconic acid or 5-ketogluconic acid from glucose and maltose. These bacteria are remarkable in that they can endure concentrations equivalent to 2% to 11% acetic acid.

The tendency today is to apply the name *Acetobacter* only to those forms that produce acetic acid, then in turn further oxidize it to $CO_2 + H_2O$. Such strains are manifestly undesirable for commercial vinegar production. Strains that produce acetic acid but do not attack it, and are also able to produce gluconic acid, are now called *Acetomonas* or *Gluconobacter*.

The chemical reaction involved in acetification is as follows:

$$CH_3CH_2OH \xrightarrow[-\,2H]{+\,\frac{1}{2}O_2} CH_3CHO \quad + H_2O$$

Ethanol — **Dehydrogenases** — **Acetaldehyde**

$$Ch_3CH_2 \cdot H_2O \xrightarrow[-2H]{+\,\frac{1}{2}O_2} CH_3COOH + H_2O$$

Hydrated acetaldehyde — **Dehydrogenases** — **Acetic acid**

Another reaction, which proceeds anaerobically, is as follows:

$$CH_3CHO + CH_3CHO \xrightarrow{+H_2O} CH_3COOH + CH_3CH_2OH$$

2 moles Acetaldehyde — **Acetic acid** — **Ethanol**

(Note: 1 mole acetaldehyde oxidized; 1 mole reduced)

Gluconic acid

Gluconic acid is widely used in the pharmaceutical industry for the production of calcium gluconate to provide calcium needed in the diet of children and pregnant women. This relatively soluble calcium salt has a distinct advantage because the gluconate portion of the molecule is completely metabolized and the calcium is transported to the necessary deposition centers in the body. Gluconate is also added to the diet of laying hens to harden the eggshells.

It is produced by the oxidation of the aldehyde group of glucose. In the simplest form it can be shown as follows:

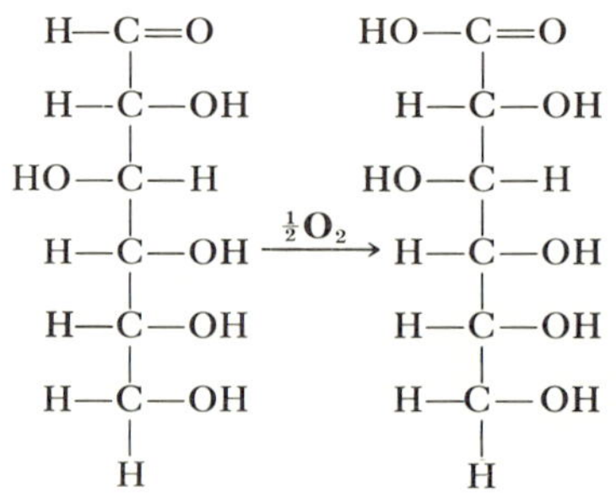

Glucose **Gluconic acid**

Certain strains of *Acetomonas (Gluconobacter)* may produce considerable quantities of gluconic acid. However, a selected strain of *Aspergillus niger* has proved the most effective and is generally used for the industrial production of gluconic acid.

Lactic acid

Lactic acid has a number of industrial uses, including food preservation, acidification of foods, nonsodium baking powders, the calcium salt used in some pharmaceuticals, and in the production of plastics.

The acid has an asymmetric carbon that gives

it the property of rotating the plane of polarized light either to the left or right depending on the positioning of the H and OH atoms in relation to the rest of the molecule.

The formula below illustrates this relationship:

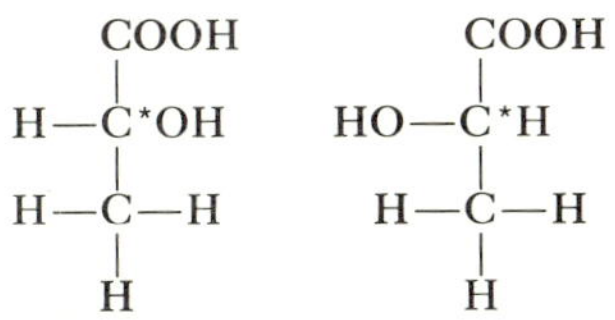

D (−) rotates light to left **L(+) rotates light to right**

*Asymmetric or optically active carbon. D and L are not used to indicate optical activity, but the molecular structure compared with glyceraldehyde.

If an equal proportion of D(−) and L(+) molecules are produced, the rotation of polarized light is neutralized and the acid is termed racemic.

The heterofermentative, or spoilage, molds or bacteria produce largely either D(−) or racemic lactic acid, while most of the useful homofermentative forms produce L(+) acids.

The L(+) acid is metabolized in the animal or human body while most of the D(−) acid is excreted.

Most of the lactic acid used in industry is produced by the bacterial fermentation of sugars. This is a relatively expensive process, but no competitive chemical synthesis is yet available.

The bacteria most commonly used are: *Lactobacillus delbrueckii* for fermenting corn sugar (dextrose); *L. bulgaricus* for fermenting the lactose in whey; and *L. pentosus (plantarum)* for fermenting the pentoses or 5-carbon sugars in sulfite waste liquor of the paper pulp industry.

The acid is harvested as the salt, calcium lactate. It is released as lactic acid by the addition of sulfuric acid. The calcium is removed as insoluble calcium sulfate.

MICROBIAL FOODS AND FOOD ADDITIVES

Use of whole cells as food

With the tremendous pressures on the world food supply, whole cells of microorganisms have been strongly considered as added food sources, especially in countries with protein-poor diets. Such microbial products can be used as additives to human food, and as fodder or feed for animals.

On the surface, the proposal for the use of microorganisms as food is most attractive. Microorganisms appear to be able to produce great quantities of high quality protein at low cost. The calculation has been made that: given for comparison a 1,000 pound steer; 1,000 pounds of soy beans seed; and 1,000 pounds of yeast cells as an inoculum for a giant fermenter, in 24 hours the steer would produce 0.9 pounds of protein; the soy beans 82 pounds; and the yeast 10,000 pounds.

However, two major problems must be solved before it becomes feasible to use microbes as a major human food source: the technical problems that will enable microbial foods to compete with conventional sources, and the great reluctance of humans to eat microorganisms.

In some cases this second difficulty has been met by mixing the microbial products with the customary foods as additives to soups, stews, ground meats, and so forth, or concealing their presence with highly flavored condiments.

The use of a microbial waste product of the brewing industry, *Saccharomyces cerevisiae*, has already been mentioned on page 240. Much of this yeast finds its way into animal feeds.

Sulfite wastes from the paper pulp industry also have been used with considerable success for producing yeast cells for use as animal feed. Most of this product becomes an additive for chicken feed.

Rapidly growing one-celled green algae (primarily *Chlorella* and *Scenedesmus* strains) have received a great deal of consideration as microbial sources of food. Being photosynthetic, they can be grown in a dilute solution of mineral salts mixture and can convert carbon dioxide to carbohydrates without the necessity of providing a sugar or similar material for cell growth. These algae have been found not only to provide protein but also to be a good source for β-carotene (the precursor of Vitamin A) and thiamine.

Food from oil

Recently, unusual strains of bacteria have been isolated that can utilize the hydrocarbons of natural gas and crude oil as the sole source of carbon for the proliferation of their cells. Inorganic salts, such as ammonium sulfate, provide the required nitrogen, while similar inorganic phosphorus and sulfur compounds provide the necessary nutrients. Using such an improbable carbon source as crude oil and simple inorganic nitrogen, phosphorus, and sulfur, these enzymatically versatile bacteria quickly produce billions of their own cells per milliliter of medium. The cells are rich in a yeastlike protein and can be a valuable source of this essential nutrient for animals or even humans. In addition to providing a rich protein source (pound-for-pound more protein than beefsteak), the cell bodies are also rich in the B-vitamins.

The recently developed biosynthetic process involving the use of these oil-utilizing bacteria promises to become a valuable source of protein for "have-not" countries. The protein extract is a light-colored to white powder, and is almost odorless and tasteless. Cost will be the major factor in the continued development of the process.

Amino acids

Microorganisms during their period of active growth may carry free amino acids within their cell membranes. This is a transitory state between synthesis of the amino acids or their release by the intracellular hydrolysis of absorbed peptides, and the synthesis of the amino acids into the various protein components.

In some nutritionally deficient mutants, a blocking of a given metabolic pathway might result in the piling up of an excess of amino acid molecules, which are then exuded into the surrounding medium. Such a mutant might well become a commercial source of a given amino acid.

Only two microbial amino acids are currently being produced commercially. The first is L-lysine. Its production is economically feasible because the cereal products in human and animal diets are often deficient in this vital amino acid. The second, glutamic acid, while not a critical amino acid, is valuable because it imparts a desirable, meatlike flavor to many otherwise bland foods.

The production of lysine is of special interest because it represents the first industrial use of bacterial production of amino acids, and because it involves the use of two deficient auxotrophs. The process was patented by Casida in 1956. It employs a deficient *Escherichia coli* mutant that lacks both the diaminopimelic acid decarboxylase and the lysine decarboxylase. As a result when grown in a medium containing a minimal amount of lysine to keep it growing, quantities of diaminopimelic (DAP) acid may be excreted into the medium. This DAP must be decarboxylated to change it to L-lysine. This is accomplished by using a strain of *Enterobacter aerogenes* that produces DAP decarboxylase.

A strain of bacteria, called *Micrococcus glutamicus* by the discoverers, exudes substantial amounts of glutamic acid. The organism is grown in a medium containing sweet potatoes, cornsteep liquor, and a small amount of biotin. The yield makes the fermentation commercially competitive with extraction of the amino acid from plant and animal proteins.

Vitamins

Many microorganisms grow well in a simple mineral salts medium with a readily utilizable carbon source. To accomplish this they must synthesize their own amino acids and the essential water-soluble vitamins. Since these vitamins tend to accumulate within the microbial cells, these whole cells become an admirable source of the B-vitamins.

The classic example is the use of debittered yeast cells compressed into tablets sold as brewer's yeast. The untreated dried yeast is used as a supplement to animal feeds, and yeast extract serves as a growth stimulator in media used for the growth of bacteria and other microorganisms.

Only a few of the commercially pure vitamins are produced by the growth of microorganisms. These come from strains that produce an excess of a given vitamin and then spill it into the medium.

In most cases, production of vitamins by microorganisms cannot compete economically with the production from other sources. One of the few vitamins for which microorganisms serve as the primary, if not the only, source is vitamin B_{12}. It apparently is produced by the microflora of the intestines and rumen, and becomes available to us from that source.

Commercially the product is obtained from one of three sources:

1. In sewage treatment substantial amounts of B_{12} are synthesized and released by the bacteria in the activated sludge method.
2. The vitamin has been obtained as a by-product of streptomycin fermentation by *Streptomyces griseus* and chlortetracycline by *S. aureofaciens.*
3. Direct synthesis of B_{12} in a medium containing glucose, corn-steep liquor, and a trace of cobalt has been carried on with special strains of *Propionibacterium freudenreichii, Pseudomonas, Streptomyces olivaceus,* or *Bacillus megaterium.*

ENZYME PREPARATIONS

This is a rapidly growing field in which much of the present knowledge is a closely guarded industrial secret, and producing organisms and operational procedures are protected by patent.

Although a number of enzymes are being sold and produced, only amylase and proteinases will be discussed.

Amylases

Amylase is a broad term that describes extracellular enzymes that hydrolyze the complex starch molecule into smaller dextrin molecules: the disaccharide, maltose, and the monosaccharide, glucose. Certain molds produce substantial amounts of amylases when growing on a starchy medium. The amylotic amylases produced by the molds, usually *Rhizopus* or *Aspergillus,* are of several types, two of which are: α-amylase, which breaks down the starch molecule at random to produce dextrans, small clusters of glucose molecules, and an occasional maltose molecule, and β-amylase, which gives almost 100% maltoses.

These enzymes may be derived as by-products of the saccharification of rice by *Aspergillus oryzae* for producing sake, or by using the same organism for direct production of the enzyme preparation from bran or other cereal products.

The enzyme, amylase, may be extracted from koji (starters prepared from saccharified rice inoculated with spores of *Aspergillus oryzae*) by precipitation of the filtrate with ethanol and drying the precipitant to a powder at 55° C or lower. The first microbial amylase preparation was prepared by Takamine in 1914 utilizing a type of koji that he produced on sterilized wheat bran and called taka-koji.

In the United States, in addition to *A. oryzae,* certain strains of *Bacillus subtilis* may be used for production of amylase.

Proteolytic enzymes

To utilize proteins as nutrients effectively, the microbial cell produces two kinds of proteolytic enzymes. The first, or proteinases, are exuded into the medium, and split the protein into the polypeptide fragments, proteoses and peptones. A small portion or all of the fragments can pass through the cell membrane. Once inside they are hydrolyzed by the second group of enzymes, the peptidases, into the component amino acids. Since these enzymes are intracellular they are released only by autolysis or a mechanical rupture of the producing cell. The combination of both enzymes is required to hydrolyze a protein into its constituent amino acids.

Because they are extracellular, the proteinases are easier to isolate and purify. Consequently they have considerably greater commercial usefulness.

Quantities of relatively pure microbial proteolytic enzymes are now being produced for commercial purposes. For most of them, selected strains of *Bacillus subtilis* are grown in huge tanks by the submerged or aerated culture method.

Some of the commercial uses are: addition to meats as tenderizers; use in the dry cleaning process for the removal of protein stains such as blood, milk, and eggs; cleaning of wounds to dissolve away fragments of dead tissue; and injection into bruises or inflammations to solubilize the accumulated blood and other materials.

The most widely publicized application of proteolytic enzymes is in the so-called enzyme detergents. The enzyme does not act on non-protein stains, and requires a pre-soak in order to allow time for the proteolysis of the stains. It does not work successfully in the presence of all detergent compounds, especially those of the dodecylbenzenesulfonate group. Since these latter are among the most efficient of the detergent compounds, the overall effectiveness of the enzyme detergents may well be less than that of some other detergent mixtures. Because the proteolytic enzymes have been found to be allergenic, serious objections have been raised by certain consumer groups which may well force these enzyme detergents off the market.

PHARMACEUTICAL APPLICATIONS

At various times the pharmaceutical industry has taken advantage of microbes and their enzyme systems as a source of useful pharmaceutical products. The two most important microbial processes are antibiotic production and the microbial conversion of sterols.

Production of antibiotics

The development of antibiotic production became a milestone in industrial microbiology. With its inception, the technique of growing huge volumes of microorganisms in a submerged aerated culture, and successfully extracting a growth product present in minute amounts became a practical reality. Today, antibiotics and many other microbial products are produced in 30,000 gallon stainless steel tanks through which air is blown to produce a submerged aerobic culture. Instead of growing as a mat on the surface, very short filaments of the mold or actinomycete are spread throughout the medium. Each fragment becomes an active antibiotic producing center.

With the discovery of penicillin and reports on the antibiotic-producing ability of actinomycetes by Waksman and his group, many of the pharmaceutical concerns established laboratories commissioned to search for new antibiotic-producing microorganisms. The search was a frantic and highly secretive one. The magnitude of the problem is indicated by the fact that Waksman and co-workers tested more than 10,000 cultures before they discovered streptomycin.

An estimate has been made that up to the mid-1950's nearly 1,000 antibiotic-producing forms had been found, and the active agent isolated, at least as a crude preparation. However, most of them proved to be too toxic, too unstable, or else were inferior to established antibiotics. As a result, although many new antibiotics have been discovered relatively few have gone into commercial production.

Today, much of the thrust of the antibiotic industry is the chemical modification of known antibiotics. This may involve the addition of side chains, modification of intramolecular bonds, or other changes that might broaden the antimicrobial range of a given antibiotic. Several new modified penicillin compounds act successfully against penicillin-resistant strains of staphylococci. Also streptomycin, which is often relatively toxic, can be converted chemically to dihydrostreptomycin, which shows substantially less neurotoxicity.

Steroids

Steroids represent a biologically active group of compounds found in the bile salts, and the sex hormones of the adrenal cortex. They include progesterone, testosterone, and cortisone.

Their basic structure consists of a tetracyclic skeleton made up of three 6-carbon rings and a 5-carbon ring attached in a stair fashion. Extraction of these compounds from animal sources for use as therapeutic agents is extremely expensive, and conversion of one compound to another may require many operations. For example, conversion of the bile salt, desoxycholate, to cortisone requires 37 steps. Some of these conversions, although extremely difficult chemically, can be accomplished easily through the magic of microbial enzymes.

With the microbial conversions available, it is now possible to use steroids extracted from some tropical plants. Most of these steroid raw materials have little biological activity. However, by adding an —OH group here, a —H there, or removing either from a critical position, tailor-made high-activity compounds can be secured. The process involves the combination of both chemical and microbial conversions. Although practically all microorganisms are capable of some of these reactions, bacteria, molds, and actinomycetes are often used.

MISCELLANEOUS APPLICATIONS

Uses of crude cultures

The retting of flax represents the classic example of a time-honored industrial procedure involving the use of naturally occurring microorganisms.

The valuable fibers of the stem of the flax plant are embedded in a pectin matrix. Mechanical separation of the fibers is most difficult to accomplish, but the flax farmers of Ireland and Belgium have long used the retting process. In late summer, the plants are pulled up by the roots and placed in bundles, which are stacked upright in small "stooks" until dried. A long trench, 3 or 4 feet deep, 6 feet or so wide, and 50 to a hundred-feet long is dug and filled with water. The bundles of flax are placed in the water and weighted down with stones. In a few days a vigorous anaerobic action begins, and the water turns black and foul smelling. After 10 to 14 days the bundles are removed and dried. During the fermentative process a strain of *Clostridium pectinovorum*, plus other bacteria, will have dissolved away much of the pectin and some of the cellulose, leaving the fibers loosely bound in a soft, woody matrix. When dry the flax is "broken" (pounded between wooden blocks), and the fibers can then be readily removed.

In the tanning industry, proteases produced by naturally occurring microbes, were long employed. In the part of the tanning process known as "baiting," hides were soaked in a suspension of fowl or dog manure. Enzymes from proteolytic bacteria and molds dissolved away fragments of flesh and aided in the removal of hair. The hides were also rendered more pliable. Today, commercial proteolytic enzymes and keratinases derived from microbes perform the same operation much more rapidly and efficiently.

Bacterial insecticides

Certain spore-forming bacteria have long been known to produce diseases in insects, especially bees and silkworms.

In 1915 Berliner in Thuringia discovered an organism, which he called *Bacillus thuringiensis*, that killed the larvae of the flour moth and other insects. The real cause of the lethal action was not established until 1927 when he found irregular rhomboid crystals in older cells that he called "resting bodies."

In the 1950's these crystals were found to be toxinlike in nature and action. Recently this thermostable insecticide was isolated and identified as an adenine nucleotide.

Steinhaus of the University of California at Berkeley, an internationally known insect pathologist, found that *B. thuringiensis* could be used successfully against the alfalfa caterpillar. Production of this organism as a commercial bacterial insecticide was, therefore, initiated. Over 120 kinds of insects are now known to be susceptible to the toxic agent of this bacillus. While still not widely used, this natural insecticide is free from the objections raised against DDT and similar synthetic insecticides. Three companies are now producing the insecticide. It is being applied to some three to four million acres.

SUGGESTED READINGS

Amerine, M. A.: The search for good wine, Science **154**:1621-1628, 1966.

Casida, L. E.: Industrial microbiology, New York, 1968, John Wiley & Sons, Inc.

Kaplan, A. M.: Industrial microbiology: concepts, challenges and motivation, Bioscience, **21**:468-470, 1971.

Rainbow, C., and Rose, A. H., editors: Biochemistry of industrial microorganisms, New York, 1963, Academic Press, Inc.

Rogoff, M. H.: Crystal-forming bacteria as insect pathogens, Adv. Appl. Microbiol., **8**:291-313, 1966.

Rose, A. H.: Industrial microbiology, Washington, D. C., 1961, Butterworth, Inc.

21 The living soil

Distribution of microorganisms in the soil varies, with the greatest numbers where the organic matter content is highest. Immediately adjacent to plant roots in the rhizosphere, is a zone of intense microbial activity. Here, nutrients are constantly exuding into the soil, and dead epithelial cells are constantly being sloughed off. Some types of roots, on the other hand, may exude substances selectively inhibitory to certain microbes.

SOIL—A MICROBIAL RESERVOIR

Soil in its development has become a culture medium in which microbes grow. The soil represents the greatest reservoir of microbes, an assortment of bacteria or other microorganisms capable of attacking almost every natural organic compound. If it were not for this amazing microbial ability, the earth would soon be covered with fallen trees, tangles of weeds, dead underbrush, and the drying carcasses of every animal from elephant to mite.

Life under such conditions could not have long continued, because the carbon dioxide, simple nitrogenous salts, phosphorus, iron, and sulfur compounds would soon have been firmly sequestered in the tissues of plants and animals, unavailable to subsequent generations. Microorganisms function as "the digestive juices of the soil."

The vital action of these soil microbes can best be visualized as we follow a growing tree through a season. As the leaves begin to burst forth, their stomata, which represent hundreds of tiny mouths, hungrily suck in carbon dioxide from the air, while the roots draw in soil water with its minerals from the moist soil. These minerals are transported to the leaf, where under the magic of photosynthesis, with chlorophyll as the machine and sunlight furnishing the energy, the mixture of gases and minerals is transformed first into simple sugar, then into the woody cellulose. Other molecules are united to make lignin, pectin (the plant-cell glue), fats, proteins, and a host of other vital compounds. The carbon dioxide, water, and simple inorganic nitrates, phosphates, and sulfates have now been tied up into complex organic compounds unavailable to other plants. However, with the first frost, the leaves begin to fall and the magic of the microbes of the soil becomes apparent. They begin the process of securing their energy and cell-building material by converting the complex cellulose back to simple carbon dioxide, and the proteins back into simple nitrates, phosphates, and sulfates.

By spring most of the fallen leaves will have disappeared—the cellulose or woody part into the air, soil air, or soil solution as carbon dioxide gas, while the nitrates, phosphates, and sulfates will be largely in solution in the soil water. Only a small portion will remain, the slowly decomposable lignins, lignocelluloses, fats, and waxes. These will slowly accumulate and eventually impart to the surface of the soil, the rich, brown color of fertile earth.

This continued action is spoken of as mineralization. Over a long period of time, the soil becomes "mature" as the surface layer of organic residue becomes deeper, and the soil itself becomes more fertile.

HISTORICAL FOUNDATION

Compared with the medical microbiologists and the fermentologists, soil microbiologists were slow in recognizing the soil as a dynamic habitat for microorganisms.

Observant naturalists did note in the soil the changes we now recognize as being caused by microorganisms. As early as the seventeenth century naturalists had observed the accumulation of brown crystalline material on the surface of soil. This was found to be mainly sodium nitrate or niter. During the campaigns of Napoleon, near the end of the eighteenth century, niter could not be secured from India. French farmers were, therefore, paid a bounty for collecting niter from their manure heaps. On recrystallization, this became sodium nitrate, the explosive component of gun powder.

The success of Pasteur in fermentology and medicine was the spark that ignited a slowly smoldering interest in soil microbiology.

In 1877 Schloesing and Müntz set up a large glass column filled with sand and limestone. Through this they passed a very slow but steady flow of sewage. After 20 days they found that the ammonia present in the sewage was converted to nitrates. The experiment was continued for 4 months. Chloroform was then passed through the column, and the nitrate-producing power ceased, indicating that the action was due to living forms.

The bacteria involved were not isolated until S. Winogradsky (1889-1892) applied the enrichment method of Beijerinck, and succeeded in isolating bacteria that produced nitrites from ammonia and nitrates from nitrites. This process is known as nitrification.

In symbiotic nitrogen fixation, bacteria growing in the roots of legumes in some fashion incorporate the inert nitrogen gas of the air into protein molecules of the host plant. Lachmann (1858) and Woronin (1866) discovered that nodules were formed on the roots of legumes. Frank (1879) concluded that nodules resulted from inoculation of the roots with bacteria. Beijerinck (1888) isolated bacteria now known as *Rhizobium leguminosarum* from the root nodules of peas.

Nonsymbiotic nitrogen fixation, the ability of free-living bacteria to convert nitrogen gas of the atmosphere into their own body protein, was also widely studied by Beijerinck. In 1893 he isolated an anaerobic spore former, which he called *Granulobacter butylicum* (now *Clostridium butylicum*), which by 1898 he had shown to be a nitrogen fixer. In 1895 Winogradsky isolated a very similar but more vigorous nitrogen fixer called *Clostridium pasteurianum. Azotobacter* (Fr. *azote* nitrogen) *chroococcum*, the most widely studied of the nonsymbiotic nitrogen fixers, was discovered by Beijerinck in 1901. It is a yeastlike cell that can easily be isolated in crude culture in a nitrogen-free medium with mannitol as a carbon source. While widely distributed, only soils that are neutral to alkaline are suitable as a habitat for this organism.

The subsequent development of the science of soil microbiology was based on the foundation laid by these and other pioneers.

SOIL AS A MICROBIAL ENVIRONMENT

Soil differs from all other environments in which microorganisms grow, such as water, intestines, and milk, but the basic growth requirements are much the same. The environment must supply moisture, oxygen (except for anaerobic forms), and a carbon and energy source (preformed organic matter for heterotrophs; carbon dioxide plus either sunlight or oxidizable inorganic compounds for the autotrophs).

The relative acidity or alkalinity (pH) is significant. The Actinomycetales seem to favor an alkaline environment, while the molds, and certainly most yeasts, grow best in acid soils.

Temperature is also important. Many bacteria seemingly grow best in the spring when the soil is beginning to warm but has not acquired the high heat of midsummer. Others continue slow growth through the chill of the winter months.

As for the necessity of oxygen in the soil, Russian workers have found that 90% to 95% of soil microorganisms are aerobic. When a heavy rain fills the soil air spaces, the soil will be temporarily water logged. When this occurs, microbial activity is curtailed until about 50% of the pore space is emptied. Rapid growth then ensues. This represents an optimium concentration for most microbes.

The growth of microorganisms is highly variable, changing from hour to hour or even minute to minute as soil moisture changes or readily available organic materials are broken down by microorganisms.

Simple bacteria

Simple bacteria in the soil, as we know them at present, belong primarily to the Pseudomonadales and Eubacteriales. In rich garden earth, billions of cells of bacteria may be present in a gram of soil.

They are a varied group, some aerobic, some facultative anaerobes, and some obligate anaerobes. Both nonspore-forming and spore-forming bacteria are found. They are mostly heterotrophs with a few autotrophs.

The whole bacterial complex is an intermeshed, interrelated operation. In many cases the waste products of one bacterium serve as the nutrients for another. A great deal of competition for nutrients apparently exists, but in spite of this, each form seems to occupy a particular ecological niche.

Higher bacteria

In terms of frequency of occurrence in the soil, members of the Actinomycetales run a close second to the simple bacteria. In many respects, because of their filamentous nature, they may resemble minaturized molds. Some even produce conidia or sporangia-like fruiting bodies. Here, however, the similarity to molds ends for they are true procaryotic forms. They grow in the soil as slender filaments, intermingled with other bacteria, and penetrating the decomposing organic matter.

In heaps of decomposing organic matter that has begun to heat, thermophilic actinomycetes are often found, along with thermophilic molds or bacteria.

The critical pH for growth is from 5.0 to 8.0, with the most vigorous growth in the alkaline range. Occasionally a few actinomycetes will be encountered in relatively acid peat bog soils. When these soils are limed to reduce the acidity, the number increases rapidly. Members of the genus, *Streptomyces*, most often encountered in soils, prefer soils with an alkaline to neutral pH. Their growth is inhibited in acid soils.

As with the molds, it is difficult to make an accurate estimate of numbers of actinomycetes in the soil. On a fairly rich medium, the more rapidly growing bacteria tend to crowd out the slower growing actinomycetes. In nutrient agar, only a few colonies are found. However, if the agar contains only 10% as much nutrient broth as usually used, most of the colonies may be actinomycetes. The most productive medium appears to be one made with soil extract and somewhat reduced nutrients. When plate counts are made, the colonies that develop may rise from mycelial fragments or asexual spores.

Unlike bacteria in general that grow best at 25% to 50% moisture, in soils, the best growth of the actinomycetes develops with 15% to 20% moisture.

The actinomycetes are capable of utilizing a wide variety of organic compounds including fats, paraffins, long-chained hydrocarbons, and phenols. When growing in the soil they impart a pleasant, earthy odor.

Myxobacteriales (slime bacteria) are apparently widespread in soil, especially soils that have received heavy applications of manure or other organic matter.

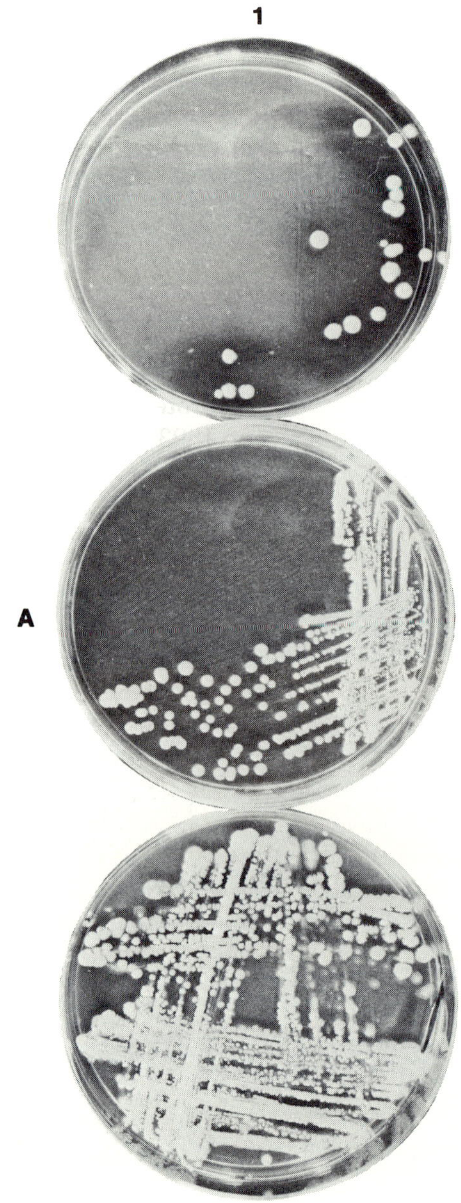

Figure 21-1
A, Comparison of colony appearance on streak plates of *Staphylococcus aureus,* **1,** plates 1 and 2, *Escherichia coli,* plate 3, and *Streptomyces* sp., **2.** Enlarged colonies of *Streptomyces* sp., **B,** and *S. aureus,* **C** (×5).

2

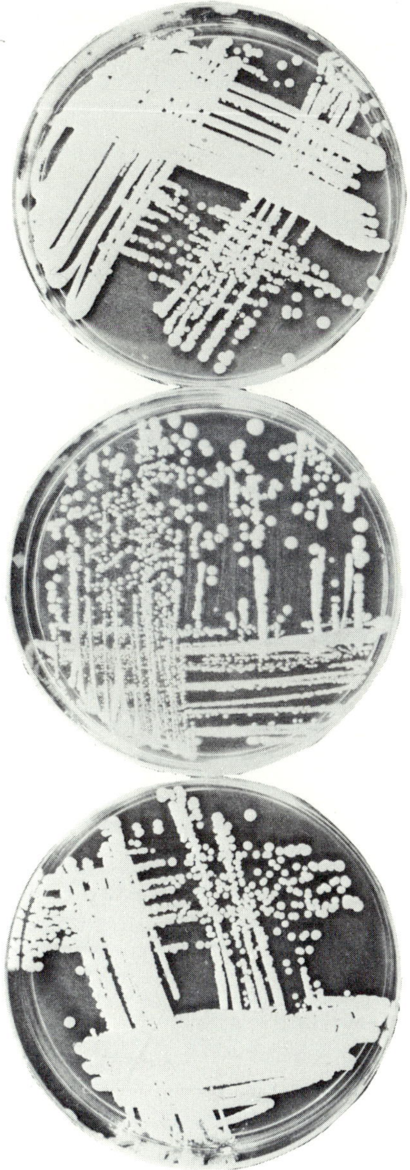

B

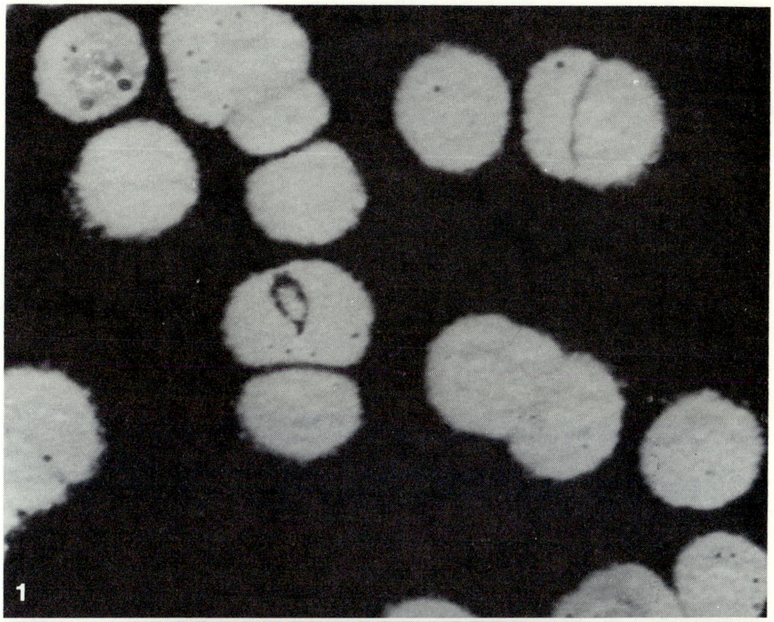

C

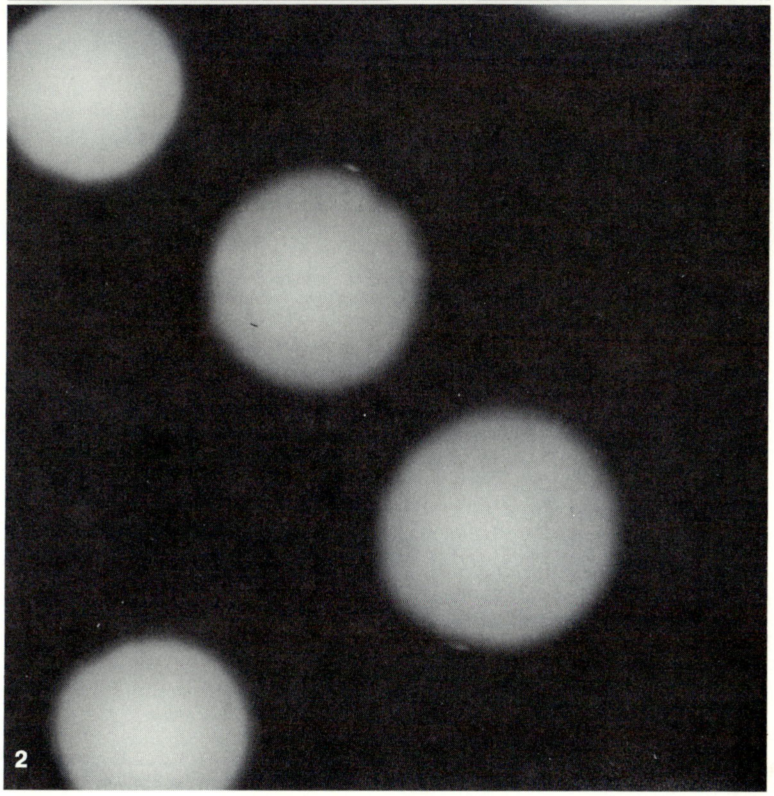

The method of their isolation is unique. A mixture of agar and water without nutrients is prepared and sterilized. To the molten agar are added living or dead cells of a culture of *Enterobacter aerogenes*. These serve as the sole carbon source of the myxobacter. The bacteria-agar mixture is poured into plates and allowed to solidify.

The plate is seeded by sprinkling finely crushed soil over the surface of the agar. After 7 to 10 days large slimy colonies develop from which mushroomlike sporing bodies rise. These sporing bodies are small but can be seen easily with a low-power dissecting microscope. The growing myxobacteria produce large amounts of slime that exude as fingerlike protrusions. The myxobacteria possess no flagella, but are actively motile by a sliding motion.

It is thought that the myxobacters play an important role in the decomposition of high cellulosic woody plant materials.

Casida's newly discovered soil bacteria

Prompted by the fact that only about 1% of the bacteria in the soil as shown by microscopic examination can be isolated by the usual plating procedures, Dr. L. E. Casida, Jr. and his co-workers at the Pennsylvania State University set out to isolate members of the unrevealed 99%.

They first developed a special system of microscopic examination that demonstrated the presence of these "unknown" organisms.

Stained with acriflavine, the organisms fluoresced in the fluorescent microscope, indicating that they were living and active.

Soil dilutions high enough to eliminate the usually found bacteria (10^{-10}, 10^{-11}) were then prepared. These dilutions were plated in special enriched media, and the plates incubated at a variety of temperatures and oxygen tensions. The investigators were gratified to find a number of slow-growing actinomycete-like forms.

Since the organisms were isolated at dilutions of 10 to 100 billion, it suggests that these are probably true representatives of the unseen 99%. As such, each gram of rich soil would thus contain billions of these forms.

Two new isolates were named, *Agromyces ramosus* (Gr. *agros* and L. *ramus* branching), and *Actinomyces humiferus*. A third type resembles the streptococci.

Fungi—molds and yeasts

Two types of fungi are most commonly demonstrable in soil, the plant pathogens and the various common saprophytic molds. The latter seem to be creatures of nutritional prosperity, growing best after heavy additions of organic matter. The soil itself is normally an impoverished medium. During periods of limited nutrients, the fungi appear to be in a quiescent state, probably in the form of conidiospores, sporangiospores, sclerotia, or similar resistant resting forms.

The three types most commonly isolated from the soil are the Phycomycetes, represented primarily by the filamentous, sporangiospore-bearing *Rhizopus*, *Mucor*, and *Absidia;* the familiar members of the Ascomycetes, the powdery blue, green, yellow, or brown *Penicillium* species, and *Aspergillus* species; and major representatives of the Fungi Imperfecti, the dark colored Dematiaceae such as *Alternaria*, *Stemphylium*, and *Hormodendrum*.

The exact growth status of molds in the soil is not completely clear. Plate counts can indicate thousands of molds per gram of soil. However, these values may merely represent counts of numbers of quiescent conidiospores or sporangiospores and not fragments of growing mycelia.

Direct microscopic examination of soils indicates that when considerable amounts of organic matter are added mold growth may become very active, with a profusion of filaments permeating the organic matter and surrounding soil. The magnitude of mycelial formation is indicated by estimates derived from the measurement of fragments of filament seen on a microscope slide. One investigator estimated the mycelia to total 10 to 100 meters in length per gram of soil.

After the readily available nutrients have been

exhausted, the fungal mycelia begin to lyse. Immediately bacteria and actinomycetes begin to colonize this readily available organic matter. The spores apparently tend to remain viable, although some lysis of mold spores by bacteria has been suspected.

In the vegetative state, molds are very vigorous decomposers of organic matter. They are also active ammonifiers releasing quantities of ammonia when growing in the presence of proteins. They are aerobic, inhibited by anaerobic conditions, but releasing quantities of carbon dioxide from the breakdown of carbohydrates.

The question of yeasts as normal inhabitants of soils has not been adequately answered. The present opinion seems to be that the primary habitat of yeasts is the flowers and fruits of various fruit trees, and vines of orchards or vineyards. Soil, on the other hand, serves as a secondary habitat or reservoir in which yeasts overwinter and in which they maintain a limited degree of activity.

Blue-green algae

Blue-green algae are common inhabitants of all soils but grow in greatest profusion on the moist surface of the alkaline desert soils of western America when a summer cloudburst leaves a transient pool of water. As the moisture evaporates and soil becomes dry once more, the blue-green algae form resistant cysts that protect them until the next rain, which may not fall for months.

Blue-green algae serve several useful functions. As photosynthetic forms they provide essential carbohydrates for the heterotrophs in the soil. A number of them, notably some members of the *Nostoc-Anabaena* group, are active nitrogen fixers. The slime that they exude also serves to glue soil particles together, lessening the likelihood of destructive soil erosion.

Green algae

The green algae, primarily as one-celled forms, are widely distributed in soils. Unlike the blue-green algae, they predominate in moist soils. In the spring in regions with considerable rainfall they may grow in such profusion that they form a grass-green film on the surface of the soil. The one-celled *Chlorococcum,* which also grows as films on tree trunks and posts, is the major algae form in these soil surface films. Some green flagellates may also be present, as well as *Scenedesmus* and similar forms. The limiting factor seems to be a high moisture content. The most beneficial role of these forms in the soil is their ability to supply readily available carbohydrates for the microflora of the soil. Like the blue-greens they form resistant cysts that carry them through periods of drought.

Protozoa

Although the natural habitat of protozoa is water, they (mostly ameba and colorless flagellates) may develop in considerable numbers in moist soils. They live primarily as bacterial predators and engulf many bacteria in a single feeding. In soils kept constantly wet and warm, such as in a greenhouse, their predatory activities may become severe enough to reduce the fertility of the soil. At times, it becomes necessary to destroy the protozoa with applications of volatile organic poisons.

The protozoa themselves are not without their predators. Certain fungi, mostly Phycomycetes, attach to and destroy both nematodes and ameba.

The soil protozoa have not been widely studied, but one investigator estimated the number of protozoa as 30,000 per gram of soil.

Like the algae, most of the protozoa form cyst-like resistant stages that carry them through periods of drought or meager nutrition.

Viruses in the soil

Phages have been found that attack bacteria including *Rhizobium, Azotobacter, Bacillus, Clostridium, Enterobacter,* and *Arthobacter.* The actinomycetes, particularly members of the genus, *Streptomyces,* also suffer from phage attacks. Recently phages that attack blue-green algae have been discovered.

METHODOLOGY

Estimation of numbers of microbes

Soon after Koch developed his plating technique, it was applied to soil samples, but the results were variable and indecisive. It is now generally conceded, however, that plate counts of bacteria represent only 1% to 10% of the total number of bacteria in a given soil.

Since the soil itself is a nutritionally impoverished medium, many bacteria will not grow on the usual rich laboratory media. For this reason, the plating medium now most frequently used for soil counts is a soil-extract sugar agar. Soil dilutions are prepared using sterile tap water, and inoculated on pour plates by the spreading rather than pouring technique. The plates are incubated for two weeks at 25° to 28° C. For growing anaerobes, plates are incubated in an atmosphere of 95% nitrogen gas and 5% CO_2.

In order to count numbers of phages and *Bdellovibrio bacteriovorans*, a plate of a somewhat richer medium is prepared and seeded heavily with the desired host organism in order to produce a solid lawn of growth. A well-shaken suspension of soil is filtered through a membrane filter with a pore size of 0.45 μm.

The filtrate is then serially diluted, and measured amounts of the dilution spread over the surface of the seeded agar with a sterile bent glass rod. If phage particles are present they are indicated as small cleared plaques after 24 hours incubation. They do not increase beyond their initial diameter.

With *Bdellovibrio*, plaques are not visible until after 2 to 4 days incubation. However, they continue to increase in size for several days.

For isolating and counting nitrifiers, a silica gel plate is prepared as a base. To this is added a measured amount of a mineral salts solution containing ammonia for *Nitrosomonas* and nitrites for the growth of *Nitrobacter*. Soil dilutions are then streaked on several plates. Silica gel is essential for growing nitrifiers because organic matter is often toxic to nitrifying bacteria. This technique is also useful for counting *Azotobacter*, and cellulose decomposers.

To count molds, large plates are used, and a medium that is acid in reaction is enriched with sugar and peptone. Algae can also be grown on a mineral salts agar medium but with only fair success.

For counting psychrophilic bacteria, plates are incubated at 0° C for two weeks or longer. For thermophiles the incubation temperature is 55° C.

Estimates of numbers of microorganisms, such as algae or protozoa that do not grow well on petri plates, can be made by the serial dilution method. Dilutions are made and dispensed into 5 tubes for each dilution. After a suitable period of incubation, tubes are examined for growth and the number estimated by the MPN method.

Direct bacterial counts may be made by spreading a measured amount of the soil suspension over a measured area of a microscope slide. Separation of living from dead bacteria is accomplished by use of the acridine stain and the fluorescent microscope. The living cells fluoresce, but the dead ones do not.

Visualization of normal growth in soil

To obtain an idea of the manner in which microorganisms normally grow in the soil, Rossi and Ricardo (1927) and Cholodny (1931) buried glass slides in the soil for periods of several hours to several days. These were then removed, fixed, stained, and examined. This technique has provided evidence of the filamentous growth of molds and actinomycetes, colonization of bacteria on organic matter or dead mold filaments, and similar phenomena.

CYCLES OF THE ELEMENTS

The elements out of which living tissues are synthesized undergo a cyclic type of molecular change in the soil due primarily to microbial action. Of these elements, the most important ones are carbon, nitrogen, phosphorus, and sulfur.

The carbon cycle

The element carbon is inert in the form of charcoal or a diamond. However, as carbon dioxide it becomes an active, vital cyclic compound. Microorganisms play an essential part in this cycle.

Since it is a cyclic process, it is necessary to break into the ring at some point in order to visualize its operation. The logical place seems to be at photosynthesis.

In the process of photosynthesis, the carbon dioxide of the air is coupled with energy supplied by the sun's rays. By way of review, the vastly oversimplified reaction can be shown as:

$$CO_2 + H_2O \xrightarrow[\text{Energy from sun's rays}]{\text{Chlorophyll}} \underset{\text{Formaldehyde}}{CH_2O} + O_2{}^{*}$$

Formaldehyde (CH_2O) does not accumulate, but serves as the transitory, intermediate building material quickly synthesized into glucose by the formula:

$$6\,(CH_2O) + \text{Synthetic enzymes} \longrightarrow \underset{\text{Glucose}}{C_6H_{12}O_6}$$

This may be further synthesized into cellulose or starch by removing a water molecule between adjacent glucose molecules as follows:

$$\underset{\text{Glucose}}{n\,(C_6H_{12}O)} - n\,(H_2O) \longrightarrow \underset{\text{Cellulose or starch}}{n\,(C_6H_{10}O_5)}$$

This is precisely what happens in a green plant, but it involves a much more complex process utilizing a series of remarkable enzyme reactions.

Carbon dioxide is immobilized in the wood, starch, or other components of the green plant, and remains quiescent until the plant decomposes as a result of microbial action, or is burned.

Under undisturbed conditions that exist in the "forest primeval" a comfortable balance exists between the bound and free carbon dioxide. During a summer night, microorganisms in the soil are busily engaged in decomposing the leaves and tiny branches that have fallen on the surface (or the residue of the leaves that still remain from the previous autumn). The Krebs cycle spinning in the enzyme system is shooting off its three carbon dioxide molecules in each revolution. Other cycles are also making their contributions. The end result is that the carbon dioxide concentration of the lower air near the ground will have increased manyfold. Added to this, the tree, in order to supply the energy necessary to transport the photosynthetically produced sugars and starches, and to synthesize them into woody material, enxymes, DNA, and so forth, must burn a portion of the stored sugars in its own enzyme furnaces. The chief of these is again the Krebs cycle. The end result is the release of energy stored from the sun, plus the CO_2 and H_2O molecules previously bound together by photosynthesis. This is known as respiration (primarily a dark process), and the tree thus returns a portion of the bound carbon dioxide into the atmosphere adjacent to itself. By sunrise, the air surrounding the tree will have become well laden with carbon dioxide. As the first rays of the sun strike the tree, the photosynthetic enzyme machines begin to operate, supplied by the carbon dioxide molecules sucked in by the wide-open stomata of the leaf. For the next few hours photosynthesis proceeds at full speed. By noon it begins to slow, and respiratory processes although active during photosynthesis begin to speed up to supply the energy required to transport the manufactured products to various parts of the plant.

It becomes apparent that if the tree had to rely solely on its own respired carbon dioxide, no growth would occur. If it had to rely on the carbon dioxide normally carried by the atmosphere, growth would be slow. The real growth impetus comes from carbon dioxide supplied by the decomposing action of the bacteria and other microorganisms in the soil beneath the branches of the tree. However, their action is not a gratuitous one. Energy for all their life processes comes from the sun's rays that fell on the plant the week or year before. This is brought back to

*Vital by-product, not considered further in discussion of carbon cycle.

the hungry soil microbes when the worn-out leaves fall. Not only does energy, but the building materials for new cells come from the photosynthetically supplied sugars, starches, celluloses, proteins, and other life products.

The ocean and bodies of freshwater have a similar but less concentrated carbon cycle.

However, man has now cast his disturbing shadow on the picture. In his frantic search for energy he has tapped the stored carbon compounds of coal in the form of fossil remains of trees and other plants that covered the earth millions of years ago. Coal burned to furnish heat, electrical power, or energy is, therefore, adding to the present level of carbon dioxide in the atmosphere. In addition an untold amount of carbon dioxide comes from the fossil fuels, gasoline and natural gas. What is disconcerting is that with the sudden addition of carbon dioxide from the distant geological past the steady-state balance has been disturbed. In the past century the carbon dioxide content of the atmosphere has risen from 290 ppm to 320 ppm; of this, 20% of the increase has occurred during the last decade or so.

The effect of increased carbon dioxide content of the air is thought by many to have a marked effect on climate. Some people have calculated that the mean temperature will increase, and the rainfall will decrease until the earth becomes a Sahara desert. Others predict that we are faced with another ice age. In any event, the active uptake of photosynthetic higher plants and algae, plus removal as insoluble carbonates in the ocean, may slow the steady increase in carbon dioxide. However, the scavenger microbes, keeping the earth free from dead leaves and branches plus animal carcasses, will still continue to return carbon dioxide to the overloaded atmosphere.

The carbon cycle is illustrated in Fig. 21-2.

Nitrogen cycle

Like carbon, the essential life element, nitrogen, is in a potentially unlimited supply. About 79% of the air we breath is in the form of nitrogen gas (N_2). As such, it is useless to us. Only when it is converted into nitrogen compounds can it be utilized for our life processes. These nitrogen compounds are incorporated into all proteins (enzymatic or structural), DNA, RNA, and the great host of the other nitrogen-containing macromolecules of life substances.

Until recently, the essential compounds of nitrogen were in limited supply. This was due primarily to the action of microorganisms. While a few were adding compounds of nitrogen, others were converting them back into gaseous nitrogen. Thus a steady or balanced state seemed to exist. As human populations increased, concern developed that the use of available fixed nitrogen compounds in crop production might exceed replenishment of the supply by natural fixation. The only great reserve of prefixed nitrogen was in the Chilean nitrate beds. The specter of their exhaustion presented a bleak picture.

However, in 1914 Haber and Bosch in Germany developed a process of uniting hydrogen and nitrogen gas (at high temperatures) into ammonia—in imitation of the lightning bolt.

The supply of fixed nitrogen has now increased to the point where concern is being expressed that we might suffer from an oversupply.

Nitrogen's tremendous biological activity stems from the ease with which it can be oxidized or reduced when the proper enzymes or catalysts are present. (This also makes it an active component of explosives.)

This range of properties is illustrated in Table 21-1.

Table 21-1. Valence states of nitrogen

NITROGEN COMPOUND	VALENCE STATE	FORMULA
Nitrate ion	+5	NO_3-
Nitrate ion	+3	NO_2-
Nitroxyl (unstable)	+1	(HNO)
Hydroxylamine	−1	NH_2OH
Ammonia	−3	NH_3

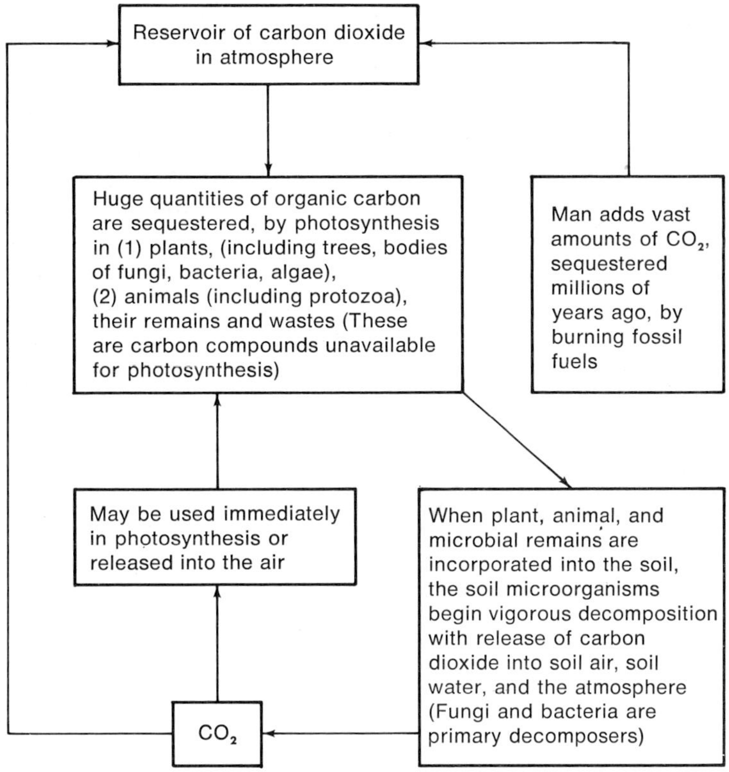

Figure 21-2
Role of microorganisms in the cyclic conversion of carbon compounds in nature.

Microorganisms participate in three ways in the cyclic change in the nitrogen of proteins and related compounds, or of nitrogen gas:

1. When proteins are attacked by bacteria, they are first hydrolyzed to amino acids. The NH_2 group of the amino acid may then be split off by hydrolysis and released as ammonia.

$$\underset{\text{Glycine or aminoacetic acid}}{NH_2CH_2COOH} + \underset{\text{Water}}{HOH} \longrightarrow \underset{\text{Hydroxyacetic acid}}{HOCH_2COOH} + \underset{\text{Ammonia}}{NH_3}$$

The ammonia thus released may then be oxidized to nitrates by the autotrophic bacteria. The first step (release of ammonia by the hydrolyis of proteins) is known as ammonification. It is an ability possessed by many common simple bacteria, actinomycetes, and molds. The second step (oxidation to nitrates) is called nitrification. The most active nitrifiers apparently are autotrophic bacteria. *Nitrosomonas* and related genera oxidize ammonia to nitrites, while *Nitrobacter* and related forms oxidize nitrites (NO_2^-) to nitrates (NO_3^-). Certain molds are thought to convert ammonia to nitrites and/or nitrates.

The nitrates in the soil are taken up by growing plants and converted back into plant protein. This process of recycling the nitrogen compounds constitutes the inner machinery of the nitrogen cycle. Like all natural processes it is not completely efficient, and a fraction of the nitrogen compounds continually disappears from circulation by denitrification or being washed into the ocean.

2. Nitrogen fixation is nature's major method

of replenishing this supply of compound nitrogen. It is accomplished by two microbial systems: nonsymbiotic nitrogen fixation, carried on by free-living bacteria, and symbiotic nitrogen fixation in which *Rhizobium* growing in nodules on the roots of leguminous plants combines the nitrogen gases of the air into the proteins in the host plant.

The primary example of symbiotic nitrogen fixation is represented by the relationship of the legumes and the *Rhizobium*. However, only about 10% of the Leguminosae possess the symbiotic relationship that leads to nitrogen fixation, and even then the soil is not appreciably enriched with the fixed nitrogen unless the plants are plowed under and undergo decomposition. (The species of Rhizobium and their host plants listed in *Bergey's Manual* appear in Table 21-2.)

The exact mechanism of symbiotic nitrogen fixation is not fully understood. However, it apparently involves the bacteroids or abnormal cell forms that develop in the nodules and the plant membranes. (See Figs. 21-3 to 21-8.) Strains of Rhizobium vary materially in their ability to promote nitrogen fixation. They may range from highly effective to a state of near parasitism.

The effect on the host plant of nodulation by effective strains of bacteria is emphasized by the example of soybean response. Soybean seeds from uninoculated plants will contain about 13% protein. (This is about the same as wheat.) However, when the plants are well nodulated by an effective Rhizobium strain, the protein content will be increased to 20% to 30%. The yield will also be increased two- to threefold.

Table 21-2. Species of Rhizobium and their host plants

SPECIES	HOST PLANTS
R. leguminosarum	Sweet pea, lentil, pea, vetch
R. phaseoli	Kidney bean, bean, scarlet runner
R. trifolii	Clover (*Trifolium* sp.)
R. lupini	Lupine, *Serradella*, *Ornithopus*
R. japonicum	Soy bean
R. meliloti	Sweet clover, alfalfa, burr clover

A symbiosis similar to the *Rhizobium*-legume relationship occurs with forest trees—*Alnus* or alder (Fig. 21-9) and the *Ceanothus* (buckthorn).

3. Denitrification is the reverse of nitrogen fixation. While the nitrogen-fixing bacteria are replenishing the supply of nitrogen compounds, another group of bacteria is reducing nitrates (NO_3) back to nitrogen gas (N_2). These are aerobic bacteria, which when growing under anaerobic conditions can utilize the oxygen in the NO_3 as the oxygen source for aerobic respiration. *Pseudomonas* species are especially active denitrifiers. Formerly denitrification was thought to be a serious threat to the economy of nature. However, we now recognize that this operation is vital in maintaining the steady state of the nitrogen cycle.

The intervention of man in the nitrogen balance in nature may be fraught with undesirable sequences. At the present time, the amount of chemically produced nitrogen manufactured each year is almost equal to the amount fixed by natural processes. These man-made nitrogen compounds are used in enormous amounts as fertilizers. One of the results is that great amounts of nitrates are carried into the streams, lakes, rivers, and even the ocean as run-off from heavily fertilized agricultural lands. When the nitrogen levels are raised in lakes, blooms of algae develop that stimulate eutrophication.

Cycle of inorganic compounds

Phosphorus, like nitrogen, is essential to all the metabolic processes of a living cell. The only natural reservoir of phosphorus is in the form of insoluble minerals in the soil, such as apatite, which is rich in tricalcium phosphate. In this form the phosphorus is useless to higher plants without the solubilizing action of soil microorganisms.

In addition to the insoluble phosphate minerals, the soil may contain fair amounts of phosphorus sequestered in the dead organic materials

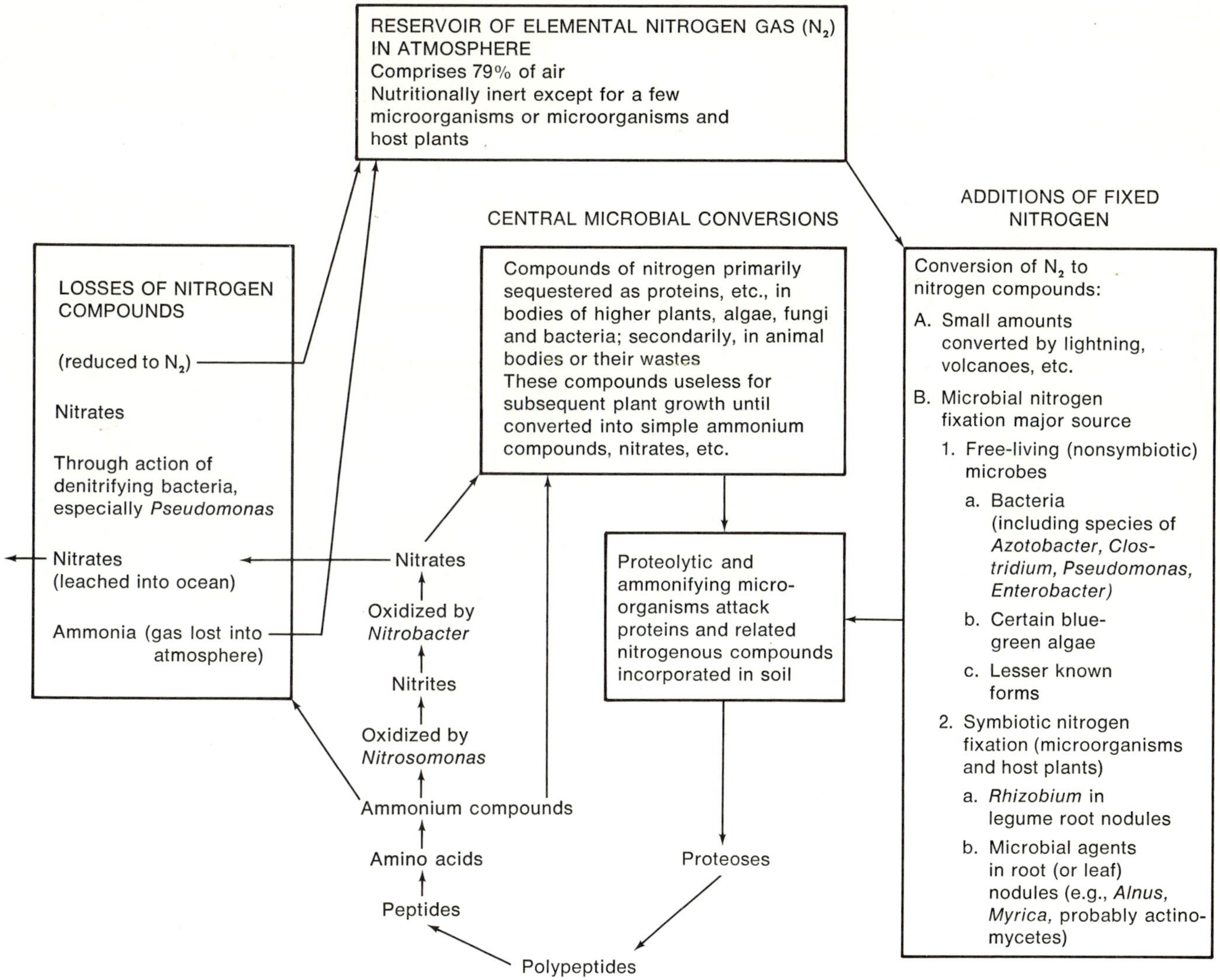

Figure 21-3
Role of microorganisms in the cyclic reuse of nitrogen compounds in nature.

in the soil. This may be in the form of plant, microbial, or animal remains. Slowly decomposable humus may hold varying amounts of unavailable phosphorus. Dead plant material contains 0.05% to 0.5% phosphorus, while the dead fungus filaments carry 0.5% to 1.0% phosphorus, and the defunct bacteria, 1.5% to 2.5%. The proportion of phosphorus is small compared with carbon and nitrogen in plant, animal, or microbial bodies. The ratio of carbon to phosphorus is about 100 to 300 : 1. With nitrogen, the ratio is 5 to 20 : 1.

However, the vital role of phosphorus is apparent when you consider that the types of phosphorus-containing organic matter in the soil include primarily DNA, RNA, phospholipids, phosphorylated sugars, coenzymes, and similar products. RNA represents the highest single concentration of phosphorus, amounting to one-fifth to one-sixth of the total.

It is, therefore, apparent that phosphorus, involved as it is in genetic messages, enzyme production, and energy-producing mechanisms, is the prime activating element in the basic processes of life.

The cyclic role of microorganisms in the life flow of vital soluble phosphate is (1) the release of the organically bound phosphates into the soil solutions in the form of soluble orthophosphate ions such as H_3PO_4, $H_2PO_4^-$, HPO_4^{--}, or PO_4^{---}, and (2) rendering soluble in a similar manner the insoluble mineral phosphates, of which tricalcium phosphate ($Ca_3[PO_4]_2$) is the most common example.

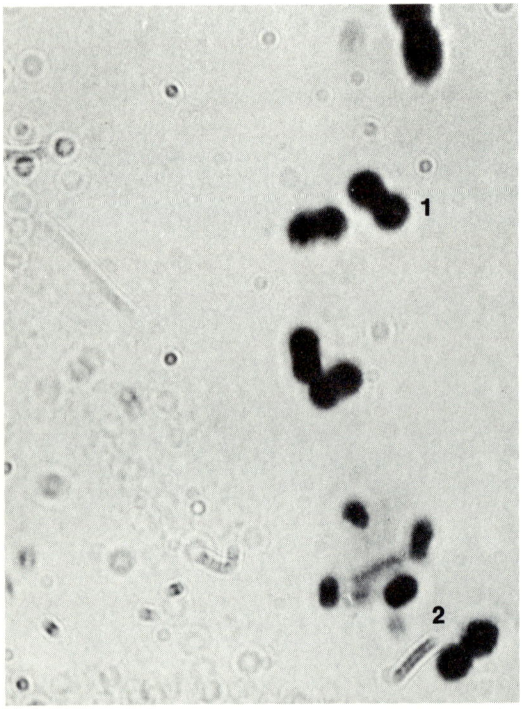

Figure 21-4
Azotobacter chroococcum cells growing in nitrogen-free mineral salts liquid medium inoculated with soil. (Note large cell size of *Azotobacter*, **1**, and the small rods, **2**. These may also be nitrogen fixers or merely forms using nitrogenous compound exuding from the *Azotobacter* cells.)

Figure 21-5
A, Root nodules on native burr clover (*Medicago* sp.) produced by *Rhizobium meliloti;*

Bacteria readily attack and solubilize the phosphorus in RNA and DNA, while other compounds such as Phytin, a calcium magnesium salt of phytic acid and a reserve sugar in various tubers and bulbs, are more slowly mineralized. As phosphate ions enter the soil solution, they may be taken up by plant-root hairs, the various microbes, or in alkaline soils changed into insoluble calcium or magnesium salts.

Solubilization of phosphates is accomplished by the production of acids of various sorts. Thus in the soil, the nitrifying bacteria will not only provide an available, soluble nitrogen compound, but they will render soluble and avail-

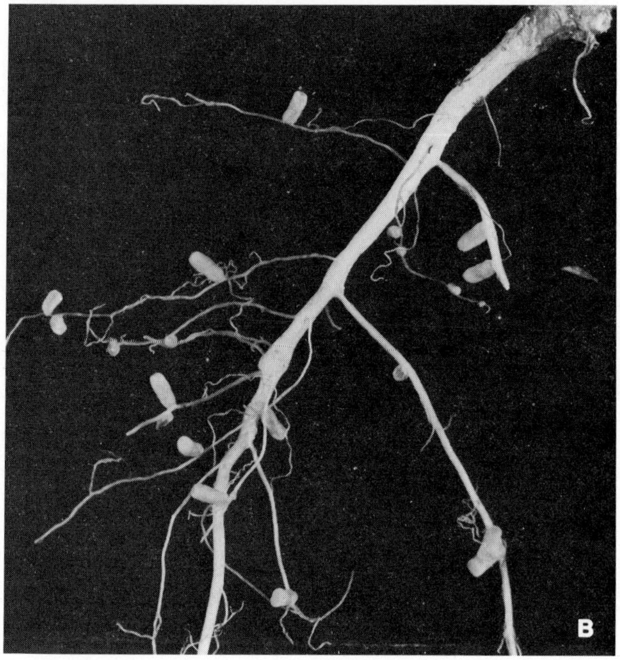

B, enlarged view of root nodules (×3).

Figure 21-6
Root nodules on native lupine (*Lupinus* sp.) produced by *Rhizobium lupini.* Note taproot location.

able to plants some of the insoluble mineral phosphates.

Thiobacillus, which oxidizes sulfur to sulfuric acid, also actively solubilizes the insoluble phosphates.

The transfer of phosphorus of the soil from soil to plant and return is essentially cyclic in nature, hence the process may be termed the phosphorus cycle.

Sulfur, like phosphorus, is a relatively rare but vital element. Unlike nitrogen with its inexhaustible supply in the air, or phosphorus with the fairly abundant concentrations of insoluble phosphate-bearing minerals in many soils, proportionately, sulfur is limited in its supply, and most of it must come from organic sources or artificial fertilizers.

Sulfur is found in organic matter in the soil (living microbes plus plant, animal, and microbial remains) in the form of three sulfur-containing amino acids: cystine, cysteine, and methionine. The B-vitamins, thiamine, and biotin together with taurine, thiourea, and other products also add small amounts.

Rainfall may add sulfur to the soil primarily as a result of the burning of sulfur-containing fossil fuels, or as fumes from some chemical processes. Sulfur added by rainfall may range from 1 lb per acre per year in unpopulated regions to 100 lb per acre per year over industrial cities.

Volcanoes also spew out "juvenile" sulfur into the atmosphere.

Under aerobic conditions, cystine and cysteine are quickly attacked and sulfates released. Methionine is not broken down by the same means.

Volatile compounds can be produced that may be lost to the atmosphere or may subsequently be oxidized to sulfate. Sulfur is largely taken up by plant roots in the form of sulfates.

Sulfur-containing fertilizers are often used. Foremost among these is ammonium sulfate. This is efficient in that nitrogen and sulfur are both made available. If animal fertilizers or sewage sludge are used, the organic sulfur-containing compounds must be decomposed in order to release the sulfates.

Elemental sulfur is sometimes added to soils as a fertilizer. Most arable soils contain a few (100 to 200, sometimes up to 10,000) *Thiobacillus* per gram. With the addition of elemental sulfur, the soil becomes an enrichment medium and the thiobacilli grow rapidly. Their major growth product is sulfuric acid. This is significant because it may make over-alkaline soils more acid or produce enough pH change to control some plant pathogens such as the potato scab.

Under anaerobic conditions in water-logged soils, *Desulfovibrio desulfuricans* may quickly reduce the sulfates to hydrogen sulfide (H_2S). The rotten-egg smell is sometimes found to emanate from such soils.

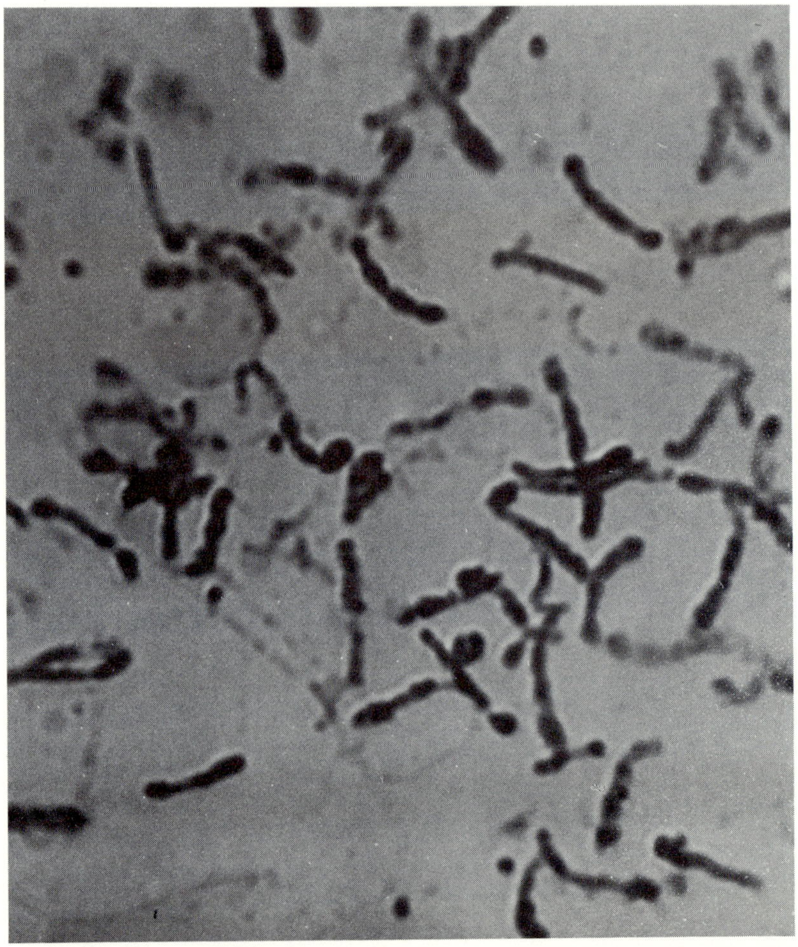

Figure 21-7
Rhizobium meliloti bacteroids (X and Y forms) as seen in exudates from crushed burr clover nodules stained with congo red (×3,000).

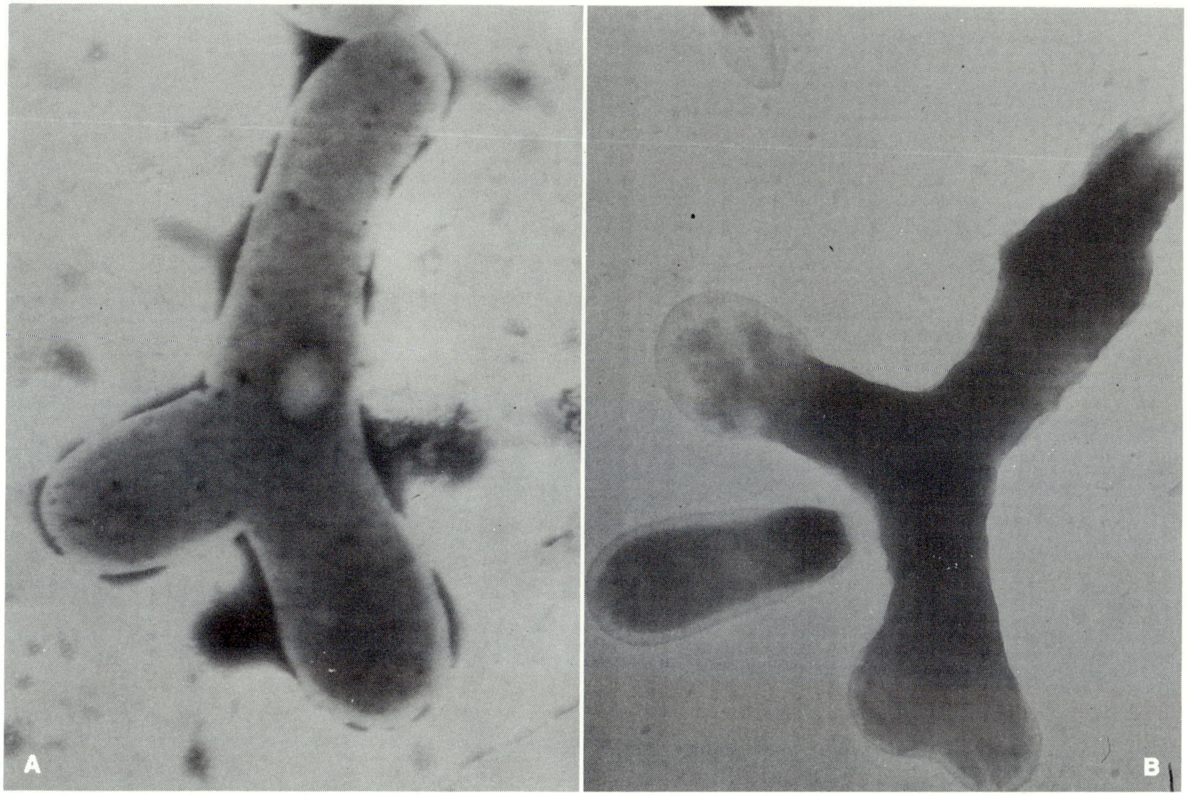

Figure 21-8
Electron micrographs of burr clover bacteroids. **A,** Fixed with gluteraldehyde; **B,** unstained (×1,200).

RESISTANT MOLECULES

Soil is the receptacle of a great variety of organic compounds—nature's contributions of plant and animal remains, and exotic compounds added by man. The voracious microorganisms searching for energy and body-building materials have versatile enzyme systems that enable them to rapidly break down most of these compounds. Certain compounds, however, are resistant. Soil microorganisms, in spite of their amazing enzymatic versatility, find that these are like a lock for which they have no key.

Naturally occurring compounds

A number of compounds that are readily oxidizable by physical means are resistant to microbial attack. For example, the lignins and other polyaromatic compounds, plus certain waxes and related substances, are only slowly broken down. These accumulate in the soil and form the organic complex termed humus. This is the material that imparts the brown or black color to highly productive soils and gives them friability and water-holding capacity.

Radioactive carbon datings on humic acid from

Figure 21-9
Root nodules associated with symbiotic nitrogen fixation in the alder, common shade tree of the genus *Alnus.* Causative agent still not identified.

solids have shown that the original plant or animal materials from which they came were incorporated into the soil as long as 1,000 to 3,000 years ago. Bodies of fungi are sometimes broken down very slowly because of the ligninlike compounds that they contain.

Some aromatic (benzene ring-like) and some aliphatic compounds are relatively resistant to microbial attack especially under anaerobic conditions. Even methane (CH_4), the main constituent of marsh gas, can be utilized by only a few microorganisms. These facts probably account for the accumulation of the great underground reservoirs of crude oil or natural gas.

Man-made synthetic compounds

In recent years man has been adding many exotic organic compounds to the soil, or to water supplies that eventually find their way into the soil.

Wide use of synthetic pesticides and herbicides has greatly increased productivity of cropped lands, but, although it appears to be a simple, effective solution to the problem of insects and weeds, it has proved to be a two-edged sword. Certain of these compounds are enzymatic enigmas to the soil microflora and may persist from year to year. Thus, while insect life or weed growth may be inhibited for more than one season, we are confronted with the threat that these toxic substances may eventually find their way into the foods we eat. Also, certain of these exotic compounds may destroy some of the vital microorganisms in the soil, thereby upsetting the balance in the microbial life and reducing soil fertility.

However, during the last decade or so a number of soil microbiologists have begun a careful study of the problem with the hope of achieving a reasonable solution. The answer seems to be the synthesis of reasonably effective pesticides and herbicides that will produce the desired controls, but in turn be destroyed in a short time by soil microorganisms and/or other environmental factors. Thus the desired control measures can be achieved without the curse of toxic persistence.

Many amateur, self-designated ecologists have shouted "abolish all synthetic pesticides and herbicides." Indeed, the word synthetic is an anathema to them. But when the cost of food doubles because it must be shared with insects or yield to noxious weeds, they probably will protest the loudest.

SUGGESTED READINGS

Alexander, M.: Introduction to soil microbiology, New York, 1961, John Wiley & Sons, Inc.

Alexander, M.: Biochemical ecology of soil microorganisms, Annu. Rev. Microbiol. **18**:217-252, 1964.

Bolin, B.: The carbon cycle, Sci. Am., **223**:125-132, September 1970.

Burges, A.: Micro-organisms in the soil, London, 1958, Hutchinson & Co. (Publishers) Ltd.

Burris, R. H.: Biological nitrogen fixation, Annu. Rev. Plant Physiol. **17**:155-184, 1966.

Deevey, E. S., Jr.: Mineral cycles, Sci. Am. **223**:149-158, September 1970.

Delwiche, C. C.: The nitrogen cycle, Sci. Am. **223**:137-146, September 1970.

Gray, T. R. G.: The ecology of soil bacteria, Toronto, 1968, University of Toronto Press.

Stewart, W. D. P.: Nitrogen fixation in plants, London, 1966, Athlone Press of The University of London.

22 Microbiology of air

EXPERIMENTAL BEGINNINGS

The first scientifically sound experiments relating to microbiology of air were those of Pasteur, when, in the early 1860's, he demonstrated the universal presence of bacteria in air.

On the basis of this, Lister in 1867 devised his method of antiseptic surgery in which he used an atomizer to spray a fine mist of carbolic acid (phenol) into the air over the area of a surgical operation; but it was soon found that contact, rather than airborne bacteria, was primarily responsible for infections. Aseptic surgery soon replaced antiseptic surgery, and the interest in airborne surgical infections subsided.

Professor Flügge of the University of Breslau in 1897-1898 concluded that air transmission of disease was of no importance. He based this on a limited series of experiments in which petri plates were exposed in front of persons who sneezed or coughed. Professor Flügge judged that all infective droplets settled out within a few feet of the "infector," therefore, diseases were not airborne. This became accepted medical dogma, and the possibility of airborne transmission of disease was essentially dismissed by physicians and public health workers for almost three decades.

Late in the World War I period, Stillman demonstrated the possible role of airborne bacteria in the transmission of pneumococcus pneumonia. *Klebsiella* and *Streptococcus pyogenes* pneumonias were also transmitted experimentally to animals by inhalation from an infected atmosphere.

In 1933, William F. Wells, an engineer in the Harvard School of Public Health, published the first of a series of papers that revived the concept of airborne infections. He developed an effective atomizer that sprayed droplets of experimental bacterial suspensions over considerable areas. The bacteria were then recovered with his centrifugal air sampler. He concluded that droplets 0.1 mm in diameter or so did settle to the ground before they evaporated. However, smaller droplets evaporated leaving "droplet nuclei" that were still infective, and that could be wafted indefinitely by air currents. This provided a logical mechanism by which diseases could be transmitted via the air. It was evident that Flügge had missed the minute particles that were carried away from his exposed plates.

Interest in the problem of airborne infections was revived and a number of other research groups were stimulated to begin reassessment.

During and after World War II the Biological Warfare Service carried out extensive studies on airborne diseases. Originally designed as a means of aggressive military action, until the experimentation was quenched by public clamor the experiments yielded a great deal of information. However, much of it found its way into "secret" files from which it may never be extracted.

Along with this, hospital epidemics caused by penicillin-resistant staphylococci and fatal epidemics of meningococcus meningitis in military establishments provided the impetus for studies directed to a further evaluation of the problem of the transmission of airborne infections.

AIR AS AN ENVIRONMENT

Today the interest in air as a major environment has provided a further incentive for the study of the microbiology of the air.

The present stage of the science of air microbiology still lags behind that of water, milk, and foods. Although a great deal of experimental work has accumulated since the 1930's, a great need exists to synthesize this accumulated knowledge into a coordinated whole. Some significant attempts have been made to do this. Examples are the report of the Second Inter-

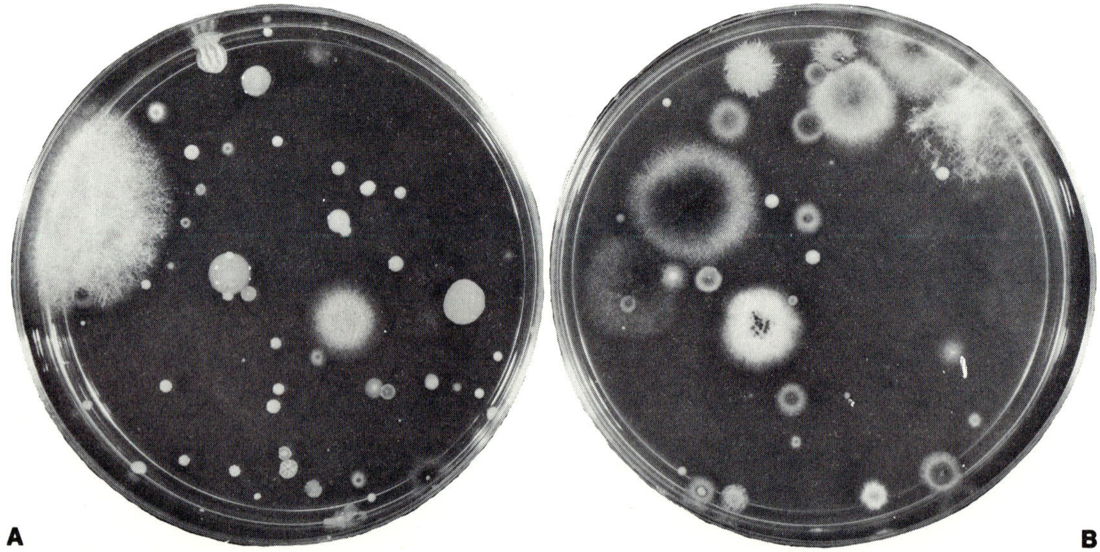

Figure 22-1
Results of gravity air sampling. Open petri plates exposed to outside air for 30 minutes. **A,** Nutrient agar, showing primarily bacterial colonies; **B,** Sabouraud's dextrose agar, showing mold colony growth.

national Conference on Aerobiology (1966), and the *Airborne Microbes,* Seventeenth Symposium of the Society for General Microbiology (1967).

In general, air is looked upon as an inert vehicle capable of carrying various bacteria, fungus spores and filaments, and droplet-enclosed virus particles. It has not been considered as an active life environment for these forms. Recently, however, at least one series of experiments suggests that bacteria carried in the clouds might actually be physiologically active and undergoing cell division.

METHODS FOR ISOLATING MICROBES FROM AIR

A variety of techniques have been devised—some simple, some elaborate—for isolating and detecting microbes in natural outdoor environments, hospital wards, military barracks, schools, or experimentally infected confined air. The microorganisms studied include vegetative and spore forms of bacteria; spores and filamentous fragments of molds; yeasts; spores, cysts, or cells of slime molds, algae, and protozoa; and animal, plant, and bacterial viruses.

Examples of some of the sampling methods or devices are described in the following paragraphs.

Gravity plates are simply standard or large-sized petri plates filled with various sterile culture media. These are left open with the agar surface exposed for varying periods. Microorganisms, dust, or other particles settle on the surface of the agar and are detected as fully developed colonies. The accuracy of this method is limited because microbes in the droplet nuclei may not settle out on the surface, but may be kept suspended by convection currents. Also, the medium used will determine only the kinds of microorganisms that will develop (Fig. 22-1).

Microscope slides covered with some sticky

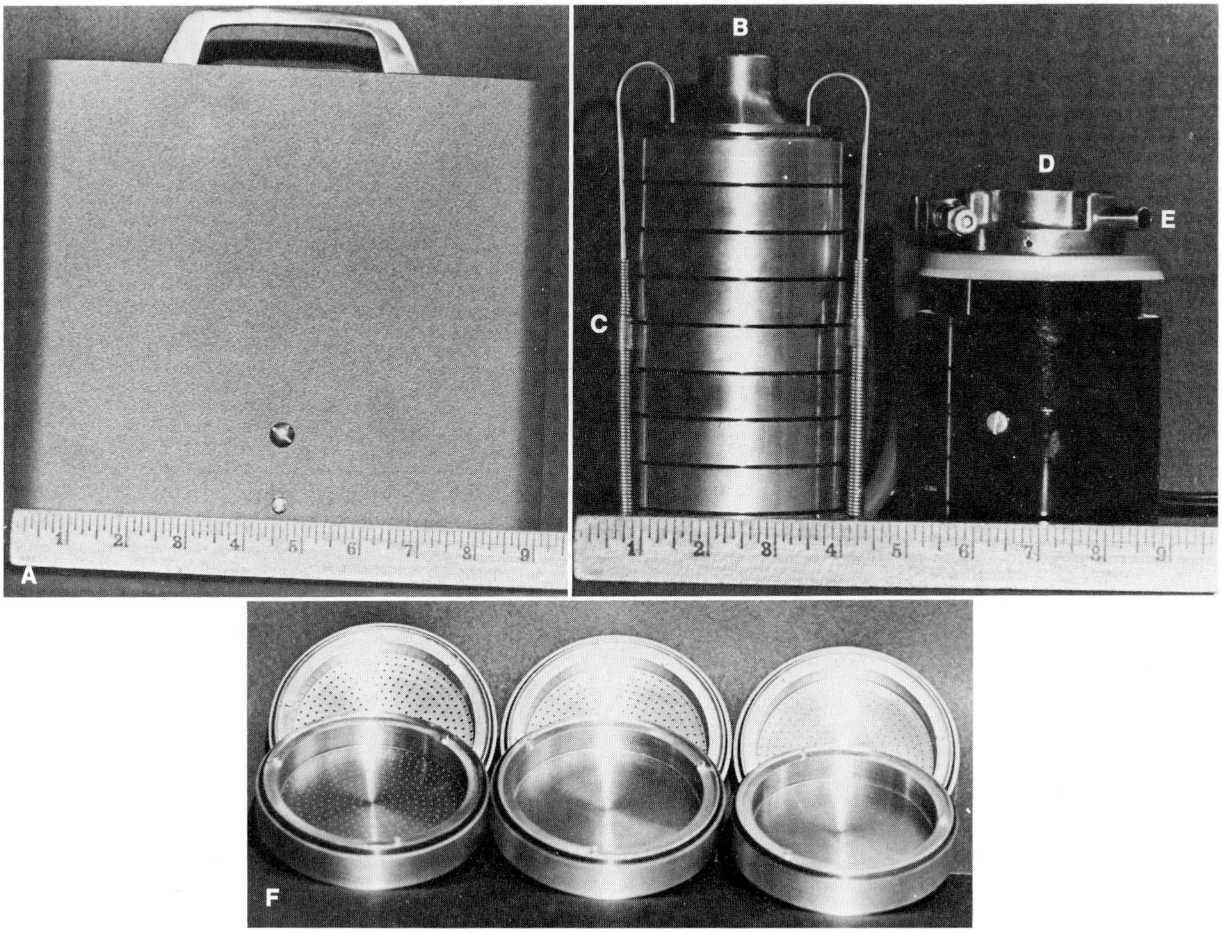

Figure 22-2
A, Andersen air sampler in carrying case. **B-E,** Air sampler ready for operation; **B,** air inlet, **C,** stacked sieve collectors, **D,** air pump, **E,** air outlet. **F,** Sieve collectors showing distribution and size of holes.

substance may be exposed directly to the air as in the gravity plate technique, or in a sampler where a jet of air is blown across the slide. This method is used primarily for the collection of mold spores.

The Wells air centrifuge uses centrifugal force to implant the organisms from the air on the collecting medium. Coupled with the air centrifuge is a special atomizer, which Wells designed, capable of spraying droplets of regulated size into the atmosphere for the study of atmospheric distribution of known microorganisms.

The Andersen sampler, developed originally by Ariel Andersen for biological warfare experiments, consists of six circular collecting devices made of metal, each of which holds the half of a petri plate filled with a suitable agar medium. This is sometimes described as the stacked sieve collector. The top collector plate has uniformly distributed large holes, the second smaller ones, and the holes become progressively smaller on each plate. The bottom collector plate has very small holes. The coarse particles (9.2 μm or larger in diameter) pass through the large holes and are trapped on the surface of the agar of the first plate. The smaller particles are carried on by the stream of air, and deposited on plates in terms of their decreasing particle size.

Stage 2 collects particles 5.5 to 9.2 μm; stage 3, 3.3 to 5.5 μm; stage 4, 2.0 to 3.3 μm; stage 5, 1.0 to 2.0 μm; and stage 6, 1.0 μm or less. The sampler is equipped with a motor that blows air through it at a controlled rate so that in a given time period a known volume of air will have passed over the plates. This device has been widely used in recent experimental studies of airborne microorganisms (Fig. 22-2).

The slit sampler is one in which air is blown through a long, narrow slit and impinges on a sticky surfaced microscope slide, or a rotating disk, or an open petri plate. Some of these devices sample large air volumes in a short period.

Liquid impingers have been designed to blow air into a liquid used to collect microorganisms. Samples of this liquid may be used to inoculate agar plates, animals, or tissue cultures.

Electrostatic devices that operate rather successfully have also been developed.

Membrane filters have been used because they remove all particulate matter. A measured amount of air is drawn through the filter, which is then placed on the surface of a suitable agar medium, similar to the water testing procedure. Colonies that develop may be counted and identified. One weakness is that the filter may not be uniformly moistened and some forms of microorganisms will fail to grow. In some cases the filters are suspended in a sterile medium to wash off the collected microorganisms. The suspension may then be used to inoculate several types of differential agar to secure a picture of the different kinds of microorganisms collected. The membrane filter has the advantage of collecting all particles larger than the pore size of the filter (usually 0.45 μm). The use of very small pore sizes makes possible the collection of viruses.

Susceptible animals have been placed in situations where they breathe air from patients suffering from a respiratory disease, or air, such as in classrooms, known to be infected. The number of infective doses per volume of air has been calculated on the basis of the number of animals that become infected in proportion to total number used in the experiment. In a few instances, human volunteers have been used for studies of airborne respiratory infections.

For high altitude air samplings from airplanes, because of the high speed of the plane, which makes accurate sample collecting almost impossible, a special device had to be developed.

AIR AS A CARRIER OF MICROBES

The study of air as a carrier of microbes represents a complex problem. Air itself may consist of the great atmospheric masses subject to motion induced by the great cyclonic weather fronts, tornadoes, great updrafts of heated air, and so forth, or may involve microclimates representing strictly localized air movements over a relatively short distance.

On the other hand, much of the experimental work is concerned with movement of air and its microbial load inside rooms, or within entire buildings. Examples of findings relating to each environment will be considered.

Microbiology of air masses

Microbiology of air masses is concerned with the transport of microorganisms, in general, by large air movements, or localized transmission of disease-producing forms in more limited areas.

A classic example of air transportation of microorganisms for great distances is represented by the dusts that arise in the Sahara desert. These may fall in visible amounts many miles away in the Mediterranean region or eastern Europe. Along with the dust particles themselves will be the microorganisms they carry.

Mold spores typical of continent land masses

have been collected over the ocean 400 miles from land. Spores from tropical soils have been collected above 10,000 feet at the edge of updrafts accompanying developing thunderheads.

The jet streams of air probably carry microorganisms for thousands of miles. Conditions are severe, so that many forms succumb to cold, desiccation, and the sun's radiation. However, many microorganisms travel on until they are swept down to the surface of the earth by rain or snow.

In connection with the space program, many air samplings have been made at various altitudes. As examples, samples collected from an airplane at 11,000 feet showed 5 to 200 microorganisms per cubic meter of air. Predominant forms were mold spores, primarily *Hormodendrum, Aspergillus,* and *Alternaria;* and bacteria, mostly *Bacillus* spores and micrococci.

Balloon samplings at 60,000 to 90,000 feet showed 1 to 5 microorganisms per 10,000 cubic feet of air. Predominating forms were *Alternaria* and *Cephalosporium* spores, and micrococci.

Escherichia coli has been collected from the air 140 feet downwind from a trickling filter sewage disposal plant (Fig. 22-3).

In terms of airborne diseases, in an epidemic of Q fever that occurred in 1951, many human cases developed downwind from sheep pens housing infected animals. The case incidence, when plotted, presented a fan-shaped distribution with the most distant cases being ten miles away from the infected animals. A persistent dry desert wind was apparently the carrier.

Spores of *Puccinia graminis,* the cause of black stem rust in wheat, have been carried northward by prevailing winds many miles over the plains of North America.

Mold spores, carried by winds, have produced allergic reactions among large urban populations.

Airborne spores of the yeastlike pathogens, *Histoplasma capsulatum* and *Coccidioides immitis* have produced infections in various caged "sentinel" animals exposed to winds blowing from areas in which the spores were endemic in the soil.

Inside air

Formerly thought to be of no significance as a carrier of pathogens, air has been proved by a great deal of evidence to be capable of transmitting diseases in buildings, sometimes for considerable distances.

In large buildings a considerable movement of air may occur. Stairwells may act like smoke stacks where the colder air at the bottom is heated and moves upward at a considerable velocity as it continues to become hotter. Elevators moving up and down act like plungers of a hypodermic syringe drawing in air from the bottom as the elevator rises, then forcing it out on the descent.

A few of the many examples of inside airborne infections follow.

The accidental transmission of the rickettsia of Q fever, *Coxiella burnetii,* through a large building housing laboratories of the National Institutes of Health was responsible for 47 cases of the disease. These occurred during a 4-month period and clearly were airborne through the ventilating system of the building.

Airflow from infectious wards in a tuberculosis sanitarium was passed through a chamber housing guinea pigs. Based on the number developing active infections among 150 animals it was calculated that 1 infective dose (10^4 organisms) was carried by each 12,000 cubic feet of air. This concentration was too low to be detectable by any mechanical sampling device but sufficient to cause infection.

The airborne nature of respiratory viruses and bacteria has been rather well authenticated. One writer likened the air in a closed classroom or theater during a cold epidemic to the water in a polluted lake.

One unusual and tragic example of an airborne virus infection occurred in a cave at Uvalde, Texas, that housed a huge colony of bats. Two persons who entered the cave contracted rabies and died. In a study that followed, caged sentinel animals, well protected from the bats, were placed in the cave. Rabies developed in 10 coyotes and 12 foxes held in the cages. A mechanical

Figure 22-3
Airborne bacteria from sewage treatment plant. **A,** Trickling filter sewage disposal tank (see Chapter 17). **B,** Close-up of boom showing formation of aerosol droplets. **C,** Petri dish air samplings collected with Andersen air sampler. Top row of plates collected 140 feet downwind from filter bed. (Note large number of coliform colonies.) (Adams, A., and Spendlove, J. C.: Science **169:**1218-1220, 1970, U. S. Army photographs.)

air sampler was then used. This was a large impinger type in which 10,000 liters of air per minute were blown against a rapidly rotating sticky disk. The rabies virus was recovered by this method.

Airborne transmission of anthrax had been demonstrated in a case in which workmen handling infected goat pelts developed pneumonic anthrax and several died. One individual, who had no connection with the plant, walked past the open door, inhaled the organism, and succumbed to the disease.

Whether a person becomes infected with airborne pathogens depends largely on the size of particles. Experiments with the Andersen sampler revealed that large particles fall out too rapidly to be inhaled. The particles 5.5 to 9.2 μm are inhalable, but are trapped in the pharynx and immobilized. The next size, 3.3 to 5.5 μm, enter the trachea and primary bronchi, the 2.0 to 3.3 μm particles penetrate the second bronchi, the 1.0 to 2.0 μm particles (bacteria carriers) move into the terminal bronchi, while particles smaller than 1.0 μm (primarily viruses) lodge in the alveoli.

SANITIZATION OR CHEMICAL DISINFECTION OF AIR

According to one estimate, the aerosol produced by a vigorous sneeze carries a million or so droplets under 100 μm in diameter, plus thousands of larger particles. The moisture carried by the finest particles evaporates immediately forming droplet nuclei that remain indefinitely suspended in the air capable of being wafted considerable distances. The larger particles quickly fall to the floor where they attach to the particles of dust.

At the peak of a cold infection, a sneeze is particularly productive because it atomizes explosively the discharges from the acutely infected nasal membranes.

Coughing is much less productive, producing mostly larger particles that quickly fall out. However, it may contribute a fair number of infective droplets. Talking is much less important because it expels few particles small enough to become droplet nuclei, unless, as someone has observed, it is accompanied by a "snort of derision."

From the viewpoint of disease transmission the infective droplets are of major importance. They are small enough to be carried deep into the lungs. The larger droplets, even though inhaled, may be trapped and inactivated by the hairs in the nose or the moist mucous membranes.

However, larger particles trapped in the dust may be carried temporarily into the air by sweeping or moving rapidly. These particles seem to be most infective as carriers of *Streptococcus pyogenes* (beta hemolytic, Type A "streps").

Use of germicidal ultraviolet rays

Wells and his group in the early 1930's were the first to apply the newly developed cold cathode–mercury vapor germicidal lamp to the sanitization of air.

This lamp, essentially a modified fluorescent lamp, has the unique property of producing rays 80% of which have a wave length of 2537 Å. This is within the maximum germicidal range of ultraviolet light: 2500 to 2600 Å.

Although the lamp would appear to be an ideal agent for air sanitization it has several objectionable features that limit its use:

1. The unprotected skin of a person exposed to its rays for any period of time shows an erythema or reddening, followed by extensive peeling. (There is no tanning effect.) The conjunctiva of the eyes temporarily develop an irritating and disconcerting redness.
2. The rays have only limited penetrating power and, as a result, are applicable only to surface sanitization.
3. Unlike visible light the lamp has only limited reflective properties.
4. After the lamp has been operating for a period of time, the unpleasant odor of ozone may develop, coupled with an irritating effect on eyes and nasal membranes. The

ozone will also damage curtains or paint in a room.

On the other hand, the germicidal lamp is highly effective against bacteria or viruses suspended in the air either as from recently expressed minute droplets, dried droplet nuclei, or smaller dust particles.

The earlier experiments indicated the substantial air disinfecting power of the germicidal lamp. Although its applications are rather limited, it has been used for the sanitization of air in hospital operating rooms; in hospital rooms occupied by persons with respiratory diseases; to provide limited invisible barriers against airborne pathogens surrounding wards for newborn babies; installed inside ducts of air conditioning systems in hospitals; installed over cosmetic packing areas to prevent airborne contamination; and in large meat storage refrigerators to prevent surface molding or formation of surface slime.

As shown by a number of surveys, installation of germicidal ultraviolet lamps in hospital operating rooms has produced marked reductions in postoperative infections.

Because of the resultant reddening of the skin and conjunctivitis that follows direct exposure, it is not feasible to install ultraviolet lights in the open hospital rooms occupied by patients. However, recessed installation above the patients causes a reduction in bacterial content of the air equivalent to 100 air changes by ventilation.

By installing germicidal lamps in a recess in the ceiling, the narrow line of irradiation immediately below will serve in effect as a narrow curtain of germicidal rays. This method has been used in large wards to divide the room into a number of invisible cubicles within which the beds of patient with respiratory diseases can be placed, effectively separating them from other patients. The nurses can pass freely from patient to patient without undue exposure to ultraviolet rays by avoiding the narrow line of irradiation.

Use of chemicals for disinfection of air

Early in World War II, military personnel became concerned with the number of outbreaks of respiratory diseases (particularly throat infections caused by hemolytic streptococci) that occurred. A special committee of experts was chosen to study the feasibility of air disinfection in barracks. In addition to a number of experimental installations of ultraviolet lamps, it was decided to use germicidal vapors as a means of air disinfection. Initially a variety of disinfectant vapors were tried including lactic acid, hypochlorous acid, iodine, and propylene glycol. These were discarded as either undesirable or ineffective. Triethylene glycol (TEG) was finally chosen as a suitable agent. This could be readily vaporized by heating and had little or no odor or other irritating properties. The major objection was that in adequate concentrations it created a faint fog in the room. Rats and mice had been exposed to the vapors for a year without any apparent harmful effects.

Added to germicidal vapors, applications of light oil to the floor were used to dispel the dust in the barracks. This combined treatment appeared to be effective in reducing the incidence of streptococcus infections, but was of only limited value in reducing viral respiratory infections. Moreover, surfaces became sticky, and the barracks had to be maintained at an uncomfortable humidity level when TEG was used. As a result, the attempt to control airborne diseases with volatile vapors was abandoned, although oiling to suppress dust levels was generally continued.

SUGGESTED READINGS

Adams, A. P., and Spendlove, J. C.: Coliform aerosols emitted by sewage treatment plants, Science, **169:** 1218-1220, 1970.

Alexander, M.: Microbial ecology, New York, 1971, John Wiley & Sons, Inc.

Dimmick, R. L., and Akers, A. B., editors: An introduction to experimental aerobiology, New York, 1969, Interscience Publishers, Inc.

Gregory, P. H.: The microbiology of the atmosphere, New York, 1961, Interscience Publishers, Inc.

Gregory, P. H., and Monteith, J. L., editors: Airborne microbes, 17th Symposium of the Society for General Microbiology, Cambridge, England, 1967, Cambridge University Press.

Lepper, M. H., and Wolfe, E. K., editors: Second international conference on aerobiology (airborne infection), Bacteriol. Rev. **30:**485-698, 1966.

Norris, J. R., and Ribbons, D. W., editors: Methods in microbiology, vol. 1, New York, 1969, Academic Press, Inc., Chap. 5.

Tyndall, J.: Essays on the floating-matter of the air in relation to putrefaction and infection, New York, 1966, Johnson Reprint Corp.

PART 3

Microorganisms and the production of disease

23 Mechanisms involved in the production of infectious diseases

INITIAL EXPOSURE TO MICROBES

From the moment the infant leaves the mother's womb, even while passing through the birth canal, the process of colonization of the body with bacteria begins. This process even continues after death as our bodies are mineralized.

The first breath of air the child breathes carries with it bacteria that may immediately begin colonization of the body. In turn, the infant's skin receives bacteria from the air, the hands of the nurse and mother, the bath water, and the clothing. The aimless fumbling attempts of the infant to place his hands in his mouth, and the oversolicitous kiss of the grandmother on the baby's mouth contribute to the colonization process.

The end result is that before the infant is a day old bacteria have already begun the process of establishing themselves. *Streptococcus salivarius* has been reported in the mouth within twelve hours of birth.

If the infant is breast fed, the feces may quickly acquire a large population of anaerobic lactobacilli. Within a short time, *Escherichia coli* will begin to establish itself in the intestines. *Staphylococcus epidermidis* appears quickly on the skin, and within a few months, the infant will have largely acquired the normal flora of the adults around it.

It would appear that the helpless infant would quickly succumb to this onslaught of potentially dangerous microbes. Fortunately, nature has provided, at least temporarily, for the safety of the infant. For the first critical six months or so, antibodies from the mother's blood, present in the infant's blood serum, provide a lifesaving degree of built-in protection. If the child is nursed by the mother, the protective action of the maternal antibodies will be prolonged. By the time the maternal antibodies have been exhausted, the baby has begun to provide its own protection.

The amount of self-protection the baby develops in the form of antibodies and active white cells is largely determined by his genetic makeup. If the infant suffers a pronounced congenital inability to produce antibodies (agammaglobulinemia), or inadequate white cell production (agranulocytosis), he may soon succumb. Fortunately, these are rare genetic accidents.

Bacteria from the environment that lodge on or in the infant's body may be destroyed or fail to develop if antibodies are present, or if the body does not provide a satisfactory environment for growth. Temperature, nutrients, and pH may all be determining factors. The bacteria may establish themselves under a mutually beneficial arrangement, may merely be tolerated, or may maintain a sort of armed truce between host and microbe. Bacteria of these types tend to remain constantly in the body, forming the so-called natural flora.

DISEASE PRODUCTION

A few microorganisms have the ability to produce disease, and their colonization or infection may become a disease condition involving a violent reaction, on the part of both the body and the disease producer, and resulting in an acute disease characterized by abrupt, marked symptoms, such as high temperature, malaise, and so forth. The condition may terminate either by complete recovery or death of the patient. In other cases, a chronic state may develop in which the pathogen produces a low-grade infection that continues for a matter of months or years. Occasionally, complete recovery may occur, or the patient may succumb.

Fortunately only a relatively small proportion

of the bacteria in our environment are disease producers. These pathogens tend to pass from the infected to noninfected persons. The mode or channel of transfer is known as an infection chain.

Many infection chains are capable of being broken by preventive measures such as the pasteurization of milk, or sanitization of eating utensils. If the organism is waterborne, chlorination may sever the infection chain and thus prevent the infection. Diseases may also be averted by control measures such as immunization.

COMPARISON OF GERM-FREE AND NORMAL ANIMALS

Animals can be grown completely free of microbes so that they will have no normal flora, no acquired antibodies, and no changes in tissue structure induced by microorganisms. For this reason they are admirably suited to studies of the effect of the normal flora as it becomes established in the body. Such animals are now readily available.

The first attempt to raise germ-free animals was carried out by Pasteur in 1885. At this early date he succeeded in growing a germ-free chick but did not pursue the study further.

It was not until 1928 that a more prolonged attempt was made to grow germ-free animals. Reynier, at the University of Notre Dame (Lobund Project) successfully grew a variety of animals in the germ-free state. Such animals are termed axenic (Gr. *a* without + *xenos* foreign), in other words without foreign life. The study of germ-free animals is termed gnotobiotics (Gr. *gnotos* known + *bios* life) and such animals may be termed gnotobiotic animals.

The original procedure of producing axenic animals was amazingly elaborate. Baby animals were taken from the mother by cesarian section under aseptic conditions. They were then maintained in a sterile chamber and fed only sterile foods. To achieve this state, a large chamber was built and equipped with a side-well filled with disinfectant. An attendant dressed in a watertight diving suit climbed down a ladder into the disinfectant, then climbed up another ladder into the sterile chamber. The technical problems, such as the simple matter of feeding a suckling mouse with sterile mother's milk, were formidable.

In the decades that have passed, tremendous strides have been made. The attendant no longer has to pass through a disinfectant bath. Instead, large, clear plastic chambers are now available. The investigator inserts his hands and arms through long rubber gloves attached to and extending through portholes into the sterile chamber. The entire device is sterilized with ethylene oxide gas.

Methods of feeding axenic animals have now been developed, and the animals, including dogs, have been successfully grown and bred for several generations in the germ-free state. The procedure has yielded germ-free insects, mice, rats, dogs, pigs, and monkeys.

In one recent case, a human infant was removed successfully by cesarian section and maintained in an axenic state for several days before introducing it gradually into the outside world.

Comparison of germ-free animals with normal animals has shown that one of the greatest differences induced by the presence of a normal bacterial flora is in the structure of the linings of the intestinal tract.

In the small intestine, the inner lining in a normal animal is well developed and in a state of so-called physiological inflammatory reaction due to the presence of the normal flora. The corresponding layer in the germ-free animal is very poorly developed. The surface area of the lining of the small intestines in the germ-free animal is markedly less than that of a normal animal. In the normal animal, the mucosa cells are renewed about twice as fast as in the germ-free animal.

In an experiment to compare the effect of pathogens, *Salmonella typhimurium* was introduced by a stomach tube through the duodenum into the small intestines of normal and germ-free mice. Killing the animals after 24 hours, the responses were then compared. The number of bacteria in the intestines was many times lower in the normal than in the germ-free animals, but the number in the lymph nodes was about the same. This suggested that the mucous membranes had about the same protective action in preventing the migration of bacteria to the lymph nodes. However, the marked reduction of bacteria in the intestines of the normal animals was found to be due to the much greater rapidity with which the normal intestines emptied as compared with those of the germ-free animals. This indicated a markedly increased rate of peristalsis in the normal animal.

When sterile starch solution was injected into the peritoneal cavity of conventional and germ-free mice, it excited the pouring-out of large numbers of phagocytic leukocytes. Both kinds of mice exuded the same kinds of leukocytes. However, after 6 hours, the number present in the normal animals was twice that found in the germ-free animals. It is thought that normal flora bacteria present in other parts of the body may enhance the increased leukocyte production observed in the peritoneal cavity.

When germ-free mice are placed in a normal colony, a considerable proportion die rather quickly. With 2-month-old mice, about 25% die, while with germ-free mice 6 months old at the time of entering the colony, 90% succumb. Bacteria that are considered avirulent for us or for experimental animals may produce fatal infections in germ-free animals.

MECHANISMS OF DISEASE PRODUCTION

Disease production or pathogenicity is a highly complex property that is genetically variable. Thus strains of a given species of bacterium may range from highly pathogenic forms to nonpathogens. This property depends on the extent to which the strain carries in its genes messages for the production of those characteristics responsible for disease production.

Highly pathogenic forms are termed virulent (L. *virulentus*, from *virus* poison). A virulent organism, therefore, has the capability of rapid and violent disease production.

Avirulent forms are essentially pathogens that have lost or do not possess the ability to produce disease.

Most strains of *Escherichia coli*, for example, are avirulent, although some strains possess considerable virulence. When these latter strains infect adults, they may produce only a relatively mild diarrhea; however, among infants or even young children fatal epidemics due to these strains have occurred.

Invasiveness implies the ability of pathogens to pass through the external defenses of the body and produce a disease.

Disease-producing properties

Pathogenicity, virulence, or invasiveness depends on a number of genetically determined properties. Chief among these are toxigenicity (the ability to produce toxins or biological poisons), capsule formation, which protects the organism against phagocytosis, and the production of enzymes that excite or assist in the process of disease production.

TOXIGENICITY. Toxins produced may be extracellular (exuded outside the cell) or intracellular (produced inside the cell and released only on the death and lysis of the cell).

EXOTOXIN. The production of exotoxin is a property of a number of pathogens. Of the exotoxins the most important are produced by twelve species of bacteria. Five of these are species of Clostridium, *C. botulinum*, *C. tetani*, and the gas gangrene producers, *C. perfringens*, *C. septicum* and *C. novyi*.

The exotoxin of *C. botulinum* is probably the most powerful poison known. One milligram of the purified toxin is equivalent to 12,000,000 LD_{50}. In other words, a series of doses, each representing 1/12,000,000 of the original milligram of toxin, would still be potent enough to kill 50% of the inoculated, susceptible animals. The designation LD_{50} means Lethal Dose–50%

and is the standard used for comparing toxigenicity, pathogenicity, viral potency, and so forth.

Six types of *C. botulinum* toxin are known. Type A is the most potent and is most often the cause of human botulism.

The botulism toxin is a true neurotoxin. The first symptoms are double vision and lack of coordination of eye muscles. These appear 18 to 96 hours after the contaminated food is eaten, and are followed by difficulty in swallowing, slurred speech, and usually death due to respiratory failure and heart arrest.

The tetanus toxin is almost equally as potent. It contains two toxins. The first, tetanospasmin, produces the extreme symptoms of muscle spasm or tetany, characterized as "lock-jaw." The second, tetanolysin, causes hemolysis of the red blood cells and heart damage.

Gas gangrene, caused by the anaerobic growth of *C. perfringens*, *C. septicum*, and *C. novyi* in lacerated tissue, involves a number of toxins, mostly necrotizing (Gr. *nekros* dead). These kill the tissues that are then attacked by proteolytic enzymes of saprophytic bacteria. Others, such as lecithinase and collagenase break down tissue components to make them more subject to proteolytic enzymatic attack. Others injure the blood cells and the heart.

The diphtheria organism, *Corynebacterium diphtheriae*, also produces a highly potent toxin that necrotizes the tissue surrounding the site of infection, usually the throat. This causes development of a false membrane that can suffocate the patient. Later, nerve and heart damage may produce death.

Streptococcus pyogenes causes "strep throat" and similar infections. Some strains cause scarlet fever because of the production of an erythrogenic toxin (Gr. *erythros* red + producing) that is responsible for the production of the scarlet fever rash.

At least four streptococcic toxins are known. Of these, streptolysins O and S hemolyze red blood cells when grown in blood broth or on blood agar plates. They are also highly toxic to white blood cells and may damage the membranes of other cells. Streptolysin O also acts as a toxic agent against the heart muscle. Intravenous injections of the toxin into laboratory animals causes heart arrest.

Staphylococcus aureus is known to produce at least six different toxic agents. Five act either against white cells (leukocidins), or red blood cells (hemolysins), or act as necrotoxins. This latter action is evidenced by the death of the skin surrounding a boil. The sixth is the enterotoxin, previously discussed in Chapter 18.

Shigella dysenteriae, the cause of the most violent of the bacillary dysenteries, produces a highly toxic neurotoxin that may induce paralysis and hemorrhage of the intestines. In the purified state, it is almost equal in potency to the botulism toxin. Fortunately, the amount produced is much lower than in botulism. It acts indirectly by cutting down the blood supply to the brain and spinal cord.

The plague organism, likewise, produces a potent necrotizing toxin. This may explain in part the high virulence of this organism.

The whooping cough organism produces an exotoxin that is specific for the respiratory cilia. This accounts for the paroxysms of coughing for which the disease is named.

Recently, what appears to be an exotoxin has been isolated from the Asiatic cholera organism, *Vibrio comma*. It seems to act specifically against the sodium-potassium pump and may explain the violent dehydration characteristic of the disease.

The exotoxins are proteins, heat labile, and highly antigenic. Because of this, they stimulate the body to produce antibodies. Active immunity can be produced by the injection of toxoids, which are detoxified toxins.

Generally the toxins react rather slowly, often requiring several days before violent symptoms appear.

ENDOTOXINS. These differ materially from the exotoxins. They are much less toxic, and are large complex phospholipid and polysaccharide molecules. They are formed inside the bacterial cell and released only by autolysis on the death of the cell. They are relatively heat stable. They are not antigenic in the sense that no antitoxins are formed. However, contact with the endotoxin develops a degree of tolerance to it.

Endotoxins are produced primarily by gram-negative bacteria and are often associated with characteristic symptoms of intestinal disturbance produced by *Salmonella* and *Shigella* infections. Injection of these toxins into experimental animals produces irreversible shock and diarrhea that rapidly becomes fatal. Small doses cause an initial decrease in white cells (leukopenia) followed by a rapid increase (leukocytosis). The temperature rises rapidly, and an increase in red cells occurs.

The toxicity is due primarily to the phospholipid fraction. In addition to the enteric bacteria, endotoxins are produced by *Brucella melitensis, Neisseria gonorrhoeae,* and *N. meningitidis,* as well as several other pathogenic bacteria.

Capsule formation. This is responsible for the virulence of a number of types of bacteria. The classic example is *Diplococcus pneumoniae.* Capsulated or smooth strains are resistant to attack by phagocytes. This gives them their marked invasive power. The noncapsulated or rough strains when injected into experimental animals are immediately ingested by the white cells and destroyed. For this reason they are avirulent or nonpathogenic.

Other bacteria that owe their invasiveness to capsule formation are Friedlander's bacillus *(Klebsiella pneumoniae),* and *Bacillus anthracis,* the anthrax organism.

Extracellular enzymes. These may impart pathogenicity or enhance the pathogenicity of other agents. Hyaluronidase, or spreading factor as it was once called, hydrolyzes hyaluronic acid or hyaluronates, which act as an intracellular glue in the host animal tissue. Several of the gram-positive bacteria produce this enzyme. This apparently enhances the spread of their toxic products.

Streptococcus pyogenes is the organism from which hyaluronidase was originally isolated, and the enzyme apparently gives the organism much of its ability to spread through host tissues. However, the organism may defeat its own purpose, because it often also has a capsule of hyaluronic acid that protects it against phagocytosis. By dissolving away its own capsule, it probably becomes susceptible to phagocytosis.

Pathogenic strains of *Staphylococcus aureus* produce the enzyme, coagulase, which causes the coagulation of blood plasma. It apparently has a dual effect, although its role is not clearly defined. By laying down a clotted fibrin barrier, the staphylococcus lesions tend to be localized, possibly a disadvantage to the bacterium. On the other hand, the fibrin coating protects the bacterium against phagocytosis.

Enzymes that lyse fibrin clots are produced by streptococci (streptokinases) and staphylococci (staphylokinases). Their role in pathogenicity is not clearly defined.

EXTERNAL AND INTERNAL DEFENSES OF THE BODY

The body is constantly assailed by pathogenic bacteria of various sorts. However, the mere contact of a pathogen with the body is rarely sufficient to produce disease. The process of disease production is a complex one, involving the interaction of a number of factors. The formula below reduces the problem to its simplest dimensions.

$$\text{Disease production (D)} = \frac{\text{(N) number of pathogens} \times \text{(V) virulence}}{\text{(R) resistance of host}}$$

Number of pathogens (N) is a factor of major importance. Few bacteria are so virulent and aggressive that a single cell gives rise to a disease condition. With many relatively virulent forms, the minimal disease-producing dose may range from a few to several hundred cells. Other less virulent forms are unable to develop a disease until the number of infecting bacteria exceeds a hundred-thousand or more. However, an infective dose of this magnitude may not be difficult to attain if you ingest food or milk in which the bacteria have been actively growing, or if you drink a glass of heavily polluted water.

Even a fleck of pus might easily carry hundreds of thousands of infective forms.

The term virulence expresses the total ability of an organism to penetrate the external defenses of the body, survive the effects of the cellular and humoral body defenses, and attack the vital organ or organs for which it has a special predilection.

The formula is written $N \times V$ rather than $N + V$, because the effect of an original small number of pathogens will be multiplied manyfold by the rapid growth of a highly virulent microbe.

Resistance of the human or animal host is the sum product of a multiplicity of factors: external defenses represented by the skin, mucosa, and mucus secretions; acidity of the stomach; natural high body temperature; lysozyme of the tears and saliva; and a variety of other protective agents. Added to this is the second line of defense, the internal defenses made up of the phagocytic cells of the blood and tissues, and the antibodies of the serum.

All of these protective factors are genetically determined in the host. High resistance to a given species or even genus of pathogenic bacteria may be inherent in a given host, although the host may be relatively susceptible to another species of bacteria. This type of selective high resistance often represents species immunity.

On the other hand, within a given species, certain individuals may show high resistance against a specific causative agent. This generally occurs because the individual has had a previous (known or unknown) contact with the pathogen that has resulted in a typical, atypical, or inapparent case of the disease. The outcome has been the presence of a high level of antibodies against the pathogen, and is termed individual active immunity. If it has been conferred by the administration of a vaccine, it represents active acquired immunity.

External defenses

External defenses represent constantly present structural or other factors of the body that prevent pathogens from penetrating into the deeper tissues and thus producing the infectious disease.

The skin has been called the "coat of armor" that protects the body. Very few bacteria can penetrate the unbroken skin. *Pasteurella pestis* and *Brucella* species are alleged to possess this power, although it is difficult to prove because an abrasion invisible to us might represent an inviting, gaping wound to the microscopic bacteria. Pathogens generally possess active enzymes that hasten the penetrating process.

Staphylococcus epidermidis is a normal inhabitant of the skin that rarely causes us any great difficulty. However, if a rose thorn or splinter of wood breaches the outer skin covering or epidermis, we may suffer from a festered sliver. When the foreign body is removed the area immediately heals.

At times nearly everyone has contact with the much more virulent *S. aureus*, often present on the skin or the respiratory mucosa. However, only occasionally do these bacteria produce boils, carbuncles, or localized mucosa infections. This again is due to the protective action of the skin. When skin infections do occur, they are often a result of growth of the bacteria around a hair follicle or in a clogged sweat gland, or because of a minute break in the skin, such as a razor nick. In hot climates, where the skin is constantly moist, bacteria such as *S. aureus* are more likely to penetrate and cause skin infections.

With lowered resistance, which accompanies the first years of puberty, susceptibility to staphylococcus infections increases. This represents a multiplicity of factors involving increased oiliness of the skin, adolescent diet habits, hormonal imbalance, and self-inoculation (perpetuated by constant squeezing of pimples). Contrary to the belief held by some, pimples do not come from impurities or bacteria in the blood. They are purely a surface phenomenon in which the ecology of the skin has been temporarily upset.

The skin itself may show an inhibitory degree of acidity, possibly engendered by bacteria growing on the surface. The presence of certain oil-like substances will cause the rather rapid death of bacteria that come in contact with the unbroken skin.

Tears serve to protect our eyes against infection. A constant flow of lachrymal fluid (tears) is secreted from the lachrymal ducts on the temporal sides of each eye. Except when the emotions stimulate a rush of tears from the accessory lachrymal glands, the ordinary flow of tears passes across the surface of the eye, drains through tiny holes at the nasal corner of each eye (the puncti), and empties through tiny tubes (the canaliculi) into the cavity at the back of the nose without overflowing the eyelids. This constant washing of the eye carries most bacteria away before they can establish themselves. In addition, the tears are rich in lysozyme, which dissolves bacterial cell walls particularly of most gram-positive bacteria.

Mucus, which is constantly being secreted, traps many of the airborne microorganisms, especially those on the larger particles of moisture or dust. Coughing and sneezing may expel a portion of the microbe-laden mucus into the atmosphere. The mucus that is sloughed off carries its load of microbes to the stomach, where the high acidity, plus certain gastric enzymes, kills many of the microorganisms.

In the nose, the hairs of the nostrils, especially those higher up in the nasal cavity, are constantly moist with mucus. These trap many of the bacteria and dust particles in the inhaled air.

The cilia, deep in the nasal tract and lining the lung passages, carry on a constant "escalator" action by which mucus and foreign objects are being steadily pushed outward by the nasal mucosa and upward by the lung passage cilia. This latter action is vitally important because it keeps bacteria and foreign matters from entering the lungs. These cilia are highly reactive, and fluid going "down the wrong throat" immediately excites the cilia to produce a paroxysm of coughing until the foreign material is carried out.

This explains why alcoholics, suffering severe alcoholic poisoning, are prone to develop pneumonia. The stupified cilia fail to sweep out the bacteria-laden droplets or mucus, allowing deep penetration into the lungs.

The mucosa that lines the respiratory, digestive, and urogenital tracts serves as a defense mechanism but is much more susceptible to infection than the skin. In addition to protecting the underlying tissue, the mucus secretions may contain antibodies of various sorts. Also saliva, nasal and other secretions contain lysozyme.

External factors such as excessive drying, low temperature, and acrid, irritating fumes may all render the respiratory mucosa more susceptible to microbial attack.

In the intestines, the mucosa may secrete coproantibodies, which are found in the feces. They are active against certain intestinal bacteria. In addition, pathogens in the intestines are subject to attack by phages, bacteriocins, and other antibacterial agents. The normal flora may keep in check certain pathogens. Thus, oral administration of broad-spectrum antibiotics that kill the normal bacteria may allow the yeast, *Candida albicans,* to gain the ascendency and produce severe intestinal infections that are sometimes fatal to infants.

The vagina, during the childbearing years, has an acid reaction maintained by lactobacilli, which represent the major members of the normal flora. The acidity controls the growth of many bacteria that might otherwise invade the tissues.

Internal defenses

It might appear logical that the stomach, intestinal tract, lungs, and so forth, being inside the body, represent internal defenses. However, since all of these tissues have direct contact with the outside environment, they represent external defenses against invasion of tissues by pathogens; hence they form part of the first line of defense.

When pathogens breach the skin or mucosa and enter the deeper tissues, they come face to face with the second defense line, the internal defenses.

The phagocytes (cell eaters) of the blood represent the initial agents that attack the invading bacteria. They are often termed the cellular defense force. The blood contains, in addition to some 5,000,000 red blood cells per cubic millimeter, about 5,000 to 10,000 white blood cells, or leukocytes, which are primarily phagocytes. These are ameboid cells that surround and ingest foreign bodies such as bacteria. Of these cells, the most active phagocytic forms are the polymorphonuclear neutrophiles (polys or neutrophiles), so called because of their lobed nuclei that form an irregular horseshoe-shaped structure in the cell (Gr. *polymorph*—many shapes, referring to the nuclear structure). The term neutrophile is applied, because in staining whole blood with a polychrome (many-colored) stain such as Wright's stain, the neutrophiles take up the neutral stain (crystal violet); cells stained with the acidic stain (eosin) are termed eosinophiles; and those taking up the basic stain (methylene blue) are called basophiles.

All of the polymorphonuclear leukocytes have two characteristics in common: irregular, multiple nuclei, and the presence of granular bodies in the cytoplasm, stained by their respective stains. Therefore, they are also called granulocytes, and normally make up 55% to 70% of the white cells and are primarily neutrophiles.

The balance of the white cells are monocytes, that is, have only a single large nucleus. The majority of the monocytes are small lymphocytes, with a much lower proportion of large monocytes.

When invading bacteria break through the skin or mucosa, the phagocytes are immediately rushed to the scene. Why this occurs is not completely understood but may involve the secretion of some chemotactic substance at the site of invasion.

The process begins with a reddening of the area due to a marked dilation of the capillaries and an increased blood supply. In some manner, the neutrophiles slither through the lumen of the capillaries, surround the bacteria, and a battle ensues. The neutrophiles are joined by the monocytes, which are also actively phagocytic.

In cases of acute infection, such as a full-blown appendicitis, the number and proportion of neutrophiles increase dramatically. The white cell count may rise to 20,000 to 30,000 with 95% of the white cells being neutrophiles. This is termed a leukocytosis (condition of marked increase in white cell production).

This acute phase of battle is usually a relatively short one. If, for example, it involves a *S. aureus* infection such as a boil, the area will be walled off by a closing of the capillaries.

The neutrophiles, initially present in great numbers, actively ingest and destroy the bacteria. However, many of the phagocytes may be killed by bacterial toxins. The neutrophiles are expendable and have a half-life of 6 hours. After 24 hours or so the monocytes move in to replace the exhausted or dead neutrophiles. The pus that has developed is made up of killed bacteria, killed white cells, living bacteria, living white cells, plasma, and cellular debris. If the battle has been won by the body, the site will have been walled off and the boil or pimple "comes to a head." If undisturbed, the monocytes complete the job of cleaning up the battle field and the infected area soon returns to normal.

If, on the other hand, at this point a pimple is squeezed vigorously, pus may be forced past the walled off capillaries, and the battle erupts afresh in the surrounding tissue. A tiny pimple may then become a large one, or a large pimple, with bacteria pressed into the deeper tissue, becomes a boil or carbuncle.

If the interference is too vigorous, the live bacteria could be forced into the lymph system. The same thing may happen if the invading organism is able to overcome the phagocytes and escape from the localized area. This is not common with *Staphylococcus aureus*, but *Streptococcus pyogenes*, because of its hyaluronidase and active toxic agents, can often escape into the lymph channels. When this occurs, the bacteria begin their journey along the lymph channels to a regional lymph node such as those found in the axilla (armpit) and the groin. The movement can be followed as a red line that moves up the arm or leg from the original infection.

Here the phagocytic battle still continues, with the neutrophiles being supported and soon replaced by the large macrophages of the reticuloendothelial system (specialized synthetic areas for phagocyte production located in the spleen, liver, thymus, and bone marrow).

The lymph nodes begin to swell due to accumulation of cells, debris, and toxic compounds. This swelling of the lymph nodes is called lymphadenitis (L. *lympha* spring water + Gr. *aden* gland + *itis* inflammation of).

At this point, the invading bacteria may pass through the blood vessel walls of the lymph node or vessels, and enter the bloodstream. When this occurs, the condition is termed a bacteremia, literally, bacteria in the blood. As the blood flows through the body it passes through the spleen and liver. These have large blood-filled sinuses or hollow structures lined with fixed phagocytic cells, which literally surround, pull out of circulation, and ingest the invading bacteria as the blood flows by. The spleen and liver also function as filters to remove worn-out red cells, dead white cells, and similar debris.

If the bacterial virulence is not sufficient to overwhelm the reticuloendothelial phagocytes, the bloodstream will soon be swept free of invaders.

The following scheme, based on an actual experiment, illustrates the clearing action of the reticuloendothelial phagocytes. Three rabbits were each given an intravenous injection of pneumococci. The first received an avirulent strain; the second, a moderately virulent strain; and the third, a highly virulent strain. Blood was drawn at intervals and cultured for bacteria. Results were as follows:

1. With the avirulent strain, bacteria were rapidly removed. After 2 hours, less than 1% remained. The blood was completely free of bacteria in 24 hours.
2. With the slightly virulent bacteria, 2% were present after 2 hours, a small number remained at 24 hours, and the blood was not completely clear until 96 hours had passed.
3. With the highly virulent bacteria, 2.5% were still present at 4 hours. By 24 hours the number had increased to $1\frac{1}{2}$ times the original inoculum, and the animal was dead in 48 hours.

If the bacteria are highly virulent, they may begin vigorous growth in the bloodstream, and a septicemia (Gr. *sepsis* decay + *haima* blood) develops. This is familiarly known as blood poisoning. It is frequently fatal and suggests that the bacteria have literally overcome all of the defenses of the body.

Septicemias of this sort are most commonly caused by the beta hemolytic streptococci, although they may occasionally be produced by the pyogenic staphylococci and some strains of *Diplococcus pneumoniae*. On rare occasions, the gonorrhea and meningitis bacteria may invade the bloodstream with serious and often fatal results. The plague organism often causes a highly fatal septicemic state. Bacteremias are common with many *Salmonella*, *Shigella*, and other enteric bacteria, with invasion of the bloodstream by way of the lymph nodes of the intestines.

Some highly pathogenic forms, such as the diphtheria and tetanus organisms, remain localized and exude into the bloodstream highly poisonous toxins that may result in fatal consequences. With the botulism organism, death can occur as a result of ingesting the toxin itself without active growth of the causative organisms.

INTRODUCING THE ANTIBODIES

The final defense mechanism is often spoken of as the humoral immune defenses. These involve the production of antibodies that are carried by the blood serum. They include antitoxins that neutralize toxins, agglutinins that clump bacteria, bacteriolysins that dissolve bacteria, and the opsonins that prepare bacteria for phagocytosis.

They may develop in the body as a result of

previous disease or a vaccine that stimulates the body to produce antibodies. The branch of science that deals with humoral immunity is called immunology. Because the antibodies are present in the serum of the blood, and their presence is detectable by laboratory tests, the practical use of such immune tests is termed serology.

This highly involved branch of microbiology is discussed in considerable detail in Chapter 24.

SUGGESTED READINGS

Burnet, M.: Natural history of infectious disease, ed. 3, Cambridge, England, 1962, Cambridge University Press.

Davis, B. D., Dulbecco, R., Eisen, H. N., Ginsberg, H. S., and Wood, W. B., Jr.: Microbiology, New York, 1967, Harper & Row, Publishers.

Kadis, S., Montie, T. C., and Ajl, S. J.: Plague toxin, Sci. Am. **220**:92-100, March 1969.

Mirand, E. A., and Back, N., editors: Germ-free biology; experimental and clinical aspects, New York, 1969, Plenum Press.

24 Host responses to pathogenic microorganisms and other foreign agents

Y. HOKAMA, Ph.D.

NATIVE IMMUNITY

ACQUIRED IMMUNITY

BLOOD GROUP SYSTEMS

As indicated in the previous chapter, animals normally possess numerous complex defense mechanisms to thwart attack by pathogenic microorganisms and harmful foreign agents, such as toxins and allergens, to which they are constantly exposed. This capacity of the host organism to withstand deleterious agents within his environment is native or innate immunity. This is, for the most part, genetically determined within a species with respect to a particular antigen. This inherited capacity may vary from species to species. On the other hand, specifically acquired immunity is an individual property, dependent on the individual's previous contact with the particular harmful agent or toxin. In this chapter we will consider the native and acquired defense mechanisms that enable the host to resist the harmful effects of pathogenic agents.

NATIVE IMMUNITY

Two general areas are considered under native immunity: active and passive defense mechanisms. The active defense mechanisms of the host involve the tissues and structures of the body, and their components, which are in part or wholly functionally directed for active resistance to harmful external or opportunistic endogenous microorganisms and their toxins. The passive defense mechanism or nonsusceptibility involves no specific tissues or structures but is, rather, an environmental situation within the host that fortuitously is unsuited for initiation of a disease process by a particular pathogenic microorganism.

Active defense mechanisms

EPITHELIAL SURFACES. The outer covering or skin of the body, and the mucous membranes composed of specialized epithelial cells lining the respiratory and gastrointestinal tracts serve as the first line of defense against infection. These surfaces act primarily as mechanical barriers to invasive and pathogenic microorganisms. In addition, substances such as fatty acids found on the surfaces of the skin are bactericidal to certain pathogenic microorganisms. An example of this is the bactericidal effect of lactic acid against some species of the genus *Salmonella.* The specialized epithelial cells of the mucous membranes secrete mucus (protein-carbohydrate complexes), enzymes, and other bactericidal factors that contribute to the host's ability to resist attack by pathogenic microorganisms and harmful agents.

TISSUE AND BODY FLUIDS. When microorganisms enter the host after trauma or injury to the epithelial surfaces, they are immediately confronted by secondary defense substances found in body fluids; some in serum, and others in tissue fluids. When isolated from animal tissues and body fluids, the compounds show antimicrobial activity in the test tube. Characteristics of some of these chemical substances are summarized in Table 24-1. Information as to whether all of these materials exert antimicrobial activity in the body itself is as yet incomplete. Nonetheless, since these molecular compounds are found distributed through the tissues their potential function as bactericidal agents cannot be excluded.

GROUP A COMPOUNDS. The antimicrobial effect, in the laboratory, of short-chain and long-chain fatty acids has been long known. Short-chain fatty acids may act as metabolic inhibitors against microorganisms and in this manner exert their killing effect, whereas long-chain fatty acids appear to exert antibacterial effects by their action as surfactants (surface acting agents). Similarly, sterols (lipids) such as bile salts (cholates, and desoxycholates of bile) act as surface-active agents, killing and lysing particularly the gram-positive capsulated pneumococci. However, there are exceptions in that the related *Streptococcus faecalis* proliferates in media containing bile salts, and in this way appears to be similar to some gram-negative organisms of the Enterobacteriaceae. This explains its presence in the intestines.

Free porphyrins exert a high degree of bactericidal activity in vitro against gram-positive microorganisms. However, the porphyrins occurring in nature are normally bound to proteins to form hemoproteins. These appear to have no antimicrobial activity. The hemoproteins are found in tissues as enzymes such as catalases of the liver and red blood cells, myeloperoxidases of the leukocytes, and hemoglobulin of the erythrocytes.

GROUP B COMPOUNDS. Numerous basic polypeptides containing high levels of basic amino acids have shown various degrees of antimicrobial activity in vitro and are found in a variety of highly specialized tissues. They affect primarily the gram-positive microorganisms. The bactericidal power seems to reside in their ability to complex or bind with the surfaces of the microorganisms. Previous studies have shown that the bactericidal activity of basic polypeptides could be neutralized by the addition of acid mucopolysaccharide, deoxyribonucleic acid (DNA), and ribonucleic acid (RNA) to the me-

Table 24-1. Antimicrobial substances in tissue and body fluids

MOLECULAR SUBSTANCES	MAJOR SOURCE	CHEMICAL PROPERTIES	EFFECT ON TYPE OF BACTERIA
Group A			
Fatty acids	Ubiquitous	Short- and long-chain fatty acids	Gram negative
Bile salts	Gall bladder, liver, intestines	Sterols	Gram positive
Heme	Red blood cells, white blood cells	Porphyrins	Gram positive
Group B			
Polylysine, Polyarginine	Thymus, tissues rich in histones	Low molecular weight; basic peptides	Mainly gram positive
Protamine	Sperm	Low molecular weight; basic peptide	Gram positive
Spermine, spermidine	Pancreas, prostate	Basic polyamines	Gram positive
Group C			
Lysozyme	Ubiquitous	Low molecular weight; basic protein (enzyme)	Mainly gram positive
Phagocytin	Leukocytes	Globulin	Gram negative
Beta lysin	Serum	Multiple enzymatic system; proteins	Gram positive
Properdin	Serum	Complex group of serum proteins	Gram negative
Opsonin	Serum	Globulins (natural antibodies?)	Gram negative
Complement	Serum	Complex group of serum proteins consisting of 9 components	Gram negative

dium, thus in this way nullifying the ability of the basic compounds to bind to the cell surface of the microorganisms in vitro. Hence, the fact that the basic polypeptides react with acidic compounds would suggest that the target on the bacterial cell may be that of an acidic residue, such as terminal sialic acid residues.

GROUP C COMPOUNDS. The chemically complex group of antimicrobial substances of body fluids are listed under group C. They include:
1. The thermostabile, low molecular weight, basic protein lysozyme found in saliva, tears, leukocytes, macrophages, and many other sources. Lysozyme primarily acts on gram-positive microorganisms in vitro. As an enzyme it hydrolyzes the complex acetylaminopolysaccharide (muramic acid). Thus microorganisms that possess muramic acid in their wall structure would be readily lysed by lysozyme. The microorganism *Micrococcus lysodeikticus* is extremely sensitive to lysozyme action and thus is readily lysed by it. The antimicrobial action of lysozyme is chiefly effective within narrow pH ranges near neutrality. The cytoplasm of alveolar macrophages and leukocytes of the blood are especially rich in lysozyme.
2. Extracting leukocytes (white cells) of blood with acid yields a cytoplasmic protein, phagocytin. In the test tube, phagocytin appears to have a wide range of antimicrobial action on both gram-negative and gram-positive microorganisms. However, its bactericidal effect is

greater on gram-negative than on gram-positive microorganisms.

3. Another compound, beta lysin probably functions as an enzyme.

4. Properdin comprises a complex group of serum proteins, some of which are heat labile. Bactericidal activity has been exhibited against gram-negative bacteria in the presence of complement and magnesium ions (Mg^{++}). Properdin, complement, and Mg^{++} ions constitute the properdin system, which also has been implicated in the inactivation of a number of viruses.

5. In normal fresh serum of animals there are proteins that enhance the phagocytosis by leukocytes of bacteria and inert particles. These proteins, called opsonins, are natural antibodies. Opsonin has no specificity, and in this respect differs from a specific antibody, but it does appear to have predilection for gram-negative microorganisms and inert particles such as carbon. Like properdin, opsonin requires complement action in the enhancement of phagocytosis by leukocytes. Opsonin is measured or titrated by determining the number of microorganisms ingested per phagocyte (phagocytic index) obtained from normal fresh serum, and then comparing this with the number of microorganisms ingested per phagocyte obtained for fresh serum of a patient. This gives the opsonic index:

$$\text{Opsonic index} = \frac{\text{Patient's phagocytic index}}{\text{Normal phagocytic index}}$$

6. Complement, designated C′, is made up of a group of nine distinct functionally interrelated protein entities. It is found in fresh blood serum of animals, and occurs in a variety of species. It generally diminishes in concentration during some of the disease processes. Differences in complement activity among various species occur, and these are attributable to the variations in the proportions of the nine C′ components.

Chemical examination of fresh serum has shown that certain components of C′ are readily destroyed by oxidizing agents and by heat, being inactivated at 56° C in ½ hour. For example, it has been known for some time that C′ can be separated into two major components, called midpiece and endpiece, by simply immersing fresh blood serum, contained in a dialyzing cellophane membrane, into distilled water. Such a process is called dialysis. The midpiece, a water-insoluble precipitate but soluble in physiological saline, consists of C′1 and C′3 forms within the dialyzing membrane. The endpiece remains in solution and consists of C′2 and C′4. The midpiece or the endpiece by itself shows no C′ activity, but when recombined, complement activity is restored. Heating of C′ at 56° C for 30 minutes inactivates C′1 and C′2, while treatment of serum with zymosan removes the C′3 component. Treatment of fresh serum with dilute ammonium hydroxide results in C′4 activation. The various residual active C′ components remaining after physical or chemical treatment are indicated in Table 24-2.

In recent studies five other C′ components have been recognized and designated C′5, C′6, C′7, and C′9. In addition, C′1 has been shown to consist of three subunits held together by calcium ions (Ca^{++}).

The sequence of action of complement can be shown diagrammatically using the blood serum from a rabbit immunized against sheep red blood cells (hemolysin) against its homologous antigen, the red blood cell sheep erythrocyte, shown at the bottom of the facing page.

The complex sequential action of complement following reaction of an antigen and antibody shows that enzymatic activity of C′ occurs at the early steps and involves C′1 following its attachment to the antibody portions of the AgAb complex. Activated C′1 has been shown to have esterase activity. Appearances of chemostatic activity associated with phagocytic migrations, anaphylatoxin release, immune adherence properties, release of histamine, and so forth, have been observed following the late phases of C′ activation. These processes appear following C′3 and/or C′5 activation. It is apparent that the complement system in association with opsonins, properdins, and specific antibodies pro-

Table 24-2. Complement components as demonstrated by selective physical or chemical activation

PHYSICAL OR CHEMICAL TREATMENT OF A SERUM	RESIDUAL ACTIVE C′ COMPONENT IN SERUM		RECOMBINATION OF C′ COMPONENTS TO GIVE ACTIVE C′
Dilution with water	a (midpiece)	C′1, C′3	a + b
	b (endpiece)	C′2, C′4	
Zymosan	c	C′1, C′2, C′4	a + c
NH_4OH	d	C′1, C′2, C′3	d + b d + c
56° C	e	C′3, C′4	e + c e + d

vides an extraordinarily complex degree of control of the immune response and for the active defense of the host against harmful agents.

CELLULAR ELEMENTS IN INNATE IMMUNITY. The phagocytes or eating cells of the blood and reticuloendothelial system comprise one of the most important active defense mechanisms in their contribution to native resistance against pathogenic microbes. These act by ingesting bacteria or other foreign particles. Metchnikoff first recognized the significance of leukocytes in native resistance to microbes. This was an outgrowth of an earlier observation of intracellular digestion by cells of the water flea.

In the animal body there are two main varieties of phagocytic cells, microphages and macrophages. The microphages include the polymorphonuclear leukocytes, which make up more than half of the white cells of the blood. Of these the neutrophiles exhibit the greatest phagocytic activity and constitute the major cells in the acute inflammatory response to infection; for example, the number and proportion rises rapidly in a case of acute appendicitis. Eosinophiles and basophiles exhibit less phagocytic activity.

The cells of the reticuloendothelial system are comprised of macrophages, either sessile or wandering. Sessile, or fixed, macrophages are found lining the capillary endothelium and sinuses of the liver (Kupffer's cells), spleen, bone marrow, lymph nodes, and other organs. They exert phagocytosis by ingesting foreign bodies from the blood as it flows past. The wandering macrophages, also known as histiocytes, include the blood monocytes that migrate through the endothelium and tissues and assist in the repair of damaged tissues by ingesting dead tissue materials and inert particles and further aid in the disposal of erythrocytes and leukocytes that

Early steps

1. Ag (sheep erythrocyte—antigen) + AB (antibody or hemolysin—rabbit serum) AgAb (complexes)
2. AgAb + C′1$^{Ca^{++}}$— — — —AgAbC′1
3. AgAbC′1 + C′4— — — —AgAbC′1,4
4. AgAbC′1,4 + C′2^{Mg}— — — —AgAbC′1,4,2

Late steps

5. AgAbC′1,4,2 + C′3— — — —AgAbC′1,4,2,3
6. AgAbC′1,4,2,3 + C′5,6,7— — — —AgAbC′1,4,2,3,4,5,6,7
7. AgAbC′1,4,2,3,5,6,7 + C′8,9— — — —AgAbC′1,4,2,3,5,9
   (end point) Lysis of the erythrocyte Ag$^{x\ (activated)}$

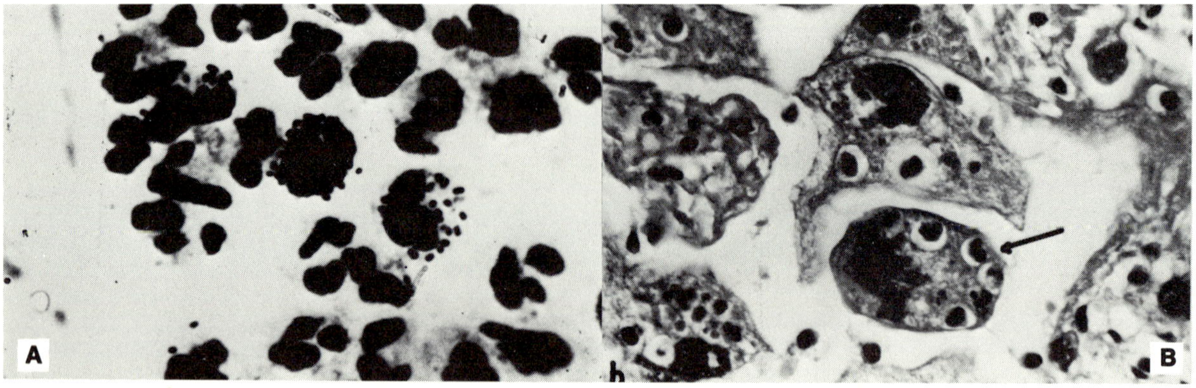

Figure 24-1
A, Phagocytosis of bacteria *(Neisseria gonorrhoeae)* by leukocytes; **B,** ingestion of leukocytes by malignant tumor cells.

have passed through into the injured areas of tissues.

The manner by which phagocytes ingest inert particles, microorganisms, dead cells, and so forth, is called phagocytosis. Another term, pinocytosis, is used in the same manner, but generally refers to ingestion of soluble proteins, or low molecular weight molecules. Phagocytosis is a complex phenomenon involving essentially three sequential steps: (1) the attraction of the phagocytes to the site of injury, known as chemotaxis; (2) the act of ingestion of the microorganism, phagocytosis itself; and (3) the ultimate disposition of the microorganisms within the phagocyte. Examples of phagocytosis of bacteria by leukocytes of the blood and the ingestion of leukocytes by malignant tumor cells are shown in Fig. 24-1.

Inflammation. The inflammatory process is a significant orderly response of the host in native resistance to the invasion of microorganisms and to traumatic tissue damages. It involves most of the effects already discussed and also brings into play other host factors that may contribute to active and passive resistance. Inflammation may be summarized by the sequence of events resulting from tissue injury as follows:

1. Tissue injury results in increased permeability and dilation of blood vessels (producing reddening around an infection site), which then initiates the deposition of fibrin that results from the activation of the clotting system.
2. Fibrin formation helps to localize the microbial agents by trapping them in the coagulum and in this way enhances phagocytosis.
3. Emigration of blood leukocytes and plasma cells occurs, and these cells contribute to phagocytosis and possible localized antibody synthesis, respectively.
4. Increase in plasma at the site ensues (as shown by swelling), and in this way increases the level of bactericidal substances, and at the same time dilutes the bacterial toxins formed.
5. Appearance of acute phase protein such as C-reactive protein may occur and aid in localized phagocytosis.
6. The physiological and biochemical states, such as increased CO_2, decreased O_2 tension, increased body temperature, and accumulation of organic acids, that occur at the site

of injury may be deleterious to the invading microbes.

Passive defense mechanisms

A variety of conditions that may be closely synonymous with innate resistance provide the indifference or passivity on the part of the animal tissue to afford a suitable environment for the ubiquitous pathogens. Some of the more significant conditions include body temperature, tissue metabolites, oxygen tension, and nonsusceptibility to toxins.

It is known that animals of different species have variations in body temperatures. Thus it is to be expected that animals with different body temperatures would exhibit different inhibition of microbes. An instance of such a phenomenon was shown by Pasteur, who demonstrated that chickens, which have high body temperatures, are normally resistant to *Bacillus anthracis*. They became susceptible when their body temperature was lowered by immersion in cold water. On the other hand, the bacillus is highly infectious for mammals in general. Viruses and bacteria require certain specific basic chemical compounds or metabolites in order to grow. Thus animals with tissues deficient in necessary metabolites would selectively inhibit growth of microbes requiring such metabolites. High oxygen tension in normal tissues prevents proliferation of anaerobic microbes, and in this way maintains passive resistance. Finally, certain species of animals possess tissues that are indifferent to bacterial exotoxins. For example, frogs are entirely resistant to diphtheria and tetanus toxins, whereas these toxins are lethal to most mammals.

ACQUIRED IMMUNITY

As already indicated, acquired immunity is an individual property, dependent on the host's previous exposure to or experience with the foreign agent. Nevertheless the ability to respond is an inherent property, and is genetically endowed within a given species.

Individuals may acquire immunity either by active or passive means. Active immunity may be acquired by a previous infection by a pathogenic microbe, or by active immunization with bacterial and viral antigens. Subsequent to these experiences, the individual develops antibodies specifically for the microorganism, and thus acquires an active immune or resistant state to the microbe. This immune state may be of short or long duration, generally dictated by the inciting agent. For example, immunity to virus infections such as smallpox and yellow fever viruses may persist for years, whereas resistance developed against pneumococci or typhoid organisms is generally of short duration.

Passive immunity (borrowed immunity) may be acquired by the transfer of human or animal serum that contains antibodies protective against a specific agent, to another individual who has no protective antibodies. An example of this is the transfusion of neutralizing antibodies against measles virus to expectant mothers who have been exposed to the viruses in early pregnancy. Passive transfer to individuals of serum containing specific antibodies to certain infectious microbes was common practice prior to antibiotic therapy. Such practices for protection from infectious diseases are generally no longer employed, but in some cases, such as the treatment against snakebites and other toxins, passive transfer of antivenoms or antitoxins is still employed. Transfer of immune cells has been attempted, and this is termed adoptive immunity.

Immunoglobulins

This section discusses numerous areas that encompass, either directly or indirectly, the broad concept of acquired immunity.

Immunoglobulin (Ig) comprises a heterogeneous group of serum proteins, which migrate in electrophoretic examination in the gamma (γ) region of the serum electrophoretic pattern. Immunoglobulins are found primarily in the liquid portion of blood and the lymphatic fluid. Blood has two major constituents: (1) the fluid portion or plasma, which may be further sepa-

Table 24-3. Immunoglobulin classification and characteristics

TYPE OF IG	ABBREVIATION	SEDIMENTATION CONSTANT, S	MOLECULAR WEIGHT	AGAR IEM ELECTROPHORETIC CHARACTERISTICS
Gamma-G sub G_a, G_b, G_c, G_d	γG	6-7	1.6×10^5	$\gamma_2 - \alpha_2$
Gamma-A	γA	6-15	1.6×10^5 plus polymers	$\gamma_1 - \alpha_2$
Gamma-M	γM	19-32	1×10^6 plus polymers	$\gamma_2 - \beta_2$
Gamma-D	γD	7	1.6×10^5	$\gamma - \beta$
Gamma-E	γE	7.7		γ_1
Gamma-L and U	γL γU	3-5	2.2×10^4	$\gamma_2 - \alpha_2$

rated into serum and fibrinogen (clotting factor); and (2) the cellular elements (the leukocytes, or white blood cells, and the red blood cells).

The recent classification of the immunoglobulins and some of their chemical and physical characteristics are summarized in Table 24-3.

The basic structural four-chain units of γA, γM, and probably γD and γE, are essentially similar. These immunoglobulins are differentiated by their antigenic variation of the heavy (H) chains. Thus, the H chain of γG is different in composition from γA or γM, and so forth. However, the light (L) chains of all the immunoglobulins thus far recognized are similar in antigenicity. There are two types of light chains, designated Types I and II. The γL and γU are similar to the L chains.

The term antibody, used in a limited sense, is generally synonymous with the γG of the IG group, but used in a broad context would be synonymous with all of the immunoglobulins. Antibody may be defined as a protein substance appearing in serum as a consequence of exposure of an individual to an antigen. A typical antibody response of a rabbit following inoculation of tobacco mosaic virus (TMV) is shown in Fig. 24-2. The initial 1.0 mg of tobacco mosaic virus and antigen given intravenously into a rabbit initiates the primary antibody response. The primary antibody response appears with the following sequence of events: (1) an initial induction period (2 to 3 days for TMV antigen); (2) a rapid synthesis with the appearance of antibodies in the blood; (3) a maximum peak; and (4) a gradual decline in the level of antibody in the blood. A second administration of TMV antigen at the time of decreasing antibody levels in the blood causes a secondary antibody response, which stimulates the primary response at an accelerated rate in all steps up to the peak antibody level, but with a slower rate of disappearance from the blood.

Of interest to the immunologist is the source of immunoglobulins and the theories to account for their specificity to antigens and to their synthesis following exposure of the individual to microbes or foreign antigens. Evidence to date suggests that the reticuloendothelial cells, particularly the lymphoid series of cells, and more specifically the plasma cells, are involved in antibody synthesis.

Of numerous theories advanced to explain antibody production, two are currently accepted by most immunologists. These are the instructive and clonal selection theories. In essence, the instructive theory proposes that the antigen that enters the host comes into contact with and enters the immunologically competent cells.

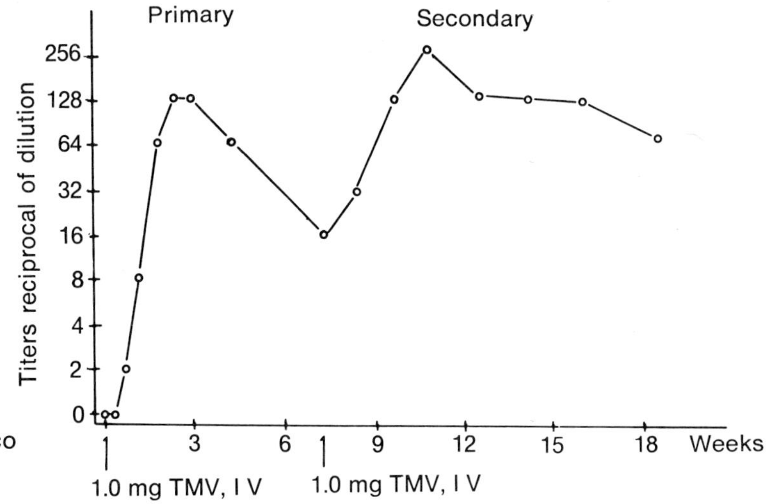

Figure 24-2
Typical antibody response of rabbit to intravenous inoculation, **IV,** with tobacco mosaic virus, **TMV,** illustrating a rise in titer.

Within the cells the antigen then instructs or dictates to the cell to produce specific globulin molecules that would fit the inducing antigen precisely in a lock and key fashion. The antigen is given the key role in the instructive theory. The clonal selection theory postulates that clones of immunologically competent cells, each containing the genetic code of the antibody complementary to one or more antigenic determinants, are present from birth. Sufficient numbers of such clones would provide an immune potential against all existing potential antigens that might be encountered. Thus, the antigen selects or affects its specific complementary clones and stimulates them to proliferate and synthesize the corresponding specific antibody. This concept stresses the significance of the role of the cells. In time both concepts may prove partly valid.

Antigens

The term antigen refers to any foreign substance which, when it enters into a host, induces or stimulates the production of immunoglobulins (antibodies). These resulting antibodies then specifically combine with the antigens as shown by various immunological tests. There are many varieties of antigens including bacteria and their fractions, viruses, proteins, carbohydrates, and, in some instances, lipids. Some antigens are designated complete and others are incomplete. The former group can induce production of antibody in the body and react with it in the test tube. The incomplete antigens or haptens are incapable of stimulating antibody synthesis in the host unless coupled with a carrier antigen but may react with the antibody in the laboratory as determined by special serologic methods.

Proteins of high molecular weights are usually good antigens capable of stimulating antibody production. Some examples, in order of decreasing molecular weights, are tobacco mosaic viruses, thyroglobulin, immunoglobulins, and serum albumins. Carbohydrates, which are basically made up of residues of hexoses and hexosamines, are generally less satisfactory antigens than proteins. Examples of carbohydrate antigens are type-specific polysaccharides of pneumococci and mucopolysaccharides of the red blood cells. The fatlike lipids are generally poor in, or have no complete antigenic activity, but many molecular weight lipids may act as haptens. Nucleoproteins are antigenic.

Immunological methods

Agglutination. Agglutination (clumping) is a term used to describe the antibody and antigen interaction that results from the combining of antigens of large particle sizes, such as bacterial or red blood cell suspensions, and their corresponding antibodies. The antibody involved in the agglutination reaction has been called agglutinin and the corresponding (homologous) antigen agglutinogen. Thus the antibody in the serum of a typhoid patient would be an agglutination against the typhoid bacteria providing the agglutinogen.

Agglutination occurs when the antibody has two reaction sites (is bivalent), each of which combines with two separate antigenic determinants. A series of such reactions and combinations of such smaller complexes results ultimately in a wire mesh-like framework that eventually builds up to larger clumps, which then flocculate out at the bottom of the test tube. Such postulated reactive combination provides the basis for the lattice framework theory of Marrack.

Agglutination tests to detect circulating antibodies to specific bacteria after an infectious episode are carried out routinely in the clinical laboratory. For example, antibody levels to typhoid or brucella infections are determined routinely with corresponding bacterial antigens as diagnostic agents. Antibody levels in an agglutination reaction are estimated by a series of twofold dilutions of the patient's serum and a constant amount of the antigen suspension. The reciprocal of the highest dilution of the serum sample showing a positive agglutination reaction is referred to as the titer of the serum sample. This procedure is known as titration.

Precipitation. This occurs between an antibody and its specific soluble antigens under optimum antibody-antigen ratios, pH, temperature, and electrolyte concentrations to give visible precipitates. The mechanisms associated in aggregation are probably similar to those for agglutination. It differs from the latter in that the antigen involved is soluble instead of being particulate (as with bacteria). Precipitation reactions have been employed to measure the concentration of numerous antibodies and their corresponding soluble antigens. The antibody in this reaction is termed precipitin and the corresponding antigen precipitinogen.

Complement-fixation. Complement-fixation (CF) refers to a serological procedure or test involving antibody and antigen, and the subsequent fixation or binding of complement to the AgAb complexes formed. Like agglutination and precipitation, it is an immunological means of determining semiquantitatively the presence or absence of antigens or antibodies.

The use of complement-fixation as a diagnostic procedure for diseases is limited because of its technically unwieldy and exacting nature. It is, nevertheless, the most frequently performed test for the diagnosis of syphilis. The complement-fixation test, in some instances, supplies information that is not obtainable by other serologic tests.

The principle of complement-fixation tests may be summarized as follows: Complement obtained from a suitable source, such as fresh guinea pig serum, enters into combination with the antibody residue of an antigen-antibody complex. The complex formed may be either of soluble or particulate antigens with their corresponding antibodies, and the complexes need not be visible macroscopically. The essence of the test is to determine whether an antibody-antigen complex is formed with the subsequent fixation of complement. Either the antibody or the antigen must be known in order to examine for one or the other.

The test is performed as follows: Patient's serum, antigen, and complement are first mixed in a test tube and incubated, usually at a cold temperature. After the proper interval, a second known antibody-antigen system is added to the same mixture. The second, or indicator system, is an easily visualized one, which consists of sheep erythrocytes mixed with rabbit serum containing antibodies against sheep erythrocytes (hemolysin). The purpose of the second system is to determine the fate of the comple-

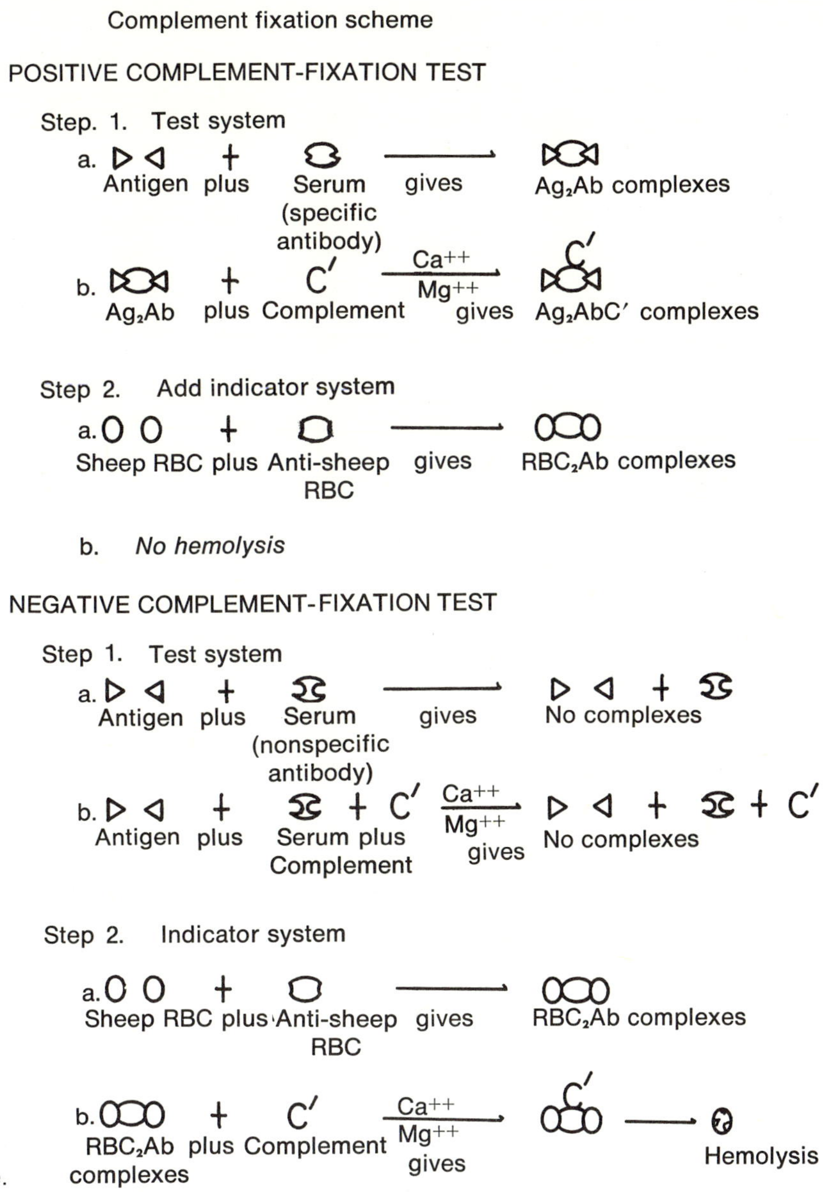

Figure 24-3
Complement-fixation scheme.

ment added to the unknown first system. If complement is fixed with the first serum-antigen mixture, then essentially none remains to react with the second or indicator system, and hence no lysis of sheep erythrocytes occurs. This result constitutes a positive complement-fixation test. If, on the other hand, no reaction occurred in the first system, serum and antigen, then free complement will be available to bind the second indicator system, and thus lysis of the sheep erythrocytes occurs. Lysis then constitutes a negative complement-fixation test. Gradations of positivity in the test occur. A summation of the complement-fixation test is shown in Fig. 24-3.

NEUTRALIZING ANTIBODIES. The term neutralizing antibodies generally refers to immunoglobulins that are induced by exposure to toxins of animals, plants, bacteria, and to viruses. These antitoxic antibodies neutralize or nullify the effect of their corresponding toxins or viruses.

The presence of neutralizing antibodies or antitoxins may be determined by mixing immune serum containing antitoxins and the corresponding toxin in a test tube, and subsequently inoculating the mixture into an animal (for example, mouse or guinea pig) susceptible to the toxin. If the animal survives, the toxin was neutralized. Appropriate controls should be carried out consisting of the same amount of toxin (without the antitoxin) in saline solution. Thus control animals given toxin alone in physiological saline solution will die.

Neutralizing antibodies may be produced in the host when virus antigens are inoculated, or when an individual recovers from a virus infection such as poliomyelitis, smallpox, measles, or mumps. These viral antibodies are able to neutralize or kill corresponding viruses against which they are formed. Viral neutralizing antibody levels may be demonstrated in monolayer tissue culture systems, or by appropriate complement-fixation procedures. Similarly, rickettsiae induce the production of neutralizing types of antibodies in the host after an overt rickettsial infection or after the administration of rickettsial antigens.

HEMAGGLUTINATION. This generally refers to the agglutination of any red blood cell (erythrocyte) by its corresponding or homologous specific antibody. However, when myxoviruses, such as the influenza viruses, attach themselves to erythrocyte surfaces and aggregate these cells under appropriate conditions, the cells clump or agglutinate. This is also known as hemagglutination. A hemagglutination test for the quantitation of influenza viruses has been devised. Agglutination occurs more efficiently at 4° C than at 37° C. Although influenza viruses have no specific antibody activity, they appear to have receptors on their cell surfaces called hemagglutinins that are capable of binding certain antigens on the surface of the red blood cell. The viral hemagglutinin has been associated with the ability of influenza viruses to infect their host. The hemagglutination of erythrocytes by viruses such as the influenza virus may be neutralized by the addition of specific antibodies to the viruses. The host antibody reacts with the hemagglutinin of the viruses and thus nullifies its ability to react with the red blood cells. Such a reaction is called hemagglutination inhibition, and has been used widely to determine the presence of antiviral antibodies in individuals suspected of influenza.

IMMUNODIFFUSION. This is a general term recently introduced to describe immunologic tests carried out in or on media such as agar and in which reactants mix solely as a result of diffusion. Two general types of immunodiffusion have been extensively employed in the examination of many precipitin-precipitinogen systems. The Oudin single diffusion is performed generally in test tubes in which a constant amount of antibody or antigen is fixed in a volume of 1% agar containing electrolytes. The agar in the tube is overlayered with a solution containing the homologous antigen or antibody. As the latter substance, at some optimum region of AgAb proportion, diffuses into the agar, a precipitate in the form of a ring will occur. The Ouchterlony double-diffusion test differs in that the antigen and its homologous antibody, placed in wells in the agar at

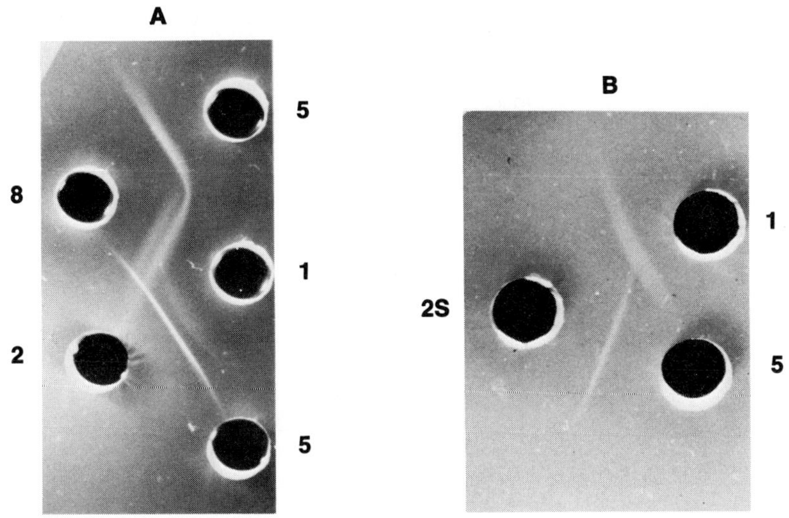

Figure 24-4
Ouchterlony double-diffusion test. **A,** Complete fusion of antigen-antibody complexes of **1** and **5** with antibody **8; B,** partial reaction of identify between antigens **1** and **5** against antibody **2S.**

an appropriate distance, usually 1 cm, are permitted to diffuse out into the agar. At some point of optimum proportion, the antigen-antibody complexes precipitate, and a line of white precipitation occurs at some distance between the two wells. Such double-diffusion examination can be seen in Fig. 24-4. A reaction of identity occurs when two antigens are identical and react with their homologous antibody resulting in complete fusion of the antigen-antibody complexes as shown in the reaction of numbers 1 and 5 with antibody 8 in Fig. 24-4, *A*. Partial reaction of identity is shown between antigens numbers 1 and 5 against antibody 2S in Fig. 24-4, *B*. A partial fusion with spurring of one can be noted. Note the crossing of the precipitation lines. Thus, double diffusion can be a useful means of characterizing relationships of several antigens and of their corresponding antibodies. Immunodiffusion can be employed for quantitation of antigens or antibodies under appropriately controlled conditions.

Immunofluorescent labelling. Accumulation of immunoglobulins or antigens at sites of tissue injury may be determined by the immunofluorescent labelling technique. For example, rabbit antibody to human γ-globulin may be chemically labelled with fluorescein isothiocyanate, a dye that gives an apple-green fluorescence when exposed to ultraviolet light. The fluorescent-labelled antibody is then added to freshly obtained and quickly frozen pathological tissues cut in 5 to 10 μm sections. If human γ-globulins are localized at the site of injury, the apple-green fluorescent staining can be seen in the region of tissue injury when examined with the special fluorescent microscope set-up. Fig. 24-5 shows an example of such an examination.

Electrophoresis and immunoelectrophoresis

Electrophoresis may be separated into two broad categories, free boundary and zonal. In free-boundary electrophoresis of serum proteins, the samples are separated in a liquid phase system of low ionic concentration, with constant pH, electric current, and voltage. The proteins are permitted to migrate through a liquid buffer medium, and the separation is dependent on their net electrical charge at the pH examined. All serum proteins are negatively charged at pH 8.6, and move toward the anode (positive electrode). Albumin, being the most negatively charged, has the greatest mobility and appears closer to the cathode (negative electrode). Free-boundary

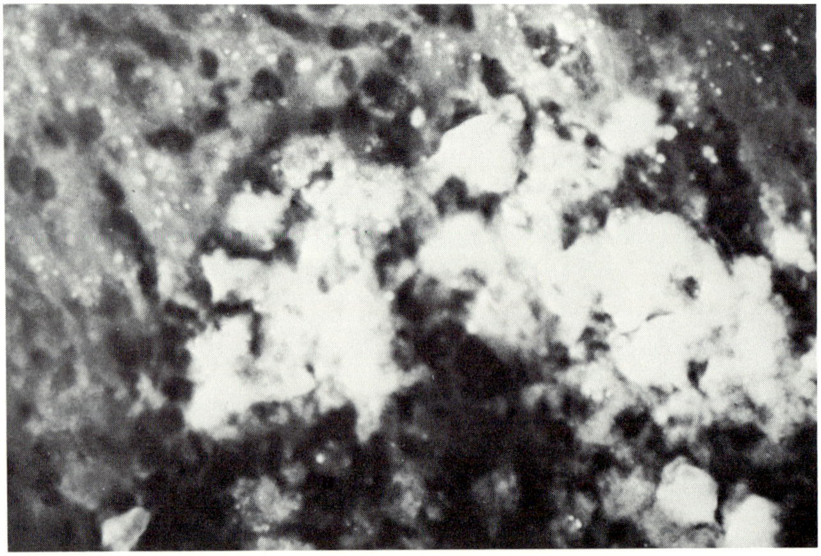

Figure 24-5
Rabbit antibody to human globulin labelled with fluorescein isothrocyanate, then applied to quickly frozen pathological tissue. Accumulation of human globulin at site of injury shown by fluorescence of antibody under ultraviolet light when examined with fluorescent microscope.

electrophoresis is a useful procedure for the determination of the isoelectric point, the pH at which a protein become electrically neutral and hence has no electrophoretic mobility.

Zonal electrophoresis utilizes a semisolid medium that serves as a bridge between the electrodes, and on which the samples are separated on the medium. The migrations of the substances are dependent upon their net electrical charge at the pH examined. Zonal electrophoresis is carried out at low ionic concentrations, 0.05 to 0.075 range, and the buffer system used generally consists of sodium barbital at pH 8.6. Some of the more common media used for zonal electrophoresis are cellulose acetate, agar, agarose, gelatin, starch, and polyacrylamide. The latter two media add a second dimension to electrophoresis in that proteins can be separated on the basis of their molecular shape and size in addition to their net electrical charge. This is called the sieving effect and is controlled by altering the concentration of the gels.

Typical electrophoretic patterns of a normal human serum and of a patient with monoclonal gammopathy of γA, multiple myeloma disease are shown in Fig. 24-6. Five major components, near the anode (positive electrode) are the albumin followed by α_1-, α_2-, β- and finally γ-globulins. This is due to an increase in the γA synthesis. Another example of zonal electrophoresis with polyacrylamide may be seen in Fig. 24-6, *C*.

Immunoelectrophoresis is a procedure that utilizes two principles for analysis of antigens: zonal electrophoresis, and double diffusion. The procedure initially consists of separating the proteins, for example, serum proteins, by electrophoresis in agar, agarose, or gelatin, and subsequently adding antiserum to the electrophorized proteins in wells cut parallel to the electrophoretic run. Fig. 24-7 shows a normal and a γA gammopathy immunoelectrophoretic pattern.

Hypersensitivity

Hypersensitivity occurs in an individual previously exposed to a foreign substance (antigen) who may develop a variety of antibodies. A subsequent contact with the same or related antigens may result in a severe hypersensitive or allergic reaction deleterious to the tissue involved. The inducing antigen may be protein or nonprotein in nature.

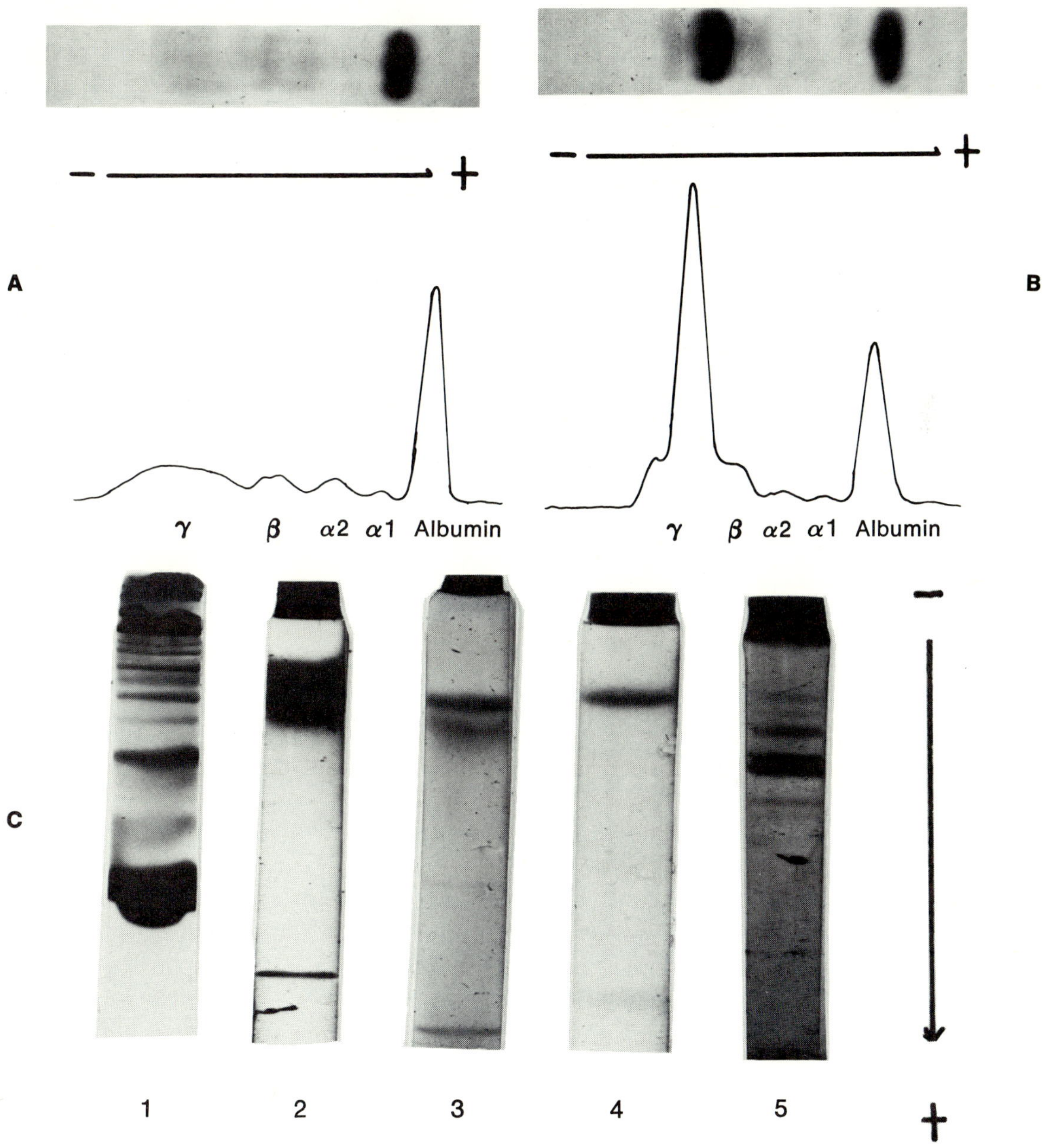

Figure 24-6
Comparative electrophoretic patterns: **A,** normal blood serum; **B,** serum from patient with multiple myeloma; **C,** various zonal electrophoretic patterns with polyacrylamide.

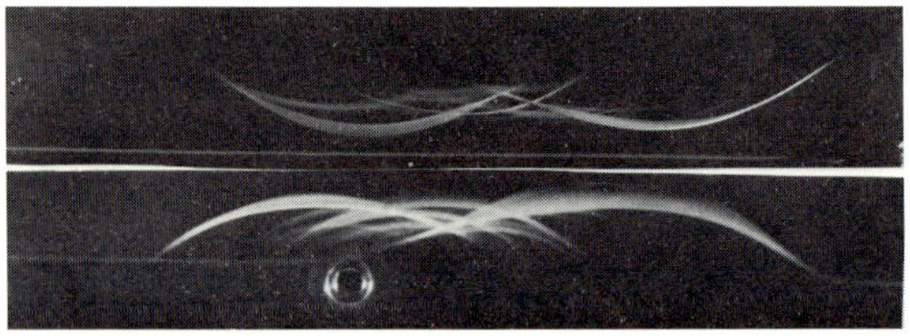

Figure 24-7
Immunoelectrophoretic patterns of normal and multiple myeloma serum.

Hypersensitivity, according to Raffel, may be classified into two types: immediate and delayed. In the immediate type, reaction to the responsible antigen follows within seconds, or minutes after contact. Although the outward manifestations may vary, the immediate hypersensitive states may have many common features. They include rapidity of response, demonstrable antibody in the blood, and no direct participation of tissue cells per se, but only secondarily as the result of antigen-antibody reaction.

Included as immediate hypersensitive states are the Arthus phenomenon, and reactions of the anaphylactic type. The Arthus reaction occurs as a result of a direct reaction and precipitation of the antigen and antibody in the blood vessels, which may then lead to subsequent possible complement action, injury, and an inflammatory reaction. The reactions of the anaphylactic type may be further categorized as evanescent cutaneous, anaphylactic, and atopic. These immediate hypersensitive states share a common feature in that they probably depend for their occurrence upon pharmacologically active intermediary substances released as a result of the initial antigen-antibody union. A number of intermediary pharmacological agents appear to be involved, including histamine, serotonin, anaphylatoxin, and heparin. These agents appear to affect target tissues such as smooth muscles and body vessels and to induce the shocklike reaction seen in the anaphylactic type of immediate hypersensitivity and associated with the release of intermediary substances such as histamine following an antigen-antibody union. Ordinarily, bacterial and viral vaccines seldom induce anaphylactic states. However, penicillin, although of low incidence, has induced anaphylaxis in individuals hypersensitive to penicillin. Serum sickness, which occurs in some instances following intravenous or intramuscular injections of large quantities of foreign serum, may be due to an antigen-antibody reaction. The individual develops fever, body rash, joint pains, and may be ill for several days. Serum sickness is another example of the anaphylactic type of hypersensitivity.

Hay fever, urticaria (hives), asthma, and certain food and drug allergies are examples of atopic hypersensitive states. Atopic hypersensitivity is restricted to human beings, and the susceptibility of the individuals appear to be genetically transmitted.

Immunoglobulin γE has been implicated in immediate type, atopic hypersensitivity. The immunoglobulin appears to have great affinity for epithelial cells.

Delayed hypersensitivity is characterized by a comparatively lengthy period between contact with the antigens and the onset of local or systemic effects. No observable reactions occur for several hours, and the peak is reached between 24 to 72 hours after contact and may persist for several days. No humoral antibody has been demonstrated, and no intermediary mediators have been associated with the delayed response. The cellular elements, such as the reticuloendothelial cells (primarily lymphocytes), have been implicated in mediating some of the effects of delayed hypersensitivity. Finally, there are no specific shock tissues as seen with the immediate type hypersensitivity.

Delayed hypersensitivity is a characteristic response of the body to various chronic infections, as well as exposure to certain plants and chemical substances. Examples of disease in which delayed hypersensitivity manifests itself are tuberculosis and many systemic fungal diseases such as the infection caused by *Coccidioides immitis.*

Transplantation

Transplantation involves the transfer of tissues or whole organs from one individual (donor) to another (recipient) or from one part of the body to another. The transfer of tissues from one part of an individual to another part of the same individual is called an autograft. Transplantation of a tissue or an organ from one person to another, both of whom are of the same inbred strain, is referred to as isograft. Since no histocompatibility differences exist between these individuals, essentially no immune response to the graft or host occurs. Examples of this in humans are transfer of organs between identical twins. A graft of tissues or organs between two individuals of the same species is referred to as allograft. Histocompatibility differences may be weak or strong, depending upon the individuals and species involved. Transplantation of tissues or organs between two individuals of different species is called xenograft, or heterograft.

Autografts and isografts are the most successful since, because of similar histocompatibilities, no immune response to the graft occurs. In contrast, xenografts and allografts are much less successful. There are as yet many limitations to successful transplantation. Thus, effective allograft and xenograft transplantations of vital organs must await further solutions to the problems of tissue rejection, immunological tolerance, and a more precise measure of assaying for histocompatibility between donor and recipient.

The antigens involved in transplantation have been designated as histocompatible lymphocytic antigens (HLA).

Autoimmunity refers to an immune response within an individual against his own tissue constituents, which then initiates a disease process. In a majority of instances, it is assumed that individuals produce antibodies to harmful agents or foreign substances as a protective measure. It is, however, becoming clearly evident that many of the diseases afflicting man are of the autoimmune nature. The immunoglobulins to self tissues are induced either indirectly by foreign agents, which have antigens (cross-reacting antigens) in common with the individual, or by direct induction of immunoglobulin synthesized by the individual's own tissues.

An example of induction of autoimmune response due to indirect causative factors may be rheumatic heart disease. Following a severe beta hemolytic streptococci infection, an individual may develop severe damage to heart tissues. It is known that common antigens are present in human heart tissue and beta hemolytic streptococcal extracts. It is suggested that after recovery from the streptococcal infection the antibody induced by the organisms may then bind or react with the common antigen in the heart tissue and thus, in this manner, initiate serious damage to the heart.

A possible direct induction of an autoimmune response leading to injury of the tissues is thyroiditis or Hashimoto's disease. It is suggested that the individual's tolerance to his own thyroglobulin breaks down. This then results in the synthesis of antibodies to one's own thyroglobulin. How tolerance to an individual's own tissue breaks down is as yet not clearly evident. The antithyroglobulins enter the thyroid and react with the thyroglobulin. The immunological reaction eventuates a disease process recognized as thyroiditis.

Examples of other possible autoimmune diseases are glomerulonephritis, systemic lupus erythematosus (SLE), and some hemolytic anemias. In many, if not all of these diseases, complement undoubtedly plays a key role in the autotissue damaging processes.

BLOOD GROUP SYSTEMS

The blood isoantigens comprise 11 groups and consist of approximately 40 different distinct antigens. The best known of these are the ABO and Rh groups because of their significance in medicine, especially in the transfusion of blood from one individual to another.

ABO blood group

Landsteiner (1900) demonstrated that when erythrocytes of certain individuals were mixed with the serum of others, a hemagglutination of the cells occurred. This reaction may be termed isohemagglutination. The antigen involved is generally spoken of as isohemagglutinogen and the corresponding antibody as isohemagglutinin. Two isohemagglutinogens, designated A and B, were recognized in the red blood cell. It was then shown that all human blood, regardless of race, could be classified into four types on the basis of the presence or absence in the red blood cells of the isoantigenic components A and B. The four blood types in the Landsteiner scheme are represented in Table 24-4.

A type O individual has no A or B isohemagglutinogen in the erythrocytes, but does possess the reciprocal anti-A and anti-B isohemagglutinin in the plasma. Type A individuals have A isohemagglutinogens in their red blood cells, and the reciprocal isohemagglutinin anti-B in their plasma. Type B individuals have B isohemagglutinogens in their red blood cells, and the reciprocal isohemagglutinin anti-A in their plasma. The individuals with AB isohemagglutinogens in their red blood cells have no reciprocal isohemagglutinins in their plasma.

The isohemagglutinogens A and B are genetically transmitted, and are established in the red blood cells of the fetus and newborn. On the other hand, the natural isohemagglutinins, anti-A and anti-B, are not demonstrable in the plasma at birth, but appear 3 to 6 months later, increase rapidly to a maximum at 5 to 10 years of age, and thereafter decrease gradually.

For the determination of the ABO blood types, only antisera for types A and B are necessary. An appropriately cleaned three-well microscope slide is labelled 1, 2, and 3. To well 1, one drop of anti-A isohemagglutinin is placed in the center; to well 2, one drop of anti-B isohemaggluti-

Table 24-4. Blood types in the ABO group according to Landsteiner

TYPE	ISOANTIGENS IN ERYTHROCYTES	ISOAGGLUTININ IN SERUMS	REACTION* WITH SERUM OF TYPE: O (ANTI-A AND B)	A (ANTI-B)	B (ANTI-B)	AB (NONE)	PERCENTAGE OF OCCURRENCE IN ADULTS, NORTH AMERICA
O	None	Anti-A (α) Anti-B (β)	−	−	−	−	45
A	A	Anti-B (β)	+	−	+	−	42
B	B	Anti-A (α)	+	+	−	−	10
AB	AB	None	+	+	+	−	3

* + isoagglutination

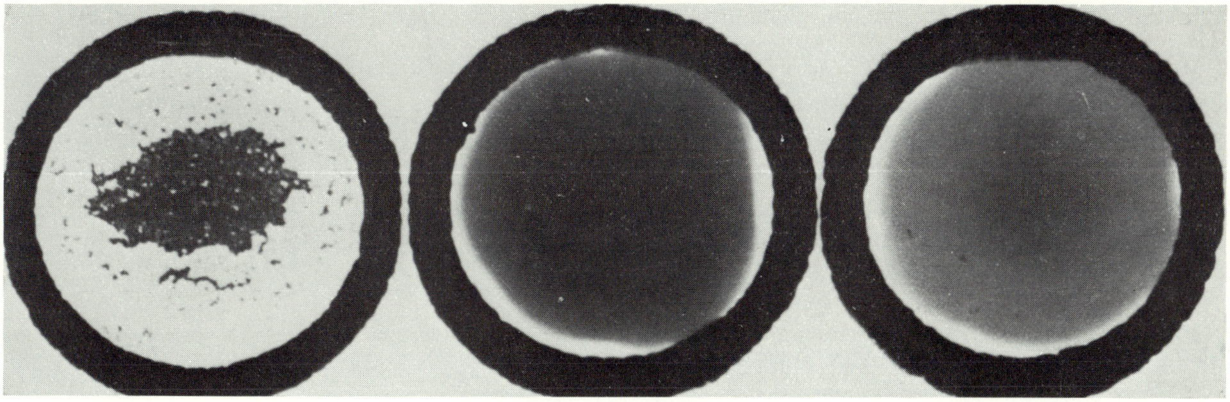

Figure 24-8
Test tube isohemagglutination test. **1,** Positive reaction; **2,** negative reaction; **3,** saline control.

nin is placed in the center; and to well 3, one drop of physiological saline is placed in the center. To each of the wells, one drop of a 5% suspension of washed red blood cells of the unknown in physiological saline is added. The slide is gently agitated for 5 minutes, then examined microscopically.

If no hemagglutination occurs in wells 1 and 2, the blood is of type O. If hemagglutination occurs in both wells 1 and 2, the blood is of type AB. If hemagglutination occurs in well 1, but not 2, the blood is of type A. If hemagglutination occurs in well 2, but not in 1, the blood is of type B. Positive, negative, and saline control patterns of an isohemagglutination test are shown in Fig. 24-8.

Compatibility of blood types must be assured before blood is transfused between donor (giver) and patient (recipient). Type O individuals are designated universal donors and may, under emergency situations, be used as donors to A, B, and AB types. Although the plasma of type O contains anti-A and anti-B isohemagglutinins, they are diluted by the recipient's blood if the O blood is given slowly. In some instances type O individuals with high levels of anti-A and anti-B isohemagglutinins are not suitable as universal donors. Type O individuals can receive only from type O donors since their isohemagglutinins anti-A and anti-B would react immediately with donor types of A, B, or AB.

Individuals of type AB are designated universal recipients since their plasma contains no isohemagglutinins.

The frequency of the ABO types for adults on the North American continent is indicated in the last column of Table 24-4. Phenotypic variations in the frequency of ABO types among different ethnic groups exist.

ABO groups are of tremendous importance in blood transfusions, transplantation, in cases of disputed parentage, and in the identification of individuals.

Rh blood group

The Rh group consists of a complex group of isoantigens found in the red blood cells, and which are genetically controlled. Unlike the ABO group, there are no natural reciprocal isohemagglutinins. Landsteiner and Wiener (1940) demonstrated that the red blood cells of Rhesus monkeys contained an isoantigen that induced antibody formation when administered into rabbits. This isoantigen, which they designated Rh, is found in 85% of American white persons. Since the Rh isoantigens are highly antigenic, Rh-positive blood given to an Rh-negative individual would induce synthesis of anti-Rh antibodies in the latter individual. A second transfusion of Rh-positive blood to the Rh-negative individual previously exposed to Rh-positive blood may induce severe transfusion reactions with hemolysis of the donor's red blood cells.

During pregnancy, a severe disease, known as erythroblastosis fetalis, may occur when an Rh-negative mother and an Rh-positive father conceive an Rh-positive fetus. The Rh-negative mother subsequently becomes immunized to the Rh-positive isoantigens of the fetus and produces anti-Rh agglutinins. The passage of the anti-Rh agglutinins, back across the placenta to the fetus, causes the disease in the fetus.

SUGGESTED READINGS

Burnet, M.: The mechanism of immunity, Sci. Am. **204**:58, January 1961.

Carpenter, P. L.: Immunology and serology, ed. 2, Philadelphia, 1965, W. B. Saunders Co.

Coons, A. H.: The application of fluorescent antibodies to the study of naturally occurring antibodies, Ann. N. Y. Acad. Sci. **69**:658, 1957.

Crowle, A. J.: Immunodiffusion, New York, 1961, Academic Press, Inc.

Dixon, F. J., Jr., and Kunkel, H. G., editors: Advances in immunology, vol. 7, New York, 1967, Academic Press, Inc.

Hirsch, J. G.: Antimicrobial factors in tissues and phagocytic cells, Bacteriol. Rev. **24**:133, 1960.

Kabat, E. A., and Mayer, M. M.: Experimental immunochemistry, ed. 2, Springfield, Ill., 1961, Charles C Thomas, Publisher.

Landsteiner, K.: Zur kenntnis der antifermentativen, lytischen und agglutinerenden werkungen des blutserums under der lumph, Centralbl. f. Bakt. **28**: 357, 1900.

Landsteiner, K., and Wiener, A. S.: An agglutinable factor in human blood recognized by immune sera for rhesus blood, Proc. Soc. Exp. Biol. Med. **43**: 233, 1940.

Pauling, L.: Theory of structure and process of formation of antibodies, J. Am. Chem. Soc. **62**:2643, 1940.

Race, R. R., and Sanger, R.: Blood groups in man, Springfield, Ill., 1958, Charles C Thomas, Publisher.

Raffel, S.: Immunity, ed. 2, New York, 1961, Appleton-Century-Crofts.

Rapp, H. J., and Borsos, T.: Complement research: fundamental and applied, J.A.M.A., **198**:1347, 1966.

Symposium: Blood groups in intrahuman species, Ann. N. Y. Acad. Sci. **97**:3, 1962.

Yachin, S.: Functions and mechanisms of actions of complement, N. Engl. J. Med. **274**:140, 1966.

Zweifach, B. W., Grant, L., and McCluskey, R. T., editors: The inflammatory process, New York, 1965, Academic Press, Inc.

25 Major bacterial diseases

STAPHYLOCOCCUS INFECTIONS

STREPTOCOCCUS INFECTIONS

GONORRHEA

SYPHILIS

TUBERCULOSIS

SALMONELLA AND SHIGELLA INFECTIONS

BACTERIAL DISEASES OF LESSER IMPORTANCE

During the last three decades the human infectious disease picture has changed radically. Antibiotics, active immunization, and improved public health measures have made many bacterial and some viral diseases rare indeed. However, a few bacterial diseases still persist, and one, gonorrhea, has become epidemic throughout the world.

The major bacterial diseases are staphylococcus infections, gonorrhea, streptococcus infections, syphilis, tuberculosis, and Salmonella-Shigella outbreaks. Of these, only staphylococcus infections are not reportable to public health agencies.

Accurate information of the number of cases of bacterial diseases in a given segment of the

Table 25-1. Public health reportable infectious diseases State of California 1967

DISEASE	NUMBER OF CASES	PERCENT OF BACTERIAL DISEASES	PERCENT OF ALL REPORTABLE DISEASES*
Gonorrhea	75,679	58.7	42.8
Streptococcus infections	33,590	26.1	19.0
Syphilis	10,436	8.1	5.9
Tuberculosis	3,846	3.0	2.2
Salmonella-Shigella combined	3,255	2.5	1.8
Total bacterial diseases	128,839	100.0	71.7
Total bacterial and viral diseases	176,679	-----	100.00

*Of the 36 diseases that are reportable, 8 are virus diseases representing 21.7% of the total number of cases.

population is difficult to obtain, and only a fraction of certain types of infection may be reported. To provide a comparison of the relative incidence of these diseases, Public Health Infectious Disease Reports of the State of California were used (Table 25-1).

STAPHYLOCOCCUS INFECTIONS

The staphylococci are represented by two species *Staphylococcus aureus* and *S. epidermidis.* They constitute a distinctive group of bacteria in the family Micrococcaceae and are frequent, if not constant, inhabitants of the skin and mucous membranes of man and animals. The individual cells are spheres 0.5 to 1.0 μm in diameter. The cells may divide in any direction and often remain attached after cell division. As a result an irregular three-dimensional arrangement of cells may develop, resembling a bunch of grapes (Gr. *staphylē* cluster). However, in infected tissues or young liquid cultures, the cells form in two's or four's or small clusters. On solid media, clusters of 50 or more cells are often seen. (Figs. 25-1 to 25-3.)

The staphylococci are small or large clusters of cocci, gram positive, nonmotile, and nonsporing. Nutritionally, they are fairly fastidious, requiring amino acids and certain of the B-vitamins for growth. Unlike the streptococci, they are strongly catalase positive. As facultative anaerobes, they grow well in the absence of oxygen in a carbohydrate-containing medium, but even better when oxygen is present. They are among the hardiest of all nonsporing bacteria, remaining alive for many weeks in stored pus, and with-

Figure 25-1
Colonies of nonhemolytic strain of *Staphylococcus aureus* growing on blood agar.

standing heat up to 60° C for 30 minutes or so. On the other hand, they are highly susceptible to gentian violet and the quaternary ammonium chloride disinfectants, but resistant to mercuric chloride and certain of the phenols.

Of the two species in the genus, *S. aureus* is a pathogen, while *S. epidermidis* is parasitic with only slight pathogenicity.

Staphylococcus aureus

On blood agar, *S. aureus* is usually beta hemolytic. Pathogenicity is apparently coupled with the production of certain enzymes or toxins, which include the following:

1. Coagulase, an enzyme that causes the rapid coagulation of citrated or oxalated human or rabbit plasma, is considered a major indicator of pathogenicity.
2. Most coagulase-positive strains also produce the enzyme deoxyribonuclease (DNase), which hydrolyses DNA (1% in a nutrient agar medium), leaving a clear zone around the colony.
3. The pathogenic strains also ferment mannitol, with acid production, and grow in a medium containing up to 10% sodium chloride. On mannitol salt agar pathogenic strains produce relatively large colonies surrounded by a yellow zone indicating acid production.
4. Pathogenicity-enhancing agents include leukocidin, toxic for white cells; staphylokinase, which slowly dissolves fibrin clots; hyaluronidase, which attacks the hyaluronic acid or tissue glue, apparently adding to the invasiveness of the organism; and lipolytic enzymes, which resist the bactericidal skin lipids.
5. A few strains produce an enterotoxin when

Figure 25-2
A, Scanning electron micrograph of *Staphylococcus aureus* fixed with gluteraldehyde and shadowed with gold. Shows three-dimensional effect (×3,000). (Courtesy JEOL [U.S.A.], INC.) **B,** Scanning electron micrograph of *S. aureus* similar to **A,** but enlarged 30,000×. (Courtesy JEOL [U.S.A.], INC.)

A

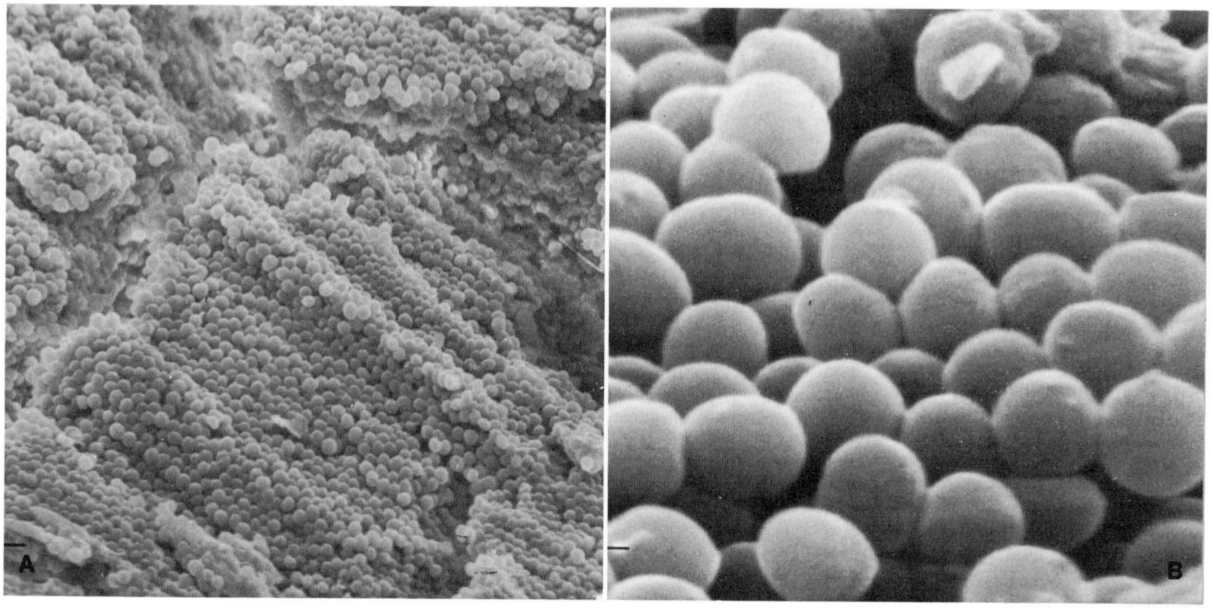

B

grown in infected foods. This toxin acts specifically on the intestines causing an acute enteritis or food poisoning.

S. aureus is carried by a third to a half of the healthy population, especially in the nose. A few persons constantly carry the organism, but most of us are transitory carriers. Most highly susceptible are infants, especially males, and diabetics.

An attack does not produce immunity even to the same strain. A person who fingers a boil on the back of his neck may develop a similar boil at a point he then scratches.

Perhaps the most common, and often the most irritating, of the staphylococcus infections are the recurrent pimples of puberty, that is, acne vulgaris, folliculitis (hair follicle infections), or hidradenitis (Gr. *hidros* sweat + *aden* gland), all associated with hyperactive sebaceous glands. The more severe skin infections, furunculosis or boils, often develop as the result of a pimple that has been too vigorously squeezed, forcing the bacteria deeper into the surrounding tissue. Sometimes boils become open suppurating abscesses.

Intradermal, or within the skin, infections include the acute exfoliative dermatitis of infants, and the impetigo of older children or adults.

Osteomyelitis, or bone infection, often chronic, is a serious manifestation of a staphylococcus attack. Staphylococcus infections of wounds or surgical operations at one time were common.

During the past decade, epidemics of staphylococcus infections caused by penicillin-resistant strains have posed a serious public health problem. Fortunately, a thorough revamping of hospital procedures has done much to alleviate this problem.

Following extensive use of oral antibiotics, a staphylococcus enterocolitis sometimes develops. This is characterized by a watery diarrhea and evidence of invasive damage to the bowel. The feces contain large numbers of staphylococci, usually resistant to several different antibiotics.

Staphylococci are dispersed most commonly

A

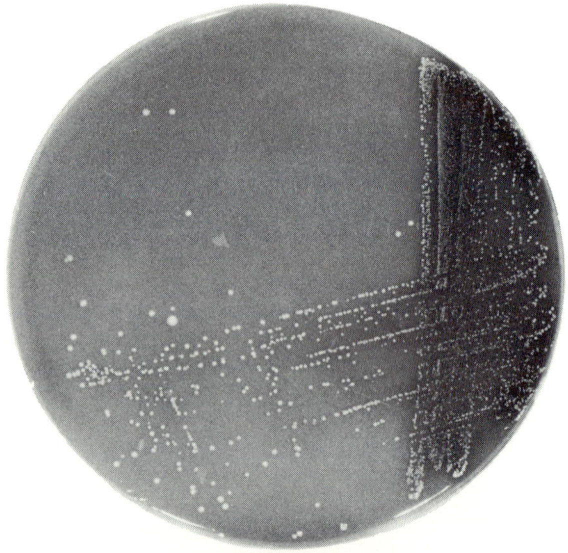

B

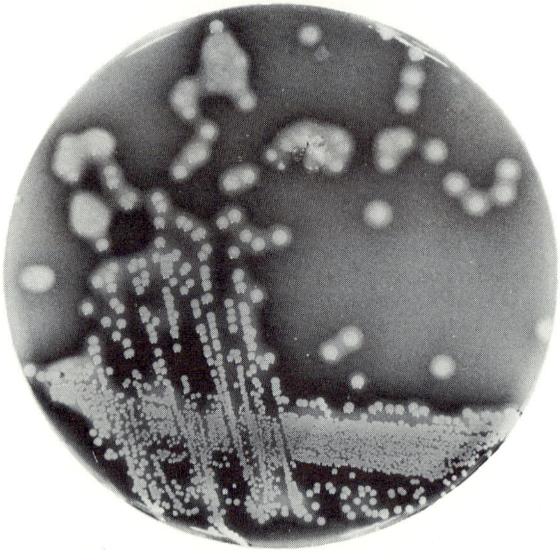

Figure 25-3
Comparative colony growth on blood agar of *Staphylococcus epidermidis*, **A,** and beta hemolytic strain of *S. aureus*, **B.** Comparison of colony growth on blood agar of **C,** *S. aureus*, **D,** *S. pyogenes* (beta hemolytic), and **E,** *S. mitis* (alpha hemolytic).

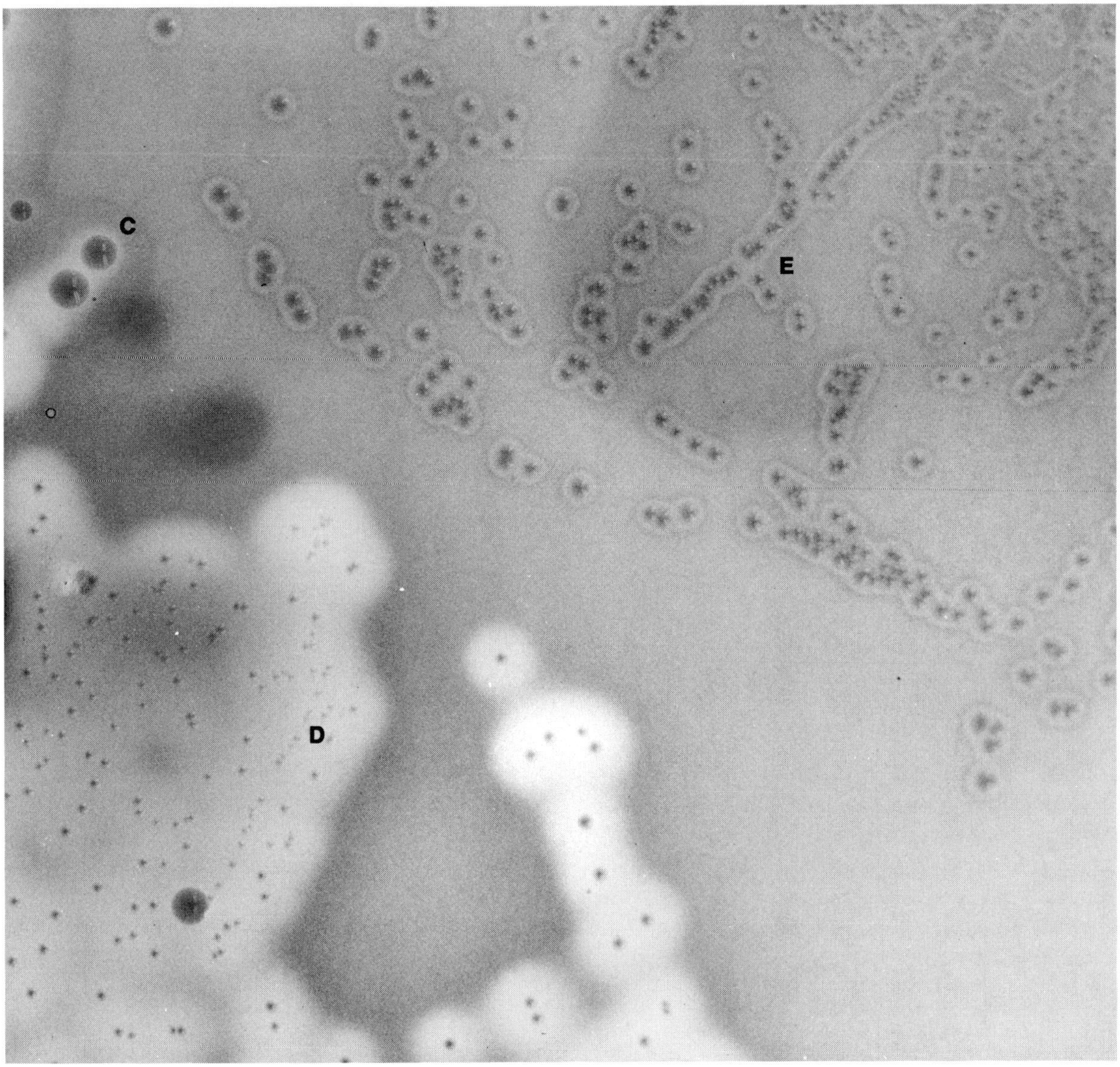
C
E
D

by contact through infected hands, kissing, and so forth, or by inanimate objects such as towels, washcloths, or clothing. Droplet transmission through the air following sneezing also occurs rather frequently.

In hospitals, the bacteria may come from a carrier site (nose or throat) or a lesion in another patient, the patient himself, or a member of the staff. Endemic hospital strains differ from the usual staphylococcus population because they are generally resistant to several antibiotics.

Staphylococcus epidermidis

In terms of pathogenicity, *S. epidermidis* is generally a relatively harmless parasite. It is coagulase negative, does not ferment mannitol, and lacks most of the pathogenicity-enhancing toxins of *S. aureus*. It is a constant inhabitant of the skin and the mucosa including membranes of the nose, throat, and intestines. The skin and nose of infants are colonized during the first few days, and this persists throughout their lives.

Only under special conditions does this organism produce infections. It can often be isolated from small, insignificant pimples. A wooden splinter may carry the bacteria through the skin barrier and a festered sliver may be the result. Stitch abscesses may sometimes be caused by *S. epidermidis*. Mild urinary infections occasionally follow catheterization, and chronic septicemias may be caused by this organism after open heart surgery or implantation of artificial valves in the circulatory system.

STREPTOCOCCUS INFECTIONS

Members of the genus *Streptococcus* represent a varied and widely distributed group of bacteria ranging from the highly virulent *Streptococcus pyogenes* to the nonvirulent and useful *S. lactis*. Many species are regularly found in the mouth and upper respiratory tract and the intestines of man and warm-blooded animals. Others are common inhabitants of milk and other dairy products, or moist plant materials rich in sugars.

The streptococci are spherical to ovoid bacteria (1.0 to 0.4 μm in diameter) that divide only on one plane producing diplococci, short or long chains of cells, generally longer in liquid media than on solid media. Diplococci or short chains of diplococci form when streptococci grow within the body tissues. (Fig. 25-4.)

Streptococci are facultative anaerobes, often making their best growth in liquid media that provide reduced oxygen. On solid media, surface growth is scanty and colonies are small to almost pinpoint in size. They are generally nonmotile, although some strains of *S. faecalis* display active motility. They are catalase negative and have no cytochrome system. Nutritionally, they are fastidious, requiring B-vitamins and certain amino acids. Although gram positive, older cultures decolorize rather easily. They do not reduce nitrates, and unlike *Diplococcus pneumoniae*, the cells are not soluble in 10% bile solution and are not sensitive to optochin.

Brown (1919) grew various strains of streptococci on blood agar and classified the observed hemolysis as:

1. Alpha hemolytic—characterized by a typical greening around the colony in which some intact erythrocytes can be observed microscopically (the green color represents a reduced form of hemoglobin)
2. Beta hemolytic—characterized by large, distinct, cell-free clear zones around the colonies
3. Gamma—an absence of any visible hemolysis

Beta hemolytic streptococci

The primary pathogen among the streptococci is *S. pyogenes* (occasionally but erroneously called *S. hemolyticus*). Rosenbach (1884) isolated the organism from suppurative lesions and gave it the name, *Streptococcus pyogenes* (Gr. *pyon* pus).

Some three decades ago, Lancefield discovered that the various strains of beta hemolytic streptococci produced in the external cell coat a group of specific carbohydrates, the C carbohydrate antigens, which make up about 10% of the dry

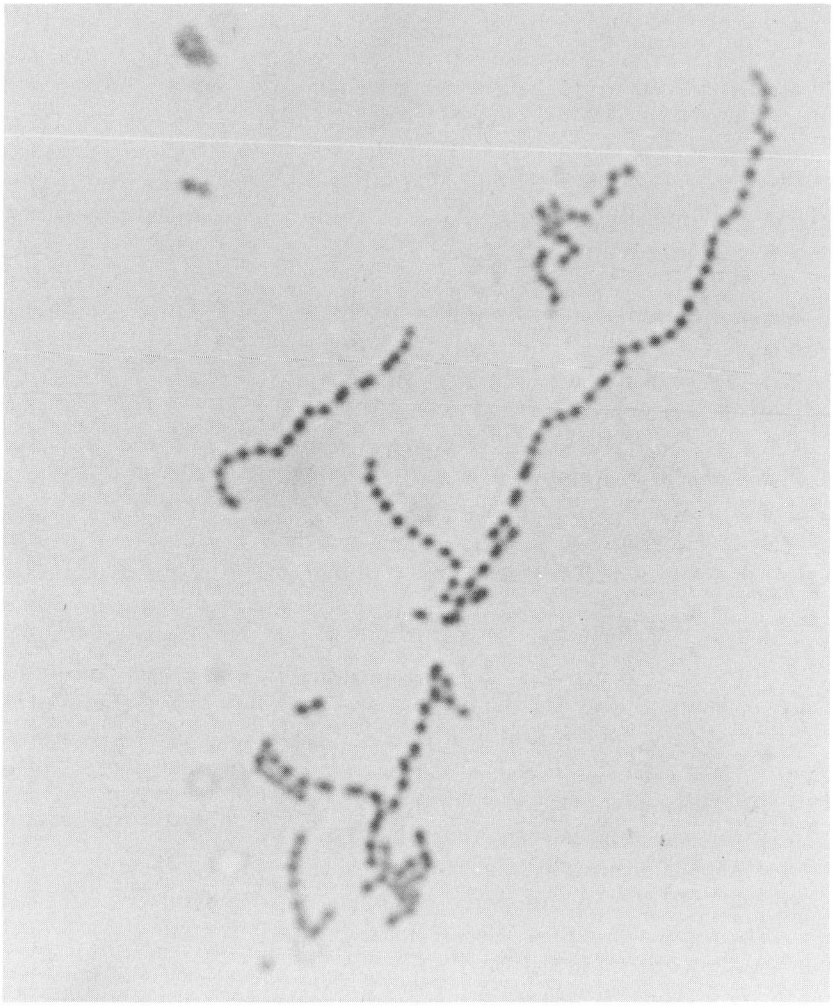

Figure 25-4
Photomicrograph of *Streptococcus pyogenes* showing typical long-chain grouping when grown in broth medium.

weight of the cell. They may be removed by boiling in dilute hydrochloric acid (pH 2.0). Rabbits can be immunized against the C carbohydrate antigen, and known immune sera used to type the C antigen of unknown strains. Practically all of the human infections (97%) are caused by Lancefield group A strains, while a small proportion (3%) are caused by group C. These remaining strains are primarily animal pathogens rarely infecting humans. Some members of groups B, C, D, H, K, O, and all of group N are nonhemolytic

The pathogenicity of group A streptococci depends on their ability to produce a number of invasive or other agents, including:

1. The hemolysins: streptolysin O, which is oxygen labile, and therefore causes hemolysis only around deep colonies in agar pour

plates or in blood broth, but not on blood agar; and oxygen stable streptolysin S, which produces the clear hemolytic zones surrounding surface colonies on blood agar. Patients recovering from streptococci infections show antistreptolysin in the blood serum.

2. Streptokinase A and B, which promote the lysis of human blood clots.
3. Erythrogenic toxins A, B, and C, which produce the scarlet fever rash.
4. Deoxyribonuclease (A, B, and C), or streptodornase, which depolymerizes the extremely viscid DNA in thick pus that results from the disintegration of leukocytes.
5. Proteinase, which splits the M protein at pH 5.5 to 6.5. It may be involved in the heart damage in rheumatic fever.
6. Hyaluronidase, which attacks the cellular glue or ground substance of host tissue.

Types of group A streptococcus infections

Infections caused by group A streptococci are of two types:

1. Suppurative, which include scarlet fever, septic sore throat, cervical adenitis, peritonsillar abscess, middle ear infection, mastoiditis, puerperal sepsis (childbed fever), meningitis, pneumonia, erysipelas and related cellulitides of the skin.
2. Nonsuppurative, which follow in the wake of streptococcus infections, and probably represent sensitivity reactions. Acute glomerulonephritis, a kidney disease, generally follows a week or so after a bout of strep throat or similar streptococcus infections. Very few streptococci are involved, and the disease may represent an autoimmune reaction although the mechanism is not clear.

Rheumatic fever, with its arthritic and cardiac symptoms, usually develops 2 to 3 weeks or months after an acute streptococcus infection. During a streptococcus epidemic, about 3% of persons infected develop rheumatic fever as a sequela. It apparently represents a sensitivity reaction (see Chapter 24). Rheumatic fever is a most important disease, producing partially disabling damage to heart valves in millions of Americans. It is chronic in nature and marked by recurrent episodes, each resulting in increased heart damage, and it leads ultimately to death in early or late middle age. While heart surgery gives relief in some cases, recurrent episodes with streptococci may reduce the efficacy of surgical intervention. Daily administration of rather large doses of penicillin to prevent the development of streptococcus infections has been used with some success.

Transmission of streptococcus infections

During interepidemic periods, a small percent of the population (less than 10%) carries the streptococcus organism in the nose or throat. During epidemics, the carrier rate is materially higher. In addition to direct contact, the organisms may be carried by droplets ejected through coughing, sneezing, or even vigorous talking, especially if the organism is in the nose or sinuses. Infected drinking glasses or other eating utensils, recently contaminated, may also transmit the disease. Dust from floors, blankets, and so forth also occasionally carries the organism.

Pathogenicity of alpha hemolytic streptococci

Some of the hemolytic or viridans-type streptococci fall within the Lancefield O-A groups and show moderate pathogenicity. However, *S. salivarius* may be either alpha or gamma hemolytic on blood agar and is normally found in human throats. The bacteria colonize the nose and throats of infants a few hours after birth and can be seen as gram-positive diplococci or short chains within superficial epithelial cells scraped from the inside of the cheek.

The closely related *S. mitis*, which is alpha hemolytic on blood agar, is also frequently found in throat cultures. The colonies resemble pneumococci but the streptococci are not bile soluble or optochin sensitive.

Alpha type streptococci produce transitory bacteremias following tooth extractions or tonsillectomies. However, the organisms soon disappear, except in individuals who have heart valve lesions resulting from a rheumatic fever type infection. The lesions serve as niduses, and bacteria grow into large colonies that cause heart valve vegetations that are often fatal. This type of infection is termed a subacute bacterial endocarditis.

The alpha and gamma hemolytic streptococci are generally resistant to most antibacterial agents.

GONORRHEA

Gonorrhea has apparently plagued man for centuries. The time or place of its origin is unknown. The causative organism was described by Neisser in 1879.

The genus, *Neisseria,* includes, in addition to *N. gonorrhoeae* (gonococcus), the causative agent of cerebrospinal meningitis, *N. meningitidis* (meningococcus), and several other slightly pathogenic or nonpathogenic forms including *N. catarrhalis, N. sicca,* and the yellow *N. flavescens.*

The *Neisseria* organisms are nonmotile, nonsporing, gram-negative cocci. They generally occur in pairs, flattened at the point of contact to give a double coffee bean appearance. The cells are usually 0.6 to 0.8 μm in diameter, although larger irregular cells are sometimes seen. The pathogenic species are fastidious, requiring enriched media in order to grow, while the other strains grow readily on nutrient agar. The most useful medium, chocolate agar, is prepared by heating blood agar (made with a heart infusion agar base) to 80° to 90°C. This releases the hemoglobin from the red blood cells and destroys the thermolabile toxic inhibitors (see Fig. 24-1, *A*).

The species of *Neisseria* can be separated on the basis of acid production as shown in Table 25-2.

Transmission and pathogenicity of *Neisseria gonorrhoeae*

Except in rare cases, gonococci are transmitted by sexual contact. In a few instances towels, clothing, and so forth, freshly infected by pus from acute cases, might become a vehicle for transmitting the disease.

The incubation period for gonorrhea is usually 3 to 5 days, although it may be as short as 1 day or as apparently long as 30 days. In the acute stage of the disease, in the male infection begins with a highly irritating, profuse urethral discharge. In untreated cases the infection may extend to the prostate, epididymis, and testes resulting in fibrosis and stricture of the urethra, and occasionally sterility.

In females, the initial infection, while producing an irritating urethral discharge of creamy pus, is less noticeable than in the male; however, the adjacent tissues, including the cervix and fallopian tubes, may be quickly involved if treatment

Table 25-2. Identifying tests for *Neisseria* species

SPECIES	GROWTH ON NUTRIENT AGAR WITHOUT BLOOD	GROWTH 22° C	ACID PRODUCTION		
			GLUCOSE	MALTOSE	SUCROSE
N. meningitidis	−	−	+	+	−
N. gonorrhoeae	−	−	+	−	−
N. catarrhalis	+	+	−	−	−
N. sicca	+	+	+	+	+
N. flavescens	+	−	−	−	−

is not begun. Sometimes acute tubal infection or salpingitis may result in symptoms resembling acute appendicitis. If the tube is ruptured an acute peritonitis may result. The uterus and Bartholin glands may also be infected.

Vaginitis is less common in adult females because the acidity may inhibit the growth of the gonococci. About 20% to 30% of infected children and adult females may also suffer a gonococcal infection of the mucosa of the anus and rectum that may be refractory to treatment.

In the acute stage, particularly with the female, the bacteria-laden pus cells may find their way into the blood stream resulting in the development of a bacteremia. In the course of this metastasis (shift of the pathogen), it may lodge in the heart, producing an endocarditis, or in the joints resulting in an acute arthritis; in rare instances, a meningitis may result. Occasionally acute eye infection develops, which if untreated may quickly cause blindness. A parotiditis, or infection of the salivary gland, sometimes occurs. Lesions of the mouth occasionally develop.

After a period of time the acute symptoms subside, generally leaving a low grade chronic infection that may persist for 5 to 15 years.

Diagnosis of gonorrhea

In acute cases in the male a Gram stain of the pus generally shows large numbers of pus cells, laden with gram-negative intracellular diplococci. With females, diagnosis by direct Gram stains is often complicated by the presence of nonpathogenic neisseria. Cultures grown on chocolate agar under reduced O_2–high CO_2 may be used. The pus cells (polymorphonuclear neutrophiles) usually carry 20 gonococci or so, although as many as 100 per cell occasionally are seen.

In chronic infections the number of bacteria per cell is greatly reduced. Nonpathogenic neisseria are not intracellular. If smears are stained with specific fluorescent antibody, a rapid diagnosis is also possible. With females or males in the chronic stage, the most satisfactory method of diagnosis is the culture method.

Repeated attacks of gonorrhea can occur, indicating that little or no immunity results from an infection.

Treatment of gonorrhea

Prior to the development of the sulfa compounds, treatment of gonorrhea was most difficult. Remarkable results were achieved when sulfathiazol or similar compounds were first used. Initially 90% to 95% of the cases responded quickly. However, sulfa-resistant strains developed immediately, and within a few years the sulfa compounds had become essentially useless. Penicillin treatment was introduced with remarkably effective results. Apparently no penicillin-resistant strains have as yet developed; however, the dosage has had to be substantially increased. This may be due, in part, to the production of penicillinase by the nongonorrheal bacteria associated with chronic cases.

A portion of patients show an allergic reaction to penicillin. However, tetracycline and some related antibiotics have been successfully substituted.

At one time, babies born of infected mothers often developed gonorrheal infections of the eyes, resulting in serious corneal damage or blindness. Routine instillation of silver nitrate solution into the eyes immediately after birth overcame this tragic condition. Today, penicillin ointment has largely replaced silver nitrate.

SYPHILIS

The exact orgin of syphilis is not known. A common belief, not well authenticated, is that the returning sailors of Columbus brought the disease with them from the New World. The organism was not seen and described until 1905, when Schaudinn and Hoffmann, using a modified Giemsa stain, were able to see the delicate spirochete in syphilitic material. They called it *Spirochaeta pallidum,* now changed to *Treponema pallidum.*

The organism occurs as a fine, protoplasmic spiral 0.25 to 0.3 μm in diameter, 6 to 14 μm in

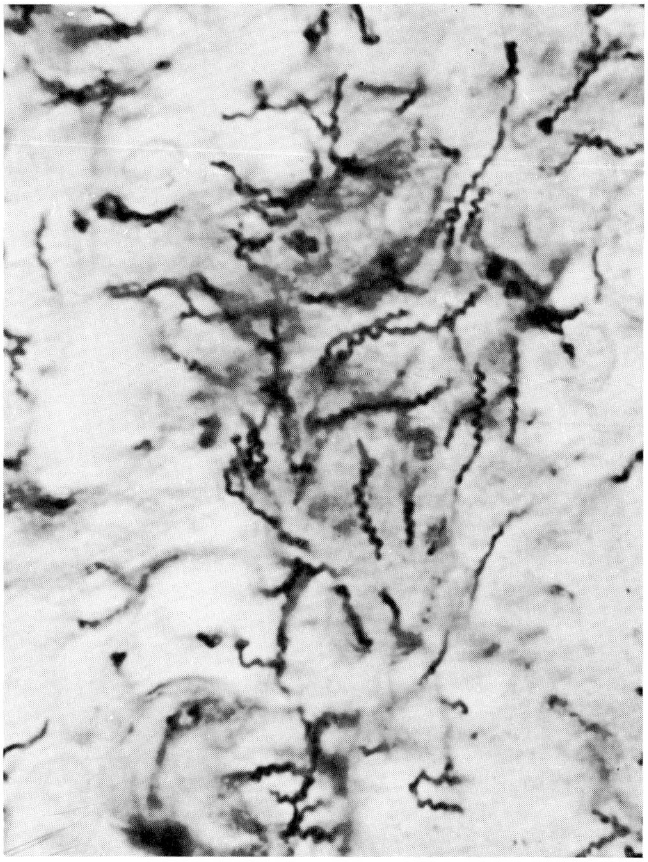

Figure 25-5
Treponema pallidum in tissue section (silver stain). (From Swatek, F. E.: Textbook of microbiology, St. Louis, 1967, The C. V. Mosby Co.)

length with 6 to 14 spirals (Fig. 25-5). Unlike the spirillum, it does not have a rigid cell. It moves by a rapidly rotating motion apparently activated by a bundle of filamentous strands that extend through the length of the cell.

The syphilis spirochete can be stained with various dyes, but its diameter is near or below the limits of resolution for the light microscope. There are special silver stains that plate the fine spirals making them resolvable. Routinely, the organism is visualized with the dark-field microscope in which side illumination makes the organism appear as a scintillating spiral against the dark background.

The organism is an obligate anaerobe that dies quickly in the presence of air. However, when the serous exudate from a syphilitic chancre is sealed in a capillary tube, the spirochetes remain mobile for a month or so if stored in a refrigerator. Cultivation on artificial media is extremely difficult.

Syphilis is transmitted primarily by sexual contact, and approximately 90% to 95% of the primary lesions are on or in the genitalia. Extragenital chancres are mostly on the lips, and are usually transmitted by kissing. Inanimate objects, such as contaminated drinking glasses, mouthpieces on wind instruments, surgical or dental instruments, can transmit the disease.

The organism apparently enters the body

through minute breaks in the skin or mucous membranes. When the spirochete enters the tissue, it begins a steady but relatively slow reproduction. After an incubation period of 15 to 30 days (average 3 weeks), the primary lesion or chancre appears as a small reddened spot, which soon becomes pimplelike. This continues to enlarge and spread, becoming infiltrated with blood. When fully developed, it is surrounded by a raised copper-colored ring resembling a scab. This is usually a half-inch or so in diameter. The center becomes sunken or indurated, and is covered with an adherent fibrinous membrane, quite easily removed to show a serosanguinous discharge. This is filled with spirochetes and is the material obtained for dark-field examinations. If the disease is untreated, the chancre, which is practically painless, usually heals within 2 to 3 weeks (up to 6 weeks), with or without scarring. The syphilitic lesion (and sometimes the disease) is called a hard chancre to distinguish it from the excruciatingly sensitive, painful, soft chancre caused by *Haemophilus ducreyi*.

About 6 to 12 weeks after the subsiding of the chancre, the organism begins to spread throughout the body and the first symptoms of the secondary or generalized stage appear. The secondary stage usually lasts for 2 years or so, with a range of 1 to 4 years. Recurrent outbreaks of symptoms may occur.

The major symptoms of secondary syphilis are fever, painful headaches, malaise, and fatigue. Swollen lymph nodes are common. About 80% of the cases develop skin eruptions, and about a third develop eruptions (gray patches) in the mouth and throat. These teem with spirochetes and become the source of many of the extragenital infections. The skin lesions may likewise be a source of infection.

After two years or so, the immunity level is raised to the point that the generalized treponemal infection is overcome. However, not all the spirochetes are destroyed, and the disease subsides into the tertiary stage. In this stage the organisms are localized in various organs and tissues, which are slowly but persistently destroyed. As the normal tissues are killed by losing their blood supply, they are replaced by gummatous tissue. A degree of tissue allergy seems to be involved in the process. Development of tertiary or late syphilis is a slow process requiring 10 to 20 years before the symptoms become apparent. Some of the symptoms involve nodular or nonnodular skin lesions, destructive bone damage, eye damage and blindness, and gumma formation in the throat resulting in a painless but persistent hoarseness. Gummas may develop in the liver or in the heart with eventual fatal results. The most tragic symptoms are those resulting from damage to the brain (general paresis) or the spinal cord (tabes dorsalis), which reduce the patient to a stumbling, shuffling, imbecile—a caricature of his earlier self. In untreated cases, about one-fourth pass through the tertiary stage without evident lesions but with a persisting positive blood test. Another fourth may even become serologically negative. However, in half the cases, the damaging lesions of tertiary syphilis occur.

Congenital syphilis is acquired when an infected mother conceives. The spirochetes do not penetrate the placenta until the fourth to fifth month. Often the fetus develops generalized syphilis, dies, and abortion occurs. If the child is born, it may carry the symptoms of primary, secondary, or late syphilis. An infected child is not likely to survive for any period of time.

Diagnosis of syphilis

Diagnosis involves the clinical appearance of a primary chancre, dark-field examination of chancre or secondary lesion exudates, and serological tests. The first serological test was the complement-fixation test developed by Wassermann. The original antigen was the water extract of the spirochete-laden liver of a syphilitic fetus. Later it was found that the lipid extracted from beef heart muscle with ethanol gave the same reaction. This is known as cardiolipin. The Kolmer modification has now replaced the classic Wassermann test; and in routine testing the complement-fixation test has largely been replaced by the easier, more rapid flocculation test such as the Kahn, Kline, Eagle, or Hinton. A rapid slide-precipitation test is used to test blood collected for transfusions. The antigen used is essentially the same as for complement-fixation tests.

Nelson and Mayer in 1949 mixed serum from an infected patient and complement from fresh guinea pig serum with actively motile *Treponema pallidum* from the syphilitic lesion of an experimentally infected rabbit. They found that the highly motile spirochetes were immobilized. This was the basis for the *Treponema pallidum* immobilization (TPI) test. The antibody appears to be different from the complement-fixing antibody of the Wassermann test. The TPI test is rather difficult and complicated, but it is used to check questionable flocculation test results. A fluorescent-antibody test, which is both sensitive and specific, has also been developed for syphilis.

Both the anticardiolipin and TPI antibodies appear about 1 to 3 weeks after the development of the chancre. Both rise for several months in untreated patients. The flocculation antibody then begins to decline but the TPI antibody remains high.

Treatment of syphilis is fortunately rather direct and effective. In 1943 penicillin was found to be specific for the spirochetes, and it has become an effective means of controlling the disease. Unfortunately, especially in females, the primary lesion may be undetected and treatment may not be initiated until the secondary symptoms are recognized. Treatment is then more complicated and sometimes less effective. For penicillin-sensitive individuals, tetracyclines and erythromycin have been used with considerable success.

TUBERCULOSIS

Tuberculosis has long plagued mankind. Once called the Great White Plague, tuberculosis has been steadily but slowly declining in importance as a killer disease since about 1870. However, it still occupies a significant position among the bacterial diseases. As a fatal disease, the greatest number of deaths occurs among indigent men over 65, who eke out a lonely impoverished existence in the slums of large cities. Among the tuberculous youth, the death rate is highest among females.

The genus *Mycobacterium*

The causative organism, *Mycobacterium tuberculosis*, is part of a rather distinct group of bacteria assigned to the family *Mycobacteriaceae* of the *Actinomycetales*. The mycobacteria range from the pathogenic *M. tuberculosis* and the deadly but insidious host dependent, *M. leprae*, through a group of atypical forms of variable pathogenicity, to the so-called saprophytes, *M. phlei* and *M. smegmatis*, which are widely distributed in soil and dust.

The mycobacteria are aerobic, nonmotile, slender, straight, or curved rods ranging from 1 to 7 μm in length to 0.2 to 0.6 μm in width. Occasionally rudimentary branching may be seen.

They generally stain with the acid-fast stain, often showing a beaded or barred appearance due to glycogen granules and polymetaphosphate bodies (volutin granules). *M. tuberculosis* cannot be stained by the conventional Gram stain procedure. The more easily stained species are gram positive. The cell walls are unique in that they contain up to 60% lipids, compared with 20% in the usual gram-negative forms and the 1% to 4% of the gram-positive bacteria.

M. tuberculosis and related atypical forms grow very slowly. Generation times are 12 to 24 hours and, for growth cultures, an incubation period of about 3 weeks is required. Colonies are small, dry, scaly, and often corrugated. In a liquid medium, growth forms as a surface pellicle. However, the addition of the surfactant Tween 80 causes the organisms to grow throughout the medium.

M. leprae apparently has not been grown successfully on artificial media. However, when an inoculum of several thousand bacteria is injected into the foot pad of a mouse, it grows slowly.

After 6 to 9 months the number in the tissue will have increased to several million.

The saprophytic strains grow rapidly (but not as fast as *Escherichia coli*) on ordinary nutrient agar. When glycerol is added to the medium, many of the cells in a week-old culture, will have become acid fast.

In a suitable liquid medium, pathogenic strains of *M. tuberculosis* will form long filamentous serpentine masses of cells. This is known as cording and is considered an evidence of pathogenicity.

Course of tuberculous infection

Unlike most acute bacterial infections, tuberculosis represents a disease in which a constant battle seems to rage between the defenses of the body and the invasiveness of the organism. This often results in an incarceration of the bacteria in a calcareous nodule in which they may remain quiescent for decades or throughout the patient's life. On the other hand, slowly developing lesions can produce an extreme destruction of tissues. In rare cases, such as miliary or "galloping" tuberculosis, the patient may succumb within weeks or a few months.

In a tuberculin-negative (nonsensitized) person, exposure to the disease by inhaling infected droplets, kissing, using infected drinking glasses or other fomites, or ingesting infected material may result initially in the formation of active lesions. Usually these are in the parenchyma of the lungs, or sometimes in the mesentary of the intestines. Sublesions may then be walled off by calcareous deposits, leaving nodules too small to be detected by X-ray examination.

With a tuberculin-positive person already sensitized to the mycobacterium, nodules may form. However, with reinfection, the hypersensitivity may increase to the point that the center of the primary nodule undergoes a characteristic caseation (L. *caseus* cheese) necrosis. The killed tissue is semi-solid with a white, cheeselike consistency. Fibrosis and calcification may produce nodules detectable with X-ray examination.

These nodules may remain inactive indefinitely, or with lower resistance or malnutrition the lesion may be reactivated, and a cavity may form in the lung parenchyma. This may become productive, releasing bacteria-laden sputum that may be spread in the patient's environment through coughing or expectoration. Within the patient's body, infections may now develop in adjacent lung tissue, or the sputum that is swallowed may infect the intestinal cell wall and particularly the mesenteric curtain with its array of lymph nodes. As the patient becomes increasingly hypersensitive to the bacteria, cavitation, involving loss of substantial amounts of lung tissue, may occur. If blood vessels become eroded, the bacteria may be carried to various parts of the body where lesions of the skin, bones, kidneys, prostate, spleen, or even the brain may develop. Tissues with the best oxygen supply are most likely to be attacked.

In rare instances, a large blood vessel in a lesion, especially if in the meninges, may rupture, flooding the blood stream with roving bacteria. The result is the disastrous miliary (L. *milium* millet seed) or "galloping tuberculosis." Small lesions, resembling millet seeds in size and appearance, will simultaneously develop in many of the organs of the body. On autopsy the lungs may be a caseous mass, and most of the organs will be seeded with lesions. Such a patient often shows shocking emaciation. Until recently, miliary tuberculosis and tuberculous meningitis were invariably fatal.

However, many people live for years with walled-off nodules in their lungs from which the bacteria are never shed. Such cases are said to be arrested.

The tuberculin test

Tuberculin (O.T.—old tuberculin), made from the growth products of the tuberculosis organism, is injected into the skin to detect sensitivity to the organism. A positive tuberculin test alerts the physician to the possibility of a reactivation of an old tuberculous lesion.

In addition to O.T., a mycobacterial cellular extract, Seibert's purified protein derivative

(P.P.D.) may also be used as a skin sensitivity test reagent. Tuberculin or P.P.D. tests are made by injecting prescribed amounts into the skin. At one time, the patch sensitivity test was widely used. However, it gave a large proportion of false negative reactions and was, therefore, discontinued.

Diagnosis of tuberculosis

Today, public health agencies, aware of the insidious potential of tuberculosis, maintain extensive chest X-ray campaigns for the detection of persons with active or inactive tuberculous lesions. Suspicious X-rays are generally followed by tuberculin tests, or positive tuberculin tests call for chest X-rays.

Final diagnosis requires identification of the bacteria by acid-fast stains or the recently developed fluorochrome stains.

Microscopic methods require a concentration in the sputum of about 10^5 to 10^6 bacteria per milliliter to give 1 cell per field. However, growth of bacteria in a suitable medium or animal inoculation will give positive results in specimens with too few bacteria to be detected microscopically. Cultures and animal inoculation are, therefore, often used in equivocable cases.

Atypical bacteria causing tuberculosis-like infections

A group of mycobacteria has recently been found that produces tuberculosis-like infections. These mycobacteria grow in cultures or in inoculated animals in a manner unlike the known species of pathogenic mycobacteria. Known as the atypicals, they are now being widely studied.

Treatment of tuberculosis

Until antibiotics became available, tuberculosis was considered incurable. However, a prolonged course of rest, adequate diet, and sunlight resulted, in a portion of the cases, in the arrest of the disease.

In 1945, streptomycin was found to act specifically on *M. tuberculosis*. However, the glowing reports of cures were short-lived because the antibiotic proved to be toxic to some people, producing serious side reactions. In addition, a proportion of the bacteria was found to have become streptomycin resistant.

A second agent, isoniazid (INH) was found to be specific against *M. tuberculosis*. When given to patients in the acute stage it tends to eliminate the bacteria; however, in some cases, resistant strains develop a few months after treatment. Fortunately these strains, which are catalase negative, also show lessened pathogenicity. Treatment must be continued for an extended period and is most effective in the control of recent, acute infections.

The most important single control measure is the discovery and treatment of active cases that can become vectors of infection. This is the basis for the chest X-ray program.

Unlike many vaccines, a tuberculosis vaccine prepared from killed bacteria is not feasible. Only living, avirulent bacteria are able to impart immunity. Using a highly virulent strain of *M. bovis* previously isolated by Nocard, Albert Calmette, of the Pasteur Institute, and his assistant, Alphonse Guérin (1921), grew the organism through a total of 230 transfers in potato-bile medium, and it became completely avirulent. The resulting preparation was termed the bacille Calmette-Guérin (BCG) vaccine. With it, 2-week-old calves could be successfully vaccinated and would develop a firm immunity. Vaccination of children born of tuberculous mothers was therefore initiated. This was a high-risk group, many of whom usually soon died of tuberculosis. By 1925, a total of 2,000 children had been vaccinated and the tuberculosis death rate declined to only 0.5 % compared with an expected mortality of 25%. The method is now widely used in some countries.

SALMONELLA AND SHIGELLA INFECTIONS

The *Salmonella* and *Shigella* species are responsible for a few major well-authenticated outbreaks and numerous unreported or even unrecognized episodes of enteritis (intestinal inflammation). The estimate has been made that for every reported case 100 to 10,000 unreported cases may occur.

Both genera have many characteristics in common. They are gram-negative rods, 0.3 to 0.7μm in diameter and 1 to 3 (rarely 4) μm in length. They are nonsporing, noncapsulated, and do not liquify gelatin, proteolyze milk, or hydrolyze urea. They do not ferment lactose, sucrose, or salicin. They are facultative anaerobes and, with one exception, reduce nitrates to nitrites.

However, the *Salmonella* species are distinct in many respects from *Shigella* and fall into two natural groups. The first is *S. typhosa*. This species generally is actively motile with many peritrichous flagella. None of the species produces indole. However, *S. typhosa* differs from other *Salmonella* species in that it ferments glucose with the production of acid only, while others produce both acid and gas from the carbohydrates they ferment.

Typhoid, caused by *S. typhosa*, is now a rare disease in this country. The few cases that do occur are often acquired outside the United States or come from carriers infected in some other country.

Salmonelloses other than typhoid and typhoid-like diseases

The salmonella infections most commonly encountered today are the food poisonings or gastroenteritides. The typhoid-related enteric fevers or paratyphoid infections are rarely seen. The paratyphoid organisms are *S. paratyphi*, the cause of paratyphoid A, *S. schottmuelleri* the agent for paratyphoid B, and *S. hirschfeldii* often called paratyphoid C. All three are natural pathogens of man causing true enteric fevers. Of the three, *S. schottmuelleri* has been also found in domestic animals.

The salmonella species responsible for the foodborne gastroenteritides are primarily natural pathogens of animals rather than man. For example, swine serve as the natural habitat for *S. cholerasuis*, while the natural host of *S. typhimurium* is mice, rats, and related forms. However, it is also found widely in other warm-blooded animals, and even has been isolated from snakes or pet turtles. *S. enteritidis*, which often produces a highly potent endotoxin, has a similar natural distribution among the warm-blooded animals.

In addition to the described salmonella species, a very large number are known only by their somatic and flagella antigens. Following isolation from infected feces or food involved in a food poisoning, final serological identification can be made at one of the salmonella typing centers.

Between 800 and 1,000 serological types have been involved in food poisonings. In addition to the feces and sometimes urine of animals or man—whole eggs and egg products (dried eggs, egg albumin), meat and meat products (especially poultry), and infected raw milk or milk products may carry the pathogens. The period of communicability following infections varies greatly. Commonly, this ranges from 3 days to 3 weeks but occasionally may persist for a few months. The disease may appear as a widespread outbreak, resulting from group consumption of contaminated food, or as sporadic cases from direct contact with animal or human carriers.

The incubation period is usually 6 to 48 hours, depending on the number of bacteria ingested. The outset is sudden beginning with acute nausea, vomiting, abdominal pains, chills, sometimes followed by fever and diarrhea. The symptoms usually last from 1 to 4 days. Disturbance of the normal flora by the oral use of broad-spectrum antibiotics may sometimes precipitate an attack.

Identification generally involves isolation of the bacteria from the feces by direct streaking on SS (Salmonella-Shigella) or desoxycholate citrate agar, or following enrichment in selenite-F or tetrathionate broth. After isolation, the organism is usually identified by serological tests and/or cultural methods.

The shigelloses

Although man is the usual reservoir of the *Shigella* species, a few strains have been isolated from dogs. As with typhoid, the initial source of all shigella outbreaks is an active case, or a convalescent or healthy carrier.

As a disease, shigellosis, or bacillary dysentery, has plagued mankind for centuries. The word dysentery means literally "sick intestines," while the adjective "bacillary" is added to distinguish it from the protozoan-caused amebic dysentery.

The causative agent (which now bears his name) was first identified by Shiga in Japan (1898) as the bacillus of Japanese dysentery.

Members of the genus *Shigella* are nonmotile and ferment sugars, producing acid but no gas. They are nonlactose fermenters, except for *S. sonnei,* which produces acid but no gas from lactose in 20 to 30 days.

Two species, *S. alkalescens* and *S. dispar,* are of questionable pathogenicity. The remaining species are pathogenic.

The shigellae are less invasive than the salmonellae, and can rarely be isolated from the blood. Bacillary dysentery is primarily a disease of the colon or terminal ileum, usually producing only superficial ulcerations in the intestinal mucosa. The disease runs an acute but unpleasant course and is generally self-limiting.

After an incubation period of 1 to 4 days the onset is marked by diarrhea accompanied by vomiting, cramps, and eventually by tenesmus (unproductive defecatory straining) accompanied by evidences of dehydration. In severe cases, the feces may contain pus, mucus, and blood. Occasionally chronic infections develop, resulting in recurrent episodes of enteritis. This condition may be diagnosed as a chronic ulcerative colitis.

In the United States, the death rate is less than 1% (usually infants or young children), but in the tropics it may reach 10% to 20%.

Shigella dysenteriae infections are usually more severe than those produced by other species. This is probably due to the production, not only of endotoxins, but also of a soluble exotoxin-like protein. The feces often contain large numbers of shigella, occasionally yielding almost pure cultures. Stool cultures may become negative 1 to 2 weeks after convalescence begins, but more commonly after a month or so.

BACTERIAL DISEASES OF LESSER IMPORTANCE

Of the lesser bacterial diseases, primary atypical pneumonia, chancroid, and meningococcus meningitis still occur with variable frequencies.

The causative agent of primary atypical pneumonia (PAP) is *Mycoplasma pneumoniae* (Eaton agent), a member of the Mycoplasmatales, or Mollicutes. The importance of this group is just now being recognized. Epidemics of PAP occur with some frequency among populations in close contact, such as in military barracks or orphanages. However, the incidence of the disease among the general population is difficult to determine because only about 5% of the cases are sufficiently severe to require medical attention. The remainder probably diagnose their own illness as "flu" or a cold. While the fatality rate is not high, a severe case may be markedly enervating.

Chancroid or soft chancre, a venereal disease sometimes called the third venereal disease, has a relatively low incidence compared with gonorrhea and syphilis. The causative agent, *Haemophilus ducreyi* (or Ducrey's bacillus) is a fastidious rod. Transmitted almost entirely by sexual contact, it produces a painful, suppurating open sore or chancre that fortunately is controllable with chemotherapeutic agents or certain antiseptics.

Meningococcus meningitis epidemics, caused by *Neisseria meningitidis,* a once dread event among the young, now rarely occur among the general population. However, several spectacular outbreaks have been reported among military personnel during the past decade.

Diphtheria is caused by *Corynebacterium diphtheriae* through the action of a highly potent exotoxin. Immunization with the diphtheria toxoid neutralizes the toxin and provides almost complete immunity.

Tetanus results from a powerful neurotoxin

produced by *Clostridium tetani.* It is now preventable by immunization with tetanus toxoid. Its efficiency was demonstrated in World War II when only 4 cases of tetanus occurred among the more than 10,000,000 toxoid-immunized American troops. Among the unimmunized German and Japanese troops, the incidence of tetanus among the wounded was 1 to 10 per 1,000.

Whooping cough or pertussis, long a dread cause of serious illness or death of infants, can now be prevented by the use of a bacterial antigen made from the causative agent *Bordetella pertussis.*

The two toxoids (diphtheria and tetanus) and the bacterial (pertussis) antigen combined as DTP provide a high degree of immunity if administered to infants and young children during the first five years of life.

Among the animal-borne diseases, individual cases of human brucellosis, tularemia (deerfly or rabbit fever), and plague (the dread Black Plague of history) may occur sporadically.

The causative agents of brucellosis are *Brucella abortus* (cattle), *B. melitensis* (goats), and *B. suis* (swine). Cases occasionally occur in rural regions as a result of contact with infected animals or drinking unpasteurized milk.

Tularemia and plague are endemic in rodent populations especially in the western United States. Biting insects or contact with infected animals occasionally cause human infections.

SUGGESTED READINGS

Blair, J. E., Lennett, E. H., and Truant, J. P., editors: Manual of clinical microbiology, Baltimore, 1970, The Williams & Wilkins Co.

Burnet, F. M.: Natural history of infectious disease, ed. 3, Cambridge, England, 1962, Cambridge University Press.

Burnett, G. W., and Scherp, H. W.: Oral microbiology and infectious disease, ed. 2, Baltimore, 1962, The Williams & Wilkins Co.

Burrows, W.: Textbook of microbiology, ed. 19, Philadelphia, 1968, W. B. Saunders Co.

Cowan, S. T., and Steel, K. J.: Manual for the identification of medical bacteria, New York, 1965, Cambridge University Press.

Davis, B. D., Dulbecco, R., Eisen, H. N., Ginsberg, H. S., and Wood, W. B., Jr.: Microbiology, New York, 1967, Harper & Row, Publishers.

26 Rickettsial and viral diseases

THE RICKETTSIACEAE

THE CHLAMYDIACEAE

VIRAL DISEASES

The rickettsiae and viruses are known almost entirely by the diseases or cytopathogenic processes that they produce, and their obligately parasitic state of existence. In this respect they differ from the bacteria, and for this reason are grouped by Bergey in the class, Microtatobiotes. The Microtatobiotes are divided into two orders, the Rickettsiales and the Virales.

The order Rickettsiales is divided into four families of which the Rickettsiaceae and Chlamydiaceae are responsible for human diseases. As a group they are very small, rodlike or coccoidal forms, barely within the limits of resolution of the light microscope. They have bacteria-like cell walls, possess both DNA and RNA, divide by transverse fission, and respond well to antibiotic therapy. For all practical purposes they are minute, nonmotile bacteria that have progressively mutated to the point where they lack the ability to produce most of the enzymes required for an independent existence. They can reproduce only in the presence of living animal tissue. Some stages will pass through bacteria-tight filters, but generally they are retained by such filters. They stain poorly by the Gram method but are gram negative. Satisfactory stains can be obtained by the Giemsa or Macchiavello methods.

Recently one species, *R. quintana*, cause of the louse-borne disease that incapacitated so many troops during World War II and was called French fever, was grown successfully on a nonliving medium.

THE RICKETTSIACEAE

The most important of the disease entities caused by the *Rickettsia* are typhus and spotted fevers.

Epidemic or louse-borne typhus

Since no cases have occurred in the United States since 1917, epidemic or louse-born typhus is discussed primarily because of its tremendous historical importance. Among the millions of American troops in World War II, protected as they were by antityphus immunization, only 64 cases developed, none of them fatal. Typhus has probably been a recurrent plague of humanity for several millenia. The earliest medical description was given by Fracastorius in 1546.

The term typhus, meaning hazy, refers to the confused mental state of the seriously ill, delirious patient, and was coined by Hippocrates. It was not applied directly to epidemic typhus fever until 1760. It should not be confused with typhoid, which means typhus-like.

Epidemic typhus is a disease of wars, famines, and misery; emerging when hordes of people, unwashed and louse-ridden, are huddled abjectly together. As the disease develops, the lice hurriedly crawl away from the fevered bodies of the sick or of the cooling corpses to the bodies of the well. The louse, its gut loaded with rickettsiae, has the unpleasant habit of chewing through the skin to feed, then when surfeited, defecating in the abraded area. History is replete with reports of the ravages of typhus.

The disease is transmitted only by the bite of head or body lice, and not from person-to-person. This is a classic example of a disease in which an epidemic can be abruptly halted by destroying the vital link in the infection chain, in this case, the infected lice.

With an effective chick yolk-sac–grown vaccine available, the disease no longer need constitute a threat. In the rare cases that now occur, administration of continuing doses of tetracyclines or chloramphenicol brings dramatic improvement in two or three days.

The causative organism is *Rickettsia prowazekii* (the species name honoring von Prowazek, who, like Ricketts, was a scientific martyr to the disease).

Endemic or flea-borne typhus

This rare and relatively mild disease, caused by *Rickettsia mooseri*, is carried by rats and transmitted by humans through the bite of the rat flea. There is no man-to-man transmission. Only sporadic cases occur, confined to endemic areas. Fewer than 100 cases per year are now reported in the United States, although formerly the number ran into several thousand.

The spotted fever group

This is a group of disease entities, usually tick-borne or mite-borne, similar to typhus. Rocky Mountain spotted fever and tsutsugamushi fever of Japan are the most important.

Rocky Mountain spotted fever, sometimes merely called mountain fever, was first described in Montana in 1873 and soon reported in Idaho. It eventually assumed epidemic proportions in the Bitter Root Valley. This prompted the United States Public Health Service to establish a laboratory in Hamilton, Montana, for the study of this and related diseases.

In 1906 Ricketts proved that the wood tick, *Dermacentor andersoni*, was the carrier and soon showed that the infective agent overwintered in the eggs of the wood tick, making it one of the reservoirs of the disease along with rabbits, field mice, and probably other wild animals. In 1909 he demonstrated the rickettsiae in blood smears of man, monkey, and guinea pig, and the gut of the wood tick. The disease was also proved to be distinctive from typhus. At first it was thought to be limited to the mountain states, but in 1931 it was reported in the East. The dog tick, *D. variabilis,* was shown to be the vector.

Rocky Mountain spotted fever ranges from mild to highly fatal. In a number of large outbreaks before chemotherapy was available, the mortality ranged from 5% to 90%.

The organism, *R. rickettsii,* as seen in infected tissue, resembles minute pneumococci occurring in pairs, about 1.0×0.2 to $0.3\ \mu m$. Gram negative, it stains well only by the Giemsa or Macchiavello methods. It grows readily in the egg-yolk sac. A killed vaccine from this source is often given outdoorsmen in high-incidence areas. Immunization must be repeated each year.

Tsutsugamushi fever, or scrub typhus, is a disease present in Japan, Korea, Formosa, India, and even northern Australia. The vector-reservoir relationship is similar to Rocky Mountain spotted fever in that the red field or harvest mites (*Leptotrombidium* species) may serve both as vector and reservoir together with field rodents. The mite is similar to the American chigger. It attacks and sucks the blood in the same fashion and area of the body.

The causative organism, *R. tsutsugamushi,* is seen as tiny bipolar staining rodlets, 0.3 to 0.5 μm long by 0.2 to 0.4 in width. It reproduces slowly in mouse cell cultures; fairly rapidly in the yolk sac.

Q fever

The Q fever disease entity differs from those previously discussed in that its transmission is not completely dependent on arthropod vectors. Indeed, the organism is almost bacteria-like in its ability to remain viable and maintain infectivity in dust and air.

The organism occurs as a diplobacillus ($0.25 \times 1.5\ \mu m$), a bipolar rod ($0.25 \times 1.0\ \mu m$), or a short, lanceolate rod ($0.25 \times 0.5\ \mu m$). Its staining characteristics are like those of other rickettsiae. It grows well in the egg-yolk sac and can be readily isolated from blood, urine, or sputum in human cases.

Continued studies have revealed that the organism is widely distributed in nature, especially in certain high-incidence, endemic areas.

It has been isolated from many different species of ticks, human body lice, a variety of small wild animals, cattle, sheep, goats, and man. A survey of Southern California made after an outbreak in 1947, showed that 10% of the dairy cattle were infected. Other surveys have shown as high as 65% of the cattle to have had the disease or have been carriers of the infective agents. The 1947 epidemic in Los Angeles County showed that a large number of persons suffering from the disease had consumed raw milk.

While milk can apparently transmit the disease, most of the infections occur among handlers of cattle and sheep, where dust from hair, dried feces, and so forth, may carry the organisms. It may be air-borne (a type of microenvironment infection) as witnessed by infections that have occurred among people with no direct animal contacts.

THE CHLAMYDIACEAE

The family Chlamydiaceae includes four important human pathogens in the psittacosis-lymphogranuloma group and the trachoma group. The chlamydia as disease agents are something of an enigma. Two decades ago they were called "large viruses" because they were generally able to pass through Berkefeld filters, grew only intracellularly, and formed inclusion bodies similar to those produced by the myxoviruses. On the other hand, they have since been found to be cell-like and possess both DNA and RNA. Also, while small, the individual body is still resolvable with the light microscope and the cell walls contain bacteria-like muramic acid. They reproduce by binary fission and are best described as viruslike rickettsia.

Parrot fever—ornithosis group

Parrot fever or psittacosis-ornithosis disease, first recognized as a distinct entity in 1880, was thought to be acquired only from infected parrots or related psittacine birds. Hence, it was called parrot fever or psittacosis. Later, it was found that a disease agent of this type infected pigeons, ducks, turkeys, and other birds, and that humans occasionally acquired infections from such birds. This disease was found to be a related, but somewhat different, entity and termed ornithosis. Nearly 130 different species of birds have been found to carry this particularly infectious type of chlamydia and a good many wild and domestic mammals may also suffer from the disease. Not only does it cause the disease in persons having contact with infected psittacine birds, but contact with pigeons may produce the disease. It has also become almost an occupational hazard with poultry men and those who work in poultry processing plants. Human-to-human infections also occur to a limited extent, especially among hospital personnel caring for infected persons.

The disease may range from subclinical through mild to severe and even fatal infections. The frequency of the disease is difficult to assess because a large number of mild cases may be diagnosed as flu or a cold.

The disease agent is primarily airborne, carried by droplets and dust from feathers and feces.

Lymphogranuloma venereum

The disease entity, known variously as lymphogranuloma venereum, lymphogranuloma inguinale, climatic bubo, or the fourth or fifth venereal disease is caused by an organism resembling in many ways the psittacosis-ornithosis organism.

Trachoma and inclusion conjunctivitis

Trachoma (granulated eyelids, granular conjunctivitis, or chronic contagious conjunctivitis) is an important chlamydia-caused disease. In some parts of the world, it is still an important cause of impaired vision or blindness in children and adults.

It is a disease of antiquity. Descriptions of it have been found on Egyptian papyri. Hippocrates wrote of it and Cicero, Horace, and Pliny, the younger, were said to have suffered from it. The current high incidence in North Africa, Egypt and the Middle East has prompted concerted control efforts by the World Health Organization. In the United States it is still a public health problem of some significance among the Indians of the Southwest.

A similar but less serious disease is inclusion conjunctivitis, inclusion blennorhea, or swimming pool conjunctivitis. This was differentiated from trachoma in 1933. By 1936 the causative organism was found capable of producing not only conjunctivitis but also a mild infection of the vagina or male urethra. It has since been found to be widespread throughout the population.

With adults, the infection may be carried from the urogenital system to the eyes by fingers, wash cloths, and so forth. Eye-to-eye transfer is relatively rare. Conjunctivitis in adults is much milder and the discharge scantier than in children. However, the eye irritation may persist for sometime. The disease may be transmitted by swimming in unchlorinated swimming pools or the waters of heavily used freshwater beaches.

VIRAL DISEASES

Although most bacterial diseases have been eliminated, many viral diseases still plague us. The common cold mars a few weeks out of each winter. That something we call intestinal flu is still a minor but annoying experience. Old fashioned influenza may still approach pandemic proportions. For many of us, cold sores or fever blisters recur with unpleasant frequency, and measles, mumps, and chicken pox, largely preventable, still attack the children of careless parents.

Hopefully, many of these diseases can be eliminated. The World Health Organization has marked smallpox for extinction – a campaign that has been partially successful but still has much to accomplish. We now have the tools to wipe out poliomyelitis if everyone cooperates in their use.

Viruses themselves have a vicious habit of mutating sufficiently to invalidate the effectiveness of a previously effective vaccine. This is especially true of influenza-causing viruses. An Influenza Surveillance Commission has been set up to check outbreaks of the disease that might be due to an antigenically different strain in time to allow the production of the new vaccine in countries to which it has not yet advanced.

The possible role of viruses as inciters of cancer is commanding a great deal of scientific attention that might lead to the control of the most feared of all killers.

Nature of the viruses of man and animal

It was not until the mid-1950's that the technique of electron microscopy had advanced to the point where the intricate details of the virus particle or virion could be visualized with any accuracy. Even so, the morphology of several of the viruses has not yet been elucidated. While the light microscope gave little information on the viral pathogens, the electron microscope, coupled with X-ray defraction, has now revealed that many of the viruses exist as cuboid or icosahedral forms with 20 faces or facets and 12 vertices.

The outer portion, or capsid, is made up of a series of essential ellipsoidal protein structures called capsomeres arranged in an orderly fashion to form the hollow cuboidal structure described above. The capsomeres each appear to be a single polypeptide molecule. Inside the capsid is a helical strand or coil of DNA or RNA. This may be single or double stranded. Some virions have, in addition, an external lipoprotein envelope.

Two other types exist as tubelike or bullet-shaped structures made up of capsomeres in an orderly arrangement with an internal helix or coil of DNA or RNA.

The DNA or RNA has a molecular weight of 2 to 160 daltons with 10 to 150 cistrons. The viruses are relatively resistant to adverse conditions. They are inactivated when the capsid protein is heat denaturated. This generally requires a few minutes at 55° to 60° C. Some are inactivated in a matter of hours at room temperature but retain their viability for many months when held at −70° C in a Dry Ice refrigerator.

When freeze-dried (lyophilized) they remain viable indefinitely at refrigerator temperatures.

They will be inactivated by X-ray or ultraviolet light, which is absorbed by the nucleic acids and causes the formation of dimers or hydrogen bonds between adjacent pyrimidine bases.

Methods of growing viruses

Unlike bacteria, which can be grown on a variety of inert nutrients, viruses have no self-replicating system and must rely on living cells to reproduce themselves. In addition, the cells must be peculiarly adapted to reproducing a given virus.

Host animals, or closely related animals, were originally the only means of replicating given viruses. Guinea pigs, mice, rabbits, and so forth, were used to grow many human and animal viruses. Unfortunately, some viruses could be grown only in humans, chimpanzees, or monkeys.

Embryonated chicks, that is, fertile eggs that have been incubated at the proper temperature to initiate the growth of the embryo, were found by Goodpasture in 1931 to be useful for growing many kinds of viruses. When the embryos had been incubated for 7 to 14 days, the virus-bearing material was inoculated into the yolk sac, amniotic sac, the chorioallantoic membrane, or even into the embryo itself. Although largely superseded by more advanced methods, the chick embryo is still widely used whenever it is desirable to produce quantities of myxoviruses and poxviruses for vaccine production or chemical studies.

Organ cultures, that is, the growth of tiny bits of embryonic organs in a highly enriched artificial medium, were developed by Harrison in 1907. Although sporadic attempts were made to grow viruses on bits of tissue, contamination plagued the investigators until it was found that antibiotics added to the nutrients effectively controlled bacterial growth.

The major breakthrough came in 1949 when Enders, Weller, and Robbins first grew poliovirus in other than cultured nerve cells. It was largely this discovery that provided the impetus for the development of the highly successful Salk and, later, the Sabin vaccines.

These culture methods involve (1) tissue cultures in which fragments of tissue are dispersed in a suitable medium and develop into tiny masses of living cells, and (2) cell cultures prepared by macerating tissues into fragments small enough to provide individual cells. When dispersed in a special medium, these grow out to form a monocell layer on which suitably diluted viruses will form plaques similar to those produced by phages in lawns of bacterial cells. From these, the concentration of virions can be estimated and pure cultures of a given virus obtained. The cells used for cell cultures may come from established lines maintained by biological supply houses or virus laboratories or developed by the individual investigator.

One of the major difficulties that slowed the development of virology was the obtaining of viruses in pure preparations free from host cell debris. This is now accomplished by cell maceration, filtration, and centrifugation for several hours in a refrigerated high-speed centrifuge at speeds sufficient to achieve up to 100,000 × gravity.

The importance of securing large numbers of virions for study becomes apparent when we consider that for successful electron microscopy, a minimum of 10^9 virions per milliliter is required.

Overview of the human virus diseases

With the amazing recent advances in virology has come an appreciation of the complications inherent in a study of viral diseases in man. Groups of symptoms thought to be characteristic of a single disease are now known to be produced by any of a number of different viruses. Also, the laboratory methods used for the diagnosis of viral diseases are often much more complicated than those available for the identification of bacterial pathogens. Therefore, routine laboratory diagnosis of viral diseases is impractical.

With these factors in mind, no attempt will be made to discuss details of specific viral diseases as was done with the bacteria.

A brief overview of the virus diseases is given below:

1. The enterovirus (intestinal virus) group includes the polioviruses, the Coxsackieviruses, and the echoviruses. These latter are the enteric cytopathogenic orphans. However, of the more than 20 types, some produce aseptic meningitis, others, the syndrome called intestinal flu, and some incite symptoms resembling the common cold.
2. With the rhinoviruses, which are responsible for the common cold, more than 100 serotypes are known. Although an attack by a given strain may make us immune for several years, other strains to which we are susceptible are always appearing.
3. Another group, the arboviruses (*ar*thropod *bo*rne), further illustrates the complexity of the problem. They have been divided into three groups based on similarities.

Group A is composed of more than 19 strains, including the viruses of eastern equine encephalomyelitis, western equine encephalomyelitis, and Venezuelan equine encephalomyelitis. These, known familiarly as "brain fever of horses," may occasionally be transmitted to man.

Group B, with more than 33 known strains, includes yellow fever, dengue or break-bone fever, and the St. Louis and Japanese encephalitides.

Group C, represented by diseases found in many parts of the world, has some 150 known serological strains. Only Colorado tick fever is common to the United States.

4. The myxovirus (and paramyxovirus) group includes the agents of influenza, with its three major types and several subtypes, parainfluenza (9 known strains), mumps, and Newcastle disease (Figs. 26-1 to 26-3).

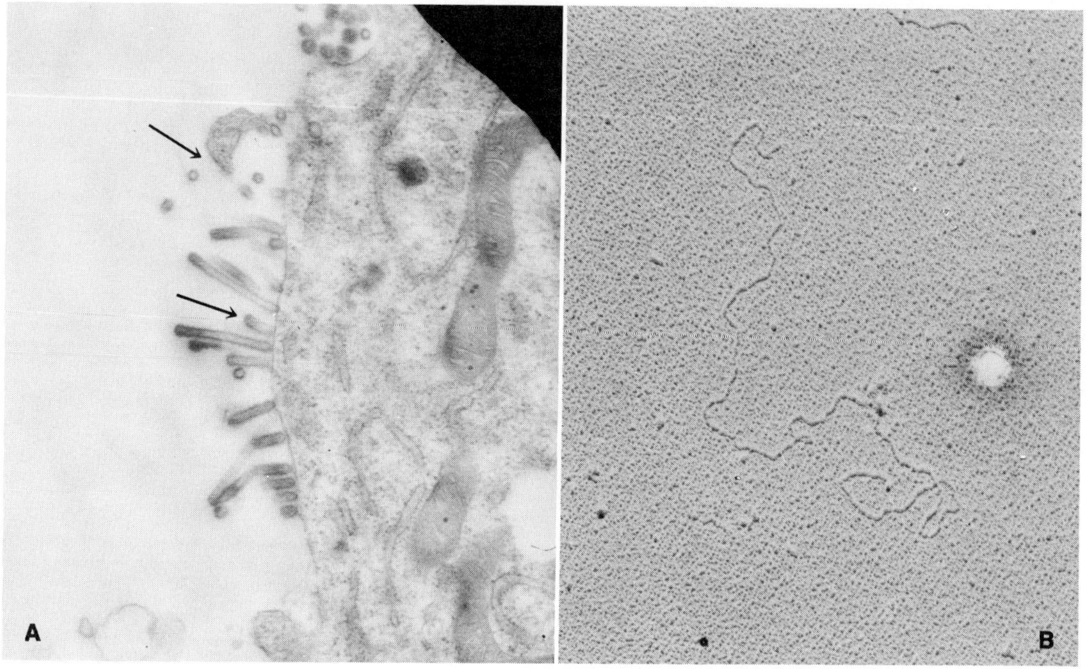

Figure 26-1
A, Electron micrograph of influenza virus replicating in human conjunctival cell cultures, showing filaments and spheres (×20,000). (Courtesy F. Chang and J. T. Seto, California State University, Los Angeles.) **B,** Electron micrograph of influenza virus RNA (×46,500). (Courtesy K. K. Li and J. T. Seto, California State University, Los Angeles.)

5. Another group includes the viruses of measles (rubeola), German measles (rubella) and exanthema subitum (a mild, afebrile disease of young children characterized by a transitory rash).
6. Rabies is of importance because, due to the endless reservoir of susceptible wild animals, it will be constantly with us. However, by immunizing our pets it can be kept out of our homes.
7. The papovaviruses include the causes of infectious warts and similar infections in animals.
8. The adenoviruses represent another important group that is responsible for acute respiratory diseases other than the common colds or influenza. These often appear in epidemic forms. More than 30 strains of this virus is known.
9. The herpesvirus group includes the viruses responsible for cold sores or fever blisters, chicken pox or shingles (an adult neurological manifestation of chicken pox), and infectious mononucleosis.
10. The poxviruses include the agents of smallpox (variola), vaccinia (the vaccine virus), and the skin disease, molluscum contagiosum.

Classification of human and animal viral pathogens

The pathogenic virions differ in a variety of respects:

1. In shape most of them are spherical, al-

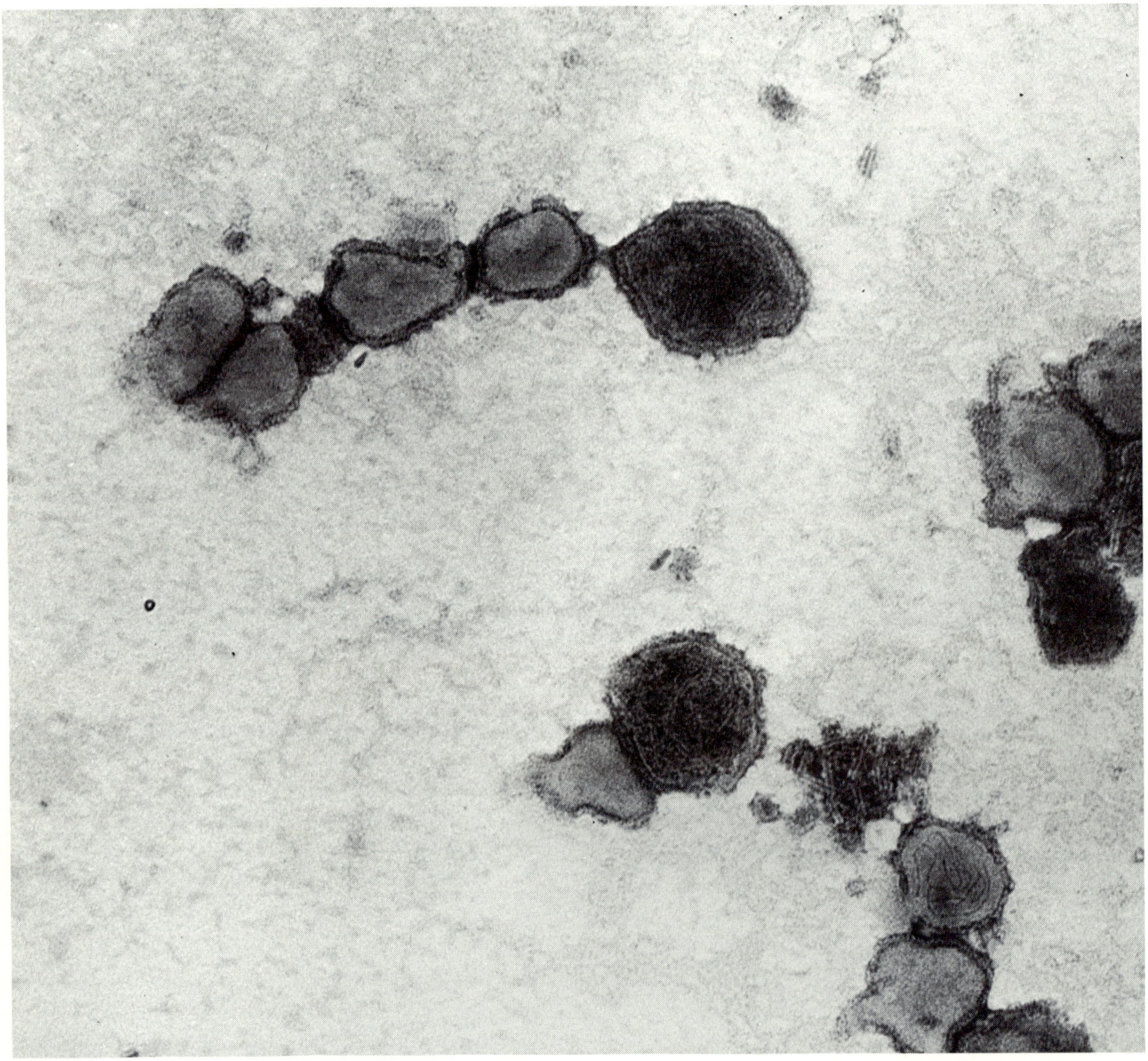

Figure 26-2
Electron micrograph of Sendai virus (a paramyxovirus) replicating in tissue cell culture (×96,000). (Courtesy Mildred Hochheiser, California State University, Los Angeles.)

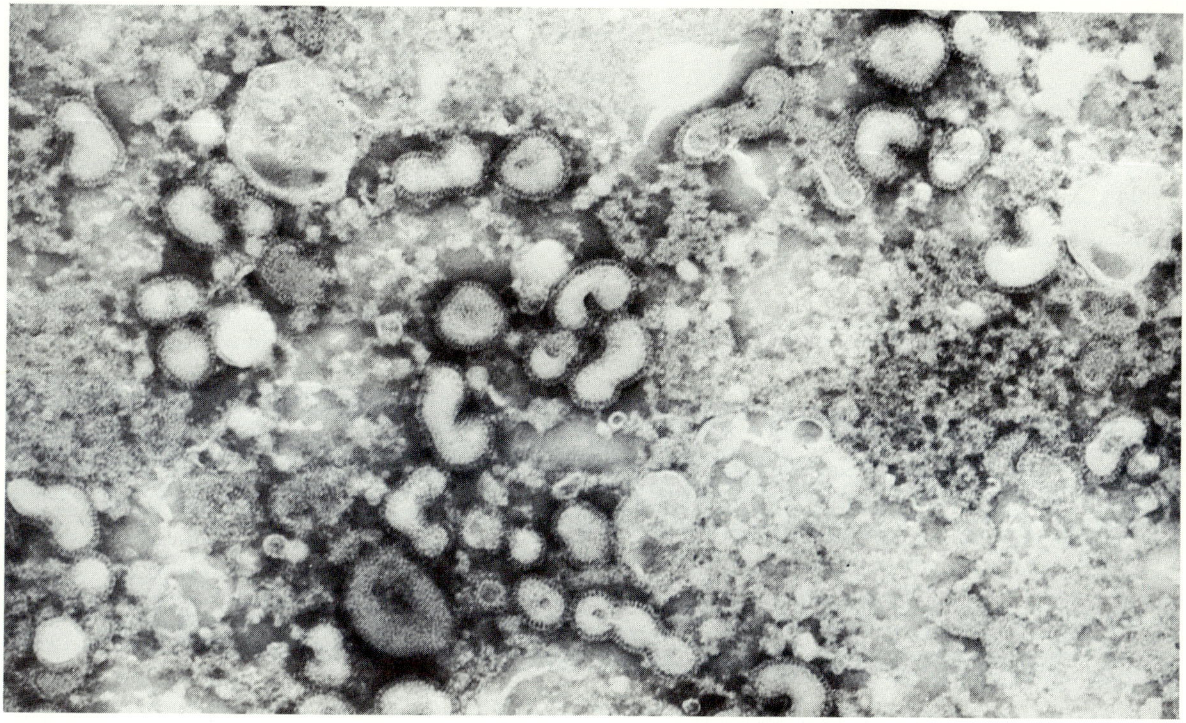

Figure 26-3
Electron micrograph of myxovirus in tissue cell culture (×30,000). (Courtesy J. T. Seto, California State University, Los Angeles.)

though a few resemble bricks; some may be filamentous; and at least one, the rabies virus, has a strange bulletlike appearance.

2. The genome of some is composed of DNA, while others are RNA viruses.
3. The DNA or RNA may be single or double stranded. With some very small RNA viruses, the RNA may also function as mRNA.
4. The virions may vary from 20 to 200 or more nm in size.
5. Some types show a distinctive envelope; others do not.

On the basis of these characteristics, this heterogeneous assemblage of viruses can be arranged into groups having similar characteristics. This classification is shown in Table 26-1.

Interferon

In the mid-1930's, several workers observed that an infection of an experimental animal with one virus often prevented a subsequent infection when another virus was introduced. This phenomenon was termed virus interference. In 1957, a compound responsible for interference was isolated from cells and called interferon. It was found to be a protein with a molecular weight of about 30,000, and nontoxic to the cells of other vertebrates.

It was also found that production of interferon was induced by double-stranded RNA viruses or certain single-stranded RNA viruses during the period when in the double-stranded replicative state. It was also later discovered that a synthetic

Table 26-1. Classification of the major viral diseases of man and animals

NAME	SHAPE	SIZE IN NM	DNA OR RNA	PRESENCE OF ENVELOPE	MAJOR DISEASES
I. Deoxyribonucleic acid viruses					
Parvovirus (L. *parvus* small)	Spherical	20	SS*	–	Virus diseases of rats and mice
Adenovirus (isolated originally from adenoid tissue)	Spherical	70-80	DS†	–	Human respiratory disease agents as designated below: adenoid-degenerative (AD); adenoidal-pharyngeal-conjunctival (APC); acute respiratory disease (ARD), respiratory infective (RI). (31 serotypes known)
Adeno-associated	Spherical	20	SS or DS	–	Found in tissue cultures of adenoviruses not connected with any known human disease
Papovavirus (name compounded from *pa*pilloma, *po*lyoma and *va*cuolating viruses)	Spherical	45-55	DS	–	Infectious warts of man; Shope papilloma of rabbits, and similar diseases of cattle, dogs, horses, monkeys, goats, etc.; polyomas (spontaneous tumors of number of animals)
Herpesvirus (Gr. *herpes* creeping; refers to nature of disease)	Spherical	150-200	DS	+	Herpes simplex (fever blisters, cold sores of man, and similar disease in monkeys); varicella (chicken pox or herpes zoster, shingles); pseudorabies, in various animals; infectious mononucleosis; cytomegalic inclusion disease, fatal to infants
Poxvirus	Brickshaped	100 × 200 × 300	DS	–	Variola (smallpox); alastrim (mild smallpox); vaccinia (vaccine virus); cowpox (distinct from vaccinia), and related diseases in sheep, swine, birds, rabbits; molluscum contagiosum (a skin disease)

*Single stranded
†Double stranded

Table 26-1. Classification of the major viral diseases of man and animals—cont'd

NAME	SHAPE	SIZE IN NM	DNA OR RNA	PRESENCE OF ENVELOPE	MAJOR DISEASES
II. Ribonucleic acid viruses					
Picornavirus (It. *pico* small + RNA)	Spherical	20-30	SS	–	Poliomyelitis; Coxsackie-viruses; echoviruses; common cold; foot-and-mouth disease of cattle; other diseases in monkeys, horses, swine, rats, cats, mice, fowl, fish
Togavirus (includes arboviruses and related viruses not arthropod borne)	Spherical	40-60	SS	+	Group of more than 200 viruses, borne mostly by mosquitoes and ticks; alternate between animal host and insect vector; some 80 strains known to infect man. Major diseases: eastern equine encephalomyelitis (EEE); western equine encephalomyelitis (WEE); St. Louis encephalitis (SLE); Venezuelan equine encephalomyelitis (VEE); Japanese encephalitis (JE); Denver fever; yellow fever; Colorado tick fever
Myxovirus (Fig. 26-4)	Spherical or filamentous	80-120	SS	+	True influenza: Type A—human, may infect swine; pandemic; Type B—small outbreaks, rarely epidemic; Type C—mild infections, often inapparent; Also equine and avian influenzas
Paramyxovirus (resembling myxovirus)	Spherical or filamentous	100-300	SS	+	Mumps, rubeola or measles; distemper of dogs, rinderpest of cattle caused by agents resembling measles virus; Parainfluenza 1, 2, 3, 4, mostly infectious to small children, Type 2, group-associated, causes mild upper respiratory disease in adults; Newcastle virus, cause of poultry diseases, occasionally conjunctivitis in humans

Continued.

Table 26-1. Classification of the major viral diseases of man and animals—cont'd

NAME	SHAPE	SIZE IN NM	DNA OR RNA	PRESENCE OF ENVELOPE	MAJOR DISEASES
Coronavirus	Spherical	80-120	SS	+	Bronchitis in chickens; about 10% of mild human colds
Leukovirus	Spherical	100-120	SS	+	Leukoses and sarcomas of chickens, hamsters, mice; some may produce tumors in other animals
Rhabdovirus (Gr. *rhabdos* rod)	Bullet shaped	70-180	SS	+	Rabies, an incidental disease in man. Reservoirs: bats, skunks, foxes, wolves, squirrels, dogs, cats
Reovirus (*r*espiratory *e*nteric *o*rphan)	Spherical	70-80	DS	−	Inapparent infections: three mammalian (cattle, horses); five avian serotypes
Unclassified viruses	Spherical	60	?	+	Rubella (German measles)
Has not been cultivated in laboratory. Apparently occurs only in man.	?	?	?	?	Infectious hepatitis; serum hepatitis

Data from Andrews, C. and Pereira, H. G.: Viruses of vertebrates, ed. 2, Baltimore, 1967, The Williams & Wilkins Company; Fenner, F. J., and White, D. O.: Medical virology, New York, 1970, Academic Press, Inc.

polynucleotide (poly I-C), which was an analog of double-stranded viral RNA, was able to induce the formation of interferon in experimental animals.

After interferon is produced by infected cells, the new protein apparently migrates to adjacent cells and affords them protection against the virus. Curiously enough, interferon does not act directly on the virus itself. Rather, it blocks the synthesis of complete viruses by the infected cell. The virus enters the cell in a normal fashion, uncoats, and synthesis proceeds up to the point where mRNA is to be translated. This, however, is blocked, not by the interferon itself but by another protein, a translation-inhibitory protein, produced by the cell as the result of the intermediation of the interferon. This apparently blocks the translation of the viral mRNA, not the cell mRNA.

The use of poly I-C in the prevention or control of certain virus diseases (not all are subject to interferon control), and also some types of tumors and possibly malaria, is being widely studied. This holds definite promise; however, the possibility exists that the compound may induce autoimmune reactions in humans. If this danger can be shown to be nonexistent, poly I-C may be of great value in the control of the common cold, or even some types of cancer.

Latent infections

Virologists believe that viruses that are carried in the host cells may serve as a major reservoir for an undetermined number of animal viruses.

In the case of the bacterial viruses, lysogeny represents a true latent state in which the phage may be sequestered in the host genome for many cell generations before reverting from the prophage to the active lytic state. This may be spontaneous, or in response to an inciting agent such as ultraviolet light.

No such clear-cut evidence of latency has been found and extensively studied in the case of the animal viruses. However, recurrent herpes simplex (cold sore or fever blister) infections may be considered a typical example of latency in an animal virus.

Recently, Dr. Frederick Robbins, a Nobel laureate in virology, has suggested that an attack of measles in childhood may be followed many years later by an attack of a chronic virus infection. Two wasting and frequently fatal diseases are suspected as having this etiology: they are multiple sclerosis and a rare brain fever, Dawson's encephalitis. Although the child apparently recovers completely from the initial attack, the measles virus remains in an inactive or latent state. What activates this virus, or causes it to produce these particular disease manifestations, is not well understood. However, measles and these possibly resultant latent infections are now preventable by vaccination.

Viruses and cancer

The subject of viruses and cancer is most difficult to treat with certainty because of the complexity of the inciting causes, including as they do, physical, chemical, and viral agents. In addition, a great amount of experimental work is in progress on the oncogenic role of viruses, so that anything written today may need to be drastically revised tomorrow.

The terminology itself confuses and complicates the picture. The commonly used term, cancer, comes from the Latin word meaning crab, and was so used because some of the early observers of skin cancers saw an outline reminiscent of the shape of a crab. The more sophisticated term carcinoma comes from the Greek and also means crab. The word tumor is derived from the Latin word *tumere* to swell. An oncologist is a scientist engaged in tumor research, and thus concerned with oncogenesis (gr. *onkos* mass + *genesis* production). Such masses or swellings may be termed neoplasms (Gr. *neo* new + *plasma* formation). A sarcoma (Gr. *sarx, sarcos* flesh + *oma* a swelling) is a tumor that begins in the connective tissue.

The development of cancer apparently begins when the usual cell-reproducing mechanism is disturbed. Under normal conditions each new body cell is essentially identical with the parent cell. During infancy and up to adulthood, cell reproduction goes on at a rapid rate. When maturity is reached, new cells are produced only when worn out cells must be replaced (since cells usually have a sort of "built-in obsolescence"), or when an injury destroys cells and the healing process requires rapid cell replacements. Crowding of one cell against another apparently activates the control mechanism.

Since only a certain number of cells in a given tissue are required at a given time, the machinery of reproduction must be turned off or on as needed. This attribute is inherent within each cell or tissue.

When this control mechanism is lost, cells continue to reproduce, often at an accelerated rate, producing neoplasms. In time, these interfere with the normal flow of blood, oxygenation of tissues, removal of growth metabolites, and may seriously crowd vital organs interfering with their normal function. Eventually this results in the death of tissue areas, and finally the individual.

These new tissue cells are often bizarre in shape and metabolism. Often they are larger than normal cells, and their growth is not hampered or slowed by the crowding effect of adjacent cells. Also, as a neoplasm increases in size a mass of cells may break off, enter a damaged blood vessel, and be carried into another part of the body to initiate another colony of neoplastic cells.

The original aberrant cell is a mutation, one so genetically damaged that it has lost the ability to turn cell reproduction off or on as needed. This means that the control genes no longer operate. A variety of agents are able to produce this genetic damage. If the genetic damage does not interfere with the control mechanism, but merely damages the genes responsible for one or more of the metabolic pathways and thus slows down the reproduction of the cell, the agent may be termed a mutagen. If, however, the cell reproduction control mechanism is inactivated and a runaway growth of the damaged cells begins, the initiating agent is termed a carcinogen or cancer producer.

Not all mutagens are carcinogenic; however, all carcinogenic agents are mutagenic. A variety of agents may be carcinogenic including (1) chemicals, (2) various kinds of radiation, and (3) viruses.

Of the chemicals, the tars and other components of cigarette smoke have been most widely publicized as carcinogens. Many types of smoke, including exhaust emissions from automobiles may have this same property. Caffeine, which mimics in its molecular structure certain components of DNA, likewise has some carcinogenic properties. Indeed, it is likely that many chemicals in our environment may also exert the same action.

X-rays, used extensively as mutagens by geneticists, may also activate the formation of fatal neoplasms. Atomic radiations may also react in the same fashion.

The role of viruses in the production of human neoplasms is still largely unsolved. Two viruses, those causing warts and molluscum contagiosum, are capable of producing neoplasms. These, however, are benign and self-limiting and are limited to the noncritical skin and subcutaneous tissues. Without doubt, certain viruses are capable of producing human carcinomas, but the problem of proving that a given virus causes a given type of cancer is almost insolvable.

To apply Koch's postulates would require that the virus always be found in a given type of cancer; that it be isolated in a pure culture; that it be injected into statistically significant numbers of other humans and produce the same type of cancer in all of them; and that it be isolated in pure culture from all of the experimental animals. Manifestly, for moral and humanitarian reasons it is essentially impossible to carry out the last two steps in this set of proofs.

However, several viruses are known that produce malignant tumors in animals. These include Rous sarcoma virus in chickens and Shope papilloma virus of rabbits.

Recently an important new experimental tool has become available. Malignant tumors can be induced by injecting certain viruses into newborn mice, hamsters, or other rodents. Especially useful are polyoma and the simian SV_{40}. While both are papovaviruses, they rarely if ever produce malignant neoplasms in nature. Also, they can be grown readily in mouse or hamster cell cultures. By utilizing such cell cultures as models, an understanding has evolved of the probable manner in which viral neoplasms are formed.

To a cell culture of hamster or mice cells are added a large number of polyoma virions, perhaps as many as 1,000 per cell. An occasional cell undergoes what is known as a "transformation." (This should *not* be confused with transformation as it occurs in bacteria.) In the rest of the cells the virus enters the cell, produces more DNA and its protein coat, and destroys the cell.

In about 1% to 5% of the cells, a mutant virion enters the cell and attaches itself to the nuclear membrane. The protein coat disappears and the DNA enters the cell. Normally in the unmutated virus, each strand of the double strand has about 5,000 bases. Since three bases are required to code for a single amino acid this means about 1700 amino acids can be produced. This is sufficient to provide for possibly 8 or 12 proteins. However, this unusual mutated virion apparently lacks the code for providing a protein coat or capsid for itself. Instead it attaches to the DNA of one of the chromosomes. Although incomplete, it does have enough bases (perhaps 1200 or so) to produce 4 to 8 proteins. In some manner, these proteins interfere with the control mechanism that turns cell division off and on. The result is that each infected cell continues to reproduce, piling up into a many-celled mass. The virion DNA provides the genetic message that keeps the cell in a transformed or uncontrolled growth state and keeps on replicating the offending virus DNA. The cell-virus association is similar in many respects to that of lysogeny in bacteria.

If a rather large dose of the transformed cells from the cell culture is injected into a mouse, it will develop a tumor, even though the virus alone would not cause a malignant tumor. However, if a mouse is given a very small dose of polyoma virus that will cause it to develop immunity, then injected with transformed cells, the immune mouse does not develop a tumor. This strongly indicates the possibility that at least for virus transformed neoplastic cells, it may be possible to develop immunity. This holds out a future hope for the immunological prevention or control of virus-initiated cancers.

Without doubt some mutant strains of human viruses that cause totally different diseases may have oncogenic properties. We do know that neoplasms can be induced in suckling rodents by the injection of some human adenoviruses. One of the herpesvirus group, the Epstein-Barr virus, has been isolated from a type of malignant lymphoma found in children in Africa.

A number of viruses found in birds produce little effect until late in the life of the birds when they may produce diseases of the blood-making system, or on rare occasions solid tumors of various sorts.

The role of viruses in human cancer is far from elucidated. The problem is like the solving of a jigsaw puzzle that moves slowly at first, then more rapidly as additional pieces are filled in.

SUGGESTED READINGS

Andrews, C., and Pereira, H. G.: Viruses of vertebrates, ed. 2, Baltimore, 1967, The Williams & Wilkins Co.

Fenner, F., and White, D. O.: Medical virology, New York, 1970, Academic Press, Inc.

Horsfall, F. L., Jr., and Tamm, I., editors: Viral and rickettsial infections of man, ed. 4, Philadelphia, 1965, J. B. Lippincott Co.

Smith, D. T., et al.: Zinsser Microbiology, ed. 14, New York, 1968, Appleton-Century-Crofts.

27 Protozoan diseases

The protozoa represent a huge and varied phylum with some 30,000 described species, occupying a variety of ecological niches in the economy of nature. Only 25 to 30 species are known to live parasitically in the human body. Of these, approximately half exist in a state of commensalism or near commensalism, living in our bodies without producing any appreciable deleterious effects. The remainder are markedly pathogenic, producing diseases of great medical and economic importance.

PROTOZOAN AND BACTERIAL DISEASES

Protozoan diseases differ materially from bacterial diseases. Most of the pathogenic protozoa are difficult to isolate and grow in pure culture. Diagnosis of the disease is reached largely by microscopic methods. In addition, immunity is highly variable and no preventive or therapeutic immunization procedures are available. Many of the protozoan diseases are chronic, giving rise to long-continued and often recurrent disabilities.

The majority of the protozoan diseases are transmitted only by specific insect vectors. Thus control of the disease may often be accomplished by elimination of the insect vector or the reservoir. Some, however, are transmitted by food or water contaminated by human feces. The four major groups of protozoan diseases are (1) amebic dysentery, (2) malaria, (3) the African sleeping sickness group, and (4) the Leishmaniases or sand fly fevers. In addition, of minor importance are Chagas' disease and the poorly defined toxoplasmosis infections.

INTESTINAL PROTOZOA

Six different species of ameba parasitize the intestinal tract.

1. *Entamoeba histolytica* is the causative agent of amebic dysentery.
2. *E. hartmanni* is a recently isolated nonpathogenic variant of *E. histolytica.*
3. *E. coli* is a common nonpathogen that resembles *E. histolytica* in a number of respects and may complicate the diagnosis of amebic dysentery. However, the cysts are somewhat larger than the *E. histolytica* cysts and have eight instead of four nuclei (Fig. 27-1).
4. *Endolimax nana,* another nonpathogen, is characterized by small cysts with one to four minute nuclei.
5. *Iodamoeba bütschlii* is characterized by irregular cysts, 5 to 20 μm in diameter with one, rarely two nuclei.
6. *Dientamoeba fragilis* is a rare form, lacking a cyst stage. Sometimes it is responsible for the production of diarrhea in children.

Another parasitic ameba of interest, although not of the intestinal tract, is *E. gingivalis.* This is found in the mouth along the gum margins and between the teeth, especially where pyorrhea is present. It is a commensal, probably feeding on bacteria and organic debris.

In addition to the amebas, members of the Mastigophora are encountered periodically in the intestines. These are *Chilomastix mesnili, Trichomonas hominis,* and *Giardia lamblia.* Only the last named organism is considered pathogenic. It may cause diarrhea in children. The disease can be cured by treatment with Atabrine.

Amebic dysentery

The development of an active case of amebic dysentery appears to be a complicated affair. The incubation period has been estimated to range from a few days to months or even years. Factors that apparently operate are initial number of cysts, virulence, bacterial pathogens present, and condition of the mucosa of the intestines.

In the development of an infection the cysts are ingested in food or water, pass through the

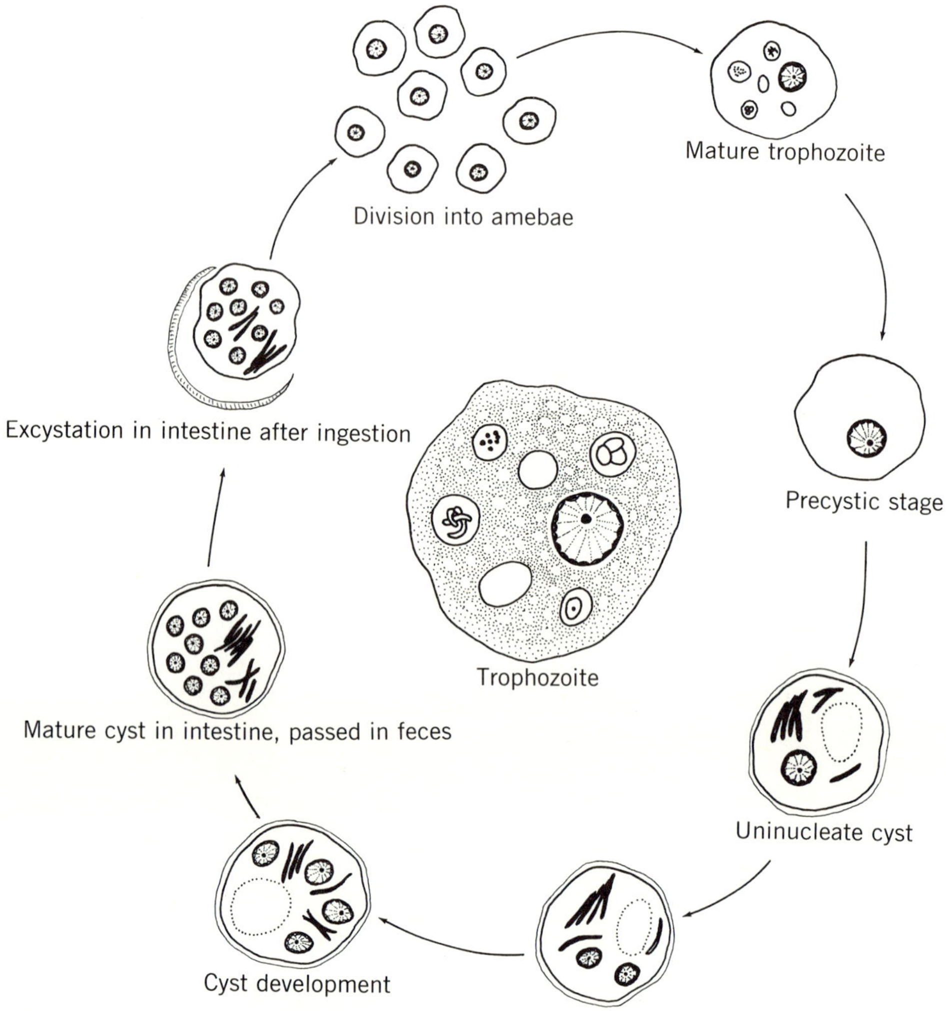

Figure 27-1
Life cycle of *Entamoeba coli.* (From Swatek, F. E.: Textbook of microbiology, St. Louis, 1967, The C. V. Mosby Co.)

stomach, and excyst in the small intestines. This process involves the formation of four small trophozoites or active protozoa from the four nuclei of the cyst. They begin feeding and soon reach full size. During this period they move from the cecum to the lower level of the large bowel. They attach and begin to erode a passageway through the protective mucosa. This continues to enlarge, forming a flask-shaped ulcer with a ragged edge.

In active infections, the person may suffer from severe diarrhea that may develop into a dysentery with considerable blood and mucus.

The organisms may metastasize into the liver, forming distinct abscesses. On rare occasions similar abscesses may even form in the lungs.

In acute infections, the patient suffers marked intestinal discomfort, diarrhea, and even dysentery with numerous stools and general malaise. Death may occur in uncontrolled acute attacks. In subacute infections, the symptoms are similar but less pronounced.

Many cases become chronic and recurrent. Their diagnosis represents a major problem for the clinician. Here the primary symptoms may be periodic bouts of diarrhea, interspersed with periods of constipation. Coupled with this may be a slight loss in weight, tendency to tire easily, mental dullness, and a general lack of physical and mental vigor.

Diagnosis of amebiasis

E. histolytica may be present in the intestines in three stages. In acute cases, organisms taken from the intestinal lesion, or present in diarrheal or dysentery feces, will be in the actively motile trophozoite or vegetative stage. These ameba may be large, 8 to 60 μm, ameboid bodies showing active motility and generally containing ingested blood cells.

The second stage is the precyst stage in which the ameba become sluggish and begin to round off. This occurs primarily in the lumen of the large intestines.

The final stage is the fully formed cyst. This is a round body, 5 to 20 μm in diameter, with one, two, or four distinct nuclei. These represent the highly resistant forms responsible for the transmission of the disease. They retain their viability for long periods in cold water, can easily be carried by infected food, are more resistant to chlorination than most nonsporing bacteria, and are not killed by the acidity of the stomach. In water treatment, because of their size they are effectively removed by filtration.

In the diagnosis of acute amebiasis the first procedure is to examine the diarrheic stools while still warm, by suspending them in warm saline and using a microscope equipped with a warm stage. Materials obtained from intestinal lesions in the lower bowel by means of the proctoscope are also extremely useful. These materials often contain trophozoites identified by their size, shape, nucleation, and active motility.

In chronic cases, formed stools may be examined for the presence of cysts. These specimens are normally suspended in an iodine solution. This imparts a distinctive diagnostic coloration to the cysts, nuclei, and other structures within the cyst itself.

Active (or chronic) cases may be treated by means of a number of specific antiamebic drugs, often combined with a broad-spectrum antibiotic for the control of concomitant bacterial pathogens.

Any attempts at direct control of the disease in a population is complicated by the fact there are many chronic carriers (estimated at 10% of the population in the United States). In the tropics the carrier incidence may be as high as 40%. The disease is thus endemic and cysts are continually present in sewage, while food itself may be contaminated by careless food handlers who are also carriers. Flies may also carry the cysts.

Control of amebiasis can be accomplished only by effective water treatment, safe food handling, and elimination of flies as possible carriers.

MALARIA

In terms of the world picture, malaria has been in the past, and in many areas still continues to be, one of the most important and devastating of the human diseases. Unlike the acute bacterial diseases, which run a violent course followed by a complete recovery, untreated malaria is characterized by recurrent bouts of chills and fever that can make an emaciated shuffling old man out of a formerly vigorous youth.

In severely infected regions, the whole population degenerates in the same fashion, thus disrupting and destroying the economic and intellectual life of the populace.

The name malaria stems from the early, prevalent belief in the miasmal theory of the nature of disease transmission. The disease was thought to be caused by malignant "night air" that came from the swamps. The victims did not recognize that the angry whine of the hungry anopheline mosquito, rather than the swamp mists, was the announcement of disease and death.

In the last few decades many areas of the earth, including the United States, have been released from the dread grip of malaria. However, in our own country, as recent as 1937, a million new cases developed annually. Most of these were in the Deep South, where the combination of swamps and subtropical climate provided the hordes of *Anopheles* mosquitoes necessary for widespread transmission of the disease.

The role of the mosquito as a transmitter of the malarial protozoan parasite has been known since the turn of the century. In 1898, Ross demonstrated that bird malaria was transmitted by mosquitoes. In the same year, Grassi, Bignami, and Bastianella demonstrated that the human disease was also carried by the anopheline mosquito.

The common *Culex* and *Aedes* house mosquitoes, for some biological reason, do not transmit malaria, while the various species of *Anopheles* do. Some 200 species of *Anopheles* are known. These vary considerably in their effectiveness as transmitters of the disease.

The anophelines can easily be recognized by their feeding habits. When the *Culex* and *Aedes* mosquitoes feed, their abdomens remain parallel to the surface of the skin; the *Anopheles* feed with the abdomen pointing almost upright. Biting occurs at dusk or early in the evening and is confined to the female mosquitoes.

When blood is drawn from an infected human by the mosquito (man is apparently the only reservoir), the malarial protozoan undergoes a cyclic sexual stage that results in production of numerous, thin, pointed sporozoites that find their way into the mosquito's saliva. On subsequent feeding of the mosquito, these sporozoites in the saliva are injected into the circulating blood of the victim where they are quickly removed from the bloodstream by the fixed tissue cells of the liver. Here they remain for many days as cryptozoites before being released from the liver back into the bloodstream.

When released they immediately enter the red blood cells. At this point, in stained blood smears, they may be recognized as a distinct "ring" structure, a fraction the diameter of the red blood cell. This is the ring stage of the trophozoite (Gr. *trophē* nourishment + *zoon* animal). From this develops the ameboid trophozoite, a considerably enlarged form. The red blood cell also begins to enlarge, reaching about one and a half times its original diameter. The parasite or trophozoite continues to grow, pigments (waste products of metabolism of hemoglobin) begin to appear, and the parasite divides into two bodies. This is known as the schizont stage. The two nuclei continue to divide, producing 12 to 24 nuclei depending on the species of parasite. The cytoplasm now surrounds each nucleus, forming the daughter cells or merozoites (Gr. *meros* part).

At this point the red cell bursts, releasing the 12 to 24 merozoites that immediately enter the red blood cells, form ring trophozoites, and the process is repeated.

The release of the merozoites is marked by the beginning of chills and fever. The length of time between these episodes is quite regular and varies with the species of *Plasmodium*.

The disease is diagnosed by means of thin or thick blood smears stained with Giemsa stain. The blood must be taken at appropriate times to increase the likelihood of finding cells that are recognizably infected.

Species of the malaria organisms

The malaria organisms belong to the class of protozoa known as the *Sporozoa*. The species of the genus *Plasmodium* (Gr. *plasma* something formed + *eidos* appearance) are as follows:

1. *P. vivax* (L. long-lived) is not as highly pathogenic as other strains, but is widely distributed. The cycle of chills and fever is 48 hours, although a 44-hour strain has been reported.
2. *P. ovale* is similar to *P. vivax* in most respects except that the infected red cell is oval in shape.
3. *P. malariae* shows a characteristic 72-hour cycle between episodes.
4. *P. falciparum* (L. *falx* sickle + *parere* to produce) is the most pathogenic of the malarial forms. The chill is less marked; the fever more prolonged and severe. The length of the cycle is 36 to 48 hours.

Treatment and control of malaria

The first chemotherapeutic agent found to be specific for the control of malaria was the bark of the cinchona tree, which contained the alkaloid, quinine.

During World War II, a number of compounds were synthesized that reacted against the parasite much as did quinine. Notable among them was the widely used Atabrine (quinacrine). These compounds suppressed the parasites in the bloodstream but did not "cure" the disease by destroying exoerythrocytic forms sequestered in the liver cells. Consequently, the disease often recurred. A new chemotherapeutic agent, primaquine, has the added advantage that it is able to affect a cure by destroying the parasites sequestered in the liver.

The major procedure for controlling the disease is destruction of the mosquito population, thus breaking the infection chain. This is accomplished by draining swamps, covering water surfaces with an oil film to suffocate the mosquito larvae, or planting guppies, the small viviparous fish, to feed on the larvae. The use of insecticides such as DDT has proved highly effective; however, mosquito strains resistant to the insecticide have developed.

INSECT TRANSMITTED TRYPANOSOMAL DISEASES

Flagellated forms belonging to the family, Trypanosomatidae, of the Mastigophora are capable of parasitizing the blood and other tissues. Consequently they are often termed hemoflagellates. The human pathogens are represented by two genera, *Trypanosoma* and *Leishmania*.

Trypanosome diseases

There are two important trypanosome diseases, both transmitted by the flies of the genus *Glossina*, known familiarly as tsetse flies. These are bloodsucking flies resembling the common housefly.

The first disease, called African sleeping sickness, is caused by *Trypanosoma gambiense* and was first described by Dutton in 1902. It is a long (15 to 40 μm), broad, ribbonlike form with an undulating membrane. It usually has a long flagellum extending out from the undulating body.

The incubation period is from 2 to 3 weeks to several months after the bite of the fly. Initially, the parasites are chiefly found in the blood. Later they move into the lymph system, the spleen is enlarged, and edema, with visible wasting, develops. Coupled with the edema is evident heart damage. The parasites now move into the spinal fluid where large numbers may be present. With the central nervous system involved, the victim shows marked apathy, somnolence, and occasionally maniacal symptoms. Generally death occurs. The disease may be several months in duration before ending fatally or before the rare recovery begins.

The reservoir is thought to be the large game animals such as the sitatunga antelope, which may harbor the flagellate without developing any symptoms. Attempts to control the disease

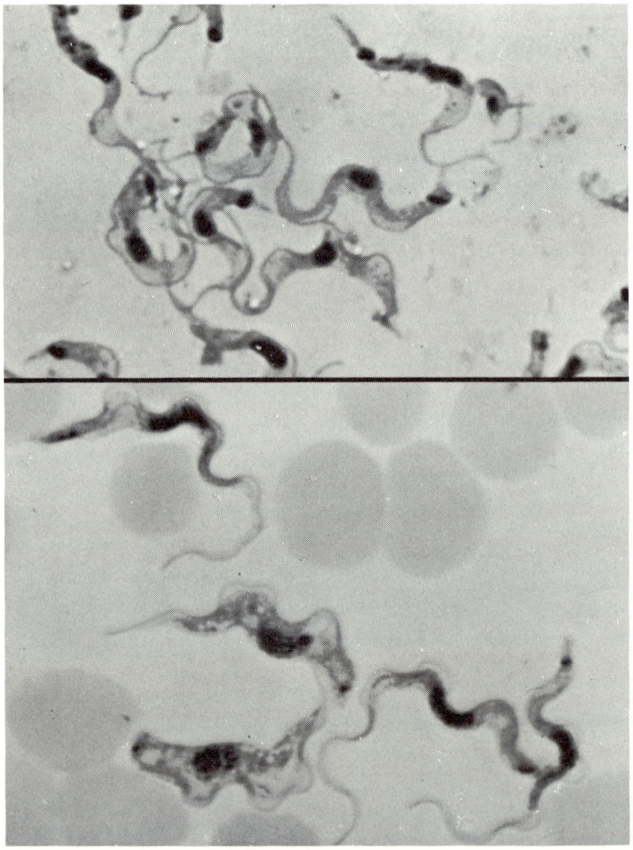

Figure 27-2
Trophozoites of *Trypanosoma rhodesiense* in blood smear. (From Swatek, F. E.: Textbook of microbiology, St. Louis, 1967, The C. V. Mosby Co.)

by removing or destroying the wild animal reservoirs have been unsuccessful. It may be that man is the primary reservoir. Trapping and poisoning the tsetse fly, and clearing areas around villages have been helpful.

Several synthetic compounds, primarily arsenicals, have been developed. Some of these have proved useful if given for prophylactic purposes or early in the course of the disease.

The Rhodesian sleeping sickness is similar in many respects. However, the parasite, *T. rhodesiense* appears to be different and the course of the disease more rapid and severe (Fig. 27-2).

The Leishmaniases

The *Leishmania* are flagellated forms, representatives of the class Mastigophora, and resemble the genus *Trypanosoma.* They are responsible for three distinct diseases, kala-azar, tropical ulcer or oriental sore, and uta or espundia.

All three are transmitted by the bite of one of the minute night-biting, bloodsucking insects, known as sand flies that belong to the genus, *Phlebotomus* (Gr. *phlebos* vein + *tomos* a cutting). The sand fly ingests blood from infected humans or reservoir animals, and develops a heavy infection in its intestines. The parasites may be numerous enough to block the pharynx. In their vigorous effort to feed, the flagellates may be forced into the body of the host.

Kala-azar (Hindu black poison) develops after an incubation period of 1 to 4 months. It is characterized by a high but irregular fever, marked enlargement of the spleen and liver, dropsy, emaciation, and hemorrhages of lips and nose. The name of the disease is derived from the dusky hue of the skin of a severely infected patient. In untreated cases the fatality rate is about 90%.

The disease is found in India, China, south Russia, parts of the Mediterranean coast line, southern Arabia, and parts of Central and South America. While it is primarily a disease of man, dogs, mice, and hamsters can be infected.

The disease is controlled by treating the patients with certain synthetic arsenicals and spraying the fly-breeding places with insecticides.

Oriental sore, caused by *L. tropica*, is characterized by large ulcers resulting from the multiplication of the organism in the macrophages of the subterranean subcutaneous tissues. The incubation period is 10 days to several months after the bite. The lesion may persist for several months, then heal spontaneously leaving a noticeable scar.

Uta or espundia, a disease resembling tropical ulcer, is widely distributed in South and Central America. The causative organism, *L. braziliensis*, produces open weeping lesions surrounding the mucocutaneous borders of the mouth and nose. The lesions may persist for months before healing. If the infection spreads to the pharynx, it may be fatal.

OTHER PROTOZOAN DISEASES

Chagas' disease

Chagas' disease, or American trypanosomiasis, differs from African trypanosomiasis in several respects. The disease is transmitted by the bite of the cone-nosed bugs (*Triatoma* species and related genera) belonging to the family Reduviidae. (They are related to the bedbugs.)

The disease is confined largely to Central and South America. Reservoirs of the disease include cats, dogs, and wild rodents. The bugs acquire the trypanosome from biting infected animals. They then bite sleeping persons, usually around the eyes and mouth. The trypanosomes are excreted from the bug, and the feces may infect the wound made by the biting insect. After an incubation period of 1 to 2 weeks, the bitten person develops an irregular fever and edema, especially of the eyelids. This is soon followed by enlargement of lymph nodes, liver, and spleen. After the initial acute stage, the trypanosome may become intracellular in the lymph-macrophage system and also in cells of the heart muscle, endocrine glands, and certain cells of the brain. In the acute stage, the mortality rate among children is very high. With adults, the disease may enter a chronic stage with slow but frequently fatal consequences.

Recently, promising chemotherapeutic agents have become available. These are most effective if administered early in the course of the disease.

Toxoplasmosis

In 1909 Nicolle and Manceaux, working at the Pasteur Institute in Tunis, isolated an unusual form from a native rodent kept for experimental purposes. This organism was banana or crescent shaped, 5 to 10×2 to 4 μm, and easily seen in Giemsa stained material. This, they named *Toxoplasma gondii*.

Its exact nature is not clear. "Doubtless it is a member of the protozoan tribe," one protozoologist expressed it; although it was once suspected of being a yeastlike form.

It is widely distributed in wild and domestic mammals, birds, and man. It represents a biological enigma. The wide spectrum of susceptible forms is different from that seen with most protozoan diseases. Also, it apparently produces a substantial degree of immunity, again unlike most protozoan diseases.

Clinical cases of the disease are extremely rare. The first human case was reported, and the causative agent identified, by Wolf and Cowan in 1937.

On the other hand, there may be a large number of subclinical manifestations, or it may produce a type of diffuse pneumonitis that might be identified as the flu or another of the virus diseases. On recovery, a solid immunity seems to be produced.

In surveys using a dye-test method, results indicated that 60% of those tested in the United States have had the disease. Skin sensitivity tests with toxoplasmin have given 17% to 35% reactions in the general population.

Studies of pork from slaughterhouses have shown that up to 22% of the meat samples examined contained the organism.

The disease is diagnosed by (1) direct examination of tissue using the Giemsa stain; (2) impression smears made from tissue or suspected lesions, dried and stained with Giemsa stain; and (3) intraperitoneal injection of infected material into mice. (After a few days the organisms are present in great numbers in stains of the intraperitoneal exudates.)

Trichomonas vaginalis infections

Trichomonas vaginalis is a flagellate that often parasitizes the female vagina, and sometimes infects the bladder. Surveys have shown that 25% to 70% of women examined harbor the parasite. An active invasion results in the production of a profuse, frothy, white discharge coupled with unpleasant irritation and itching.

The invasion of the parasite may be controlled by the use of vinegar or lactic acid douches or localized applications of terramycin, Aureomycin, or other proprietary compounds under the direction of a physician.

The organism may invade the male urethra and prostate, producing a prostatitis. This may be complicated by secondary nonspecific bacterial infections.

SUGGESTED READINGS

Burrows, W.: Textbook of microbiology, ed. 18, Philadelphia, 1963, W. B. Saunders Co.

Hegner, R. W., and Engemann, J. G.: Invertebrate zoology, ed. 2, New York, 1968, The Macmillan Co.

Manwell, R. D.: Introduction to protozoology, New York, 1961, St. Martin's Press, Inc.

Smith, D. T., et al.: Zinsser Microbiology, ed. 14, New York, 1968, Appleton-Century-Crofts.

28 Fungus diseases

Molds, yeasts, and related simple fungi common to our environment are represented by thousands of species. Of these, fewer than a hundred are capable of causing human diseases, and of these less than a dozen can cause fatal infections. Only a few, the common dermatophytes (Gr. *derma* skin + *phyta* plants) such as the ringworm organisms, can be readily transmitted from person-to-person, or from animal to man. However, epidemics of these diseases sometimes develop. A few others, notably *Candida albicans,* the thrush organism, are normal inhabitants of the body, only rarely causing infections. The remainder are inhabitants of the soil or decaying organic matter and are often endemic to some geographical area.

FUNGI AS THE CAUSE OF DISEASES

The diseases of man caused by fungi fall into two rather distinct groups:

1. Superficial—those that attack primarily surface structures of the body such as skin, hair, and nails.

2. Deep-seated or systemic diseases that attack the lungs, brain, and other organs. These fungi are primarily free-living forms in nature and are acquired from the environment. In many cases, deep-seated infections may produce death.

   The most important of the deep-seated fungus diseases of man are blastomycosis (often termed American blastomycosis), coccidioidomycosis or coccidioidal granuloma, cryptococcosis, and histoplasmosis.

THE DERMATOPHYTES

Keratin (Gr. *keras* horn) is a protein made up of long, fibrous molecules, which forms hair, fingernails, horns of animals, and the hard, outer layer of the skin. It is insoluble and relatively inert.

Very few microorganisms are enzymatically equipped to attack this protein; it, therefore, imparts to the skin its armorlike ability to resist the penetration of most microorganisms. However, a distinct group of fungi have evolved that are able to utilize keratins. These are termed dermatophytes, and they are responsible for the superficial infections of the skin and hair known as ringworm, athlete's foot, and so forth.

Certain of these forms lead a saprophytic existence in the soil, and places such as the moist wood floors of showers, with periodic excursions to the human or animal skin, hair, or nails. Others are primarily inhabitants of the human or animal body with rare excursions into saprophytic life in the soil, or suitable moist habitats in the human or animal environment.

The saprophytes can be rather easily isolated by burying hair or fingernail clippings in the soil and culturing the resultant organism on Sabouraud's or similar agars. Keratinophilic fungi have been isolated from soils all over the world.

Ringworm

Only the most common types of ringworm fungi will be considered.

The small-spored *Microsporum* is responsible for the most of the ringworm of the scalp, which often becomes epidemic among children. Two species are generally involved, *Microsporum audouini* (Fig. 28-1) and *M. canis*. The first produces about half the ringworm of the scalp in children. This disease is termed tinea capitis (L. *tinea* worm, hence ringworm; and L. *caput* head).

M. canis, often acquired from the family dog (as the name implies), is responsible for most of the remaining ringworm infections in children. The organism may also cause fungus infections of the smooth or glabrous skin.

The nature of the organism, with its resistant spores, is conducive to the development of epidemics. It can be spread by direct contact, use of the common towel or washcloth, wearing of borrowed infected hats or caps, the infected backs of theater or other chairs, and so forth. Skin infections can be transmitted by autoinoculation from a ringworm on the scalp.

In the classroom, it may be detected by using Wood's ultraviolet lamp or "black light" to examine the heads of children. The infected areas fluoresce in a distinctive green fashion; *M. canis* is less uniformly fluorescent.

After the age of puberty, susceptibility to these organisms decreases greatly.

Among adults, ringworm of the scalp and smooth skin sometimes assumes epidemic proportions. The disease is less common than the childhood infections and occurs more frequently in Europe than in the United States. The organisms usually responsible are *Trichophyton* species (Gr. *trichos* hair). The two most common species are *T. tonsurans* and *T. schoenleini*. These organisms both grow into the hair, which may break off near the base leaving a bare patch covered with stubs of hair. They are transmitted in the same manner as with children and, in addition, the barber's clippers, shears, and comb may be involved.

A less common form, *T. verrucosum* (L. *verrucosus* resembling a wart) infects the scalp, beard, and smooth skin. Unlike the other two,

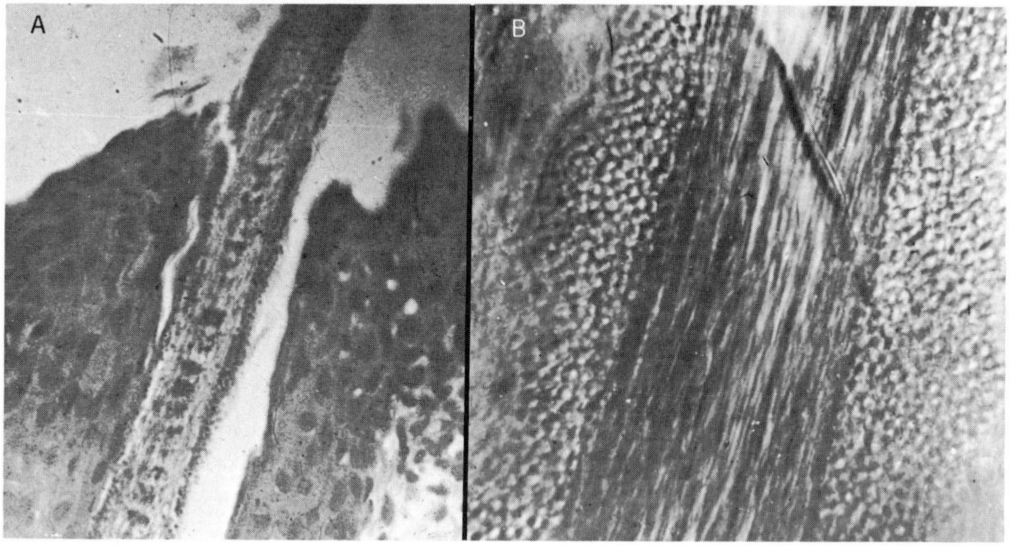

Figure 28-1
Hairs infected with *Microsporum audouini,* showing spores along hair shaft. **A,** Low power magnification; **B,** high power. (From Swatek, F. E.: Textbook of microbiology, St. Louis, 1967, The C. V. Mosby Co.)

this organism grows inside or on the outside of the hair.

Athlete's foot

T. mentagrophytes (L. *mentagra* skin eczema) is the most common cause of athlete's foot or tinea pedis. Most commonly it invades the area between the toes but may extend to the bottom of the foot, or produce ringworm of the smooth skin on other parts of the body or the hair follicles of the scalp or beard. The trichophytons are more prone to elicit inflammatory reactions of the skin than do the *Microsporum* species.

Athlete's foot is spread from improperly sanitized shower floors, wood floors of bath houses, and so forth, or by infected shoes. If shoes are known or suspected to be infected, they may be sanitized by moistening pledgets of cotton with formalin, placing one in each shoe, and storing in a tight shoe box for 24 hours.

Epidermophyton floccosum (L. *floccosus* tufted, as in wool) is the primary cause of tinea cruris, (also called eczema marginatum, dhobie itch, and "jockey's itch"), which is marked by a reddening of the regions of the inside of the thighs and the adjacent moist areas in the inguinal regions. This is followed by an exfoliation of the skin. The infection may also appear between the toes and the fingers. The organism does not infect the hair or hair follicles.

Chronic infections of the fingernails and toenails may also occur, resulting in progressive damage to the nails, which become irregular and finally distorted. Several species of dermatophytes may be responsible for this condition.

Diagnosis or identification of the causative dermatophytes is accomplished by the microscopic examination of skin, fingernail scrapings, or short pieces of hair. These are mounted on a slide in a few drops of 20% to 40% potassium or sodium hydroxide solution. This will dissolve the keratin, leaving visible pieces of mycelia or spores (Fig. 28-2).

Final diagnosis depends on the growing of the

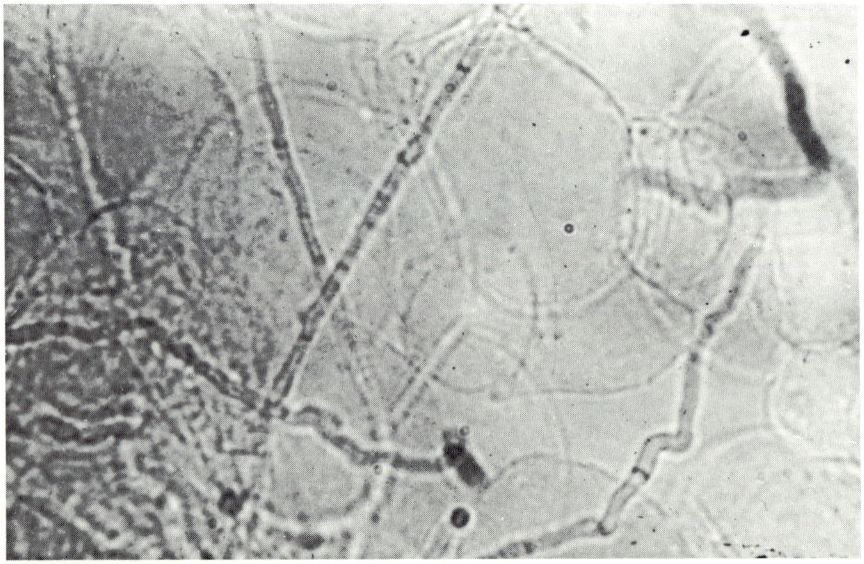

Figure 28-2
Potassium hydroxide mount, showing fungus mycelium in skin scales. (From Swatek, F. E.: Textbook of microbiology, St. Louis, 1967, The C. V. Mosby Co.)

causative organism on Sabouraud's dextrose agar with chloramphenicol and Actidine added to suppress bacterial growth, (pH 5.5). The resultant colony is then identified by mycelial and spore morphology.

Treatment is sometimes difficult, but undecylenic acid or zinc undecylenate are apparently rather effective when used as foot powders or sprays. Soaking the feet in a 50 ppm solution of sodium hypochlorite may be helpful. Often infections tend to heal spontaneously after a period of time.

DEEP-SEATED FUNGUS INFECTIONS
Blastomycosis

A vast majority of the cases of blastomycosis occur in the Mississippi River Valley in the United States, and the provinces of Ontario and Quebec in Canada. It has, therefore, been looked upon as indigenous to North America, and designated North American or merely American blastomycosis. However, suspected cases were reported from Mexico (1953), and Venezuela (1954), and in 1964 well-authenticated reports of blastomycosis came almost simultaneously from the Congo, Uganda, and the Republic of South Africa.

The organism, *Blastomyces dermatitidis*, was first isolated in Baltimore in 1898 by Gilchrist and Stokes. The organism is dimorphic; that is, in tissues it appears as a budding, yeastlike form but, on Sabouraud's agar it may initially produce a white, cottony, mycelial growth, a mixture of one-celled forms and mycelial threads. Subsequently, the surface growth is made up primarily of "prickles" of closely packed mycelia.

The disease may develop as a primary infection of the skin, beginning as a small papule usually at the site of a scratch or other skin abrasion. This may enlarge, forming an ulcerated surface. From this, the organism may enter the bloodstream and many small abscesses may occur throughout the body. A second manifestation is a lung infection resembling tuberculosis. Although the record of deaths due to this cause is fairly accurate, the disease is nonreportable, so it is difficult to determine the exact number of atypical infections it may produce.

The exact habitat of *Blastomyces dermatitidis* has not been determined conclusively. All indications are that the organism is endemic to the soil, particularly in areas where men or animals have lived. However, when soil samples are placed on the surface of a plate showing active mycelial growth, lysis of the mycelium may occur, making the soil suspect as a normal habitat.

Coccidioidomycosis

Coccidioides immitis appears to be restricted to the western hemisphere. In the United States the indigenous area appears to be the arid Southwest. It has also been reported from Guatemala, Honduras, Venezuela, Paraguay, and Argentina. In endemic areas the disease occurs frequently enough to consider it as a major disease. Fiese, for example, estimated in 1958 that some 35,000 cases developed in the highly endemic San Joaquin Valley. In the high-incidence Maricopa and Pima counties of Arizona 5,000 to 10,000 new cases are estimated each year. In these areas and in El Paso County, Texas, practically every inhabitant has acquired the infection within a few years of birth or after taking up residence in the area. In about 80% of the cases, the infection is so mild that no medical attention is required, and it is probably passed over as the flu. In fact, it is often called "desert fever." Of those requiring medical attention, the mortality rate is very low (averaging 61 deaths per year during 1952-1963).

Persons who have had the disease generally show positive skin tests when the antigen, coccidioidin, is injected into the skin. In endemic areas, a high percentage of the inhabitants show a positive skin test. Lower animals, such as cats, dogs, cattle, coyotes, horses, swine, and rodents also show a high incidence of infection in the endemic areas.

Cryptococcosis

The distinctive feature of cryptococcosis is that the causative organism is a true, imperfect yeast showing no filamentous growth.

The organism was first isolated from a fatal case in 1894 by Busse in Germany. The older literature calls the disease European blastomycosis, torula meningitis, or torulosis. The causative organism is *Cryptococcus neoformans*. (Gr. *kryptos* hidden). It was earlier called *Torula histolytica*.

It is worldwide in its distribution. In the United States an estimate has been made of 200 to 300 cases of cryptococcal meningitis per year. However, for every case of this type there are a great many undiagnosed, subclinical, pulmonary cases. In New York City alone an estimated 5000 to 15,000 cases per year may occur.

In addition to humans, the disease can be found in horses, cattle, dogs, cats, and various wild animals. There have been no proved cases of man-to-man or animal-to-man infections.

The ultimate source of the organism is the soil. However, pigeon feces, and perhaps that of other birds, may serve as an enrichment to stimulate the growth of the soil form. This may be due to their high urea, purine, and total nitrogen content. Birds, because of their high body temperature, do not suffer from the disease.

The organism is yeastlike, round to ovoid, and 5 to 10 μm in diameter. When seen in tissue exudates, it has a large gelatinous sheath that differentiates it from other mycoses. It grows easily on Sabouraud's agar producing a white to tan smooth colony.

It may attack any part of the body, but seems to have a predilection for the central nervous system, producing symptoms resembling tuberculous meningitis or brain tumors.

It may produce primary lung infections or deep skin abscess. From there it may metastasize into various organs of the body in addition to the central nervous system. Amphotericin B seems to be specific as a therapeutic agent.

Histoplasmosis

Histoplasmosis was discovered in 1906 in livers and spleen sections of natives from the Canal Zone who had died of what was thought to be leishmaniasis. Darling, because he assumed these agents were related to the protozoa, called the organism, *Histoplasma capsulatum.*

It has since been found to be worldwide in distribution, although it is endemic only in certain geographical areas. It requires moist soils and temperatures not much higher than 100° F. In the United States, it is confined primarily to the Ohio River Valley, especially along the river bottoms. It could not survive in the hot, dry, climate of the southwestern United States.

Histoplasmosis is generally a mild disease. However, a very large percentage of people residing in the endemic region show positive skin reactions when tested with histoplasmin. On autopsy, many people from these areas are found to have characteristic small, nontuberculous, calcareous nodules in the lungs. This suggests a high incidence of inapparent or undiagnosed infection.

The causative organism apparently is a saprophyte and is transmitted to humans by dust or spore inhalation as a true airborne disease. Dogs are almost as susceptible as humans. It has been found in amphibians, birds, insects, and mammals in general. There is no evidence that infected animals may serve as reservoirs.

In addition to the inapparent or undiagnosed, common, low-grade, pulmonary infections that appear to be so widely distributed, the disease is occasionally encountered in a more severe form, which then leads to its diagnosis. When severe it may be in the form of (1) a systemic, febrile disease with enlargement of the spleen; (2) a similar disease but characterized by lymph node enlargement; (3) pulmonary involvement; and (4) an ulcerative type of skin infection.

The disease is diagnosed by culturing on Sabouraud's and blood agar. On the first medium it develops a distinctive mycelial type of growth, while on sealed blood agar plates it grows in a yeastlike form. Also, in patients with the disease the skin test will be positive.

In most cases, treatment with amphotericin B, coupled with surgical removal of localized lesions, especially in the lungs, may effect a cure.

NORMAL FLORA INFECTIONS

Candidiasis

The organism, *Candida albicans* (L. *candidus* dazzling white + *albicans* white) is a yeastlike form that is apparently a common inhabitant of the body, kept under control by the other normal flora. It can often be isolated in small numbers from the mouth, vagina, or feces of persons in normal health.

Formerly known as *Monilia albicans,* or more recently *Oidium albicans,* it was first isolated from a case of thrush or "white mouth" by Langenbeck in 1839. In smears from the mouth or other sources it appears as a small, oval, budding yeast cell. These cells resemble the commercial yeast but are somewhat smaller. Occasionally, short, filament-like pseudomycelia may be seen.

SAPROPHYTIC MOLD INFECTIONS

In very rare instances molds, normally considered strict saprophytes, have been involved in human infections. A few examples are discussed below:

Phycomycosis is a disease caused by *Mucor, Absidia,* or *Rhizopus* species. A number of fatal cases have been reported, especially in Japan. These have been mostly rapidly fatal brain infections, although the lungs or other organs have sometimes been involved. These cases have all developed in persons suffering from advanced stages of diabetes mellitus or multiple myeloma. Spontaneous infections in swine, dogs, and horses have been reported.

Infections have developed as a result of inhalation of spores.

Cladosporiosis is caused by *Cladosporium* or *Hormodendrum* species. Only a few cases have been reported, all in Japan. These led to fatal brain abscesses. Infection was due to pulmonary inhalation.

Geotrichosis, caused by *Geotrichum candidum,* has been reported. This has been characterized as a chronic pulmonary infection.

Aspergillosis, caused primarily by *Aspergillus fumigatus,* is a fairly common disease in poultry, especially pigeons. Strains of the organism vary greatly in virulence, and the virulent strains produce a fulminating, rapidly fatal pneumonia. It may become epidemic with brooder chicks, resulting in high mortality and severe economic losses. It is transmitted primarily by the feeding of moldy grain.

Man occasionally contracts the disease, usually as an infection of the outer ear. It may sometimes extend into the inner ear. Lung infections are rare but may be mistaken for pneumonia. It is almost an occupational disease, occurring most often among pigeon fanciers.

SUGGESTED READINGS

Ajello, L.: Comparative ecology of respiratory mycotic disease agents, Bacteriol. Rev. **31**:6-24, 1967.

Burrows, W.: Textbook of microbiology, ed. 18, Philadelphia, 1963, W. B. Saunders Co.

Converse, J. L., and Reed, R. E.: Epidemiology of coccidioidomycosis, Bacteriol. Rev. **30**:678-694, 1966.

Dubos, R. J., and Hirsch, J. G., editors: Bacterial and mycotic infections of man, ed. 4, Philadelphia, 1965, J. B. Lippincott Co.

Smith, D. T., et al.: Zinsser Microbiology, ed. 14, New York, 1968, Appleton-Century-Crofts.

29 Disease of plants

Plant diseases effect everyone in some way. They may strike directly at the livelihood of the farmer or gardener, and they have a direct bearing on the availability of the food we eat. In some instances, they have proved ruinous to a whole country. The classic example is the Great Potato Famine of Ireland in 1845-1846.

IMPORTANCE AND NATURE OF PLANT DISEASES

The annual cost of crop losses in the United States due to plant diseases amounts to several billion dollars.

Not all disease conditions in crops and plants are due to microorganisms. However, the microorganisms represent a formidable and destructive group. Included are the bacteria, fungi, ranging from the Fungi Imperfecti and the Phycomycetes to some of the Basidiomycetes, the strange protozoan Plasmodiophorales, and the viruses.

BACTERIAL DISEASES OF PLANTS

Burrill of the University of Illinois in 1878 first recognized that bacteria can cause diseases in plants when he studied fire blight in pears. The seventh edition of *Bergey's Manual* lists 193 species of bacteria as plant pathogens.

Major bacterial diseases

A few of the most commonly encountered types of bacterial plant diseases are discussed in the following paragraphs.

1. Soft rots are caused by *Erwinia* species and are most commonly seen in cabbage, carrots, potatoes, lettuce, and celery. The condition may develop in growing potatoes or carrots while still in the soil if they have been damaged in some manner. The soft rot is characterized by the developing of a soft slimy area of decay, often with an offensive odor. The damage results from the hydrolysis of the pectin compounds that serve as the "glue" in the middle lamella of the cell walls thus holding the cells together. As the pectin compounds are hydrolyzed, a slimy disintegration of the tissue occurs, and because of this property, the name *Pectobacterium* was once proposed to replace *Erwinia*. The most important species is *E. carotovora* (the carrot eater), which seems to be widely distributed.

If a sprout from an infected potato cutting grows into a plant, its growth will at first be slowed, and the stem will then turn red to purple and finally black as the plant succumbs. This is termed black leg in potatoes.

2. Fire blight practically wiped out the Bartlett pear orchards in some areas before practical control measures could be developed. Fire blight, so called because the dry leaves and stems appear to have been killed by fire, can also develop in certain apples and some stone fruits. The causative organism, *E. amylovora*, appears to be carried by honeybees from infected to noninfected blossoms, or from the sweet exudate that forms in droplets on the infected stems. Wind-blown rain may also be an important carrier. Pruning tools that have been used to cut diseased branches may infect healthy trees. After the initial infection the bacteria apparently move through the intracellular spaces; the cells quickly lose water and death occurs.

The bacteria seem to overwinter in infected twigs. When spring growth begins, the sugary exudate leaks from the area of new active infection, and the disease is rapidly spread.

Some varieties of pears are quite resistant to infection, but the fruit is inferior in quality. They are often used as the root stock, and Bartlett scions are grafted on them. If infection of any of the grafts occurs, they may be cut off aseptically and the dead material burned.

3. Leaf spot or black arm of cotton was first described in 1891. When cotton plants are infected, leaves develop spots, the stalk turns black, dies, and the young bolls dry up and drop off. The causative organism is *Xanthomonas malvacearum*. Control measures consist of soaking seeds in sulfuric acid solution before planting, or using varieties that are resistant to the bacteria.

4. First observed in 1892, bacterial blight of beans is characterized by dark to black spots on leaves and on the pods of the beans. Three different species of bacteria, *Xanthomonas phaseoli*, *Pseudomonas syringa*, and *Corynebacterium flaccumfaciens*, produce similar manifestations of the disease.

Diseases caused by Mycoplasma

The *Mycoplasma* represent the unique group of microorganisms briefly discussed in Chapter 7.

In 1967 the first indication appeared that mycoplasmas might cause certain diseases in plants. At that time, electron microscope studies were reported on aster yellows. This disease, known and studied for years, occurs also in potatoes, onions, carrots, and similar plants. It is transmitted by a leafhopper and was thought to be a virus disease. However, the electron micrographs showed typical mycoplasma bodies, present in large numbers, in thin sections of diseased flower stalks, but absent from normal, uninfected stalks. The same bodies were also seen in the salivary glands of the infected leafhopper, but not in the uninfected insect. Relatively large numbers were also observed in extracts from infected plants and leafhoppers. As a result of the studies, it was found that aster yellows could be controlled with tetracycline.

In only one case, white leaf of sugar cane, described in Taiwan, was it possible to isolate and grow the mycoplasma outside the plant.

Although 40 mycoplasmal infections of plants have been described, it is impossible to determine how many different strains may be involved.

FUNGAL DISEASES OF PLANTS

Fungal diseases of plants are produced by a large and varied group of forms, with representatives from all four of the major classes of fungi; Phycomycetes, Fungi Imperfecti, Ascomycetes, and Basidiomycetes.

Some of the plant diseases caused by the Phycomycetes are black wart of potato, brown leaf and stem spot of corn, crown wart of alfalfa, root rot of peas. *Pythium*, which is the cause of the "dampening-off" disease of seedlings, results in a severe economic loss in nurseries. Other diseases caused by the Phycomycetes include the late blights of a number of crop plants, the downy mildews often seen on grasses, and the soft rot of sweet potatoes.

The Fungi Imperfecti, which grow familiarly as spoilage molds on fruits, vegetables, jellies, and so forth, generally produce powdery, dark masses of growth. Some of the common plant diseases caused by these fungi are gray neck-rot of onions; *Fusarium* infections of cabbage, tomatoes, and potatoes; early blight of potatoes and tomatoes (Figs. 29-1 and 29-2) which causes seemingly healthy plants to suddenly turn brown and die; onion smudge, in which a black, powdery growth appears on the onion during storage; anthracnose of green string beans, which may be recognized by the distinct, black spots on the pod; and white rot of onions.

Many of the Ascomycetes resemble the Fungi Imperfecti in their growth habits except that they may occasionally develop a perfect or ascospore state. The plant pathogens range from relatively simple forms, with only meager mycelial growth, to forms with an involved life cycle. Some of the representatives include the cause of peach leaf curl, and the powdery mildews of plants such as roses and grasses.

Two diseases caused by the Ascomycetes, Dutch elm disease and chestnut blight, practically wiped out the stands of these trees in northeastern United States. Apple scab, ergot of grains, and brown rot of stone fruits are also caused by the Ascomycetes.

These fungi often produce spores in profuse

Figure 29-1
Tomatoes infected on vine with early blight, a common fungus disease caused by *Alternaria solani.*

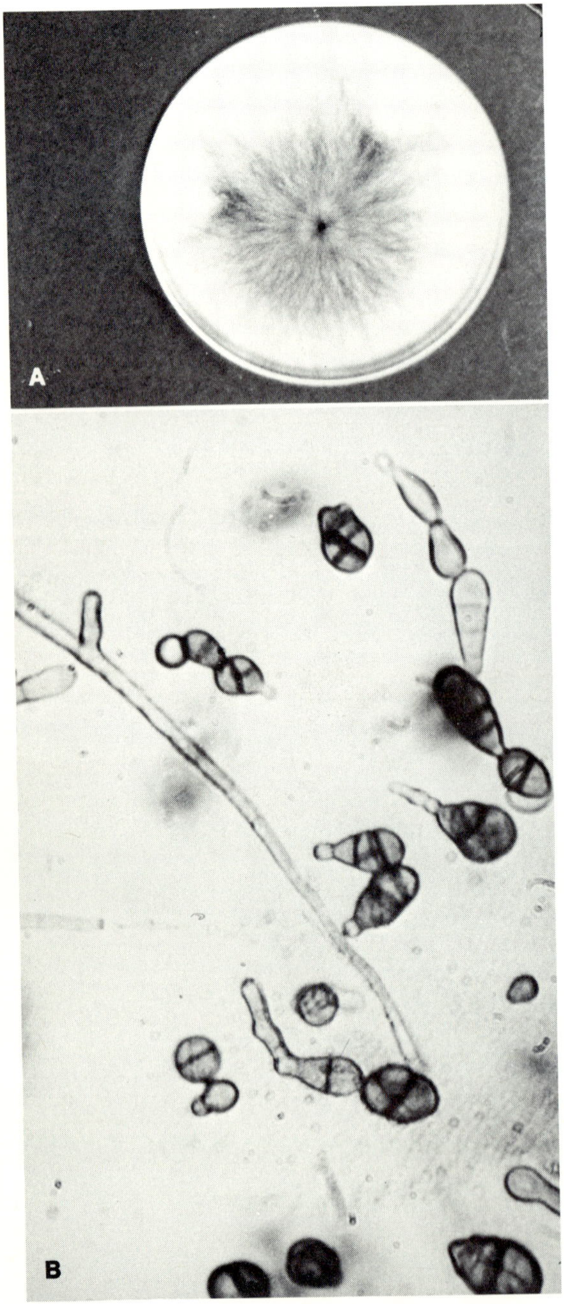

Figure 29-2
A, Colony of *Alternaria solani* isolated from infected tomato on Sabouraud's agar;
B, spores and mycelia of *A. solani* (×2,000).

amounts; these may be windborne and scattered for considerable distances.

The basidiomycetes, familiar as mushrooms and shelf fungi, are active as wood-rotting forms. They are often not considered microorganisms. However, their spores may be small enough to fall within the microbial range. Important among the diseases produced by these forms are the smuts of corn, wheat, oats, and barley. Bunt or stinking smut (so-called because an infected head of wheat smells like rotting fish) is caused by *Puccinia graminis*, whose life cycle is shown in Fig. 29-3.

Rusts and white pine blister may have a life cycle in which the development takes place in part in an intermediate host plant. The wild gooseberry or the barberry and white pine blister disease, is an example of this.

Other basidiomycetal diseases include onion smut, rhizoctonia disease of potatoes, and white-heart rot of certain species of trees used for lumber.

DISEASES OF PLASMODIAL ORIGIN

The plasmodia are rather rare forms represented by a single family and eight genera. Some are parasites of algae and some of the filamentous molds. Most of them are parasitic on higher plants. Only two species are of economic importance.

Biologically, the plasmodia represent an anomalous group, considered by some to be fungi and by others as protozoa. The exact position is not clear. Some earlier workers placed them with the slime molds (myxomycetes); others considered them to be related to the fungi, Chytridiales, which produce motile zoospores and lack a true mycelium.

In nature they exist in the soil, or in water as resting spores that germinate to form motile swarm cells with two flagella, one short, the other long. When these swimming spores contact a hair root of the host they lose their flagella and change into the myxameba stage. This blob of naked protoplasm enters the host cell and

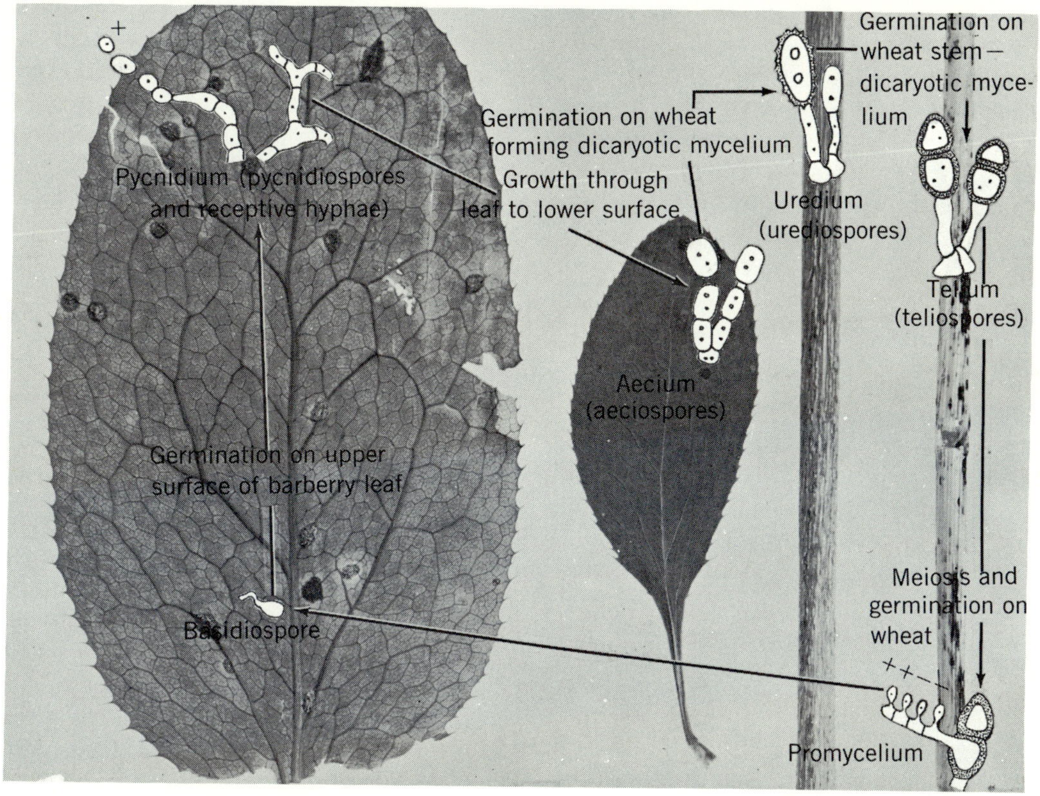

Figure 29-3
Life cycle of *Puccinia graminis,* a basidiomycete causing the important plant disease, wheat rust. (From Swatek, F. E.: Textbook of microbiology, St. Louis, 1967, The C. V. Mosby Co.)

begins to proliferate, forming a multinucleate plasmodium around the host cell nucleus. As the plant cell divides, a portion of the plasmodium is found in both mother and daughter, both of which may be greatly enlarged. Eventually a large number of spores are formed and are released from the hypertrophic root tissue. These spores may immediately germinate into motile zoospores and continue the infective cycle, or may remain in the ungerminated state for several years.

Economically, the most important plasmodial disease is the clubroot of the Cruciferae. The organism, *Plasmodiophora brassicae,* discovered by Woronin (1877), causes infections in cabbage, rutabagas, turnips, radishes, and many other members of the mustard family, both cultivated and wild. In addition, it has also been reported in plants as divergent as red clover, rye grass, and poppies.

In clubroot of cabbage and related forms, the roots develop long, spindlelike enlargements. Sometimes, gall-like enlargements develop on the roots near the surface. The initial infection causes wilting, but death of the plant may be due to bacterial soft rots that develop after the infected root breaks down.

The spores released from the disintegrating

infected roots may remain in the soil for several years, making a prolonged crop rotation necessary if the disease is to be controlled by this method. However, the disease is usually controlled by draining the soil to prevent high moisture content, and by applying lime to produce an alkaline reaction. The disease may be spread by water draining from infected fields, may be wind borne, or may be carried on the feet of domestic animals.

VIRAL DISEASES OF PLANTS

Long before bacteria or viruses were known, a number of abnormalities that we now recognize as viral diseases were described. As early as 1576, Charles l'Ecluse wrote of the striking and often beautiful variegation in tulips, then called "color break." We now know that this is actually a symptom of a mosaic-like viral disease, transmitted by aphids.

Potato leaf roll or curl disease was described just before the American Revolution and was attributed to a "degeneration" of the stock. We now recognize that viruses were the cause of this "degeneration."

Today, more than 300 kinds of plant viruses are known. Originally, each new virus was thought to be limited to a small number of host plants, but it soon became apparent that some were able to infect a broad spectrum of plants. Each new virus brought a serious problem, and in a few instances they almost wiped out an entire crop industry before the cause and mode of transmission became known and control measures could be applied.

As examples of the effect of viral diseases, grass leaves may show longitudinal streaks; great masses of foliage may rapidly turn yellow (yellows); a metallic bronzing may appear on leaves; leaves or stems may show bizarre growth distortions or malformations (as in curly top in sugar beets); punched out brown areas may appear in leaves (as in cherry leaf tatter); a mass of small stems or branches may develop from a branch tip, producing "witches brooms"; fruits may become small and distorted (as in cucumbers); peaches and other stone fruits may be misshapen and develop an insipid flavor; root tumors may form (as in clover root tumor disease); roots may become short and bunchy (corn-stunt virus); or the death of roots may occur.

Within the plant tissue, serious damage is often visible. In addition, irregular masses, often called X bodies, may be seen. These inclusions are filled with virus particles and are analagous to the inclusion bodies characteristic of many animal diseases.

Plant viruses are made up of two components, proteins and RNA. The protein in the coat represents between 65% and 95% of the constituents of the virus. Plant viruses tend to fall into two general classes, the anisometric and the isometric.

The anisometric viruses are made up of an outer filament or tube of protein, the capsid. This is composed of elliptical protein units, the capsomeres, which form a hollow tube. This might be visualized as a paper tube around which kernels of corn have been pasted in an orderly array. Inside the tube is a strand of RNA, which carries the genetic information of the virus. Some viruses resemble a bacillus in shape, being somewhat longer than they are round. Many others are long, flexible threads. For example, the tobacco rattle virus is 20 to 22 nm in diameter by 300 to 315 nm in length.

The isometric forms, once thought to be spherical, have been found to be either completely symmetrical, hollow, cuboid forms with 20 sides, or symmetrical, hollow microcrystalline bodies. The cuboid forms are generally small, about 10 to 60 nm in diameter.

Plants may be infected in a variety of ways. Biting insects, such as beetles, grasshoppers, or earwigs, may serve as mechanical carriers due to viruses adhering to their biting parts. This mode is modified to some extent in beetles, such as the turnip flea beetle. Since it has no saliva, it periodically regurgitates some of the chewed material from the foregut to aid in digestion. This may actively carry a virus, if infected plant tissue has been ingested.

Most of the plant viruses can be transmitted by aphids. One single species of aphids, *Myzus persicae*, has been shown to be able to transmit at least 50 different viruses. In many cases, the virus is merely transmitted mechanically by the stylus or sucking-mouth part. With some aphids, the viruses are swallowed and pass into the gut. They then pass through the gut wall, into the bloodstream, and into the saliva, which thus becomes infective. With a few of the insects, mainly leafhoppers, the virus may reproduce within the gut, making the insect an active reservoir of the disease. These insects have been known to remain infective for as long as 100 days. With a few leafhoppers, the plant virus may be carried over in the eggs, at least for several generations. In a few cases, the plant virus may greatly reduce the life span, or actually kill the insect carrier. Mites may also carry a number of different plant viruses. Nematodes or eelworms in the soil have been incriminated as transmitters of viral infections through the roots of the host plant. The Chytrid fungi in the soil produce a motile spore stage shown to be able to transmit some of the plant viruses.

Man himself may be a carrier of plant viruses. Cutting tools such as pruning shears, lawn mowers, hoes, and so forth, plus shoes and clothing may mechanically transmit some viruses. In the process of grafting fruit trees, a scion or bud may be grafted into a root stock that carries a latent virus. This will infect and eventually kill the new graft. Wind-borne leaves and soil as well as irrigation water may serve as transmitters of plant viruses. The tobacco in cigarettes has also been found to carry the tobacco mosaic virus.

The viral infection of plants resembles, in some respects, phage infection in bacteria. When the plant virion enters a susceptible cell, the protein coat is shed, and the process of replication of the viral RNA begins. This may be intranuclear or within the cytoplasm.

At this point, inclusion bodies may be seen. Viral synthesis seems to be largely confined to the nucleolus. A slow cell-to-cell transfer of virus begins as the disease progresses. The streaming of the cytoplasm hastens this process. In other cases, the virus enters the phloem and is carried rapidly through the plant.

Measures used for the control of viral diseases may be direct or indirect. Since many of the diseases are insect transmitted, destruction of the insect vector may prove an effective control measure. If a soil becomes infected, crop rotation can permit a lapse of several years between susceptible crops and may thus free the soil from infection. Elimination of infected seeds may also reduce the frequency of disease. With fruit trees, pruning to remove infected branches may prove effective.

MECHANISMS RESPONSIBLE FOR DISEASES IN PLANTS

Bacterial or fungal plant pathogens may damage or kill the host plant by one of several mechanisms. Often a plant disease may involve an interplay of two or more of these mechanisms. The toxins are generally the most damaging of the pathogenetic agents, often interrupting some vital metabolic action, causing localized tissue damage or the death of the entire plant.

Some pathogens produce enzymes that destroy plant substances, such as the pectins that glue the plant cells together, or break down the hormones that govern plant growth. This may result in stunted plant growth or the formation of bizarre shapes of stems or other structures.

Some bacteria, in particular may synthesize great quantities of slimes. These may physically block the vessels that conduct water or plant growth solutions. Some slime producers also produce toxins that accentuate the damage to the infected host plant. The toxins are of particular interest because they can be readily isolated and their mode of action studied. These toxins differ considerably from the toxins that produce disease processes in animals. The plant toxin molecules usually have a lower molecular weight, and they rarely stimulate antibody formation when injected into experimental animals.

A unique relationship has been noted in soybeans. *Rhizobium japonicum* is the symbiotic nitrogen-fixing organism that forms nodules on the plant roots. Effective strains induce active nitrogen fixation. This greatly stimulates plant growth and may increase the protein content of the soybeans by as much as two and a half times. Some bacterial strains, however, while stimulating normal nitrogen fixation, also produce a toxin that causes chlorosis in the growth of new leaves on the soybean. The toxin has been isolated from both the nodules and the chlorotic young leaves. It is called rhizobitoxine, and has been found to produce typical chlorosis in the seedlings of many different plants.

Another unique relationship is exhibited by the phytopathic fungus, *Helminthosporium victoriae,* that causes blight only on oats belonging to the Victoria strain. This particular oat variety was imported from South America for crossing with North American strains to produce a cultivated variety resistant to smut and crown rust. Such rust-resistant strains were developed and released about 1955. Soon after, "Victoria blight" assumed epidemic proportions and the strange selectivity of *H. victoriae* was discovered.

Some years earlier, workers had found that this fungus produced a toxic substance now termed victorin. Mild alkali hydrolysis yielded 5 to 6 common amino acids plus a nitrogen-bearing terpene, victoxime. (This latter compound alone is not toxic; evidently for toxicity, presence of the amino acids is required.) Its toxic action appears to injure the cell membrane, causing leakage of cell electrolytes. Early blight of potatoes, tomatoes, and tobacco is caused by *Alternaria solani.* The toxic agent, alternaric acid, in low concentrations causes wilting of many kinds of seedlings.

SUGGESTED READINGS

Dowson, W. J.: Plant diseases due to bacteria, ed. 2, Cambridge, England, 1957, Cambridge University Press.

Maramorosch, K., Granados, R. R., and Hirumi, H.: Mycoplasma diseases of plants and insects, Adv. in Virus Res. **16**:135-193, 1970.

Owens, L. D.: Toxins in plant disease: structure and mode of action, Science **165**(3888):18-21, 1969.

Smith, K. M.: Insect virology, New York, 1967, Academic Press, Inc.

Smith, K. M.: Plant viruses, ed. 4, London, 1968, Methuen & Co. Ltd.

Walker, J. C.: Plant pathology, ed. 3, New York, 1969, McGraw-Hill Book Co.

30 Vistas in microbiology

With the great amount of new microbiological knowledge now available and the urgent need for applying this knowledge to improve the quality of our life and environment, it seems fitting to end this book with a guess, in a prophetic vein, suggesting some of the ways in which microorganisms may aid in making the world a better place to live. Along with the explosion of new microbial knowledge and the population explosion has come an explosion in popular interest in improving the environment in which we live. These three may provide the ingredients for a whole series of significant new developments in which microbes may play a major role.

MICROBIOLOGY AND INFECTIOUS DISEASES

We now have the means at hand to eliminate many, if not most, of the infectious diseases. We have already seen some remarkable results. Cholera is unknown in the Western World. In the United States, typhoid and diphtheria are almost unknown, and summers are no longer saddened by epidemics of poliomyelitis.

Elimination of specific common diseases

Viral diseases. We now possess the fundamental knowledge required to banish several specific viral diseases from among the populations of the world. The World Health Organization has marked smallpox for extermination. Methods now available can do the same for measles, mumps, chickenpox, and poliomyelitis.

On the other hand, influenza will probably continue to be a perennial threat because of the apparent ease with which the influenza virus mutates into an immunologically different strain. When this occurs, everyone becomes equally susceptible to the new strain. As a result, great pandemics may occur, such as the devastating one of 1918-1919, or the severe outbreak of Hong Kong or Asian flu of the early 1950's. To prevent the recurrence of such a catastrophe, the World Health Organization has established the Influenza Surveillance Commission. Its purpose is the rapid detection of any new strains that carry the threat of becoming pandemic. Hopefully, the new virus strain can be isolated and a vaccine prepared for wholesale immunization in time to prevent the worldwide spread of the infection.

Bacterial diseases. Our present knowledge has been effective in eliminating a number of killer bacterial diseases from the more advanced countries of the world. Already mentioned are cholera, typhoid, and diphtheria. Plague is almost forgotten, except for occasional sporadic outbreaks. The remarkable manner in which tetanus was prevented in the Armed Forces of the United States in World War II is proof of the success of a campaign to control a bacterial disease by immunization.

New developments can be expected that may contribute to the partial or complete control of gonorrhea, syphilis, and the nontyphoidal salmonella-shigella infections. More effective control of cholera may be expected in the high-incidence endemic areas.

Protozoan diseases. These diseases represent a more complicated problem, especially those that are insect borne and have wild animal reservoirs (unlike viral and bacterial diseases, no wholesale immunization is feasible). However, as an example, in the last few decades elimination of the mosquito has practically wiped out malaria in our own country. The World Health Organization is sponsoring a rigorous campaign to control malaria in those parts of the world in which it is still a major problem. Other protozoan infections, such as African sleeping sickness and the leishmaniases will receive continuing or increasing attention.

Virus-caused cancer. The problem relating to the control of the cancer caused by viruses is still far from solved. However, a solution seems possible within the next decade. If this occurs, it may become feasible to immunize the population against a substantial proportion of the cancer attacks that now kill so many Americans.

Immunization through interferon. Although a breakthrough has developed that makes possible the artificial stimulation of the production of interferon as a nonspecific agent in the control of virus or bacterial diseases, its potentialities are still questionable. However, it may provide the means of eliminating, or at least controlling, the currently uncontrollable common cold and similar respiratory diseases caused by many different strains of related viruses.

Role of the microviruses. The recent discovery of viruses that apparently function without the usual protein coat, hence are microviruses, may provide the breakthrough necessary to enable us to control infectious hepatitis, some virus cancers, and several other diseases in which the nature of the specific agent is not yet completely understood.

Future of antibiotics. The antibiotics are no longer the panacea of all infectious diseases, as originally thought. However, they are still highly useful, especially the broad-spectrum types. The genetic development of the "resistant factor," which confers resistance against several antibiotics particularly with the salmonella-shigella group, now complicates the use of antibiotics. Furthermore, the development of specific antibiotic resistance, as seen with the staphylococci, adds further difficulty. The success of chemically modified penicillin suggests the necessity of a continual modification of the chemical structure of known antibiotics to keep abreast of the newly developing resistant strains.

MICROORGANISMS AND THE FOOD SUPPLY

Although we have not reached the state of starvation to which Malthus doomed the world, the need for an increased food supply and the presence of many undernourished nations or parts of nations, are still very much with us.

During 1970, according to a recent survey, the world's food supply was increased 2%; yet, during the same period the population increased at the same rate, leaving no net gain.

Microorganisms, if properly used, hold great promise as a means of increasing our world food supply.

Increased productivity of farmlands

Microorganisms and organic matter, rather than reliance on synthetic inorganic fertilizers, hold the key to increased agricultural productivity. True, the inorganic fertilizers may produce a temporary yield increase, but the end result is a slow depletion of the organic matter content of the soil, bringing increased erosion and a steady decrease in the future productivity.

Solid waste disposal has the potential of providing a vast new supply of vital organic matter. This will be discussed later.

As we learn more about the great number of minute, slow-growing bacteria in the Casida group (see Chapter 21) we may discover a new means of further stimulating crop production.

Recent studies with *Rhizobium* have shown that, by using the transfer of DNA from highly effective strains to less effective strains (transformation), it has been possible to create artificial symbiotic nitrogen-fixing bacteria, superior to any strains found in nature. Think what this could mean to areas of India, for example, where the diet is depressingly low in proteins, if it became possible to grow soybeans containing three times the protein content of rice.

Whole microbes as food sources

Whole microorganisms have tremendous potential as sources of food materials rich in proteins and vitamins. Use of crude oil as a carbon source for growing high-protein bacteria has already been discussed. Yeasts, as waste products of the brewing industry, are now being widely used as feeds for poultry, cattle, and hogs, and to a lesser extent as supplementary human foods. This source can be greatly expanded.

Certain of the green algae have proved their value as food sources. In the future this reservoir may be effectively tapped by using ponds of treated sewage effluent for growing algae. These algae could be harvested for use of animal feeds, or refined for use as human foods, probably disguised in conventional foods but serving to increase the nutritional value. These added surface areas of algae-laden water would also increase the removal of CO_2 from the atmosphere and thus improve the quality of the air.

Biological insect control

With the known toxic dangers associated with DDT and other insecticides, the use of biological methods holds great promise. We already have a precedent in the use of *Bacillus thuringiensis* (see Chapter 20). Use of this control method has been limited, primarily by economics. A governmental subsidy for the application of this method, in place of highly toxic pesticides, might increase its feasibility.

Viruses that attack specific insects also show great promise. Again development of their potential will depend on available research funds to provide the basic information required for their effective application.

Biodegradable pesticides and herbicides

One of the chief objections to the use of synthetic pesticides and herbicides is that they are still present on crops at harvest, and may remain in the soil itself for many years. The problem is similar to that of nonbiodegradable detergents.

A great deal of interest is now being shown in biodegradable pesticides and herbicides. These compounds remain active long enough to control the insects or weeds but are then quickly converted into innocuous compounds. In most cases, the microorganisms in the soil effect this decomposition, although a slow decomposition by the oxygen of the air may also be involved.

MICROBIAL IMPROVEMENT OF OUR ENVIRONMENT

Water supplies

Some water supplies pass all tests for pollution yet are high enough in nitrates to make them a hazard to health. Bacteria are available that can readily reduce the nitrates to harmless nitrogen gas, which is then returned to the atmosphere leaving no residue (see Chapter 19). Bacteria hold the key to this conversion and may be widely used for this purpose in the future. Pilot plant operations have already established the feasibility of this method.

Sewage disposal

Water pollution control procedures have made remarkable strides in the past two decades or so. However, waters are still being extensively polluted, decreasing the total amount of water available for future use. This has forced us to face the prospect of directly reusing our water supplies. This involves three major difficulties:

1. Final treatment of sewage effluent to make it potable for reuse involves the use of chemicals, which continue to accumulate, thus limiting the number of times water can be recirculated.
2. With each reuse, waters rapidly increase in nitrate content. A bacterial method for reduction of the nitrates may be required in the future.
3. No method of treatment fully guarantees the successful removal of all viable enteric viruses. When this is accomplished, and it is now entirely feasible, a major objection to reuse of water will have been met.

With sewage treatment, the settleable solid material (sludge) poses a major problem. Sludge is usually decomposed anaerobically in huge sludge digestion tanks. One of the products is a substantial amount of marsh gas, high in methane. This resembles natural gas in many ways, and in larger cities is often used as fuel for combustion engines. It is also used, in some cases, to provide fuel for electric power plants. Here microorganisms are providing a usable product, often wasted by burning in an open flame. This wasteful procedure adds carbon dioxide to the atmosphere without putting any of the potential energy to practical use. It is possible that in the future, smaller communities may use their sewage disposal plant as a source for domestic gas.

Solid wastes

Someone has said that "we have scientific know-how to send a man to the moon but we are being buried in our own garbage." This, unfortunately, is true. However, microorganisms hold the greatest promise for the successful solution of the problem.

The bulk of the decomposable portion of garbage remaining after the cans, bottles, and similar materials are removed, is in the form of cellulose (paper, wood, and so forth). This is generally buried because burning pollutes the air.

However, microorganisms could provide the means of alleviating this problem while providing benefits that will greatly improve our environment. Three microbial approaches suggest themselves:

1. Conversion of nonmetallic solids into methane would be equivalent to sludge digestion on a vastly larger scale, with many times more raw material available each day, and a correspondingly higher gas yield.

That this suggestion is feasible is borne out by pilot plant studies in Iowa during the depression years. Small tanks, equivalent to sludge tanks, were provided the cooperating farmers. These were filled with shredded cornstalks or corncobs, covered with water, and seeded heavily with a crude culture of methane-producing bacteria. The tank was closed and in a short time large quantities of methane gas developed. This was trapped in special tanks, over water, and piped to the gas stove. The yield was sufficient to provide cooking gas for the home, and required only periodic replenishment of the plant material.

Such installations utilizing the solid wastes of a city could provide great quantities of gas with the properties of natural gas.

2. A second solution is one dear to the heart of the ecologically minded. Many amateur gardeners place all their grass clippings, trimmings, and even household garbage into a compost pit, where the microorganisms quickly convert it into a high humus organic fertilizer. This could also be accomplished by a city, developing a sort of high capacity mechanized compost heap for solid wastes. This is being seriously considered at the present time.

One of the objections is that we would have far more compost than we could use. However, there are many areas in need of the surplus compost. One is our farm lands, which are constantly being "mined" of their organic matter. This not only reduces productivity but also lowers their water-holding capacity and ability to resist erosion. Wholesale quantities of humus could be used to improve our soils.

A second area of great need is also apparent. In middle America, following the great drought of the Dust Bowl days, millions of tons of rich, productive surface soil were either swirled away by the winds, or washed by floods into the Gulf of Mexico. The end result is that millions of acres were depleted in the process. Many of these acres had been marginal grass or forest lands and should have remained in that state. However, the false economic boom of the post-World War I period resulted in many acres of marginal land being ravaged and denuded by the plow and rapacious farming practices.

Efforts have been made, with some success, to reclaim these lands. However, great amounts of rich organic humus fertilizer, which might be made available by composting solid wastes, would work wonders with these soils. Many marginal areas might be converted into forest or lush grass lands that could then, through greatly increased photosynthesis, help stem the constant and alarming increase in carbon dioxide in the air resulting from the burning of fossil fuels.

These same vast supplies of humus might do much to remove the blight from the many acres of land denuded by the strip coal mining practices. Here again, this might aid in the establishment of a vital, photosynthetically active, plant cover.

3. A third microbial solution is the saccharification of paper and other cellulosic wastes into sugar. This product could then be fermented by yeasts to produce great quantities of ethyl alcohol, which could be used as a substitute for or additive to gasoline as a fuel for automobiles.

That this is not a wild dream is again borne out by experiences of the depression days. On an experimental basis, farmers in some areas of Iowa traded their cheap corn to an industrial alcohol plant for a blend of ethanol and gasoline. This proved to be a satisfactory fuel. However, as soon as the price of corn advanced, the practice was no longer feasible. It may be that with great quantities of cheap carbohydrates available in garbage, alcohol from this source might once more be a desirable automotive fuel. In addition, great quantities of yeast suitable for animal feed or human foods would automatically become available.

In addition to the few applications suggested, we may well see other uses of microorganisms that go beyond our wildest dreams.

The microbes have enormous enzymatic versatility and an amazing ability to reproduce. The future use of them will be limited only by our own ingenuity, and not by the capabilities of the microbes.

Name index

Subject index

A

B

C

D

E

F

N

O

P

S

T

U

V